Advances in Intelligent Systems and Computing

Volume 384

Series editor

Janusz Kacprzyk, Polish Academy of Sciences, Warsaw, Poland
e-mail: kacprzyk@ibspan.waw.pl

About this Series

The series "Advances in Intelligent Systems and Computing" contains publications on theory, applications, and design methods of Intelligent Systems and Intelligent Computing. Virtually all disciplines such as engineering, natural sciences, computer and information science, ICT, economics, business, e-commerce, environment, healthcare, life science are covered. The list of topics spans all the areas of modern intelligent systems and computing.

The publications within "Advances in Intelligent Systems and Computing" are primarily textbooks and proceedings of important conferences, symposia and congresses. They cover significant recent developments in the field, both of a foundational and applicable character. An important characteristic feature of the series is the short publication time and world-wide distribution. This permits a rapid and broad dissemination of research results.

More information about this series at http://www.springer.com/series/11156

Stefano Berretti · Sabu M. Thampi
Praveen Ranjan Srivastava
Editors

Intelligent Systems Technologies and Applications

Volume 1

Springer

Editors
Stefano Berretti
Dipartimento di Ingegneria
 dell'Informazione (DINFO)
Università degli Studi di Firenze
Firenze
Italy

Praveen Ranjan Srivastava
Indian Institute of Management Rohtak
Rohtak, Haryana
India

Sabu M. Thampi
School of CS/IT
Indian Institute of Information Tech.
 and Management – Kerala (IIITM-K)
Trivandrum
India

ISSN 2194-5357   ISSN 2194-5365 (electronic)
Advances in Intelligent Systems and Computing
ISBN 978-3-319-23035-1   ISBN 978-3-319-23036-8   (eBook)
DOI 10.1007/978-3-319-23036-8

Library of Congress Control Number: 2015946580

Springer Cham Heidelberg New York Dordrecht London

Printed on acid-free paper

Springer International Publishing AG Switzerland is part of Springer Science+Business Media
(www.springer.com)

Preface

Intelligent systems refer broadly to computer embedded or controlled systems, machines and devices that possess a certain degree of intelligence with the capacity to gather and analyze data and communicate with other systems. There is a growing interest in developing intelligent technologies that enable users to accomplish complex tasks in different environments with relative ease. The International Symposium on Intelligent Systems Technologies and Applications (ISTA) aims to bring together researchers in related fields to explore and discuss various aspects of intelligent systems technologies and their applications. ISTA'15 was hosted by SCMS School of Engineering and Technology (SSET), SCMS Group of Institutions, Kochi, India during August 10–13, 2015. The Symposium was co-located with the Fourth International Conference on Advances in Computing, Communications and Informatics (ICACCI'15).

In response to the call for papers, 250 papers were submitted to the symposium. All the papers were evaluated on the basis of their significance, novelty, and technical quality. Each paper was rigorously reviewed by the members of the program committee. This book contains a selection of refereed and revised papers from intelligent techniques and applications track and the special track on intelligent image processing and artificial vision.

There is a long list of people who volunteered their time and energy to put together the conference and who warrant acknowledgment. We would like to thank the authors of all the submitted papers, especially the accepted ones, and all the participants who made the symposium a successful event. Thanks to all members of the Technical Program Committee, and the external reviewers, for their hard work in evaluating and discussing papers. The EDAS conference system proved very helpful during the submission, review, and editing phases.

We are grateful to the General Chairs and members of the Steering Committee for their support. Our most sincere thanks go to all keynote and tutorial speakers who shared with us their expertise and knowledge. Special thanks to members of the organizing committee for their time and effort in organizing the conference.

We thank the SCMS School of Engineering and Technology (SSET), SCMS Group of Institutions, Kochi for hosting the event.

We wish to express our thanks to Thomas Ditzinger, Senior Editor, Engineering/Applied Sciences Springer-Verlag for his help and cooperation.

August 2015

Stefano Berretti
Sabu M. Thampi
Praveen Ranjan Srivastava

Organization

Honorary Chairs

Lotfi A. Zadeh   Founder of Fuzzy Logic, University of California
Berkeley, USA
William A. Gruver   Simon Fraser University, Canada

ICACCI Steering Committee

Ravi Sandhu	University of Texas at San Antonio, USA
Sankar Kumar Pal	Indian Statistical Institute, Kolkata, India
Albert Y. Zomaya	The University of Sydney, Australia
H.V. Jagadish	University of Michigan, USA
Sartaj Sahni	University of Florida, USA
John F. Buford	Avaya Labs Research, USA
Jianwei Huang	The Chinese University of Hong Kong, Hong Kong
John Strassner	Software Labs, Futurewei, California, USA
Janusz Kacprzyk	Polish Academy of Sciences, Poland
Tan Kay Chen	National University of Singapore, Singapore
Srinivas Padmanabhuni	Infosys Labs, India & President at ACM India
Suzanne McIntosh	New York University and Cloudera Inc., USA
Prabhat K. Mahanti	University of New Brunswick, Canada
R. Vaidyanathan	Louisiana State University, USA
Hideyuki TAKAGI	Kyushu University, Japan
Haibo He	University of Rhode Island, USA
Nikhil R. Pal	Indian Statistical Institute, Kolkata, India
Chandrasekaran K.	NITK, India
Junichi Suzuki	University of Massachusetts Boston, USA
Deepak Garg (Chair)	IEEE Computer Society Chapter, IEEE India Council & Thapar University, India

Pascal Lorenz	University of Haute Alsace, France
Pramod P. Thevannoor (Vice Chairman)	SCMS Group of Institutions, Kochi, India
Axel Sikora	University of Applied Sciences Offenburg, Germany
Maneesha Ramesh	Amrita Vishwa Vidyapeetham, Kollam, India
Sabu M. Thampi	IIITM-K, India
Suash Deb	INNS India Regional Chapter
Arun Somani	Iowa State University, USA
Preeti Bajaj	G.H. Raisoni COE, Nagpur, India
Arnab Bhattacharya	Indian Institute of Technology (IIT), Kanpur, India

General Chairs

Soura Dasgupta	University of Iowa, USA
Jayanta Mukhopadhyay	Indian Institute of Technology (IIT), Kharagpur, India
Axel Sikora	University of Applied Sciences Offenburg, Germany

Publication Chair

Sabu M. Thampi	IIITM-K, India

Technical Program Committee

Program Chairs

Juan Manuel Corchado Rodriguez	University of Salamanca, Spain
Stefano Berretti	University of Florence, Italy

TPC Members/Additional Reviewers

Girijesh Prasad	University of Ulster, UK
Hanen Idoudi	National School of Computer Science - University of Manouba, Tunisia

M.V.N.K. Prasad	IDRBT, India
Wen Zhou	Shantou University, P.R. China
Marcelo Carvalho	University of Brasilia, Brazil
Yoshitaka Kameya	Meijo University, Japan
Lorenzo Mossucca	Istituto Superiore Mario Boella, Italy
Rodolfo Oliveira	Nova University of Lisbon, Portugal
Vamsi Paruchuri	University of Central Arkansas, USA
Maytham Safar	Kuwait University, Kuwait
Wei Tian	Illinois Institute of Technology, USA
Zheng Wei	Microsoft, USA
Uei-Ren Chen	Hsiuping University of Science and Technology, Taiwan
Son Doan	UC San Diego, USA
Roman Jarina	University of Zilina, Slovakia
Sanjay Singh	Manipal Institute of Technology, India
Ioannis Stiakogiannakis	France Research Center, Huawei Technologies Co. Ltd., France
Imtiez Fliss	ENSI, Tunisia
Alberto Nuñez	University Complutense of Madrid, Spain
Sandeep Reddivari	University of North Florida, USA
Haibin Zhu	Nipissing University, Canada
Philip Branch	Swinburne University of Technology, Australia
GianLuca Foresti	University of Udine, Italy
Yassine Khlifi	Umm Al-Qura University, KSA, Saudi Arabia
Abdelhafid Abouaissa	University of Haute Alsace, France
Jose Delgado	Technical University of Lisbon, Portugal
Sabrina Gaito	University of Milan, Italy
Petro Gopych	Universal Power Systems USA-Ukraine LLC, Ukraine
Pavel Kromer	VSB - Technical University of Ostrava, Czech Republic
Antonio LaTorre	Universidad Politécnica de Madrid, Spain
Suleman Mazhar	GIK Institute, Pakistan
Hidemoto Nakada	National Institute of Advanced Industrial Science and Technology, Japan
Yoshihiro Okada	Kyushu University, Japan
Hai Pham	Ritsumeikan University, Japan
Jose Luis Vazquez-Poletti	Universidad Complutense de Madrid, Spain
Rajib Kar	National Institute of Technology, Durgapur, India
Michael Lauer	Michael Lauer Information Technology, Germany
Anthony Lo	Huawei Technologies Sweden AB, Sweden
Ilka Miloucheva	Media Applications Research, Germany
Sunil Kumar Kopparapu	Tata Consultancy Services, India

Jhilik Bhattacharya	Thapar University, India
Lai Khin Wee	Universiti Malaya, Malaysia
Mu-Qing Lin	Northeastern University, P.R. China
Monica Mehrotra	Jamia Millia Islamia, Delhi, India
Shireen Panchoo	University of Technology, Mauritius
Angkoon Phinyomark	University of Calgary, Canada
Kashif Saleem	King Saud University, Saudi Arabia
Shajith Ali	SSN College of Engineering, Anna University Chennai, India
M. Emre Celebi	Louisiana State University in Shreveport, USA
Qurban Memon	United Arab Emirates University, UAE
Manoj Mukul	BIT, India
Fathima Rawoof	K S School of Engineering & Management, Bangalore, India
Kaushal Shukla	Indian Institute of Technology, Banaras Hindu University, India
Marc Cheong	Monash University, Australia
Senthilkumar Thangavel	Amrita School of Engineering, India
M.V. Judy	Amrita Vishwa Vidyapeetham, India
Michael McGuire	University of Victoria, Canada
Dhiya Al-Jumeily	Liverpool John Moores University, UK
Kambiz Badie	Iran Telecom Research Center, Iran
Vasudev Bhaskaran	Qualcomm Inc., USA
Antonis Bogris	TEI of Athens, Greece
Minas Dasygenis	University of Western Macedonia, Greece
Swathi Kurunji	Actian Corporation, USA
Hailong Li	Cincinnati Children's Hospital Medical Center, USA
Jian Lu	University of Massachusetts Lowell, USA
Punit Rathod	Indian Institute of Technology Bombay, India
Ajmal Sawand	Paris Descartes University, France
Dimitrios Stratogiannis	Wireless and Satellite Communications Group, Greece
Zhong Zhang	University of Texas At Arlington, USA
Runhai Jiao	North China Electric Power University, USA
Yunji Wang	University of Texas at San Antonio, USA
Michael Affenzeller	Upper Austria University of Applied Sciences, Austria
Andrea Omicini	Alma Mater Studiorum-Università di Bologna, Italy
Kuei-Ping Shih	Tamkang University, Taiwan
Guu-Chang Yang	National Chung Hsing University, Taiwan
Massimo Cafaro	University of Salento, Italy
Swati Chande	International School of Informatics and Management, India

Grammati Pantziou	Technological Educational Institution of Athens, Greece
Sheng-Shih Wang	Minghsin University of Science and Technology, Taiwan
Laurence T. Yang	St. Francis Xavier University, Canada
Georgios Kambourakis	University of the Aegean, Greece
Kazuo Mori	Mie University, Japan
Hadj Bourdoucen	Sultan Qaboos University, Oman
Thanh Long Ngo	Le Quy Don University, Vietnam
Abdelmadjid Recioui	Universitry of Boumerdes, Algeria
Christian Schindelhauer	University of Freiburg, Germany
Yuh-Ren Tsai	National Tsing Hua University, Taiwan
Minoru Uehara	Toyo University, Japan
Bin Yang	Shanghai Jiao Tong University, P.R. China
Shyan Ming Yuan	National Chiao Tung University, Taiwan
Meng-Shiuan Pan	Tamkang University, Taiwan
Mohan Kankanhalli	National University of Singapore, Singapore
Philip Moore	Lanzhou University, P.R. China
Sung-Bae Cho	Yonsei University, Korea
Eraclito Argolo	Universidade Federal do Maranhão, Brazil
Tushar Ratanpara	Dharmsinh desai University, India
Bulent Tavli	TOBB University of Economics and Technology, Turkey
Andre Carvalho	University of Sao Paulo, Brazil
Valentin Cristea	University Politehnica of Bucharest, Romania
Boris Novikov	Saint Petersburg State University, Russia
Prasheel Suryawanshi	MIT Academy of Engineering, Alandi (D), Pune, India
Kenneth Camilleri	University of Malta, Malta
Atilla Elçi	Aksaray University, Turkey
Stephane Maag	TELECOM SudParis, France
Jun Qin	Southern Illinois University Carbondale, USA
Mikulas Alexik	University of Zilina, Slovakia
Ali Hennache	Al-Imam Muhammad Ibn Saud Islamic University, Saudi Arabia
Mustafa Man	University Malaysia Terengganu, Malaysia
Misron Norhisam	Universiti Putra Malaysia, Malaysia
Hamid Sarbazi-Azad	IPM & Sharif University of Technology, Iran
Dhaval Shah	Institute of Technology, Nirma University, India
Luiz Angelo Steffenel	Université de Reims Champagne-Ardenne, France
Ramayah Thurasamy	Universiti Sains Malaysia, Malaysia
Chi-Ming Wong	Jinwen University of Science and Technology, Taiwan

Shreekanth T.	Sri Jayachamarajendra College of Engineering, India
Kazumi Nakamatsu	University of Hyogo, Japan
Mario Collotta	Kore University of Enna, Italy
Balaji Balasubramaniam	Tata Research Development and Design Centre (TRDDC), India
Chuanming Wei	Boradcom Corporation, USA
S. Agrawal	Delhi Technological University (DTU) Formerly Delhi College of Engineering (DCE), India
Mukesh Saini	University of Ottawa, Canada
Ciprian Dobre	University Politehnica of Bucharest, Romania
Traian Rebedea	University Politehnica of Bucharest, Romania
Zhaoyu Wang	Georgia Institute of Technology, USA
Chakravarthi Jada	RGUKT Nuzividu, India
Eduardo Rodrigues	Federal University of Rio Grande do Norte, Brazil
Dinesh Sathyamoorthy	Science & Technology Research Institute for Defence (STRIDE), Malaysia
Tomonobu Sato	Hitachi, Ltd., Japan
Peng Xia	Microsoft, USA
Bei Yin	Rice University, USA
Ahmed Almurshedi	Universiti Teknologi Malaysia, Malaysia
Amitava Das	CSIO, India
Povar Digambar	BITS Pilani Hyderabad, India
Joydev Ghosh	The New Horizons Institute of Technology, India
Son Le	Aston University, UK
Wan Hussain Wan Ishak	Universiti Utara Malaysia, Malaysia
Edward Chu	National Yunlin University of Science and Technology, Taiwan
Shom Das	National Institute of Science & Technology, India
Akshay Girdhar	Guru Nanak Dev Engineering College, Ludhiana, India
Akash Mecwan	Nirma University, India
Prasant Kumar Pattnaik	KIIT University, India
Ramesh R.	Asiet Kalady, India
Mostafa Al-Emran	Al Buraimi University College, Oman
Weiwei Chen	University of Southern California, USA
Adib Chowdhury	University College of Technology Sarawak, Malaysia
Josep Domingo-Ferrer	Universitat Rovira i Virgili, Spain
Ravi G.	Sona College of Technology, India
Govindarajan Jayaprakash	AmritaVishwavidyapeetham University, India
Vinayak Kulkarni	MIT Academy of Engineering Pune, India

Nandagopal Jayadevan   Amrita School of Engineering, India
  Nair Lathika
Tonglin Li   Illinois Institute of Technology (IIT), USA
N. Mathan   Sathyabama University, India
Hu Ng   Multimedia University, Malaysia
Sindiso Nleya   Computer Science Department, South Africa
Marcelo Palma Salas   Campinas State University (UNICAMP), Brazil
Muhammad Raheel   University of Wollongong, Australia
Mohammed Saaidia   University of Souk-Ahras. Algeria, Algeria
Jose Stephen   Centre for Development of Advanced Computing,
                                India
Mohammed Mujahid   Hafr Al-Batin Community College (HBCC), Saudi
  Ulla Faiz   Arabia
Karthik Srinivasan   Philips, India
Shrivishal Tripathi   IIT Jodhpur, India
Hengky Susanto   University of Massachusetts at Lowell, USA
Haijun Pan   New Jersey Institute of Technology, USA
Bhupendra Fataniya   Sarkhej Gandhinagar Highway, India
Afshin Shaabany   University of Fasa, Iran
Ashutosh Gupta   Amity University, India
Pablo Cañizares   Universidad Complutense de Madrid, Spain
Georgios Fortetsanakis   University of Crete, Greece
Tilahun Getu   École de Technologie Supérieure (ETS), Canada
Filippos Giannakas   University of the Aegean, Greece
Navneet Iyengar   University of Cincinnati, USA
Jamsheed K.   Amrita Vishwa Vidyapeetham, India
Rupen Mitra   University of Cincinnati, USA
Krishna Teja   University of Cincinnati, USA
  Nanduri
Hieu Nguyen   INRS-EMT, Canada
Joshin Mathew   Indian Institute of Information Technology an
                                Management - Kerala, India
Sakthivel P.   TCS, India
Lili Zhou   China Telecom Inc., P.R. China
Scott Kristjanson   Simon Fraser University, Canada
Yupeng Liu   BROADCOM, USA
Sreeja Ashok   Amrita Vishwa Vidyapeetham, Kochi, India
Parul Patel   Veer Narmad South Gujarat University, India
Mona Nasseri   University of Toledo, USA
Indhu R.   CDAC, India
Pranali Choudhari   Fr. C. Rodrigues Institute of Technology, India
Anastasia Douma   University of the Aegean, Greece
Aswathy Nair   Amrita School of Arts and Science, Kochi, USA
Zakia Asad   University of Toronto, Canada
Kala S.   Indian Institute of Science Bangalore, India

Pınar Kırcı	Istanbul University, Turkey
Lee Chung Kwek	Multimedia University, Malaysia
Azian Azamimi Abdullah	Nara Institute of Science and Technology, Japan
Archanaa Rajendran	Amrita, India
Marina Zapater	Universidad Politécnica de Madrid, Spain
Piyali Das	University of Cincinnati, USA
Divya G.	Asiet Kalady, India
Medina Hadjem	Université Paris Descartes, France
Pallavi Meharia	University of Cincinnati, USA
Priyanka Shetti	Amrita Vishwa Vidyapeetham, India
Xiaoqian Wang	UTA, USA
Mingyuan Yan	Georgia State University, USA

Contents

Part I
Intelligent Techniques and Applications

Butterfly Mating Optimization

**Chakravarthi Jada, Anil Kumar Vadathya, Anjumara Shaik,
Sowmya Charugundla, Parabhaker Reddy Ravula
and Kranthi Kumar Rachavarapu**

Abstract This paper presents a novel swarm intelligence algorithm named as Butterfly Mating Optimization (BMO) which is based on the mating phenomena occurring in butterflies. The BMO algorithm is developed with novel concept of dynamic local mate selection process which plays a major role in capturing multiple peaks for multimodal search spaces. This BMO algorithm was tested on 3-peaks function and various convergence plots were drawn from it. Also, BMO was tested on other benchmark functions to check and discuss thoroughly its capability in terms of capturing the local peaks. Various comparisons were made between BMO and GSO, a recent swarm algorithm for multimodal optimization problems. BMO was also tested on a function with varying dimensionality at higher level. Finally based on various assumptions through simulations, possible future work is discussed.

Keywords Butterflies · Patrolling · Mating · UV distribution · Multimodal Optimization · Glowworms · Local mate selection · Bfly

1 Introduction

Nature is a hidden place for magical powers used in many ways in the world everyday. It has been a great source of inspiration for many inventions. Since the time of evolution it has solved many complex problems efficiently keeping in view all the constraints. The property of emergence through collective behavior is remarkably observed in nature. For instance, properties of water are not present in *Hydrogen* and *Oxygen* alone, but the specific combination of them emerges into water, which made life possible on Earth. In the similar way many groups of organisms together achieve a specific task by their collective behavior often described as *Swarm Intelligence*,

C. Jada · A.K. Vadathya(✉) · A. Shaik · S. Charugundla · P.R. Ravula ·
K.K. Rachavarapu
Rajiv Gandhi University of Knowledge Technologies, Hyderabad 504107, India
e-mail: {chakravarthij,anil.rgukt,anju.june.25,charugundlasowmya,
        prabhakar.ravula,rkkr.2100}@gmail.com

© Springer International Publishing Switzerland 2016   3
S. Berretti et al. (eds.), *Intelligent Systems Technologies and Applications*,
Advances in Intelligent Systems and Computing 384,
DOI: 10.1007/978-3-319-23036-8_1

which has made them emerge into efficient tools and make their survival possible [1]. For instance, colony of ants emerged as the foragers of food; school of fishes as the best self defenders; flock of birds as the migrators and foragers; glowworm swarm as the prey attractors, etc. In all these natural swarms, though the achievement of task is due to collective behavior the progress of each organism is decentralized by considering its own social and cognitive behavior. In nature what we see the so called randomness in Bee's dance, swaying of plants, butterfly movement is not really random but it has got a meaning in its own sense. One of these randomnesses is seen in the butterfly's movement. If we can understand its behavior and deduce a metaphor, it can lead us to solve many engineering problems like cooperative multi-agent systems and optimization problems. This paper explains an engineering metaphor based on butterfly mating phenomenon.

This paper is organized as follows: Prior related work is discussed in Section 2. Section 3 explains the butterfly mating mechanism used in this paper. Butterfly metaphor - the main contribution of the paper is explained in Section 4. Extensive simulation results followed by the discussions are presented in the Section 5. BMO test on higher dimensional function is discussed in Section 6. The paper concludes with few remarks in Section 7.

2   Literature Survey

Many of the engineering metaphors proposed by mimicking the swarm intelligence were applied to diversified fields such as classification, localization and clustering etc. Most of these metaphors fall under the roof of Optimization field, in which one class of problems deals with finding the global optimum and the another with finding multiple optima (not necessarily equal) [2], [3]. The advantage of having multiple optima is obtaining insight into the multimodal function landscape and an alternative solution can be chosen if the behavior of constraints in the search space makes previous optimum solution infeasible to implement. In these engineering models, Muller et al. [4] presented an optimization algorithm based on the bacterial chemotaxis to find solutions to multimodal functions. Deneubourg et al. [5] investigated the pheromone laying and following behavior of ants as an example of stigmergy using the Double Bridge Experiment. This inspired the development of one of the many ant algorithms proposed by Bilchev and Parmee [6] to solve continuous optimization problems. Marco Dorigo and Thomas Stutzle [7], have formalized Ant Colony Optimization (ACO) into meta-heuristic for combinatorial optimization problems and solved Travelling Salesman Problem (TSP). Krishnanand and Ghose [8] used the concept of luciferin released by Glowworm for attracting other glowworms and proposed a Glowworm Swarm Optimization (GSO) algorithm. By incorporating the concept of variable decision domain range, the GSO algorithm captured local peaks for multimodal search spaces and was used to detect multiple source locations with applications to collective robots. Particle Swarm Optimization (PSO) [9], [10] is a population-based search algorithm where each particle exhibits a taxis behavior towards favourable regions in the searching process when its velocity is dynamically adjusted according to the history of itself and its neighbours. Parsopoulos and Vra-hatis [11], found the local optima of multimodal functions sequentially by combining PSO variant with constriction factor along with repulsion technique.

Similar to the above mentioned swarm behavior and metaphors, the communi-cation strategies of butterflies have scope for inspiration in deducing optimization algorithms. The communication in butterflies is mainly for mating; they either adopt patrolling or perching for their mate location. In patrolling, the male butterflies fly in search of females using color and odor to screen females. On the other hand, in perching, males spend long time sitting on a prominent height so that they could survey and intercept passing females using their movement and size [12]. The next section concentrates only on the patrolling mating mechanism of butterflies in detail and the experimental simulations done previously which are taken as the basis in developing the proposed algorithm.

3 Patrolling Mating Mechanism

In Patrolling, the male butterflies are mobile. They fly in search of females con-tinuously using color and odor as the visual and olfactory parameters for the ini-tial approach and during courtship [12]. For mate locating purpose, butterflies use pheromones as the trait for shorter distances and production of iridescence (falls in the region of Ultraviolet) for longer distances. The male butterflies in general are capable of reflecting UV and the females can absorb or reject this reflected UV. The eye spots present on the dorsal fore-wings of males reflect UV and the females detect it using the receptors which can perceive higher frequency colors. After the accep-tance from female the male joins it for copulatory action or heads away if the female refuses [13], [14]. Sowmya et al. [15] conducted extensive experimental simulations of patrolling with various mate selecting schemes and the results strongly promote the localization of agents in the search space. They also stated that the localization of search space can be improved if interactions were made among males and females individually unlike their simulations in which mating was done between a male and female only. This paper takes that work as a basis to deduce a butterfly engineer-ing metaphor named as **"Butterfly Mating Optimization (BMO)"** for capturing multiple peaks in multimodal functions explained in the next section.

4 Butterfly Mating Optimization

4.1 Description of BMO Algorithm

As mentioned in the Section 3, while patrolling, butterflies mainly use UV reflectance (by males) and absorbance (by females) mechanism for mating. This proposed "But-terfly Mating Optimization" assumes that there is no differentiation among males and females and every butterfly reflects UV it has got and absorbs UV that it receives from all the remaining butterflies simultaneously. Hence this algorithm suggests a meta-butterfly model which replaces "*Butterfly in the natural environment*" with "*Bfly in the search space*". BMO aims at solving multimodal optimization problems. In this each Bfly chooses its mate anywhere in the search space adaptively in each iteration, we call this mate as a local mate i.e. *l-mate*. This adaptive selection process of *l-mate* plays a key role in the BMO algorithm. Our algorithm starts with a well random dispersion of Bflies on the search space. Then each Bfly updates its UV, distributes to and accesses UV from all the others, chooses *l-mate* and moves towards that. The algorithm's description is given below:

4.1.1 UV **Updation Phase**

In this, the UV of each Bfly is updated in proportion to it's fitness i.e function value at the current location of Bfly, as given below.

$$UV_i(t) = max\{0, b_1 * UV_i(t-1) + b_2 * f(t)\} \qquad (1)$$

UV is updated at time index 't', giving more importance to the current fitness and less to the previous UV, accordingly choose the b_1, b_2 values such that $0 \le b_1 \le 1$ and $b_2 > 1$.

4.1.2   **UV Distribution Phase**

Here, every Bfly distributes its UV to remaining Bflies such that the nearest Bfly gets more share than the farthest one. To distribute in this way, the following approach has been followed: An ith Bfly having UV_i reflects it's UV value to the jth Bfly at a distance d_{ij} which is given by

$$UV_{i \to j} = UV_i \times \frac{d_{ij}^{-1}}{\displaystyle\sum_k d_{ik}^{-1}} \qquad (2)$$

where $i = 1, 2, \ldots, N$; N is the number of Bflies; $j = 1, 2, \ldots, N$ & $j \ne i$; $UV_{i \to j}$ is UV absorbed by j^{th} Bfly from i^{th} Bfly; d_{ij} is euclidean distance between i^{th} and j^{th} Bfly; $k = 1, 2, \ldots, j, ., N$ & $k \ne i$; d_{ik} is euclidean distance between i^{th} and k^{th} Bfly.

4.1.3   *l-mate* **Selection Phase**

If every Bfly chooses another Bfly which has maximum UV as its *l-mate* and moves towards it, this leads to a rendezvous of all Bflies to a single peak. So, if this *l-mate* choosing process is made adaptive, disjoint groups of Bflies can be formed. Keeping this in view, BMO proceeds with following approach of *l-mate* selection.

Initially an ith Bfly arranges for itself all remaining Bflies in descending order based on the amount of UV it has received from them. Now, if every Bfly chooses the

1st Bfly in its descending order as its *l-mate* and move towards that, then this behavior will lead to some kind of localization and sensing of peaks which is discussed in Section 5. But to capture local peaks simultaneously an ith Bfly should also consider the UV of remaining Bflies; by sequentially comparing UV_i with UV values of Bflies arranged in the descending order, then it chooses the first encountered Bfly which satisfies below condition as it's *l-mate* i.e., which has got more UV than it has.

$$UV(i^{th} \ Bfly) < UV(j^{th} \ Bfly) \tag{3}$$

where $i = 1, 2, \ldots, N$; $j = 1, 2, \ldots, N - 1$, j is index of Bflies in the descending order of ith Bfly.

4.1.4   Movement Phase

Each Bfly is moved in the direction of its *l-mate* according to Eqn. 4.

$$x_i(t + 1) = x_i(t) + B_s * \left\{ \frac{x_{l\text{-}mate}(t) - x_i(t)}{\|x_{l\text{-}mate}(t) - x_i(t)\|} \right\} \tag{4}$$

where B_s is Bfly step size; $x_i(t)$ is the position of ith Bfly in a time index t. The pseudo code for the BMO algorithm is given below.

Butterfly Mating Optimization
Randomly initialize Bflies;
$\forall \ i$, set $UV_i = UV(0)$;
Set maximum number of iterations $= iter\_max$;
Set $iter = 1$;
while ($iter \leq iter\_max$) do:
{
    for each Bfly i do:
       UV Updation; %using Eqn. 1
       UV Distribution; %using Eqn. 2
    for each Bfly i do:
       Select *l-mate*; %using $l - mate \ selection \ phase$
       Update position; %using Eqn. 4
    $iter = iter + 1$;
}

4.2   Behavior of l-mate Selection Phase

In BMO algorithm, *l-mate* locating range is dynamic i.e. changes w.r.t each iteration. This is analogous to the variable decision domain range $r_d(t)$ in GSO algorithm [8]. However in BMO *l-mate* locating range varies according to UV distribution and *l-mate* selection phase. This dynamic behavior can be understood illustratively from Figure 1, 2, in which the static placements of Bflies at time index t were considered.

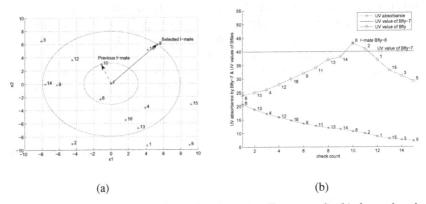

(a)   (b)

Fig. 1 (a). Dynamic behavior of *l-mate* locating range (Incremental), (b). *l-mate* locating strategy plot

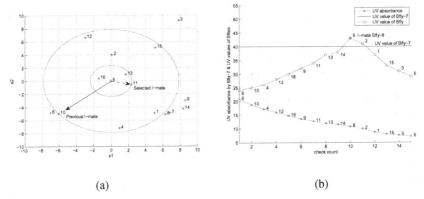

(a)   (b)

Fig. 2 (a). Dynamic behavior of *l-mate* locating range (Decremental), (b). *l-mate* locating strategy plot

According to *l-mate* selection phase, previously Bfly-10 was the *l-mate* of Bfly-7 (located at $(0, 0)$). Figure 1.(b) shows that Bfly-8 is the next *l-mate* of Bfly-7 and hence there is an *increment* in the *l-mate* locating range from Bfly-10 to Bfly-8 as shown in Figure 1.(a). Similarly, Figure 2.(b) shows that Bfly-11 is the next *l-mate* for Bfly-5 (located at $(0, 0)$) after Bfly-10, and hence there is a *decrement* in the *l-mate* locating range from Bly-10 to Bfly-11 as seen in Figure 2.(a).

4.3 GSO Algorithm vs BMO Metaphor

In GSO, glowworms use information in the variable decision domain range $r_d(t)$, controlled by the threshold number of neighbours, to select their mate based on luciferin they expose. This luciferin is updated in proportion to fitness of the glowworm represented at its current location. Overall, the agent which has higher luciferin attracts more agents towards it than others, enabling the agents to move towards the favourable places in the search space. BMO algorithm is somewhat similar to GSO but with significant differences in the mate selection. In BMO adaptive *l-mate* selection phase, discussed in Section 4.1, is such that a Bfly selects the nearest Bfly which passes highest UV towards it as its *l-mate* which results in capturing of local peaks. However, a few exceptions are observed to this *l-mate* selection which are discussed in Section 5.1.

5 Simulations and Results

BMO algorithm has been validated using various multimodal benchmark functions and for illustrative purpose 3-peaks function is considered for presenting extensive simulation results. This function $f_1(x_1, x_2)$ has 3 peaks and 2 valleys in $[-3, 3]^2$, here BMO is aimed at capturing all the three peaks located at $(0.06, 1.625)$, $(-0.45, -0.55)$, $(1.35, 0.05)$ as shown in Figure 3.

$$f_1(x_1, x_2) = 3(1-x_1)^2 e^{-[x_1^2+(x_2+1)^2]} - 10(\frac{x_1}{5} - x_1^3 - x_2^5)e^{-[x_1^2+x_2^2]} - \frac{1}{3}e^{-[(x_1+1)^2+x_2^2]}$$

(5)

The algorithmic parameters are set as 50 Bflies with step size of 0.02, $b_1 = 0$ and $b_2 = 3$ for 350 iterations.

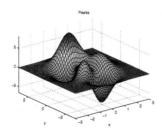

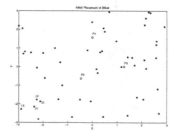

Fig. 3 3-D Surface plot of f_1 **Fig. 4** Initial deployment of Bflies

Initially the first ranked Bfly in the descending order after UV distribution is directly selected as *l-mate* [**case(1)**] and simulations were carried out using this *l-mate* selection process. Figure 4 shows the initial deployment for case(1) along with peak locations, Figure 5 shows emergence of Bflies at various iterations ($t = 50, 80, 350$) and Figure 7.(a) shows the Bflies positions at final iteration on f_1's contour. Here we observe that the Bflies were emerging into disjoint groups creating an atmosphere of localization by forming groups nearer to the peaks thereby sensing the peak locations which is a crucial result of case(1). This localizing can also be observed from Figure 8.(a) which shows the UV converged values of all Bflies vs iterations. However, because of this blind selection of *l-mate*, Bflies were getting struck after forming the groups and as a result none of the peaks is captured. To solve this problem, UV comparison of that particular Bfly with the remaining ones is introduced which is already discussed in Section 4.1. Considering this proposed

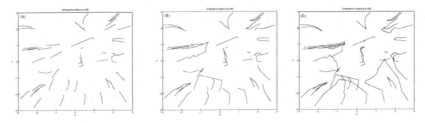

Fig. 5 Emergence plot at various iterations

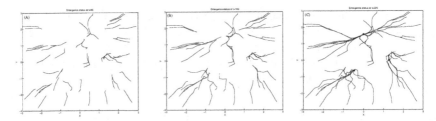

Fig. 6 Emergence plot at various iterations

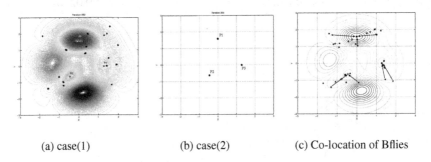

(a) case(1)   (b) case(2)   (c) Co-location of Bflies

Fig. 7 Final location of Bflies

l-mate selection process [**case(2)**] simulations were carried and results are compared with the case(1). For comparison, same initial deployment of Bflies and algorithm parameters were considered for both cases. Figure 6 shows the emergence of Bflies at various iterations ($t = 65, 100, 220$) with the proposed *l-mate* selection phase. Finally we observe that all the peaks were captured as shown in the Figure 7.(b). Figure 7.(c) shows the co-location and direction of Bflies after they enter into the contours of the peaks supporting the idea that they would ultimately reach the aimed peaks. Figure 8.(b) shows the variation of UV for all Bflies and its convergence. Figure 8.(c) shows the UV convergence of 4 Bflies (18, 25, 26, 37), shown in Figure 4, in both the cases. In the case(1) Bflies were stuck at some value of UV and for case(2) they clearly converged to the peak(P1).

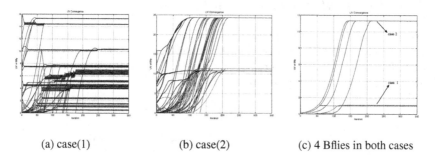

(a) case(1)   (b) case(2)   (c) 4 Bflies in both cases

Fig. 8 UV convergence

5.1 BMO vs GSO

As explained in the Section 4.3, the main difference between GSO and BMO arises in the selection of mate. For comparison, we consider the same initial deployment of agents for both the metaphors and for the same number of iterations. The Figures 9, 10 show the emergence plots of GSO and BMO respectively capturing all the peaks of f_1. In GSO, we can see that agents were reaching to the nearby peaks unlike in BMO where Bflies 2, 40 reached P1 instead of P3; Bflies 22, 31 reached P1 instead of P2. This is because here decision of choosing a mate is not restricted like in GSO to some range. Agents 7, 30 initially started towards a Bfly(i) near P2, which is the nearest one with highest UV but later diverted towards P1 this is a fall out of UV distribution Eqn.(2); where if the distance d_{ij} is very small then much of the UV from i^{th} Bfly (nearly 80-90%) is absorbed by j^{th} Bfly thus leaving less amount to others, so they may get attracted by some other Bfly which gives them more UV. In addition in BMO we can see that agents were almost forming line (queuing) after they reach near to the peaks, which is also due to the UV distribution Eqn.(2). Figure 11 shows the $r_d(t)$ and *l-mate* distance variation for all iterations for agent 50, where we can observe that $r_d(t)$ is more oscillatory in GSO than in BMO in two cases. An advantage of BMO is that the tuning of number of parameters is less when compared to GSO.

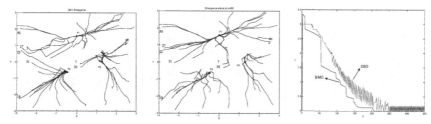

Fig. 9 Emergence plot of GSO

Fig. 10 Emergence plot of BMO

Fig. 11 $r_d(t)$ variation for agent 50

5.2 BMO for Other Standard Test Functions

To verify if BMO algorithm captures larger number of peaks, we present the results
that have validated BMO using two benchmark test functions f_2, f_3 given in equations 6, 7. Schwefel function $f_2(x_1, x_2)$ has 16 peaks in $[-150, 150]^2$ and Rastrigen
function $f_3(x_1, x_2)$ has 100 peaks in $[-5, 5]^2$.

$$f_2(x_1, x_2) = -x_1 \sin(\sqrt{|x_1|}) - x_2 \sin(\sqrt{|x_2|}) \tag{6}$$

$$f_3(x_1, x_2) = 20 + x_1^2 + x_2^2 - 10\cos(2\pi x_1) - 10\cos(2\pi x_2) \tag{7}$$

To test the Schwefel function, 500 Bflies were deployed in the search space with
the step size of 0.5, $b_1 = 0$, $b_2 = 3$ for an iteration count of 400. Figure 12, 13,
14 show the surface plots of Schwefel function, emergence of all Bflies, and final
co-location of Bflies respectively. We observe that BMO was able to capture 95%
of peaks of the Schwefel function. To test the Rastrigen function, 1000 Bflies were
deployed in the search space and was run for 120 iterations for a step size of 0.009,
$b_1 = 0.1$, $b_2 = 3$. Figures 15, 16, 17 show the surface plots of Rastrigen function,
emergence of Bflies and final co-location of all Bflies at last iteration respectively.
It is observed that BMO was able to capture 94% of peaks. In the emergence plots
of both Schwefel function and Rastrigen function, we observe that the Bflies were
almost forming queues while entering into the contour and are co-locating at the
peaks due to the adaptive *l-mate* selection process.

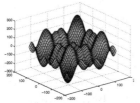

Fig. 12 3-D plot of Schwefel f_2

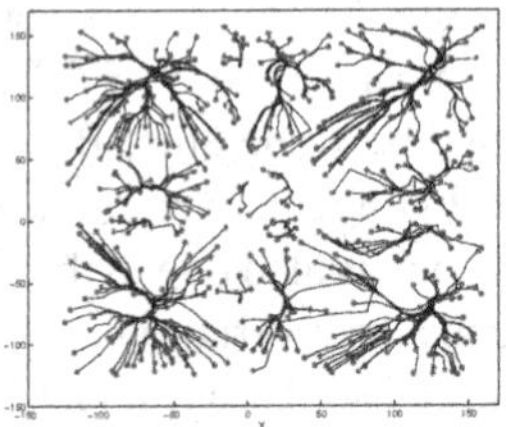

Fig. 13 Emergence plot of all Bflies

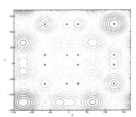

Fig. 14 Co-location of Bflies at final iteration

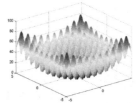

Fig. 15 3-D plot of Rastrigin f_3

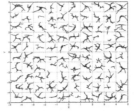

Fig. 16 Emergence plot of all Bflies

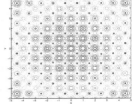

Fig. 17 Co-location of Bflies at final iteration

Table 1 Number of Bflies required for a peak-capture of 50% of f_4 function for $m = 1, \ldots, 5$

m	Number of Peaks	n_m	$n_m/n_{(m-1)}$
1	3	6	
2	9	41	6.83
3	27	150	3.66
4	81	500	3.33
5	243	2000	4

6 Application of BMO to Higher Dimensional Functions

In the previous section, we have carried out experiments for two dimensional cases. Here we show the efficacy of BMO in higher dimension spaces as well. We use the Equal peaks function $f_4(x_1, \ldots, x_m)$ for this analysis, where a m-dimensional space has $(2k + 1)^m$ peaks in $\prod_{i=1}^{m}[-k\pi, k\pi]$ range. Algorithmic parameters are kept constant for all trials as $b_1 = 0$, $b_2 = 0.3$ and $stepsize = 0.1$ and $k = 1$

$$f_4(x_1, x_2, \ldots, x_m) = \sum_{i=1}^{m} cos^2(x_i) \qquad (8)$$

Figures 18.(a) and (b) show the number of bflies co-located at each peak for dimension, $m = 4$, with 91 peaks for 500 and 2000 bflies respectively. As expected, as the number of bflies(n) increases, the number of peaks captured also increases. Figure 18.(c) shows variation in the average of distance from bflies to their finally co-located peaks as the iterations progress. We can see that as n increases the rate of convergence also increases. To characterize the dependence of the required number of bflies on the dimensionality of the problem, the number of bflies n_m required for 50% peak-capture for m = 1,...,5 are obtained and enumerated in Table 1. The values of $n_m/n_{(m-1)}$ show that the number of bflies needed to capture a given fraction of the total number of peaks does not increase exponentially.

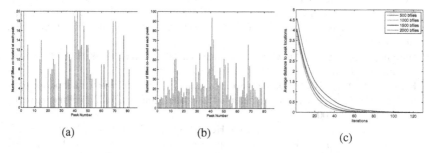

(a)   (b)   (c)

Fig. 18 Number of Bflies co-located at each peak for $m = 4$ and (a) $n = 500$ (b) $n = 2000$ (c) Average of distance from each bfly to the finally co-located peak as the iterations progress for m = 4 for n = 500,1000,1500 and 2000

7   Conclusions and Future Works

In this paper we have presented a novel Butterfly metaphor - Butterfly Mating Optimization (BMO) algorithm for capturing all the peaks of a continuous multimodal function. Initially the need for multimodal functions has been discussed followed by the study of previously existing related work. Next the Patrolling Mating mechanism has been explained followed by the phase wise detailed description of algorithm. Extensive results were presented which proved the convergence of algorithm pictorially. The comparison of BMO with GSO was done and the results were presented. Finally the validation of BMO was shown by using it on the standard benchmark functions for capturing the peaks also at higher dimensionality. In future, there is a scope to work on the following proposals: Consideration of other attracting traits of butterflies and using them in the algorithm; Mathematically proving the UV convergence of Bflies and explaining the formation of disjoint groups clearly; Analysing the parameter variation (no. of agents, no. of iterations, step size) and also the comparison of BMO and other existing algorithms; Clearly analysing the 'queueing' of Bflies while reaching the peaks; Application of BMO to signal source localization by either mobile robots or flying vehicles which can mimic the behavior of Bflies in BMO; Application in soft computing field for data clustering and classification etc.

References

1. Bonabeau, E., Dorigo, M., Theraulaz, G.: Swarm Intelligence: From Natural to Artificial Systems. Oxford University Press (1999)
2. Goldberg, D.E., Richardson, J.: Genetic algorithms with sharing for multi-modal function optimization. In: Grefenstette, (ed.) Genetic Algorithms and their Applications, ICCGA 1987, pp. 41–49 (1987)
3. Lung, R.I., Dumitrescu, D.: Roaming optimization: A new evolutionary technique for multimodal optimization. Studia Univ. Babes Bolyai, Informatica **XLIX**(1), 99–109 (2004)
4. Muller, S.D., Marchetto, J., Airaghi, S., Koumoutsakos, P.: Optimization based on bacterial chemotaxis. IEEE Transactions on Evolutionary Computation **6**(6), 16–29 (2002)
5. Deneubourg, J.L., Aron, S., Goss, S., Pasteels, J.M.: The self-organizing exploratory pattern of the Argentine ant. Journal of Insect Behaviour **3**(2), 159–168 (1990)
6. Bilchev, G., Parmee, I.C.: The ant colony metaphor for searching continuous design spaces. In: Fogarty, T.C. (ed.) AISB-WS 1995. LNCS, vol. 993, pp. 25–39. Springer, Heidelberg (1995)
7. Dorigo, M., Stutzle, T.: Ant Colony Optimization. A Bradford Book. The MIT Press, Cambridge (2004)
8. Krishnanand, K.N., Ghose, D.: Glowworm swarm optimisation: a new method for optimising multi-modal functions. International Journal of Computational Intelligence Studies **1**(1), 93–119 (2009)
9. Clerc, M.: Particle Swarm Optimization. Hermes Science Publications (April 2006)
10. Kennedy, J., Eberhart, R.: Particle swarm optimization. In: Proceedings of the Fourth IEEE International Conference on Neural Networks, pp. 1942–1948. IEEE Service Center, Perth (1995)

11. Parsopoulos, K.E., Plagianakos, V.P., Magoulas, G.D., Vrahatis, M.N.: Stretching technique for obtaining global minimizers through particle swarm optimization. In: Proceedings of Particle Swarm Optimization Workshop, pp. 22–29 (2001)
12. Rutowski, R.L.: Sexual Selection and the Evolution of Butterfly Mating behaviour. Journal of Research on Lepidoptera, 125–142 (1984)
13. Andersson, J., Borg-Karlson, A.K., Vongvanich, N.: Wiklund, C.: Male sex pheromone release and female mate choice in a butterfly. Journal of Experimental Biology, 964–970 (2007)
14. Robertson, K.A., Monteiro, A.: Female Bicyclus Anynana butterflies choose males on the basis of their dorsal UV reflective eyespots. Proceedings of the Biological Sciences/The Royal Society, 1541–1546 (2005)
15. Sowmya, C., Shaik, A., Jada, C., Vadathya, A.K.: Butterfly communication strategies: a prospect for soft-computing techniques. In: Proceedings of International Joint Conference on Neural Networks (ICJNN), pp. 424–431 (July 2014)

A Hybrid Firefly Algorithm and Social Spider Algorithm for Multimodal Function

Samiti Gupta and Sankalap Arora

Abstract The fast growing complexity of optimization problems has motivated researchers to search for efficient problem solving methods. In this paper, the concept of hybridization is introduced to solve the optimization problems which make the use of concept of exploration and exploitation over search space efficiently. The proposed algorithm is formulated by combining the biological processes of Firefly Algorithm (FA) and Social Spider Algorithm (SSA). The proposed algorithm is tested on various standard benchmark problems and then compared with FA and SSA. The results show that the proposed algorithm performs better than the FA and SSA on most of the benchmark functions.

Keywords Swarm Intelligence · Firefly Algorithm · Social Spider Algorithm · Hybridization

1 Introduction

Swarm Intelligence (SI) Algorithms is artificial intelligence technique used for solving large and complex optimization problems. SI algorithms are based on how the agents in the population interact with each other, locally and with their environment which results into "intelligent" global behavior. SI algorithms are self organized, decentralized and autonomous in nature [1, 4, 16, 33]. Conventional methods uses standard operating functions to react to predetermined stimuli but swarms have ability to adjust to new situation. Conventional systems can shift the locus of adaption from one part of system to another while in the swarm systems, the individual randomness lead to frequent novelty, which leads to evolution.

Researchers formulate optimization algorithms in order to find efficient ways to mimic evolution of a self-organizing system. Artificial Bee Colony algorithms (ABC) [9], Particle Swarm Optimization (PSO) [17], Social Spider Algorithm (SSA) [18], Genetic Algorithm (GA) [20], Firefly Algorithm (FA) [29] are some

S. Gupta(✉) · S. Arora
Computer Science Department, DAV University, Jalandhar, Punjab, India
e-mail: {samitigupta,sankalap.arora}@gmail.com

© Springer International Publishing Switzerland 2016
S. Berretti et al. (eds.), *Intelligent Systems Technologies and Applications*,
Advances in Intelligent Systems and Computing 384,
DOI: 10.1007/978-3-319-23036-8_2

17

SI based algorithms. ABC categorizes the bee in hive into three types: scout bees, employed bees and onlooker bees. The algorithm uses employed and onlooker bees for exploitation and the scout bees for exploration. PSO represents the movement of organisms in bird flock or fish school. It searches the space of the objective functions by adjusting the boundaries of individual particles over particle's position and velocity. SSA describes the foraging behavior of social spiders towards the food source position. Spiders identify their prey by notifying the vibrations on their web. GA is a heuristic approach to find optimum solutions to optimization problems using evolutionary operators, such as inheritance, mutation, selection and crossover. FA is a nature-inspired metaheuristic optimization algorithm inspired by the flashing behavior of fireflies [2, 31, 32].

In this paper research finding for firefly algorithm (FA) and social spider algorithm (SSA) are investigated separately for creating a new hybridized algorithm which uses the concept of intensification (exploitation) and diversification (exploration) efficiently. The proposed algorithm will be compared with FA and SSA, and then investigated for performance parameters viz. (i) Complexity, (ii) Time consumption, (iii) Convergence under various test bed benchmark functions [2, 4, 29].

The paper is organized in following manner: in next section the related work on swarm intelligence and nature-inspired algorithms are discussed. Then FA and SSA are elaborated by idealizing and imitating the foraging behavior of fireflies and social spiders respectively. In next section presents proposed algorithm. The subsequent part of the paper introduces benchmark functions used for testing performance of FA, SSA and proposed algorithm with the experimental setting. Then it follows by the results and comparison of algorithms. Finally, the paper will be concluded and propose some future scope.

2   Related Work

Swarm intelligence mimics the methods in nature to solve optimization problems. There are different metaheuristic algorithms used to drive a search for the optimal solution such as Particle Swarm Optimization (PSO), Ant Colony Optimization (ACO), Genetic Algorithm (GA), Firefly Algorithm (FA), Social Spider Algorithm (SSA) and Artificial Bee Colony Optimization (ABC).

PSO was developed by Kennedy and Eberhart which represents the behavior of organisms as in a flock of birds or school of fishes [7, 17, 23]. PSO uses the positions of swarm particle in search space to represent the feasible solutions of optimization problem. A particle is attracted towards current global best location and its own best location and can move randomly. PSO finds the global optimal value among all possible best solution until stopping criteria is matched.

ACO is inspired by the group foraging behavior of ants. The ants find shortest path from their colony to food sources. The ants communicate using a volatile chemical substance called pheromone [11, 21]. When an ant finds a food source, it deposits certain amount of pheromone along the path which starts evaporating with the passage of time, therefore decreases its attractive strength. The pheromone density is more on shorter paths than on longer ones, so the shorter paths are covered more quickly. Pheromone evaporation helps in achieving exploration, hence limiting exploitation. The path selection is made by other ants by lying down the pheromone trails, providing positive feedback. Using this feedback, the algorithm leads the ants to find the shortest path in the graph representing the problem to solve.

GA mimics the processes of natural evolution to solve optimization problems. GA is more robust than the conventional artificial intelligence. GA is based on the behavior of chromosomes within a population of individuals [4, 20, 22]. The chromosomes are coded with finite length vector of variables or components called genes. The idea used by GA is to do selective reproduction of offspring better than the parents by amalgamating the information from the chromosomes.

FA mimics the behavior of fireflies which uses flashing light to attract the other fireflies towards them [26, 27, 29]. The flashing light is produced by a process of bioluminescence devised in such a way that it is associated with objective function to be optimized [28, 30, 31, 32]. FA is used for solving various optimization problems such as Travelling Salesman Problem (TSP) using discrete distance between two fireflies and the movement scheme [14].

SSA formulates the search space as a web on which each position represents a feasible solution to the optimization problem and all feasible solutions to the problem have corresponding positions on this web [12, 18]. Each spider on the web holds fitness value which is based upon the objective function [12, 13], and represented by the potential of finding a food source at the position. SSA is used for solving constrained optimization problems.

ABC depends upon the intelligent foraging behavior of bees namely: employed bees, onlooker and scout. The employed bees are equal in number to food sources. The employed bees do waggle dance once they choose their food source. Onlooker bees watch the dance of employed bees and select one of their food sources depending upon dance and move towards that source, after that it evaluates it nectar amount [9, 10]. The abandoned food source is determined and is exchanged with the new food source located by scouts.

3   Firefly Algorithm and Social Spider Algorithm

3.1   Firefly Algorithm

The functioning of fireflies relies on flashing light that is produced by a physiological process called bioluminescence [1, 29]. The flashing light is an outcome of chemical reaction when an oxygen molecule combines with calcium, Adenosine Tri-phosphate (ATP) and the chemical luciferin in the presence of an enzyme luciferase. Firefly emits light in order to attract potential prey and to attract mating partners. To be unappetizing to predators, fireflies produce defensive steroids from their bodies. Hence, the flashing light also safeguards fireflies from their enemies.

The light intensity follows Inverse-square Law, it decreases with increase in distance. Moreover, air also acts as a light absorbent, that decreases the light intensity as the distance increases. These two combined elements make fireflies visible at a very small distance.

FA follows three basic rules: (i) All fireflies are unisex, they attracted towards each other despite of their sex; (ii) Attractiveness is comparable to the brightness, therefore the less bright firefly move towards brighter firefly. It will move randomly if there is no brighter one; (iii) Brightness of a firefly is determined by a objective function.

There are two important points in firefly algorithm: (i) The change in light intensity; (ii) Construction of the attractiveness. For easy understanding, the attractiveness of firefly is examined by its brightness which is further related to the objective function. The brightness of firefly at specific position x can be written as $I(x) \infty f(x)$. As the distance increases, the brightness of the firefly decreases because light is absorbed in media like air, rain etc. Therefore, light intensity $I(r)$ obeys inverse square law $I(r) = I_s/r^2$ where I_s is the source intensity. The light intensity decreases for a particular absorption factor γ with distance. That can be explained as $I = I_0 e^{-\gamma r}$ where I_0 is initial light intensity. In order to neglect singularity at $r = 0$ in the expression I_s/r^2 the effects of both inverse square law and absorption is merged and can be approximated using Gaussian form as shown in Equation 1:

$$I(r) = I_0 e^{-\gamma r^2} \ . \tag{1}$$

At some time, a function is needed which decreases at a slower rate uniformly. That can be represented as in Equation 2:

$$I(r) = \frac{I_0}{1 + \gamma r^2} \ . \tag{2}$$

As the attractiveness of firefly is directly corresponds to light intensity visualized by other fireflies and is defined in Equation 3:

$$\beta(r) = \beta_0 e^{-\gamma r^2} \ , \tag{3}$$

where β_0 is value of attractiveness when $r = 0$.

The distance between firefly i and firefly j at x_i and x_j respectively, is calculated using Equation 4:

$$r_{ij} = \| x_i - x_j \| = \sqrt{\Sigma_{i=1}^{d}(x_{i,k} - x_{j,k})^2} \ , \tag{4}$$

where $x_{i,k}$ is the kth component of the spatial coordinate x_i of the ith firefly. The movement of less brighter firefly i moves towards the more brighter firefly j is determined by Equation 5:

$$x_i = x_i + \beta_0 e^{-\gamma r_{ij}^2}(x_j - x_i) + \alpha \left(rand - \frac{1}{2} \right), \tag{5}$$

where second term is due to attraction and third term α is randomization parameter. The pseudo-code of firefly algorithm (FA) are summarized in algorithm 1:

Algorithm 1. Pseudo code of firefly algorithm (FA).

1. *Define the objective function f(x) to be minimized or maximized $x = (x_1 \dots \dots \dots x_d)^t$*
2. *The initial population of fireflies is computed $x_i (i = 1, 2, 3 \dots \dots n)$*
3. *The light intensity I is calculated that is in direct proportionate to objective function f(x)*
4. *Define absorption coefficient of light γ*
5. *while (t < maximum_generations)*
   for i = 1: n (all n fireflies)
      for j=1: i (all n fireflies)
      if ($I_i > I_j$),
         Move less brighter firefly i towards more brighter firefly j in all d dimensions;
         end if
         Attractiveness varies with distance r by means of $e^{-\gamma r^2}$
         Calculate the new solutions and re-evaluate the   light intensity
      end for j
   end for i
      The fireflies are ranked based on light intensity and discover the current best.
   end while

3.2   Social Spider Algorithm

SSA mimics the foraging behavior of spiders to perform optimization over the search space [18]. Spiders use their vibrations to determine the position of prey and as a protective warning mechanism for themselves. There are two proposed models namely information sharing (IS) model and producer scrounger (PS) model [5, 8]. The searching pattern of SSA is controlled using IS model which is based on the individual searching who look for an opportunity to join other individuals simultaneously.

The search space of SSA is devised in the form of web where each position is related to feasible solution of the problem and corresponds to the fitness value of the objective function. The quality solution (fitness) of the problem corresponds to the potential of finding a food source at the position. The vibration is transmitted over the web when the spiders move to new position and other spiders can sense it, this is how collective social knowledge is shared between them.

At the starting of the algorithm, a pre-defined number of spiders are set to the random positions on the web. Each spider s in the search space holds the information in its memory: (i) their respective position on the web, (ii) fitness of the current position of s and (iii) target vibration in the last iteration. The intensity of the vibration generated by the spider is correlated to the fitness value of the position. The vibrations generated by spider is defined by two properties viz. the source position and the intensity of the vibration of the source. The source position over the search space is defined by the objective function and the intensity of vibration varies in the range $[0, +\infty)$. The position of spider a at time t is defined as $P_a(t)$. The intensity of the vibration of spider a sensed by spider b at time t is defined as $I(P_a, P_b, t)$. The vibration intensity generated by spider s on the source position is defined as $I(P_s, P_s, t)$. This is directly correlated to the fitness of the source position $f(P_s)$ and is define in Equation 6 and Equation 7:

$$I(P_s, P_s, t) = \left(\frac{1}{C_{max}} - f(P_s) \right) \text{ for maximization}, \tag{6}$$

$$I(P_s, P_s, t) = \left(\frac{1}{f(P_s)} - C_{min} \right) \text{ for minimization}, \tag{7}$$

where the value of constant C_{max} is selected such that it is confidently large among all the possible fitness values of the maximization problem and the value of constant C_{min} is selected such that it is confidently small among all the possible values of the minimization problem. Equation 6 is used for maximization problem and Equation 7 is used for minimization problem. Both the equations ensure that, the vibration intensities are the positive values and guarantee the better fitness values. The vibration generated by spider attenuates over time and distance.

(i) Attenuation Over Distance. $D(P_a, P_b)$ Defines the distance between spider a and spider b. The maximum distance between two points is defined as D_{max} in the search space and is problem independent. For simplicity, the Equation 8 is used:

$$D_{max} = \left\| \bar{x} - \underline{x} \right\|_p, \tag{8}$$

where $\bar{x}$ is defined as upper bound and $\underline{x}$ is the lower bound of the search space. p indicates p-norm method i.e. it is used to calculate the distance. The distance between spider a and spider b is calculated in Equation 9:

$$D(P_a, P_b) = \left\| P_a - P_b \right\|_p. \tag{9}$$

In order to compute distance in this paper, Manhattan norm or 1-norm is used. The vibration attenuation over distance is calculated in Equation 10:

$$I(P_a, P_b, t) = I(P_a, P_a, t) \times \exp\left(-\frac{D(P_a, P_b)}{D_{max} \times r_a} \right). \tag{10}$$

The r_a is a user controlled parameter that controls the attenuation rate over distance defined in the range $(0,1)$.

(ii) **Attenuation Over Time.** In order to prevent pre-mature convergence, the affect of the previous vibrations is properly attenuated. The attenuation of vibration over time is defined by following Equation 11:

$$I(P_a(t), P_a(t), t+1) = I(P_a, P_a, t) \times r_a .\qquad (11)$$

The value of r_a attenuates all the previous vibrations. For simplicity of the parameter tuning, the same parameter is used as in case of vibration attenuation over distance. The position of spider a may change to $P_a(t+1)$ at time$(t+1)$ but the source position remains at $P_a(t)$.

The working of SSA is divided into three phases: initialization, iteration and final. In each pass of SSA, it begins with the initialization phase, then searching is executed in an iterative procedure and in the final phase, the algorithm terminates when the stopping criteria is matched and output the results found.

In the initialization phase, the objective function and its feasible region are defined. The value of parameters used in SSA is assigned an initial population is created in the search space. During the simulation of SSA, the total number of spiders remains constant and fixed memory location is given to each spider to store information.

The positions of spiders are generated in the search space randomly and fitness values is calculated and stored. The target vibration of each spider is set at its current position in the population and the vibration intensity is zero.

In the iteration phase, all spiders on the web move to a new position and their fitness values are calculated. The fitness values of all the artificial spiders are evaluated on different positions on the web. By using Equation 7, spiders generate vibrations at their positions. Once all the vibrations are generated, the propagation process of these vibrations using Equation 10 is calculated. Each spider s will receive different vibrations V generated by other spiders. The received information involves the source position of the vibration and its attenuated intensity. The strongest vibration $vbest$ from V is selected for each spider and compare its intensity with the target vibration intensity $vtar$ stored in its memory. If the intensity of $vbest$ is more than the $vtar$, than the s will store $vbest$ as $vtar$, otherwise the original value of the $vtar$ remains same. Then spider s performs a random walk towards $vtar$. This random walk is equated in Equation 12:

$$P_s(t+1) = (P_{tar} - P_s) \odot (1 - R \odot R) ,\qquad (12)$$

where $\odot$ represents element-wise multiplication. $P_s(t+1)$ is the position of spider s at time $t+1$ after random walk and P_{tar} is the vibration source position of the target vibration $vtar$. R is a vector of random numbers generated from zero to one and 1 is vector of ones, whose length is equal to the dimension of the objective function. An artificial spider jump away process is introduced in order to prevent spider s for getting trapped in local optima. After the random walk step, there is a small probability to decide whether the spider s will follow its present target or jump away from its current position. The probability is calculated in Equation 13:

$$p_j = \frac{r_j}{\exp\left(\frac{D(P_s, P_{tar})}{D_{max}}\right)} \ , \qquad (13)$$

where r_j is a user-defined jump away rate parameter. If the spider s is selected to jump away, a new random position is generate and assigned as the new position of s in the search space. Finally, the algorithm attenuates the intensity of the stored target vibration using Equation 11 and with this the iteration phase finishes.

The iteration phase will stop when the stopping criteria is matched. The stopping criteria can be defined as the maximum iteration number reached, the maximum CPU time used, the maximum number of iterations with no improvement on the best fitness value, or any other appropriate criteria. When the iteration finishes, the algorithm outputs the optimal value. The above three phases comprises the complete algorithm of SSA. The pseudo-code of the algorithm is summarized as follows:

Algorithm 2. Pseudo code of social spider algorithm (SSA).

1: *Values to the parameters are assigned.*
2: *Create the population of spiders in the search space and allocate memory to them.*
3: *For each spider vtar is initialized.*
4: **while** *stopping criteria not reached do*
    for *each spider s do*
      The fitness value of s is evaluated.
      A vibration is generated at the position of s.
    end for
    for *each spider s do*
      The intensity of the vibrations generated by other spiders is evaluated.
      The strongest vibration vbest is selected from V .
      if *The vibration intensity of vbest is larger than vtar then*
        Store vbest as vtar.
      end if
      A random walk is performed towards vtar.
      A random number r is generated from [0,1).
      if $r < p_j$ *then*
        A random position is assigned to s.
      end if
      The intensity of vtar is attenuated.
    end for
  end while
5: *Output the best solution with the best fitness found.*

4   The Proposed Algorithm

The main aim of using the concept of hybridization is to combine the different biological processes of two different metaheuristics algorithm in order to overcome the weakness of the individual components. Secondly, in order to achieve global optimality in a short and acceptable time. Thirdly, to use the concepts of exploration and exploitation effectively, this may lead to the optimal performance of an algorithm.

In the proposed hybrid algorithm, the process of FA and SSA is combined in such a way that they are analogous to the objective function to be optimized. FA exploits the search space efficiently that lead to better exploration. FA deals with local attraction which is stronger than the long distance attraction [31, 33]. The whole population of FA algorithm can automatically subdivided into multiple subgroups and each group searches its own local best value. Among all the possible local values, there is always global best solution which is the true optimal solution of the problem. SSA uses random walk scheme to solve the global optimization problems. Each spider in SSA analyzes the received information of the propagated vibration over the web to determine the food source position. Spiders perform random walk step towards its target that explores the solutions faster and the global optimality can be more accessible. FA draws attention of researchers by solving many of the NP-hard problems such as load dispatch problems [3], antenna design optimization [6, 34], digital image compression [15], job shop scheduling [19], scheduling problems [24], classification and clustering [25] and many more whereas SSA has potential to solve real world problems. But there are some factors which need to be improved like the complexity of the algorithm, computational time and convergence rate in order to prevent the algorithm to trap in the local optima for achieving the better optimal solution.

The improvements are made in the proposed algorithm in which the convergence rate is faster than the FA and SSA. The concept of probability p is used in this algorithm to achieve the exploitation and exploration efficiently. The value of p varies between 0 and 1. The less bright agent is attracted towards the brighter one, performing the local search using Equation 16 and the concept of random walk is used for global search using Equation 17. The concept of elitism is included so as to allow best values to pass onto the next generation in the population.

The less bright agent i is attracted towards the brighter agent j.   The attractiveness of agent is given as in Equation 14:

$$\beta(r) = \beta_0 e^{-\gamma l^2} , \qquad (14)$$

where β_0 is value of attractiveness when $l = 0$ and γ is a fixed light absorption factor. The distance between agent i and agent j at x_i and x_j respectively, is calculated using the Cartesian distance in Equation 15:

$$l_{ij} = \left\| x_i - x_j \right\| = \sqrt{\sum_{i=1}^{d}\left(x_{i,k} - x_{j,k}\right)^2} , \qquad (15)$$

where $x_{i,k}$ is the kth component of the spatial coordinate x_i of the ith agent.

The local movement is mathematically formulated in Equation 16:

$$x_i + \beta_0 e^{-\gamma l_{ij}^2}(x_j - x_i) + \alpha(rand - \tfrac{1}{2}), \tag{16}$$

where second term represents the attractiveness and third term explains the randomization with α used as a randomization parameter. The random global walk is formulated as following Equation 17:

$$x_i(t+1) = (x_{tar} - x_i) * (1 - rand * rand), \tag{17}$$

where x_i is a agent that performs random walk towards its target x_{tar} at time t + 1, 1 is a vector of ones and *rand* is vector of random generators generated from 0 to 1. The pseudo-code of proposed algorithm is devised below in algorithm 3:

Algorithm 3: Pseudo code of proposed algorithm.

1. *Define objective function f(x) for maximization and minimization problem.*
2. *Generate the initial population in the search space randomly with their fitness value calculated. $x_i (i = 1, 2, .., n)$*
3. *Assign the values to the parameters of proposed hybrid algorithm.*
4. **while** *stopping criteria not matched do*
   for *i = 1: n (all n agents)*
     for *j=1: i (all n agents)*
       if *(p < rand)*
       do local search
       $$x_i = x_i + \beta_0 e^{-\gamma r_{ij}^2}(x_j - x_i) + \alpha(rand - \tfrac{1}{2})$$
       else
         do global search
         $$x_i(t+1) = (x_{tar} - x_i) * (1 - rand * rand)$$
       end if
       Evaluate new solutions.
       If new solutions are better, update them in the population.
     end for j
   end for i
   end while
5. *Output the best solutions found.*

5   Benchmark Functions and Experimental Setting

In order to evaluate the performance of FA, SSA and the proposed hybrid algorithm, 10 benchmark functions are used. The benchmark functions [2], are listed in Table I with their range in the search space and optimal value. These functions are divided into two categories:

(i) Group I: Unimodal minimization functions such as sphere, cigar, step and schwefel's problem 2.22. These functions are used for fast converging performance.

(ii) **Group II:** Multimodal minimization functions such as rastrigin, alpine, michalewiz, ackley and griewank functions. These functions avoid pre-mature convergence, having large number of local points and can be used to test the ability of algorithm to jump out of local optima.

Table 1 Various Testbed Benchmark Functions

Benchmark Function	Formula	Dimension (n)	Search Space	Optimal Value
Sphere Function	$f_1(x) = \sum_{i=1}^{n} x_i^2$	30	(-100,100)	0
Cigar Function	$f_2(x) = x_1^2 + \sum_{i=2}^{n} x_i^2$	30	(-10,10)	0
Step Function	$f_3(x) = \sum_{i=0}^{n} (\lfloor x_i + 0.5 \rfloor)^2$	30	(-100,100)	0
Schwefel's Problem 2.22	$f_4(x) = \sum_{i=1}^{n} |x_i| + \prod_{i=1}^{n} |x_i|$	30	(-10,10)	0
Schaffer Function	$f_5(x) = (x_0^2 + x_1^2)^{0.25} (50(x_0^2 + x_1^2)^{0.1} + 1)$	2	(-100,100)	0
Michalewiz Function	$f_6(x) = -\sum_{i=0}^{n} \left((\sin (x_i) sin^{20} \left(\frac{i x_i^2}{\pi}\right)\right)$	30	(0,π)	0.966n
Rastrigin Function	$f_7(x) = \sum_{i=1}^{n} (x_i^2 - 10 \cos(2\pi x_i) + 10)$	30	(-5.12,5.12)	0
Alpine Function	$f_8(x) = \sum_{i=0}^{n} |x_i \sin(x_i) + 0.1 x_i|$	30	(-10,10)	0
Griewank Function	$f_9(x) = \frac{1}{4000} \sum_{i=1}^{n} x_i^2 - \prod_{i=1}^{n} \cos\left(\frac{x_i}{\sqrt{i}}\right) + 1$	10	(-600,600)	0
Ackley Function	$f_{10}(x) =$ $-20\exp\left(-0.2\sqrt{\frac{1}{n}\sum_{i=1}^{n} x_i^2}\right)$ $-\exp\left[\frac{1}{n}\sum_{i=1}^{n} \cos(2\pi x_i)\right] + 20 + e$	30	(-32,32)	0

All the algorithms are implemented in C++ based platform (QT) under Microsoft Windows 7 operating system. In each run, maximum numbers of 1000 function evaluations are used as termination criteria. In order to produce statistically significant results, each function is repeated for 100 independent runs for FA and SSA, proposed algorithm. The mean of FA, SSA and proposed algorithm is recorded. For FA and the proposed algorithm, the population size is 50 and the values of the parameters are $\alpha = 0.2$, $\gamma = 1$ and $\beta_0 = 1$. For SSA, the population size is 50 and value of vibration attenuation rate r_a is 0.9 and the jump away rate r_e is 0.05.

6   Simulation Results

From the simulation result comparison in Table II, the following points are noted:

(i) For Group I functions, the proposed algorithm shows outstanding results in f_1, f_2 and f_4. FA and proposed algorithm outperforms SSA in f_3. The performance of SSA is best in f_5.

(ii) For Group II functions, the proposed algorithm is at superior position in 4 out of the total of 5 multimodal minimization functions in f_7, f_8, f_9 and f_{10}. In f_6 FA performs better than other algorithms.

The proposed hybrid algorithm possesses a dominating position in the overall comparison of the simulation results.

The complexity of proposed hybrid algorithm is much less than the FA and SSA. The following results are evaluated using the 10 different benchmark functions in the given simulation environment identified in Section IV. The mean value in bold font in Table II indicates the superiority one algorithm over the two algorithms. The performance of the proposed hybrid algorithm is outstanding as compared to the FA and SSA in terms of convergence and avoiding premature optima.

Table 2 Simulation Results Of Fa, Ssa And Proposed Algorithm

Benchmark Function	Firefly Algorithm	SSA Algorithm	Proposed Algorithm
Sphere (f_1)	4.5871E-05	3.8912E+04	**4.5572E-05**
Cigar (f_2)	8.3135E-07	7.3604E+02	**1.7778E-07**
Step (f_3)	**0.0000E+00**	3.8688E+04	**0.0000E+00**
Schwefel's Problem 2.22 (f_4)	9.4198E-03	8.3456E+01	**1.3166E-04**
Schaffer (f_5)	3.9775E-03	**0.0000E+00**	2.8700E-03
Michalewiz(f_6)	**-5.4523E+00**	-2.5949E+00	-4.4152E+00
Rastrigin (f_7)	3.7410E+01	9.4307E+01	**1.6500E-04**
Alpine (f_8)	2.7180E-01	4.2880E+01	**0.0000E+00**
Griewank (f_9)	1.0556E-01	1.3853E-02	**3.3366E-05**
Ackley (f_{10})	1.1283E-03	2.0523E+01	**8.7579E-04**

7   Conclusion

The concept of hybridization is introduced to enhance the performance of FA and SSA. In FA the chances of getting trapped in local optima is still there, the solutions are still changing as the optima approaches. It is possible to improve the quality of solutions by reducing the randomness in local search. The rate of convergence in SSA is very slow as it needs large amount of time to explore the search space. The proposed algorithm had made the improvement by removing these shortcomings and used the concept of exploitation and exploration effectively and efficiently. The proposed algorithm shows its superiority by converging fast and it does not get trapped in local optima. Another strength of proposed algorithm is that it successfully avoids premature converge which contributes in its excellent performance. Future research on FA can be modified to deal with multi-objective optimization problems. It would be interesting to identify real-world applications, so that proposed algorithm deals with them effectively and efficiently.

References

1. Gandomi, A.H., Yang, X.-S., Alavi, A.H.: Mixed variable structural optimization using firefly algorithm. Computers & Structures **89**(23), 2325–2336 (2011)
2. Adorio, E.P., Diliman, U.: Mvf-multivariate test functions library in c for unconstrained global optimization (2005)
3. Apostolopoulos, T., Vlachos, A.: Application of the firefly algorithm for solving the economic emissions load dispatch problem. International Journal of Combinatorics **2011** (2010)
4. Blackwell, T.M.: Particle swarms and population diversity. Soft Computing **9**, 793–802 (2005)
5. Barnard, C., Sibly, R.: Producers and scroungers: A general model and its application to captive flocks of house sparrows. Animal Behavior **29**(2), 543–550 (1981)
6. Chatterjee, A., Mahanti, G.K., Chatterjee, A.: Design of a fully digital controlled reconfigurable switched beam concentric ring array antenna using firefly and particle swarm optimization algorithm. Progress In Electromagnetics Research B **36**, 113–131 (2012)
7. Clerc, M.: Particle swarm optimization, vol. 93. John Wiley & Sons (2010)
8. Clark, C.W., Mangel, M.: Foraging and Flocking strategies: Information in an uncertain environment. The American Naturalist **123**(5), 626–664 (1984)
9. Karaboga, D., Basturk, B.: A powerful and efficient algorithm for numerical function optimization: artificial bee colony (ABC) algorithm. Journal of global optimization **39**(3), 459–471 (2007)
10. Karaboga, D., Basturk, B.: On the performance of artificial bee colony (ABC) algorithm. Applied soft computing **8**(1), 687–697 (2008)
11. Dorigo, M., Birattari, M., Stutzle, T.: Ant colony optimization. IEEE Computational Intelligence Magazine **1**(4), 28–39 (2006)
12. Cuevas, E., Cienguegos, M., Zaldvar, D., Perez Cisneros, M.: A swarm optimization algorithm inspired in the behavior of the social-spider. Expert Systems with Applications **40**(16), 6374–6384 (2013)
13. Cuevas, E., Cienfuegos, M.: A new algorithm inspired in the behavior of the social-spider for constrained optimization. Expert Systems with Applications **41**(2), 412–425 (2014)

14. Jati, G.K.: Evolutionary discrete firefly algorithm for travelling salesman problem. Springer, Heidelberg (2011)
15. Horng, M.-H.: Vector quantization using the firefly algorithm for image compression. Expert Systems with Applications **39**(1), 1078–1091 (2012)
16. Kennedy, J., Kennedy, J.F., Eberhart, R.C.: Swarm intelligence. Morgan Kaufmann (2001)
17. Kennedy, J.: Particle swarm optimization, pp. 760–766. Springer, US (2010)
18. James J.Q, Victor O.K.: A Social Spider Algorithm for Global Algorithm. Technical Report No. TR 2013-004, The University of Hong Kong, October 2013
19. Khadwilard, A., et al.: Application of firefly algorithm and its parameter setting for job shop scheduling. In: First Symposius on Hands-On Research and Development, pp. 89–97 (2011)
20. Davis, L. (ed.): Handbook of genetic algorithms, vol. 115. Van Nostrand Reinhold, New York (1991)
21. Dorigo, M., Birattari, M.: Ant colony optimization, pp. 36–39. Springer, US (2010)
22. Mendes, R., Kennedy, J., Neves, J.: Watch thy neighbor or how the swarm can learn from its environment. In: Proceedings of the IEEE Swarm Intelligence Symposium (SIS), pp. 88–94. IEEE, Piscataway (2003)
23. Poli, R., Kennedy, J., Blackwell, T.: Particle swarm optimization. Swarm intelligence **1**(1), 33–57 (2007)
24. Sayadi, M., Ramezanian, R., Ghaffari-Nasab, N.: A discrete firefly meta-heuristic with local search for makespan minimization in permutation flow shop scheduling problems. International Journal of Industrial Engineering Computations **1**(1), 1–10 (2010)
25. Senthilnath, J., Omkar, S.N., Mani, V.: Clustering using firefly algorithm: performance study. Swarm and Evolutionary Computation **1**(3), 164–171 (2011)
26. Farahani, S.M., Abshouri, A.A., Nasiri, B., Meybodi, M.R.: A Gaussian firefly algorithm. Int. J. Machine Learning and Computing **1**(5), 448–453 (2011)
27. Palit, S., Sinha, S., Molla, M., Khanra, A., Kule, M.: A cryptanalytic attack on the knapsack cryptosystem using binary Firefly algorithm. In: 2nd Int. Conference on Computer and Communication Technology (ICCCT), 15–17 September 2011, India, pp. 428–432 (2011)
28. Łukasik, S., Żak, S.: Firefly algorithm for continuous constrained optimization tasks. In: Nguyen, N.T., Kowalczyk, R., Chen, S.-M. (eds.) ICCCI 2009. LNCS, vol. 5796, pp. 97–106. Springer, Heidelberg (2009)
29. Yang, X.-S.: Firefly algorithms for multimodal optimization. In: Watanabe, O., Zeugmann, T. (eds.) SAGA 2009. LNCS, vol. 5792, pp. 169–178. Springer, Heidelberg (2009)
30. Yang, X.-S.: Firefly algorithm, Levy flights and global optimization. Research and Development in Intelligent Systems XXVI, pp. 209–218. Springer, London (2010)
31. Yang, X.-S.: Firefly algorithm, stochastic test functions and design optimisation. International Journal of Bio-Inspired Computation **2**(2), 78–84 (2010)
32. Yang, X.S.: Chaos-enhanced firefly algorithm with automatic parameter tuning. Int. J. Swarm Intelligence Research **2**(4), 1–11 (2011)
33. Yang, X.-S.: Swarm Intelligence Based Algorithms: A Critical Analysis. Evolutionary Intelligence **7**(1), 17–28 (2014)
34. Zaman, M.A., Matin, A.: Nonuniformly spaced linear antenna array design using firefly algorithm. International Journal of Microwave Science and Technology **2012** (2012)

Visualization – A Potential Alternative for Analyzing Differential Evolution Search

P.R. Radhika and C. Shunmuga Velayutham

Abstract This paper is a preliminary investigation towards employing Visualization as a potential alternative to analyze Differential Evolution search as against the typical theoretical and empirical analyses. The usefulness of scatter plots and difference vector visualization has been observed on six Differential Evolution variants. Simulation analysis reiterated their potential beyond analyzing mere convergence. It has also been observed that scatter plots and difference vector visualization can be employed to detect premature convergence and stagnation.

1 Introduction

Evolutionary Algorithms (EAs) are population-based, iterative, search algorithms modelled on natural evolution. Genetic Algorithm (GA), Evolutionary Programming (EP), Evolution Strategies (ES) and Genetic Programming (GP) are the popular instances of EAs. By virtue of the very *evolution* metaphor, a typical EA produces a very complex search dynamics. The EA population, which is the unit of evolution, along with the variation-cum-selection operations form the very source of an EA's complex search behavior. Therefore, understanding the population comprising of individuals and its evolution is crucial in understanding and harnessing the true potential of EAs.

Theoretical and empirical analyses are the primary means, in the EA literature, to gain deeper insight about various aspects of an EA search. Visual analytics, defined in [1] as "the science of analytical reasoning facilitated by visual interactive interfaces" is a very interesting alternative. The population of an EA is a complex multidimensional data and hence will lend itself naturally for a visual analysis. Inspite of the fact that post mortem off-line visualization of an EA search are frequently

P.R. Radhika · C.S. Velayutham(✉)
Amrita Vishwa Vidyapeetham, Coimbatore, India
e-mail: radhikapr88@gmail.com, cs_velayutham@cb.amrita.edu

© Springer International Publishing Switzerland 2016
S. Berretti et al. (eds.), *Intelligent Systems Technologies and Applications*,
Advances in Intelligent Systems and Computing 384,
DOI: 10.1007/978-3-319-23036-8_3

31

used in literature, they often fall short in bringing out the complex evolutionary characteristics depicted by the changing population. So, visual analysis of Evolutionary Algorithm data has recently emerged as an alternative method of analysis among EA researchers [2-10].

Differential Evolution (DE) is a simple yet powerful Evolutionary Algorithm proposed by Rainer Storn and Kenneth Price in 1995 [11,12]. DE is a population based algorithm most suitable for global optimization in the continuous search domain. The algorithm has become popular because of its speed, robustness and good convergence properties and differs from other Evolutionary Algorithms by virtue of its *differential mutation* operation.

Interestingly, as far as the authors know, there has been no attempt in the literature in employing visualization to analyze DE search. So, this paper attempts a preliminary investigation in exploring visualization as a potential alternative to analyze a typical DE search. The current work shows the potential of two dimensional scatter plot of population cloud as well as visualization of difference vector evolution in analyzing the search characteristics of DE. By virtue of its exploratory nature, the current work, rather than attempting to gain deeper insight of a DE run, explores the potential and possibilities of visual analytics of DE using scatter plots and difference vectors by way of an example.

This paper is organized as follows. Section 2 provides a brief overview of Differential Evolution algorithm followed by an overview and discussion of the related works in Section 3. Section 4 explores visualization of DE as an alternative method of analysis and Section 5 concludes the paper.

2   Differential Evolution

The main stages in a typical DE search are population initialization, mutation, recombination/crossover and selection. The initial population is selected randomly from the entire solution space and consists of NP D-dimensional vectors. These NP vectors constitute the first generation and this step is called initialization. New vectors are produced by adding a difference of two population vectors (or sum of difference of two pairs of vectors), to a third vector, known as the target vector $X_{i,G}$, where i ∈ 1,2,...,NP and G is the current generation. This step called differential mutation is the main step in DE and produces an intermediary population of NP mutant vectors. The crossover step then builds trial vectors by crossing the parameters of each vector of the current generation with the parameters of the respective mutant vector. The final step of DE is selection where the new population members get chosen for the coming generation. The trial vector gets chosen for the next generation only if it has a lower objective function value than the target vector otherwise the target vector remains in the population for the next generation (assuming minimization problem). Until a pre-specified termination condition is achieved, the entire process of mutation, crossover and selection is iterated.

Based on different mutation strategies and crossover schemes, it is possible to generate several DE variants. The variants are typically written in the form *DE/x/y/z*

Table 1 DE Variants

Variant Name	Equation
DE/rand/1	$V_{i,G} = X_{r_1^i,G} + F(X_{r_2^i,G} - X_{r_3^i,G})$
DE/rand/2	$V_{i,G} = X_{r_1^i,G} + F(X_{r_2^i,G} - X_{r_3^i,G} + X_{r_4^i,G} - X_{r_5^i,G})$
DE/best/1	$V_{i,G} = X_{best,G} + F(X_{r_1^i,G} - X_{r_2^i,G})$
DE/current-to-rand/1	$V_{i,G} = X_{i,G} + F(X_{r_1^i,G} - X_{r_2^i,G}) + K(X_{r_3^i,G} - X_{i,G})$
DE/current-to-best/1	$V_{i,G} = X_{i,G} + F(X_{r_1^i,G} - X_{r_2^i,G}) + K(X_{best,G} - X_{i,G})$
DE/rand-to-best/1	$V_{i,G} = X_{r_1^i,G} + F(X_{r_2^i,G} - X_{r_3^i,G}) + K(X_{best,G} - X_{i,G})$

where DE means Differential Evolution, x indicates how the individuals are chosen for perturbation, y denotes the number of pair of individuals chosen and z denotes the type of crossover used. The variants used in this paper are given below.

where $X_{best,G}$ is the best member of the current generation and $r_1^i, r_2^i, r_3^i, r_4^i$ and r_5^i are mutually exclusive random indices.

In this work, binomial crossover is used with each of the above said mutation strategies.

$$U_{i,G}^j = \begin{cases} V_{i,G}^j & \text{if } (rand_j[0.1] \leq C_r) \vee (j = j_{rand}) \\ X_{i,G}^j & \text{otherwise} \end{cases} \tag{1}$$

The crossover occurs between a mutant vector and a predetermined target vector (parent) to produce a trial vector $U_{i,G}$. Here j_{rand} is a random index chosen anew for each individual to ensure that at least one parameter from the mutant vector gets a place in the trial vector. C_r, (crossover probability) is the rate of occurence of crossover between the parameters of $X_{i,G}$ and $V_{i,G}$.

The data regarding the population, obtained from random run of each of the above variants is visualized in this paper.

3 Related Works

Visualization of EA search data has been attempted, and is being attempted, by many researchers. This section presents a brief review of related works.

Lutton and Fekete [2] proposed the usage of Scatterdice/ GraphDice tool which explores the advantages of using visual analytics for visualization of population data in an EA. The ScatterDice tool enables the user to navigate in a multidimensional dataset by making 2D projections using scatterplot matrices. Thus it becomes easy to visualize the exploration capability, convergence and diversity of the population. GraphDice tool provides a way to adapt a general purpose visualization tool to the analysis of an EA.

Bedau and Bullock [3] proposed a method for visualizing population data of evolving systems using wave diagrams which helps in identifying mutations that prove to be significant across generations. Pohlheim [4], presented different visualization techniques for various time frames and datatypes and proposed multidimensional scaling techniques for higher dimensions. Kerren and Egger [5] have developed EAVis tool that provides several views that enables user to observe the evolution of each generation of the EA and thereby gain valuable insights regarding the working of the algorithm.

The usage of self-organizing maps for GA visualization [6] helps in visualizing genotypes, phenotypes, state and course of evolution. This method uses unsupervised competitive learning and has generalization capacity. Self-organizing maps reduce the dimension by using self-organized neural network and it works well for multidimensional scaling and also helps to discover hints about initialization and population diversity.

James McDermott [7] has proposed a method for visualizing evolutionary search spaces in which the distances on search spaces are used to project them into two dimensions. Various types of distances like syntactic, semantic and mutation based distances have been used. The goal is to obtain useful and attractive visualizations. These methods enable comparison of the structure of search space and see the differences between various fitness functions. Multidimensional scaling is the method used to project the search space into two dimensions.

The method of visualization of Genetic Algorithm proposed by William Shine and Christoph Eick [8] processes information generated by running a genetic algorithm and enables visualization of fitness by means of contour maps. A map can be created for each generation and these maps can then be combined to visualize evolution by forming movies that can be viewed using animation viewer. VIS System [9] examines the details of a GA run and supports the representation and examination of individuals, populations, and entire runs. In short, it helps in making all the details of a run easily accessible. GAVEL [10] is another offline visualization system proposed by Emma Hart that provides a way for us to understand how the mutation and crossover operators help in reaching the optimal solution, origin and history of each of the alleles. GAVEL enables us to view the complete ancestry tree of a solution and helps to determine ancestry at both individual as well as gene level. It also helps us in understanding how much of the search space has been explored.

Visualization is now considered to be one of the most effective methodologies for data analysis across multiple domains. Reddivari [13,16] has proposed a framework that examines usage of visual analytics in Requirements Engineering and has developed a tool ReCVisu for quantitative visualization. Jankun-Kelly [14] has introduced P-Set Model of Visualization Exploration that helps in description as well as representation of information. Katifori [15] presented a study on the effectiveness of visualization methods for supporting users in current ontology browsing for Information retrieval. Seifert [17] has proposed a visualization for faceted browsing

Table 2 Benchmark Functions

Function name	Description
Ackley's function	$$f(x) = -20 \exp\left[-\frac{1}{5}\sqrt{\frac{1}{n}\sum_{i=1}^{n} x_i^2}\right] - \exp\left[\frac{1}{n}\sum_{i=1}^{n} \cos(2\pi x_i)\right] + 20 + e, \quad -32.768 \le x_i \le 32.768$$
Schwefel's function	$$f(\mathbf{x}) = 418.9829d - \sum_{i=1}^{d} x_i \sin(\sqrt{|x_i|}), \quad -500 \le x_i \le 500,$$

with an aim to help users to find their way through large document collections by visually constructing complex boolean search queries for narrowing down the search space. Simons [18] explored the usefulness of visualization of interactive evolutionary search that represents both the design problem and design solution in upstream software design. While visual analytics of EA search data is certainly not new, visualizing a real parameter optimization algorithm like DE has not received much attention and this paper attempts a preliminary investigation in this direction.

4 Visualizing Differential Evolution

In this paper, the following paramaters have been employed for the DE variants. The population size is set as 100, scale factors F (and K) as 0.9, C_r as 0.5 and maximum number of generations as 1000. The benchmark functions used in all simulation experiments are given in Table 1. For the sake of easier analysis, we consider only 2 dimensions for both functions. While Ackley's function has global minimum value of 0 located at origin, Schwefel function's global minimum value of 0 is located at (420.9687,420.9687). The functions and their contour plots are displayed in Figures 1-4. Following [2], we collect the following information from a DE run which forms the 'population cloud': the generation numbers, the parameters x[0] and x[1] (as we consider only 2D cases), the fitness values and the difference vector (not the mutant vector) that is generated for each of the target vector.

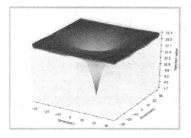

Fig. 1 Ackley's function

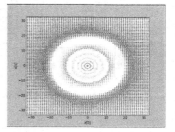

Fig. 2 Contour plot for Ackley's function

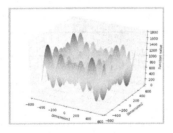

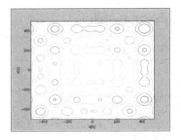

Fig. 3 Schwefel's function

Fig. 4 Contour plot for Schwefel's function

4.1 DE Visualization Using Scatter Plots

A DE run generates huge amount of data captured in population cloud. As the population cloud encompasses four information entities (excluding difference vector), a scatter plot matrix is employed to visualize different scatter plots employing all possible combinations of information entities in the population cloud. Figures 5-8 show the scatter plot matrices to facilitate visualization of different DE variants' run on the benchmark functions. To avoid redundancy, only the lower triangle of scatter plot matrix is shown.

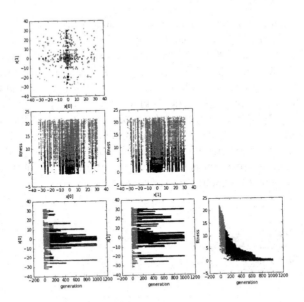

Fig. 5 Scatter plot matrix of *DE/rand/1/bin* showing successful convergence for Ackley's function

Fig. 6 Scatter plot matrix of *DE/best/1/bin* showing premature convergence for Ackley's function

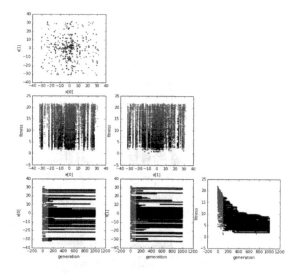

Fig. 5 shows the scatter plot matrix for a run of *DE/rand/1/bin* minimizing Ackley's function. The scatter plot for x[0] vs x[1] shows that the parameter values in the first 100 generations (shown in red) are spread out in the search space and after that (points in black) they start converging towards the global minimum. The convergence pattern is due to the unimodality of the function. The widely spread parameter values in the initial generations is by virtue of the exploratory nature of *DE/rand/1/bin*.

While generation vs fitness scatter plot displays the convergence trend of *DE/rand/1/bin*, the rest of the plots show the explored points in search space in terms of fitness and generation.

Fig. 6 shows the scatter plot matrix for *DE/best/1/bin* minimizing Ackley's function. The greedy nature of *DE/best/1/bin* is evident in the scatter plots especially in the x[0] vs x[1] scatter plot and generation vs fitness scatter plot. As can be viewed in the latter scatter plot, the convergence occurs at a value between 0 and 5 but never truly converges to 0, the global optimum.

Fig. 7 shows scatter plot matrix for *DE/rand/1/bin* solving Schwefel's function. The x[0] vs x[1] scatter plot not only shows dense cloud of parameters near (420.9687,420.9687), the global optimum signalling successful convergence, but also local optima. This is evident from distinct bands in scatter plots between parameters and fitness as well as between parameters and generation.

Fig. 8 shows scatter plot matrix for *DE/current-to-rand/1/bin* solving Schwefel's function. As can be observed from the generation vs fitness scatter plot, stagnation has occured for this DE variant between 0 and 200.

As is shown above, scatter plots have the potential to bring out more information than mere successful convergence. Various regions is the search space visited by DE (including its local optima) evident from x[0] vs x[1] scatter plot providing insight about the search dynamics.

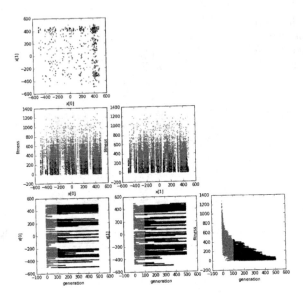

Fig. 7 Scatter plot matrix of *DE/rand/1/bin* showing successful convergence for Schwefel's function

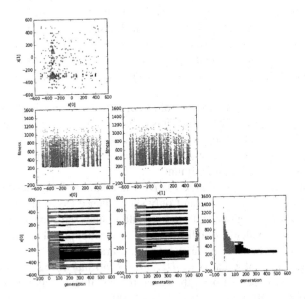

Fig. 8 Scatter plot matrix for DE *current-to-rand/1/bin* showing stagnation for Schwefel's function

More information about these regions can be inferred from scatter plots between parameters and fitness as well as parameters and generations also. Scatter plots as shown also prove to be good diagnostic tools so as to detect premature convergence and stagnation that often hinders the success of DE's (in general to EA's) search.

4.2 Visualizing Difference Vector Evolution

The defining charateristic of DE, that differentiates it from other EA's, is its differential mutation operation. Difference vectors, which are the fundamental components of differential mutation, are expected to be large in the initial phase of DE search. By virtue of larger difference vectors, target vetcors when perturbed, can achieve larger jumps in search space facilitating exploration. Towards the end of the search phase, difference vectors should be smaller to facilitate exploitation of the already explored search space.

Thus visualizing the evolution of difference vectors will provide valuable insight about the dynamics of DE search. Towards this, we recorded difference vectors which create *NP* mutant vectors against each target vector in the population for every generation. All the difference vectors obtained throughout the run of a DE variant is plotted by transporting them to the origin resulting in difference vector distribution/evolution.

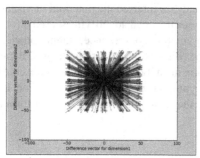 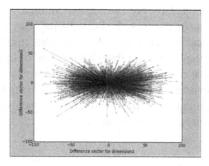

Fig. 9 Difference Vector of *DE/rand/ 1/bin* for Ackley's function

Fig. 10 Difference Vector of *DE/rand/2/ bin* for Ackley's function

Figures 9-12 show the difference vector distribution obtained from a random run of *DE/rand/1/bin*, *DE/rand/2/bin*, *DE/current-to-best/1/bin* and *DE/rand-to-best/1/bin* for Ackley's function. As is evident from the figures, different DE variants show different distribution (or evolution pattern) of difference vectors. Despite the fact that understanding a DE variant from its difference vector is beyond the scope of this paper, it is evident that visualizing difference vectors may provide rich information about the DE search characteristics. Similar to scatter plots, the difference vector visualization too has the potential of detecting premature convergence and stagnation.

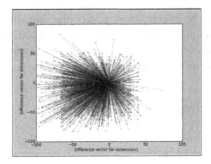 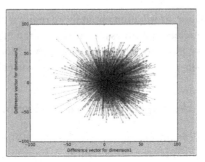

Fig. 11 Difference Vector of *DE/current-* *to-best/1/bin* for Ackley's function

Fig. 12 Difference Vector of *DE/rand-to-* *best/1/bin* for Ackley's function

5 Conclusion

This paper is an attempt to investigate the potential of visualization to analyze the Differential Evolution search against the typical theoretical and empirical analyses. Towards this, the potential of scatter plots and visualization of difference vector evolution were investigated in understanding a DE run. Visualization of six DE variants (*DE/rand/1/bin*, *DE/best/1/bin*, *DE/rand/2/bin*, *DE/current-to-rand/1/bin*, *DE/current-to-best/1/bin* and *DE/rand-to-best/1/bin*) that solves two 2D benchmark functions have been attempted. Simulation analyses inferred the potential of visualization in analyzing the DE variants' search space traversal, identifying search efforts in local as well as global optima, detecting premature convergence, stagnation etc.

Despite the fact that visualization of higher dimensions poses challenges, it can still be argued that the potential of visualization in analyzing DE search, as observed in simulation analysis, still holds. Moving beyond this preliminary investigation, by employing visualization to critically analyze the search behavior of DE will form a definite part of our future work.

References

1. Thomas, J.J., Cook, K.A.: A visual analytics agenda. IEEE Computer Graphics and Applications **26**(1), 10–13 (2006)
2. Lutton, E., Foucquier, J., Perrot, N., Louchet, J., Fekete, J.-D.: Visual analysis of population scatterplots. In: Hao, J.-K., Legrand, P., Collet, P., Monmarché, N., Lutton, E., Schoenauer, M. (eds.) EA 2011. LNCS, vol. 7401, pp. 61–72. Springer, Heidelberg (2012)
3. Bedau, M.A., Joshi, S., Lillie, B.: Visualizing waves of evolutionary activity of alleles. In: Proceedings of the 1999 GECCO Workshop on Evolutionary Computation Visualization, pp. 96–98 (1999)
4. Pohlheim, H.: Visualization of evolutionary algorithms-set of standard techniques and multidimensional visualization. In: Proceedings of the Genetic and Evolutionary Computation Conference, vol. 1, pp. 533–540 (1999)

5. Kerren, A., Egger, T.: Eavis: A visualization tool for evolutionary algorithms. In: 2005 IEEE Symposium on Visual Languages and Human-Centric Computing, pp. 299–301. IEEE (September 2005)
6. Romero, G., Merelo, J.J., Castillo, P.A., Castellano, J.G., Arenas, M.: Genetic algorithm visualization using self-organizing maps. In: Guervós, J.J.M., Adamidis, P.A., Beyer, H.-G., Fernández-Villacañas, J.-L., Schwefel, H.-P. (eds.) PPSN 2002. LNCS, vol. 2439, pp. 442–451. Springer, Heidelberg (2002)
7. McDermott, J.: Visualising evolutionary search spaces. ACM SIGEVOlution 7(1), 2–10 (2014)
8. Shine, W.B., Eick, C.F.: Visualizing the evolution of genetic algorithm search processes. In: IEEE International Conference on Evolutionary Computation, pp. 367–372. IEEE (April 1997)
9. Wu, A.S., De Jong, K.A., Burke, D.S., Grefenstette, J.J., Loggia Ramsey, C.: Visual analysis of evolutionary algorithms. In: Proceedings of the 1999 Congress on Evolutionary Computation, CEC 1999, vol. 2. IEEE (1999)
10. Hart, E., Ross, P.: GAVEL-a new tool for genetic algorithm visualization. IEEE Transactions on Evolutionary Computation 5(4), 335–348 (2001)
11. Storn, R., Price, K.: Differential evolution-a simple and efficient adaptive scheme for global optimization over continuous spaces, vol. 3. ICSI, Berkeley (1995)
12. Storn, R., Price, K.: Differential evolution-a simple and efficient heuristic for global optimization over continuous spaces. Journal of Global Optimization 11(4), 341–359 (1997)
13. Reddivari, S., et al.: Visual requirements analytics: a framework and case study. Requirements Engineering 19(3), 257–279 (2014)
14. Jankun-Kelly, T.J., Kwan-Liu, M., Gertz, M.: A model and framework for visualization exploration. IEEE Transactions on Visualization and Computer Graphics 13(2), 357–369 (2007)
15. Katifori, A., et al.: Visualization method effectiveness in ontology-based information retrieval tasks involving entity evolution. In: 2014 9th International Workshop on Semantic and Social Media Adaptation and Personalization (SMAP). IEEE (2014)
16. Reddivari, S., et al.: ReCVisu: A tool for clustering-based visual exploration of requirements. In: 2012 20th IEEE International Requirements Engineering Conference (RE). IEEE (2012)
17. Seifert, C., Jurgovsky, J., Granitzer, M.: FacetScape: a visualization for exploring the search space. In: 2014 18th International Conference on Information Visualisation (IV). IEEE (2014)
18. Simons, C.L., Parmee, I.C., Gwynllyw, R.: Interactive, evolutionary search in upstream object-oriented class design. IEEE Transactions on Software Engineering 36(6), 798–816 (2010)

Implementation of Mixed Signal Architecture for Compressed Sensing on ECG Signal

S. Gayathri and R. Gandhiraj

Abstract Persistent health monitoring is the key feature in present day wearable health monitoring system. The focus is on reducing the power consumption associated with transmission of large data content by reducing the bandwidth required. Signals sampled wastefully at Nyquist rate increases power dissipation drastically when RF Power amplifier (inside the body area networks of wearable device) transmits sensed data to personal base station. Compressed Sensing (CS) is an emerging technique that condenses the information in the signal into a lower dimensional information preserving domain before sampling process. CS facilitates data acquisition at sub-Nyquist frequencies. The original signal is reconstructed from the compressively sampled signal by solving an undetermined system of linear equations. In this paper a scalable hardware for CS in ECG signal is modeled. The factors determining the quality of reconstruction of a Compressively Sampled ECG signal is studied in both time and wavelet domain using the modeled hardware.

Keywords Compressed Sensing · Electrocardiogram · Sub-Nyquist frequency · Signal Reconstruction · Wavelet domain

1 Introduction

World Health Organization statistics shows that at present cardiovascular diseases is the major cause for current increased death rate [1]. Consistent ECG monitoring of the patients is an effective method for diagnosis and treatment. Earlier portable ECG Holter monitor systems where used for this purpose. But this bulky device needed to be carried by patients for recording ECG signals for long period of time (e.g. for two weeks). This inconvenience was alleviated by wearable body area

S. Gayathri(✉) · R. Gandhiraj
Amrita School of Engineering, Amrita Vishwa Vidyapeetham, Coimbatore, India
e-mail: gayathri.suresh20@gmail.com, r_gandhiraj@cb.amrita.edu

© Springer International Publishing Switzerland 2016
S. Berretti et al. (eds.), *Intelligent Systems Technologies and Applications*,
Advances in Intelligent Systems and Computing 384,
DOI: 10.1007/978-3-319-23036-8_4

43

networks (WBANs). The Body area networks (BAN) inside this wearable devices contains bio-signal sensors equipped with ultra-low power radios which communicates to a BAN personal base station, and finally the healthcare provider. Power dissipation increases drastically when RF power amplifier transmits sensed data to the personal base station. This led to the idea of data compression before transmission so that the duty cycle of the power amplifier, and there by power dissipation is reduced.

This paper deals with compressed sensing (CS) method [2], [3], [14] of ECG signals which facilitates ECG data compression before transmission. Compressed sensing is applicable to any signal which has sparse representation in a suitable basis. This is one of the criterions for proper reconstruction of ECG Signal from the compressed measurements. The objective of this paper is to study quality of reconstructed ECG signal by analyzing its compressibility and reconstruction error.

The three important performance metrics for compressed sensing are [6], [14]:

1. Compression Ratio (*CR*): Ratio of original length of input signal to length of compressed output signal.

$$CR = \frac{N}{M} \tag{1}$$

Where *N*= length of input signal and
M= length of compressed output signal.

2. Signal to Noise Ratio (*SNR*): Defines the error between the original signal and the reconstructed one.

$$SNR = 20\log 10 \|xorg\| / \| xorg - xrec \| \tag{2}$$

Where x_{org} is the original signal and x_{rec} is the reconstructed signal.

3. Percentage Root-mean-square Distortion (PRD): This index is the inverse of SNR.

$$PRD = | xorg - xrec | / | xorg | \tag{3}$$

Where x_{org} is the original signal and x_{rec} is the reconstructed signal.

The main objective of applying compressed sensing on a signal is to get a high CR and SNR and low PRD.

All hardware realizations for compressed sensing are accomplished by a trade off in CR and SNR. Very few hardware structures exist for Compressed Sensing. This paper discusses about the hardware for CS on ECG signal in time and wavelet domain and studies the effect of ECG signal sparsity in CS.

The paper is organized as follows: Section 2 briefly explains CS and the types of existing CS hardware structures for Bio-signals. Section 3 discuss in detail about CapMux hardware [13] which is the basic hardware structure adopted in our design and then proceed to the proposed design for ECG signal CS. Section 4 presents the hardware results and corresponding observations. The paper is concluded in section 5.

2   Compressed Sensing

Compressed sensing is a new concept which has recently gained wide importance in signal processing community. The importance of CS is due to fact that it allows a signal to be acquired and accurately reconstructed from fewer data samples than required by Nyquist-rate sampling. Compressed sensing relies on maximum rate-of-information of a signal rather than maximum rate-of-change of a signal as prescribed by Nyquist sampling [3], [14], [15]. Compressed sensing has been instrumental in research for low-power data acquisition methods.

2.1 Formal Definition of CS

The compressed sensing system performs two fundamental computational operations: encoding (or compression) and decoding (or reconstruction) [2].

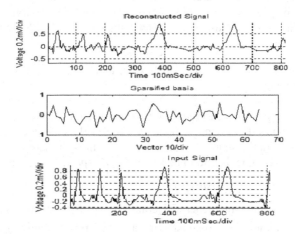

Fig. 1 Compressed Sensing of ECG Signal

Fig 1 shows a diseased ECG bio-signal after the acquisition, compression, and reconstruction stages of the CS system. In figure the analog input vector, x, comprises 810 samples whereas and the compressed analog output vector, y, has 64 samples after CS compression. Each frame is period normalized to 256 length and a compression factor of C=4 is achieved with reasonable reconstruction error. The compressed decoding of the compressed output is done by using convex optimization algorithm to achieve reconstructed ECG signal of length 800.

Compressed Sensing is a non-adaptive sampling scheme represented in matrix form as:

$$y = \emptyset\, x \tag{4}$$

Where x is the analog time-domain input sample vector of length N, y is the compressed analog output vector of length M, and ϕ is the M x N measurement matrix. CS is considered non-adaptive because ϕ remains constant [4].

In some cases, the analog input vector, x can be expanded as:

$$x = \psi \alpha \tag{5}$$

Where α is the N-sample coefficient sequence and ψ is the N x N sampling basis or sparsifying matrix. In Fig.1, the ECG signal was sampled and compressed in the time domain. The representation basis matrix, ψ, for the time-domain is simply the identity matrix. Sometimes it is useful to combine both the Equations (3) and (4) to obtain:

$$y = \Theta x \tag{6}$$

Where $\Theta = \psi \phi$.

CS captures M << N measurements from N samples using random linear projections, i.e. by projecting the input signal onto a random sparsifying matrix.

The two criteria's a signal must satisfy so as to undergo proper CS are [4], [14]:

- The sampled signal, x, should be sparse in some basis.
- The two matrices, sampling matrix, ψ, and the measurement matrix, ϕ should be incoherent.

The sparsity of a signal is measured based on the difference between the signal's information rate and rate of change. Sparsity is defined as the percentage of zero elements contained in the signal. For signal x of length N, if there are only K non-zero values, then the signal is said to be k sparse with sparsity given as [4]:

$$Sparsity(\%) = (N-K/N) \times 100 \tag{7}$$

Increased signal sparsity in higher CS compression. Sparsity in some basis is extraordinarily common for bio-signals.Signal incoherence condition between ϕ and ψ basis leads to the spreading out of measurements from sensing basis to measurement basis. For example the sparse Dirac or delta function in time domain is a constant continuous value in frequency domain. The coherence between sampling and measurement matrices is given by [3], [4]:

$$(\phi, \psi) = n\max 0 \leq j, <n< \psi j, \phi k> \tag{8}$$

Where, ψ is a column of ψ and ϕ is a row of ϕ. If ψ and ϕ are unit vectors, then $\mu \in [1, n]$.

Coherence measures the largest correlation between any column of ψ and row of ϕ. For proper compressed sensing the coherence factor (ϕ, ψ) should be minimum. The minimum number of compressed measurements for accurate compression and reconstruction is given by:

$$M \geq C.\mu^2 (\phi, \psi) .k\log N \tag{9}$$

Where C is a small constant.

It is a well-known fact that random matrices are largely incoherent with any fixed basis [2], [3]. Therefore the measurement matrix can be populated with random values from many different probability density functions including Bernoulli, Gaussian and Uniform distributions. Decoding or signal recovery is another key challenge in the implementation of a complete CS system. Reconstruction can be done by solving the equation represented in matrix form as [3], [4]:

$$x = \phi y \qquad (10)$$

But there are N unknown values in the reconstructed signal x , which has to be determined from N (<M) known values in the measured signal y. This leads to an underdetermined linear equation which has many possible solutions. The set of solutions is reduced by the condition that the solution to this linear equation should also be sparse. Using $l1$ or $l0$ norm (sparsity measure) the following minimization leads to signal recovery with considerable accuracy:

$$\min \quad \alpha \quad \text{subject to } y = \emptyset \varphi \alpha$$
$$\min \quad \|\hat{\alpha}\| \text{ subject to } y = \emptyset \varphi \alpha \qquad (11)$$

2.2   ECG Signal CS in Time Domain

Reconstruction of a 256 point data frame using M measurements, assuming that the sparsifying basis is time domain will succeed only if M is about 112 or the ECG Signal is heavily tresholed that only QRS remains in the signal. Since all portions of ECG Wave contain critical information for diagnostics, practical value of time domains sparsifying domain is very limited [5],[15].More thresholding may improve the performance, but at the expense of increase in hardware at the encoder. This is so because a signal that is k parse requires 3.5k to 4k measurements in the sparsifying domain (here time domain). Raw ECG signal is only 60 to 90% sparse depending on thresholding with 40 to 10% non-zero values.

2.3   ECG CS in Wavelet Domain

From [10] it was observed that ECG signal is highly sparse in wavelet domain. By applying wavelet transform ECG Signal can be decomposed into number of sub band signals containing the frequency and time characteristics. By Wavelet transformation ECG signal is represented as a linear superposition of wavelets [12], [15]. Wavelets are created by the scaling and translation of the mother wavelet given by:

$$\varphi_{a,b} = \frac{1}{\sqrt{a}} \left(\frac{t-a}{b} \right) \qquad (12)$$

Where a denote the scaling and b the translation.

Wavelet transformation on a signal gives approximation coefficients and decomposition coefficients depending on the level of decomposition. In wavelet domain most of the energy of the signal will be carried by the approximation function (scaling function of the chosen wavelet system). For ECG signal most of the information is present in the approximation coefficients itself. So rest of the decomposition coefficients can be safely discarded. Even in the approximation coefficients many of them are close to zero. For instance a standard ECG wave, if one data frame covers one period it will have only about 10 to 12 non zero coefficients within 24 approximation coefficients in biorthogonal 4.4 level 4 wavelets or 22 coefficients in Daubechies level4 wavelets. So typically a total of 16 coefficients from all levels will reconstruct the data with a PRD<5%.There are many wavelets like Haar, Daubechies, Biorthogonal out of which bior 4.4 level 4 and Daubechies level 4 gives the best result for ECG signals (found out by simulations in [6]). This shows that CS reconstruction has to use wavelet domain as the sparsifying domain. Let ψ be the connecting matrix between data values and wavelet coefficient values. i.e. $Xn=\psi C$ where Xn is the 256 point data frame; ψ is the 256 X 289 sparsifying matrix and C= 289 X 1 coefficient matrix in case of biorthogonal 4.4 level 4 wavelets.

2.4   Types of Existing CS Hardware for Bio-Signals

There are two types of Bio-Signal CS Encoders BioCS [4] [5] namely: Digital BioCS Encoder and Analog BioCS Encoder.

2.4.1   Digital BioCS Encoder

Digital bio CS formed by [8] used an 8-bit SAR ADC followed by 50 16-bit MACs integrated in 90 nm. The power dissipation obtained was 1.9 μW for power supply of 0.6 V, for EEG signals with compression factors from 10X to 40X.The disadvantage of this system is that the basic functionality of CS is violated. The advantage of CS was that the power consuming ADC system had to sample only the compressed measurements (which is less than those obtained via Nyquist sampling) thus leading to power reduction. In Digital Bio CS encoder ADC samples at Nyquist rate leading to higher power dissipation.

In Digital Bio CS (Fig. 2) system the conditioned bio-signal is digitized by ADC at Nyquist rate [7]. The N samples from ADC i.e. x is then applied in parallel to M multiply-and-accumulate (MAC) stages thus generating M compressed output samples, y [7].

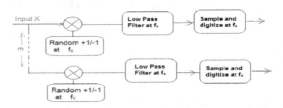

Fig. 2 Hardware of Bio-CS.

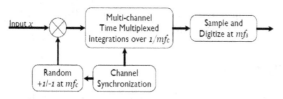

Fig. 3 Capmux Hardware Implementation [13]

The MDAC, in Fig 3, uses switched capacitor operation to first multiply the input signal, x with the measurement matrix φ using a C-2C ladder, followed by integration. The compressed signal y is then digitized and transmitted. M time's repetition of these components leads to severe power dissipation an increase in size of the system [13].

2.4.2   Analog BioCS Encoder
In Analog BioCS Encoder [13] the bio-signal is compressed by passing N samples, x, through M multiply and integrate stages. A single multiply integrator stage, named analog CS operator consists of switched capacitor Op-amp and multiply circuitry. The multiplication and integration step is done by a multiplying digital-to-analog-converter (MDAC).

3   Improvised CS Encoder Hardware Design

Existing hardware architectures discussed require active analog components in each branch which consume quiescent current. The low pass filters are constructed from high performance, low noise op-amps. As the number of channels are increased for larger number of independent measurements, the quiescent current increases proportionally. For low power sensing applications, total quiescent current can exceed the active power of the ADC, leading to higher power dissipation and thus higher errors in compressed sample acquisition. In the proposed method CS sensing errors are reduced by sharing a single multiply integrator stage (analog CS operator) with N sample values of input bio-signal. We observed that existing hardware structures uses M (number of CS measurements) integrator units for effective CS operation. The demerits of such a system have also been noticed. In this paper we imbibe the principle of Capmux architecture by [13] for implementing a scalable power optimizable hardware for CS on ECG Signal. Here we replace the M (values taken by CS from N values of input signal) analog processing units operating with frequency f_c with a single analog processing unit working with a frequency of operation Mf_c where f_c is the "chipping rate" of the input signal. The basic idea is that the single analog processing unit (here op-amp integrator unit) is time shared M times in $1/f_c$ time instead of working with M analog processing units in $1/f_c$ time. Fig.2 and Fig.3 depicts the two situations explained above.

We use mixed signal model for implementing this hardware. The Control circuitry was programmed in an Arduino Duo board. The CS operation takes place in the analog processing units. The reconstruction of the CS values acquired from mixed-signal hardware is done in MATLAB.

3.1 Working Principle

Consider the CS principle which gives compressed M values from N (>>M) input signal values. We have [Y] = [φ] [X] where X is the input vector of length N, Y is the compressed vector of length M, and φ is the measurement or sensing matrix. The key assumption here is that the input signal is a slow varying signal and will have a constant value in every T time. Here T= 256/fc since we are taking 256 input signal values within one time frame. This T time slot is further divided into M=16 even slots. This is for the column wise matrix multiplication.

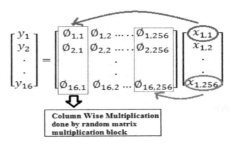

Fig. 4(a) Matrix Representation of CS

The complete matrix representation of CS is shown in Fig 4(a). In Fig 4(a) an input signal x is divided into fixed time frames of 1/fc. Within one time frame 256 values of x is taken and compressed to 16 CS values by projecting x onto a sensing matrix of dimension 16x256. As depicted in Fig.4 (a) CS operation is the column wise multiplication of input signal entries with sensing matrix. Here the sampling matrix, (i.e. the matrix representation of the domain in which input signal is sparse) is identity matrix since the signal is sparse in time domain itself.

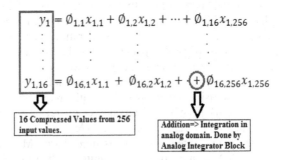

Fig. 4(b) System of Linear equations from CS matrix

During the first T time slot, the first input signal value in first frame is multiplied with each entries of the first column of sensing matrix φ. Each multiplication (say ∅1,1x1,1) happens in every t=T/16 time slot. The entire column multiplication is done in T time slot. Similarly the second input value is multiplied with second column in next T slot. These values are added with the earlier column multiplication values. Finally after 256th T slot (i.e. 1/fc time slot) all the column multiplication values for the corresponding input signal value is added together to form a linear equation. This linear equation is shown in Fig.4 (b). The addition in analog domain corresponds to integration which will be performed by an op-amp integrator.

3.2 Proposed Hardware

The input for the hardware is an ECG Signal which is usually sampled at 250 – 300 Hz. It is sparse in wavelet domain. But sparsity of a quasi-periodic signal like ECG will stand exposed fully for compressive sensing only if the samples used for compressive sensing come from an integer number of average cycle of the signal. This is ensured in this work by sampling in every 3.25 ms and using 256 such sample values as the data frame for compressive sensing. The chosen sampling rate of 3.25ms ensures 256 samples will cover one cycle of healthy ECG that has 72 beats per minute rate. The entries of the CS sensing matrix are randomly chosen. The entries used here are Bernoulli distributed ± 1 entries with an overall scaling factor of $1\sqrt{N}=1/16$ for gain normalization. The entries of this matrix are stored as a single dimensional array in the controller hardware implemented in Arduino Duo (ATMEL Microcontroller Platform). Since ECG is a low frequency signal the signal value is almost constant over one sampling interval (i.e., over 3.25 ms in the current scenario). So a sample and hold process is not required. Proposed system assigns one capacitor each for each measurement value as value holder. Thus M capacitors hold and update the measurement values as new sample values are presented to the system. As explained earlier, the proposed system uses a single Op-amp Integrator in a sharing mode [9][13]. The M value holding capacitors are multiplexed (or polled) in the integration path once in every sampling frame for integration and updating of signal value. This calls for one time-slot for each capacitor within a sample frame. This slot has a width of ~ 50 μs (3.25 ms/ 64) for a system with M = 64. Analog multiplexers with suitable address and inhibit control are used in this work to implement the required 64-channel multiplexing unit.

Bernoulli random matrix sensing involves linear combination factors +1/16 and -1/16. For -1/16, the ECG signal is passed through a switch to a unity-gain buffer amplifier input and for +1/16 the ECG signal is inverted and passed to the same buffer amplifier input. The switch outputs can be connected together at buffer amplifier input this way since the switches are operated in a non-overlapping mode. The buffer amplifier output is connected to the multiplexing integrator. These two switches are implemented using analog switches .The inversion of ECG signal is done in a unity gain inverting amplifier [11]. The first half of this available time-slot, i.e., 25μs in this work is used to set up the signal switch states for channelizing the input signal after reading the relevant measurement matrix entry and to select and connect the correct capacitor in the integration path. During this time capacitors must not lose the charges they have already accumulated. Hence all the analog MUX units are put in the inhibited state during this time. Further, the Op-amp feedback path is kept shorted during this half of slot-time to prevent the Op-amp going into open loop mode. Moreover, both the positive signal line and inverted signal line are kept open during this half of time-slot. The second half of the time-slot is used for carrying out the integration. First the shorting switch across Op-amp feedback path is opened, then the address lines of MUX units are updated, then the proper MUX unit is de-inhibited. A short time delay for allowing the integrator op-amp to slew and settle at the currently connected capacitor voltage is given after this. At the end of this delay, the proper signal channelizing switch is closed to pass on the input for integration. The capacitor channels get the 64 compressed values as charges on them.

The voltages acquired by the 64 integrator capacitors are to be sampled first and then reset to zero at the end of a data frame containing 256 samples. The sampled measurement values (M of them) will be passed on to the data communication subsystem. However, in this work these values are to be taken to MATLAB environment for software reconstruction and performance evaluation. The low cost Arduino Due Board (microcontroller platform) was used as data acquisition system.

4   Results and Analysis

The hardware described in the last chapter was assembled on project boards for testing. The Arduino Due sketch (i.e., code) was written to accept any of the three values for number of measurements – M. Thus the same hardware could be used to implement a 256/64 or 256/48 or 256/32 CS system. An ECG test signal was generated by second order RC shaping of 5 monostable pulse outputs [9]. Each of these pulses when shaped forms the P, Q, R, S, T components which will be added and amplitude adjusted in the ECG generator. The ECG Test Signal Generator assembled along with mixed signal hardware provides input to the proposed hardware. Arduino Software v. 1.5.6r was used to upload the code to the Arduino Due Board. CoolTerm 1.4.3 was used to read the analog samples delivered by the Arduino Board. The ADC in the Arduino Due was configured for 10-bit resolution [10]. Values captured from CoolTerm will be in the range 0 – 1023. These values are loaded into MATLAB workspace. The values are converted to corresponding voltage levels in MATLAB. Three different ECG test signals were employed interesting.

Fig.5 shows the ECG Test signal generated and its inverted version. The signal contains all PQRST components. The input signal undergoes CS operation as explained earlier and produces 64 compressed output values at the end of each frame.

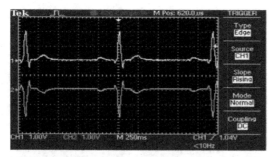

Fig. 5 Oscilloscope Screen Capture of ECG_PQRST signal and its inverted version

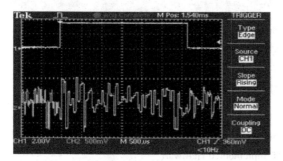

Fig. 6 Output at the Multiplexing Integrator at the end of one data frame

Fig.6 shows the system output – that is, the 64 capacitor voltages during the 257th sampling interval at the end of a data frame containing 256 samples. Sampling of these 64 capacitor voltages will give the compressively sensed measurement vector. This output is taken from the Op-amp used in the integrator. Channel1 shows the end-of-data frame indicator signal coming from Arduino due.

The first one, designated as ECG_PQRST, contains all the waves of standard ECG – P, Q, R, S and T waves – and matches with healthy ECG closely. The second test signal – ECG_QRS – was obtained by suppressing the P and T waves. This increases the sparsity of the signal both in time- domain and in terms of wavelet coefficients. The third test signal suppressed S and T too, leaving only the R wave. This signal will be the sparsest as the information level is lowest.

Detailed hardware testing (shown in Fig. 7) was done for three cases with M = 32, M = 48 and M = 64 and for each value of M all the three signals were applied. Further, for each signal, the non-sampled version as well as sampled and held version (from DAC0 pin of Arduino Board) was applied.

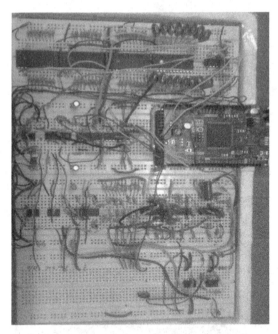

Fig. 7 Snapshot of implemented Hardware

The CS values imported to MATLAB workspace were used for reconstruction to determine the SNR and PRD metrics. The reconstruction was done in wavelet domain with Basis Pursuit reconstruction algorithm. This is a kind of Convex optimization reconstruction method where the basic concept is that since the signal x which was compressively sensed is sparse, the reconstructed signal x $=\Theta y$ should also be sparse. The sparse solution to underdetermined system of linear equations $x = \Theta y$ can be summarized as min x 1 subject to $y = \phi\psi x$ Where x 1 is the l1 norm (measure of sparsity) of the reconstructed signal given by x 1 $= xii$.

The PRD realized from hardware, though satisfactory in many cases, is much higher than the ideally expected value due to various sources of error that corrupt the compressed measurements. These sources of error are listed below:

- Offset introduced by Op-amp used for inverting the applied input and deviation from unity gain in the above amplifier.
- Mismatch in the ON resistances of the two switches used for routing the input signal as per measurement matrix entries and offset introduced by the unity gain buffer amplifier in the signal route.Mismatch in value of 64 capacitors, Loss of charge in capacitors due to MUX switching operations and Dumping of charge into integrator capacitors from control lines of switches and MUX unit.
- Loss of charge in integrator capacitors due to slewing of integrator Op-amp and Quantization error in ADC.

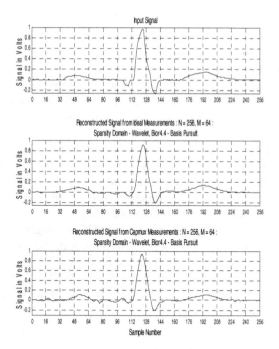

Fig. 8 Ideal and practical reconstruction of ECG_PQRST using 64 measurements

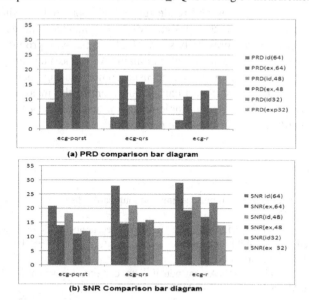

Fig. 9 (a) PRD Comparison, and (b) SNR Comparison between hardware and simulation for different ECG signals.

Fig. 8 shows the shows the results of reconstruction for ECG_PQRST signal using 64 compressed measurements. First wave is the input signal. Second plot is the reconstructed signal by using a measurement vector that is prepared by multiplying the ECG signal values by the measurement matrix in workspace. These are the measurements expected from a hardware that is ideal. Third plot is the signal reconstructed using measurements received from the hardware. Second and third plots show that the proposed hardware is performing very well in this case.

Fig. 9 shows the comparative study of SNR and PRD values of different ECG Test signals. The ideal case representing the results of mathematical CS model in MATLAB is denoted by (id, M value) in figure. The Experimental case representing the Hardware result is denoted by (ex, M value) in figure. From the PRD comparison bar diagram we observe that PRD is lowest (3%) for ECG-R test signal (signal containing only R spike of an ECG. i.e. highly sparse signal) in MATLAB simulation. The experimental hardware result is poor compared to mathematical model, but is a satisfactory hardware result with a PRD of 12%. Similarly the SNR comparison gives high SNR for MATLAB Simulation with ECG-R test signal input. This is due to the fact that ECG_R has maximum sparsity and gives better CS reconstruction.

5   Conclusion

An efficient mixed signal hardware design for Compressed Sensing on ECG signal was proposed and the design was implemented successfully. The reconstruction error of compressively sensed ECG Signal with proposed hardware is lesser when reconstructed in wavelet domain. As immediate future work can evaluate the reconstruction performance using other sensing matrices that use binary entries rather than Bernoulli entries and identify the measurement matrix that yields best result in hardware implementation and prepare a PCB version of the proposed hardware and evaluate the performance.

References

1. Alwan, A.: Global status report on noncommunicable diseases 2010. World Health Organization (2011)
2. Candes, E., Romberg, J., Tao, T.: Robust uncertainty principles: Exact signal reconstruction from highly incomplete frequency information. IEEE Trans. on Inform. Theory **52**(2), February 2006
3. Candès, E., Wakin, M.B.: An introduction to compressive sampling. IEEE Signal Processing Magazine **25**(2), 21–30 (2008)
4. Donoho, D.L.: Compressed sensing. IEEE Trans. on Inform. Theory **52**(4), April 2006
5. Dixon, A.M., Allstot, E.G., Gangopadhyay, D., Allstot, D.J.: Compressed sensing system considerations for ECG and EMG wireless biosensors. IEEE Transactions on Biomedical Circuits and Systems **6**(2), 156–166 (2012)

6. Hosseini Khayat, S.S.: ECG signal compression using compressed sensing with nonuniform binary matrices. In: 16th CSI International Symposium on Artificial Intelligence and Signal Processing (AISP), February 2012

7. Chen, F., Chandrakasan, A., Stojanović, V.: A signal-agnostic compressed sensing acquisition system for wireless and implantable sensors. In: Custom Integrated Circuits Conference (CICC), pp. 1–4. IEEE, September 2010

8. Chen, F., Chandrakasan, A., Stojanovic, V.: Design and analysis of a hardware-efficient compressed sensing architecture for data compression in wireless sensors. IEEE Journal of Solid-State Circuits **47**(3), 744–756 (2012)

9. Allen, P., Holberg, D.: CMOS Analog Circuit Design, ser. The Oxford Series in Electrical and Computer Engineering. Oxford University Press, USA (2002)

10. Margolis, M.: Arduino cookbook. O'Reilly Media, Inc. (2011)

11. Johns, D.A., Martin, K.: Analog Integrated Circuit Design. John Wiley and Sons, Inc. (1997)

12. Mishra, A., Thakkar, F., Modi, C., Kher, R.: Comparative Analysis of Wavelet Basis Functions for ECG Signal Compression Through Compressive Sensing. International Journal of Computer Science and Telecommunications **3**, 23–31 (2012)

13. Charbiwala, Z., Martin, P., Srivastava, M.B.: CapMux: A scalable analog front end for low power compressed sensing. In: International Green Computing Conference (IGCC), pp. 1–10. IEEE (2012)

14. Gayathri. S, Gandhiraj. R.: Analysis of ECG Signal compression with compressed sensing method. In: International Conference on Advance Engineering &Technology (ICAET), Bangaluru, March 23, 2014

15. Avinash. P., Gandhiraj. R., Soman. K.P.: Spectrum Sensing using Compressed Sensing Techniques for Sparse Multiband Signals. International Journal of Scientific and Engineering Research **3**(5), May 2012

The Medical Virtual Patient Simulator (MedVPS) Platform

Prema Nedungadi and Raghu Raman

Abstract Medical Virtual Patient Simulator (MedVPS) is a cutting-edge eLearning innovation for medical and other health professionals. It consists of a framework that supports various patient cases, tailored by interdisciplinary medical teams. Each virtual patient case follows the critical path to be followed for a specific patient in a hospital. MedVPS takes the student on a journey that enables the student to interview, examine, conduct physical, systematic and ultimately reach a diagnosis based on the path that is chosen. After the interactions, the student must decide whether each response is normal or abnormal and use the virtual findings to identify multiple probable diagnoses or reexamine the virtual patient with the goal to narrow down to the correct disease and then provide treatment. We present the architecture and functionality of the MedVPS platform and include a pilot study with medical students.

Keywords E-learning · Medicine · Virtual patient · Simulation · Patient · case simulation · Medical simulation

1 Introduction

Simulations provide a practical approach to acquiring and maintaining task-oriented and behavioral skills across the spectrum of medical specialties. Simulations offer clinicians a safe and credible means to acquire skills and practice managing manage clinical scenarios.

P. Nedungadi(✉) · R. Raman
Amrita CREATE, Amrita University, Coimbatore, India
e-mail: prema@amrita.edu

R. Raman
Amrita School of Business, Amrita University, Coimbatore, India
e-mail: raghu@amrita.edu

© Springer International Publishing Switzerland 2016
S. Berretti et al. (eds.), *Intelligent Systems Technologies and Applications*,
Advances in Intelligent Systems and Computing 384,
DOI: 10.1007/978-3-319-23036-8_5

59

A recent study estimates the premature deaths associated with preventable harm to patients was estimated at more than 400,000 per year [1]. This may be due to a lack of professional competence defined as, "the habitual and judicious use of communication, knowledge, technical skills, clinical reasoning, emotions, values, and reflection in daily practice for the benefit of the individual and community being served" [2]. The causes for preventable errors include performing the wrong actions, improperly performing right actions, or not performing an action based on medical evidence. Diagnostic errors can also result in wrong or delayed treatment. Simulations offer learning of uncommon cases, and multiple disease symptoms of various diseases including accurate visual and auditory inputs. Simulations can be used for learning technical procedures, clinical skills or skills required in emergency care. The clinical realism of a simulation and the feedback provided within a simulation [3] helps improve learning and performance [4].

Simulations have many advantages such as improved learning [5,6,7] in student learning due to making medical errors without harming a patient [8] and learning in a realistic context [9,10]. While rich simulation technologies have been developed in India for other domains in Engineering, Biotechnology [11] and for high school sciences [12], a complete simulation based learning and assessment platform for medical simulations has been lacking.

At present virtual patient cases offer reasonable learning about the procedures and the types of tests to perform but do not have accurate images, simulations and audio for accurate learning of patient symptoms and conditions. Other individual simulations and animations of various functions provide limited learning about the particular organ or exam.

Our Medical Virtual Patient Simulator (MedVPS) integrates real life images, audio, animations and simulations of the entire learning process into a virtual patient case to provide an end-to-end learning and assessment tool for medical students.

2   The MedVPS Platform

MedVPS consists of case based simulations that integrate medically accurate simulations, 2D and 3D animations, videos and assessments into virtual patient cases. It allows features such as the deliberate practice and feedback for skills development, exposure to difficult to visualize procedures, protocols and case studies using interactive virtual patient cases.

A typical MedVPS case consists of seven modules (Fig. 1), History Taking, Physical Examination, Systemic Examination, Investigation, Diagnosis, Treatment and Evaluation. Each module is further categorized as detailed examination procedures. Students need to select only the relevant questions or procedures that they believe are relevant to the virtual patient. MedVPS logs all student actions along with the time taken and mark it as relevant /irrelevant for the diagnosis of the virtual patient.

The web-based, responsive, mobile first design for User Interface renders the simulation seamlessly on computers and various mobile and tablet devices with Internet connectivity. It uses Scalable Vector Graphics for UI and Web Graphics Library for 3D Views and to make the animations and simulations responsive on various screen sizes. MedVPS is built using the Python and Django framework. The simulation runs on Html5 Canvas, with JS and CSS3. MedVPS is based on International Standards such as ICD-10-CM (International Classification of Diseases, 10th Revision, Clinical Modification), ATC (Anatomical Therapeutic Chemical classification system for drugs) and INN (International Nonproprietary Names for generic drugs) and hence can be easily customized and localized. MedVPS is designed to work on highly modular and reusable medical activities, and can be configured to run various virtual cases.

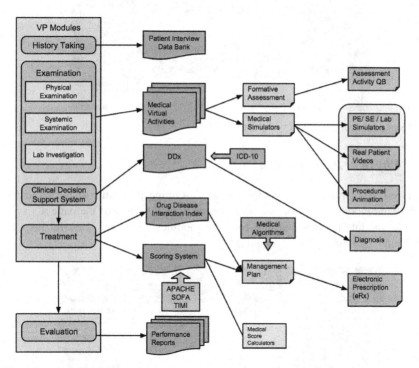

Fig. 1 MedVPS Platform

3 Virtual Patient Modules

3.1 History Taking

The 'medical history' is a structured assessment conducted to generate a comprehensive picture of a patient's health and health problems. This module helps the user ask such questions under each category. The potential list of questions are provided under each sub-section and can be accessed from the left pane of the simulator. The user can select the questions to understand the patient history.

The key to reaching an accurate diagnosis is to extract a detailed description of patient's symptoms and medical conditions. A typical sequence comprises of the introduction, understanding the complaint, history of the illness, systemic enquiry, past medical history, allergies and personal and family history.

Fig. 2 Screen showing the History Taking Module

3.2   Physical Examination

The physical examination helps the doctor determine the status of the patient's health. This module helps the user perform an overall review of the patient's system. The module is classified into four sub modules/segments - General Inspection, Vitals, Examination of Eye and ENT.

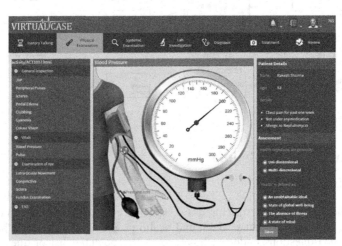

Fig. 3 Screen showing the Physical Examination Module

The user can click on the desired segment and access the tools/simulator to examine the patient. The system provides suitable tools such as thermometer, blood pressure meter (sphygmomanometer), and stethoscope under each sub-section to do the physical examination of the virtual patient. The doctor can choose the relevant physical examination he/she needs to perform with the virtual patient based on the history taking that is already performed.

3.3 Systemic Examination

Systemic examination reviews the major systems of the body like the central nervous system, the respiratory system, the cardiovascular system and the gastrointestinal system. The student should both determine and perform the activities relevant for the given virtual case to diagnose correctly.

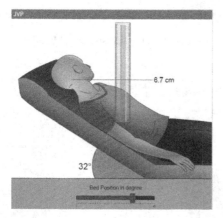

Fig. 4 Screen showing JVP being taken in the Systemic Examination Module

3.4 Investigation

The student can select various tests such as a blood test, x-ray, ECG and so on that need to be performed on the patient. The result of test is shown upon selecting the required lab test.

3.5 Differential Diagnosis

A major challenge facing the medical fraternity is the provision of quality medical care through proper diagnosis based on the symptoms. The Differential Diagnosis (DDx) module ranks diseases based on the probability of the disease symptoms of the virtual patient. Students can review the shortlisted diagnosis, go back and review the patient as needed until they arrive at the correct diagnosis.

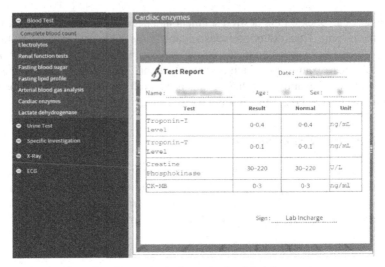

Fig. 5 Screen showing a Test Report in the Investigation Module

Fig. 6 Screen showing the Differential Diagnosis Module

3.6   Treatment Module

The treatment module allows the user to prescribe several drugs on the virtual patient. Depending on the drug, drug-to-drug interactions, drug-to-food interactions are displayed on screen to guide the student administer the correct combination of drugs. An important goal is to incorporate medical score calculators to the system to help doctors calculate the effectiveness of the treatment and survival rate for a treatment.

4 Pilot Study and Findings

The pilot study on MedVPS was conducted with 15 participants who were third year MBBS students from the Government Medical College at Trivandrum.

The workshop was designed to incorporate beginners to advanced users of computer technology. MedVPS being user friendly could be used easily and effortlessly. The whole program was catered to hold the interest of the participants, by evoking their curiosity, and making them aware of the different potentials and possibilities of the MedVPS project. The workshop was for two hours and had three sessions.

MedSim helps in learning how to manage infrequently occurring medical conditions [Please rate the following questions]

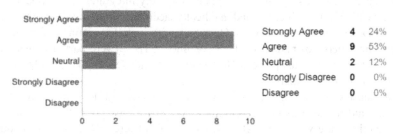

Strongly Agree	4	24%
Agree	9	53%
Neutral	2	12%
Strongly Disagree	0	0%
Disagree	0	0%

MedSim accelerates their progress in learning curves for skills development (e.g. diagnostic skills). [Please rate the following questions]

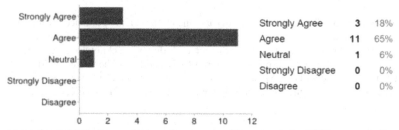

Strongly Agree	3	18%
Agree	11	65%
Neutral	1	6%
Strongly Disagree	0	0%
Disagree	0	0%

How do you rate the online performance of the simulator? [Please rate the following questions in regards to the topics that medsim demonstrated well]

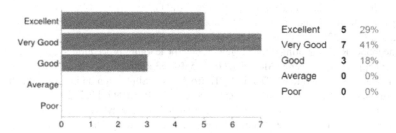

Excellent	5	29%
Very Good	7	41%
Good	3	18%
Average	0	0%
Poor	0	0%

Fig. 7 A Few Responses from the Online Feedback

The first session consisted of a brief fifteen minutes introduction to the medical simulation project and on what the workshop intended to achieve. This was followed by a fifteen minutes demo of five activities taken from more than the sixty activities developed for the project.

The second session was a sixty minutes hands-on one where the participants used the MedVPS system to diagnose virtual patients. The MedVPS cases used in the study were in the Respiratory and the Cardiology domains.

The final session was a thirty minutes one where the participants discussed their thoughts and gave their feedback orally and filled in an online survey.

5   Conclusions

Unlike other medical simulations, MedVPS uniquely integrates case based virtual patients with realistic images and medically accurate simulations to provide a learning experience for the student. Our challenges included getting real reference images and videos for patient medical conditions. These were partially overcome by re-creating images using Scalable Vector Graphics (SVG) with guidance and approval by medical experts.

We focused on learning, authenticity and assessment in developing MedVPS. Students said it was a game like environment, where they got authentic information that they rarely see in real patients. In the differential diagnosis module, students were able to learn from their errors and go back to previous modules and re-examine the patients to arrive at the proper diagnosis.

Acknowledgement This work derives its inspiration and guidance from Amrita University's Chancellor Amma. The project is funded by the Department of Electronics and Information Technology (DeitY), Government of India. We acknowledge the efforts of Rathish G and Dr. Romita and others at Amrita CREATE team, the subject guidance from faculty at Amrita Schools of Medicine and Trivandrum Medical College. We thank our partner, CDAC Trivandrum, for organising the pilot studies with the medical students.

References

1. James, J.T.: A new, evidence-based estimate of patient harms associated with hospital care. Journal of patient safety **9**(3), 122–128 (2013)
2. Epstein, R.M., Hundert, E.M.: Defining and assessing professional competence. Jama **287**(2), 226–235 (2002)
3. Kneebone, R.: Evaluating clinical simulations for learning procedural skills: a theory-based approach. Academic Medicine **80**(6), 549–553 (2005)
4. Kneebone, R.L., Scott, W., Darzi, A., Horrocks, M.: Simulation and clinical practice: strengthening the relationship. Medical education **38**(10), 1095–1102 (2004)
5. Shapiro, M.J., Morey, J.C., Small, S.D., Langford, V., Kaylor, C.J., Jagminas, L., Jay, G.D.: Simulation based teamwork training for emergency department staff: does it improve clinical team performance when added to an existing didactic teamwork curriculum? Quality and Safety in Health Care **13**(6), 417–421 (2004)

6. Grantcharov, T.P., Kristiansen, V.B., Bendix, J., Bardram, L., Rosenberg, J., Funch-Jensen, P.: Randomized clinical trial of virtual reality simulation for laparoscopic skills training. British Journal of Surgery **91**(2), 146–150 (2004)
7. Gaba, D.M., Howard, S.K., Fish, K.J., Smith, B.E., Sowb, Y.A.: Simulation-based training in anesthesia crisis resource management (ACRM): a decade of experience. Simulation & Gaming **32**(2), 175–193 (2001)
8. Ziv, A., Ben-David, S., Ziv, M.: Simulation based medical education: an opportunity to learn from errors. Medical teacher **27**(3), 193–199 (2005)
9. Khan, K., Pattison, T., Sherwood, M.: Simulation in medical education. Medical teacher **33**(1), 1–3 (2011)
10. Donaldson, L.: 150 years of the Annual Report of the Chief Medical Officer: On the state of public health 2008. Department of Health, London (2009)
11. Raman, R., Achuthan, K., Nedungadi, P., Diwakar, S., Bose, R.: The VLAB OER Experience: Modeling Potential-Adopter Student Acceptance. IEEE Transactions on Education **57**, 235–241 (2014)
12. Nedungadi, P., Raman, R., McGregor, M.: Enhanced STEM learning with online labs: empirical study comparing physical labs, tablets and desktops. In: 2013 IEEE Frontiers in Education Conference, pp. 1585–1590. IEEE, October 2013

Intelligent System Based on Impedance Cardiography for Non-invasive Measurement and Diagnosis

Pranali C. Choudhari and M.S. Panse

Abstract Impedance cardiography has become a synonym for indirect assessment of monitoring the stroke volume, cardiac output and other hemodynamic parameters by monitoring the blood volume changes of the body. In this method the changes in the impedance within a certain body segment are recorded as a measure of the physiological changes happening within that segment, thus making it suitable for assessment of physiological parameters. The accurate computer aided diagnosis of the cardiovascular diseases is gaining impetus. Accuracy and non - invasiveness of the diagnostic system have become the need of the hour. This paper presents an intelligent system for measurement of haemodynamic parameters and diagnosis of diseases with the help of impedance cardiography. The vital parameters are calculated with the bioimpedance signals obtained non-invasively along the thorax. For the diagnosis the signals are recorded across the radial artery. The multivariate analysis has been used to obtain a diagnosis index based on various time as well as frequency domain parameters.

Keywords Impedance cardiography(ICG) · Radial artery · Haemodynamic · Non-invasive · Analysis of Variance (ANOVA) · Multivariate analysis

1 Introduction

Electrical impedance measurement in human body for cardiac system analysis were first applied in Impedance Plethysmography by Jan Naboer in 1940 and in Impedance Cardiography by W.G. Kubicek in 1966 [1], to assess central and peripheral blood flow from heart.

P.C. Choudhari(✉) · M.S. Panse
Department of Electronics Engineering, Veermata Jijabai Technological Institute, Matunga, Mumbai, India
e-mail: pranalic75@gmail.com

© Springer International Publishing Switzerland 2016   69
S. Berretti et al. (eds.), *Intelligent Systems Technologies and Applications*,
Advances in Intelligent Systems and Computing 384,
DOI: 10.1007/978-3-319-23036-8_6

Biological tissues such as muscle, bone etc, and biological fluids such as blood, urine, cerebrospinal fluid etc. are neither good conductors of electricity like metal nor they are bad conductors like wood. This intermediate property of the biological matter makes it feasible to measure electrical conduction through them or impedance offered by them. Measurement of this impedance in various tissues tells about the capacity of electrical conduction of that tissue. ICG has been mainly used to determine the stroke volume and thereby the cardiac output. The evaluation of the SV from the bioimpedance signal was first done using Kubicek equation [1]and then Sramek [2] equation. It is now established that Bernstein equation [3][4] can give best results in such analysis.

In bioimpedance measurement, alternating current of 4 mA (50 KHz – 100 KHz) is passed through the thorax in a direction parallel with the spine between the beginning of the thorax (the line at the root of the neck) and the end of the thorax.

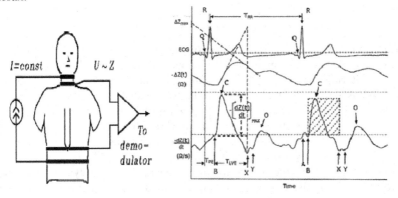

Fig. 1 Kubicek's Arrangement of electrodes at the recording of the electrical impedance for SV estimation[1] and Timing relation between ECG, and ΔZ, dz/dt of ICG.

Two voltage electrodes are placed on neck and chest. Current transmitted by outer electrodes seek the path of least resistance i.e. the blood filled aorta Fig 1.

Several researchers have validated the accuracy and reproducibility of the measurements of SV and CO using ICG against the existing invasive accurate methods such as Fick's [5], Doppler [6] and PATD[7]. Peacock et al. [8] showed the impact ICG data on diagnosis and therapeutic planning for dyspnea patients. Impedance cardiography (ICG) is a simple, noninvasive technique that is commonly used to calculate the stroke volume, cardiac output, and systemic vascular resistance (SVR). In addition, ICG readings can be taken by performing postural changes to obtain data pertaining to ventricular-vascular coupling [9,10]. ICG has been mainly used by researchers for early detection of cardiovascular diseases. Most of the methods implemented so far make use of maximal change in the dZ/dt waveform, wave morphology to detect the risk of a cardiac disease [11,12]. This essentially needs an expert to characterize the changes in the morphology and detect the amount and type of risk associated with the cardiovascular system.

Statistical methods have been used to mainly find the correlation between the obtained haemodynamic parameters of the study group and the control group. These findings have been further used devise a method of differentiating the pathological cases from the normal ones. Also, most of the existing diagnosis methods based on ICG have acquired the impedance signals across the thorax, where the velocity as well as volume of blood is maximum [13,14]. This paper presents a non-invasive method of diagnosis of diseases using impedance cardiography applied at the radial pulse. The application of electrodes along the wrist would be less traumatic to the patient in comparison to the thoracic electrodes and would also be free from respiratory artifacts and misinterpretation of impedance changes due to extra vascular lung water in cases of pulmonary infections.

2   Acquistion System

ICG signal is acquired by injecting high frequency 4mA current through the subject's hand using silver coated metal band electrodes. Due to change in the impedance the voltage of the acquired signal is varying. This change in voltage is ICG signal which is further processed in MATLAB. Input sinusoidal signal of 4mA, 100 KHz is generated using circuit shown Fig.2. It consists of timer IC which generates square wave of desired frequency followed by exponential integrator circuit using OPAMP to smooth square wave into sinusoidal signal. In order to get constant current at output irrespective of applied load transconductance amplifier is used at output stage. Output of transconductance amplifier is injected through hand and output from two voltage electrodes is taken and processed in MATLAB after amplification using instrumentation amplifier. The standard method of implementing ICG is to apply an AC current across the thorax and measure the voltage changes. Hence the Thoracic segment is used for obtain the various a haemodynamic parameters, while the wrist segment is used for acquiring the signal for diagnosis.

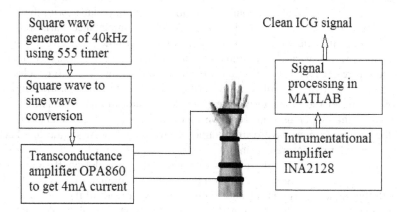

Fig. 2 ICG acquisition circuit .

3   Calculation of Haemodynamic Parameters

The most widely used method of diagnosis of the cardiovascular diseases is the ECG analysis. But this method is not suitable in proper evaluation of the functioning of the cardiovascular diseases in some scenarios. Moreover, it requires specialized environments, costly equipment, and specially trained medical personnel to obtain and/or interpret the data. Thus, there is a need to develop a system which is simple to operate, interpret and cost effective, for assessing the cardiac function of an individual. The characteristic points of the ICG signal are direct indicators of the various cardiac activities such as :

A: Atrial wave. The A-wave of the impedance cardiogram is the small negative deflection from baseline that occurs during the PEP and before the opening of the aortic valve and is strongly correlated with atrial contraction.
B: Aortic opening which marks the beginning of the systolic interval.
C: Maximum aortic flow (dz/dt) which appears after the R wave of the ECG. The shape, depth and duration of this wave help in determining dz/dtmax and the ventricular ejection time. The major determinant of the shape of the C wave is the aortic pulse pressure wave.
X and Y : Aortic and Pulmonary valve closing respectively.
O: Mitral valve opening.

The functioning and timings of the cardiovascular valves can be accurately identified with the help of the ICG signal. The analysis module has been divided into three sections viz., denoising, parameter estimation and time and frequency domain analysis for diagnosis.

3.1   Denoising

The major sources of artifacts in case of a bioimpedance signals are respiration and movement of the patient during acquisition. There is fair amount of overlap in the spectrums of the signal as well as the artifact. The signals that are obtained from the wrist have a smaller amount of variation in the impedance with the blood flow, making the artifacts more dominant. Also, the impedance variations obtained at thorax include the variation in impedance due to the other thoracic fluids such as lung water in case in pulmonary infections, change in the air content in lungs due to respiration, in addition to the aortic blood flow. Therefore, selection of the artifact removal algorithm is very critical, in order to maintain the temporal relations between the characteristic features for accurate diagnosis. The system presented in this paper uses an adaptive algorithm for baseline wander removal from the bioimpedance waveform, obtained at the radial pulse of the left hand, using the wavelet packet transform.

The algorithm computes the energy in scale of the wavelet coefficients. The energies in the successive scales are compared and the branch of wavelet binary tree with the higher energy is selected. This will be followed till the energy difference in the subsequent levels exceeds the preset threshold value. The wavelet packet decomposition of the signal was done using Daub-4 wavelet [15].

Table 1 gives the various parameters calculated for 25 normal subjects within an age group of 25 to 45 years, their formulae, mean and standard deviation and the standard normal values as per the medical data books. The data has been presented for normal subjects deliberately, so as to demonstrate the effectiveness of system in measurement of the haemodynamic parameters.

Table 1 Haemodynamic parameter calculated with help of ICG across the Thorax

Parameter	Formulae	Calculated values with mean and standard deviation	Standard Values
Heart Rate	$HR = \dfrac{60}{CC}$	73.9036 ± 9.1964	60-100 beats / min
LVET	Interval from B to X point in the ICG	0.3639 ± 0.0151	0.330 ± 0.020 sec
Stroke Volume (Kubicek)	$SV_{Kub} = \rho \cdot L^2 \cdot T_{LVET} \cdot \dfrac{\left(\dfrac{dz}{dt}\right)_{max.}}{Z_0^2}$	73.0153 ± 16.01	60 - 80 ml /beat
Stroke Volume (Bernstein)	$SV_{Bern} = V_{EPT}\sqrt{\dfrac{(dZ/dt)_{max.}}{Z_0}} LVET$	131.968 ± 15.125	136 ml/beat
Cardiac Output(CO) (Kubicek)	$CO\left(\dfrac{litres}{min}\right) = SV \times HR$	5.4055 ± 1.2153	4.0 - 8.0 L/min
Cardiac Output(CO) (Bernstein)	$CO\left(\dfrac{litres}{min}\right) = SV \times HR$	9.7637 ± 1.1845	8 -12 L/min

3.2   Time and Frequency Domain Analysis

The system presented in this paper can be used for vital parameter measurement as well diagnosis of diseases. A data set of 28 patients with 7 normal subject, 7 with cardiac abnormality, 7 with hypertension and 7 subjects with diabetes were selected. The data was obtained at wrist at a clinic in Thane, Mumbai. A significant difference was observed in the heart rate pattern of normal and abnormal subject. Hence Poincare plots were plotted to find the scatter pattern of the Heart rate values at various beats[16].

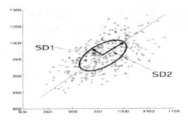

Fig. 3 Poincare plot for CC intervals.

The CC intervals (in msec) of the ICG signal are represented as two auxiliary vectors which can be defined as

$$x^+ = (x_1, x_2, x_3, \ldots\ldots\ldots\ldots, x_{N-1}) \text{ and } x^- = (x_2, x_3, x_4, \ldots\ldots\ldots\ldots, x_N)$$

The plot of x^+ vs x^- is the Poincare plot. Fig 3 shows the Poincare plot and its various statistical parameters such as SD1, SD2, centroid and the area. The CC intervals are scattered in the form of an ellipse around a centroid with SD1 as horizontal deflection from mean and SD2 as vertical deflection from mean.

$$SD1 = \sqrt{Var\left(\frac{x^+ - x^-}{\sqrt{2}}\right)} \; ; \; SD2 = \sqrt{Var\left(\frac{x^+ + x^-}{\sqrt{2}}\right)} ; \; Area \; of \; the \; Ellipse : \pi \times SD1 \times SD2$$

Fig 4 shows the Poincare plot for the different groups of subject. The plot showed a significant difference in the centroid and the area of the ellipse. Thus, a visual classification of the diseases was obtained.

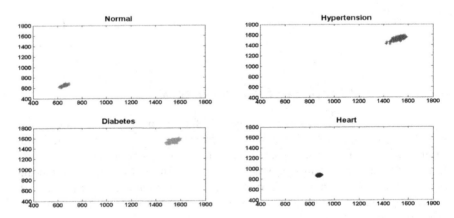

Fig. 4 Poincare plots for subject from various groups (Top left : Normal; Topright: Hypertension; Bottom left: Diabetes and Bottom right : Heart Patient.

Further, the frequency domain characteristics were explored to see if the ICG signals of normal and abnormal subject differed. Hence the Power Spectral Density (PSD) of Heart Rate signals, for each of subjects, was obtained using the Welch spectral estimation method. The zero mean HR signal was smoothed using a moving average filter of order 10. The window length for estimation was 128, with 64 overlap samples and 1024 points in the Fast Fourier Transform.

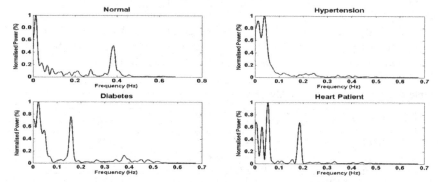

Fig. 5 PSD plots for subject from various groups (Top left : Normal; Topright: Hypertension; Bottom left: Diabetes and Bottom right : Heart Patient.

The PSD plots showed peaks at three prominent frequencies : 0.04 Hz (Very Low Frequency VLF), 0.15 Hz (Low Frequency LF) and 0.4 Hz (High Frequency HF). The amplitudes of the PSD plots at these frequencies, area of the plot from 0-0.04 Hz, 0.04 Hz to 1.5 Hz and 1.5 Hz to 0.4 Hz were used as markers for differentiation between the various subjects. Fig. 5 shows the PSD plots for four subjects each from different group.

Thus, based on the Poincare and the PSD plots the following parameters were identified for classification of the disease:

1.	SD1	6.	LF Area
2.	SD2	7.	HF Area
3.	Area of Ellipse	8.	Centre frequency in VLF (F_{VLF})
4.	Centroid of Poincare	9.	Centre frequency in LF (F_{LF})
5.	VLF Area	10.	Centre frequency in HF (F_{HF})

3.3 ANOVA and Multivariate Regression

Though the above shortlisted parameters could significantly classify the subject correctly, the decision mainly depended on the skills of the technician who, would interpret the data. Thus, it was necessary to identify the most significant parameters and develop a diagnosis index which could easily and accurately classify the subject into various groups. One way ANOVA was performed for each parameter with $\alpha = 0.05$. The null hypothesis is that the means of all groups are equal

- $H_0 = \mu_1 = \mu_2 = \mu_3 = \mu_4$

The alternative hypothesis is that at least one of the means is different from another.

Table 2 depicts the statistical parameters obtained for the above shortlisted classification parameters using one way ANOVA. There are four types of diseases, thus the degrees of freedom between the group = 4-1 =3, while degrees of freedom within the group = 28-4=24. From the F distribution table, the $F_{critical}$ value for 3 degrees of freedom between the group and 24 degrees of freedom within the group is 3.0088. After the one way ANOVA F test, if the obtained F value is more than the $F_{critical}$, then the null hypothesis can be rejected.

Table 2 Statistical parameter of ANOVA for the given data set.

Parameter	$F_{obtained}$	P	$F_{critical}$
SD1	23.454	0.000009 => 0	3.0088
SD2	8.619	0	3.0088
Area of Ellipse	18.79	0	3.0088
Centroid	13.43	0	3.0088
VLF Area	22.412	0	3.0088
LF Area	18.501	0	3.0088
HF Area	8.968	0	3.0088
F_{VLF}	12.834	0	3.0088
F_{LF}	12.288	0	3.0088
F_{HF}	5.184	0.007	3.0088

This would imply that there is a definite difference in the means of the said parameter within the different groups. This fact can be effectively utilized for classification of the subject into normal and pathological cases. Table 2 suggests that the Null hypothesis should be rejected in every case, thus indicating that all the parameters are significant in deciding the category of the patient. But, including all the parameters for classification would make the system fairly complicated. Thus, 7 important parameters were identified based on the following condition

$$Parameter\ is\ significant\ if\ \begin{matrix} P = 0 \quad and \\ |F_{obtained} - F_{critical}| \geq 6 \end{matrix}.$$

Thus, SD1, Area of the ellipse, centroid, VHF Area, LF area , F_{VLF} and F_{LF} were recognized as the significant parameters. Matlab was used to perform the multivariate regression to obtain a diagnosis index based on the selected parameters. The relation thus obtained for the diagnosis index was able to classify the subjects into three categories only viz., Normal, Diabetes and Heart Patient or Hypertension. The index values for Heart and Hypertension patients were overlapping. Hence a second level of regression was performed between the parameters of the third category (after 1st level of classification) , VLF Area and F_{VLF}. Thus, the complete classification was performed using two indices: $Index_1$: Level 1 Classification and $Index_2$: Level 2 Classification.

$$Index_1 = 0.0163 * centroid + 0.1197 * VLF\ area + 54.6211 * F_{LF} - 0.2018$$
$$* Sd1 - 0.0021 * Area\ of\ ellipse - 0.0601 * LF\ area$$
$$- 182.7115 * F_{VLF}$$
$$Index_2 = 0.0048 * LF\ area + 260.0203 * F_{VLF}$$

Table 3 shows the criterion for classification obtained with the help of the two indices.

Table 3 Disease Classification Criteria

Sr. No	Type of Subject	Condition
1.	Normal	$Index_1 < 10$
2.	Heart Patient	$10 < Index_1 < 35$ and $Index_2 > 13$
3.	Hypertension	$10 < Index_1 < 35$ and $Index_2 < 13$
4.	Diabetes	$Index_1 > 35$

4 Conclusion

This paper presents an intelligent system based on impedance cardiography which can be used for bedside parameter estimation as well as easy diagnosis. Elimination of the thoracic electrodes would make the measurement less traumatic to the subject. The prime contribution of this system is the statistically developed diagnosis index which is capable of identifying four different categories of ailments, in comparison to the existing methods which have used wave morphologies and frequency domain characteristics to categorize mainly heart diseases. A mathematical derived constant can be obtained that could be multiplied with radial pulse impedance to get a correct estimate of cardiac output. The statistically developed diagnosis index developed in this paper can be used successfully for disease classification using ICG.

References

1. Kubicek, W.G., Karnegis, J.N., Patterson, R.P., Witsoe, D.A., Mattson, R.H.: Development and evaluation of an impedance cardiac output system. Aerospace Medicine 37(12), 1208–1212 (1966)
2. Sramek, B.B.: BoMed's electrical bioimpedance technology for thoracic applications (NCCOM): Status report, May 1986 Update, pp. 19–21. BoMed Ltd., Irvine (1986)
3. Bernstein, D.P.: Impedance cardiography: pulsatile blood flow and the biophysical and electrodynamic basis for the stroke volume equations. Journal of Electrical Bioimpedance 1, 2–17 (2010)
4. Henry, I.C., Bernstein, D.P., Banet, M.J.: Stroke volume obtained from the brachial artery using transbrachial electrical bioimpedance velocimetry. In: 34th Annual International Conference of the IEEE EMBS San Diego, California, USA (2012)

5. Barde, P., Bhatnagar, A., Narang, R., Deepak, K.K.: Comparison of non-invasive cardiac output measurement using Indigenous impedance cardiography with invasive fick method. International Journal of Biomedical Research 3, 11 (2012)
6. Schmidt, C., Theilmeier, G., Van Aken, H., Korsmeier, P., Wirtz, S.P., Berendes, E., Hoffmeier, A., Meissner, A.: Comparison of electrical velocimetry and transoesophageal Doppler echocardiography for measuring stroke volume and cardiac output. British Journal of Anaesthesia 95(5), 603–610 (2005)
7. Sharma, Singh, A., Bhuvanesh, K., Anil, K.: Comparison of transthoracic electrical Bioimpedance cardiac output measurement with thermodilution method in post coronary artery bypass graft patients. Annals of Cardiac Anaesthesia 14-2 (2011)
8. Peacock, W.F., Summers, R.L., Vogel, J., Emerman, C.E.: Impact of impedance cardiography on diagnosis and therapy of emergent dyspnea: the ED-IMPACT trial. Academic Emergency Medicine 13(4), 365–371 (2006)
9. DeMarzo, A.P., Kelly, R.F., Calvin, J.E.: Impedance cardiography: a comparison of cardiac output vs waveform analysis for assessing left ventricular systolic dysfunction. Prog. Cardiovasc. Nurs. 22, 145–151 (2007)
10. Antonicelli, R., Savonitto, S., Gambini, C., et al.: Impedance cardiography for repeated determination of stroke volume in elderly hypertensives: correlation with pulse echocardiography. Angiology 42, 648–653 (1991)
11. Demarzo, A.P.: Using impedance cardiography to detect subclinical cardiovascular disease in women with multiple risk factors: a pilot study. Prev. Cardiol. 12, 102–108 (2009)
12. Ferrario, C.M., Jessup, J.A., Smith, R.D.: Hemodynamic and hormonal patterns of untreated essential hypertension in men and women. Ther. Adv. Cardiovasc. Dis. 7(6), 293–305 (2013)
13. Jindal, G.D., Nerurkar, S.N., Pedhnekar, S.A., Babu, J.P., Kelkar, M.D., Deshpande, A.K., Parulkar, G.B.: Diagnosis of peripheral arterial occlusive diseases using impedance plethysmography. J. Postgrad. Med. 36, 147 (1990). [serial online] [cited 2015 Jun 4]
14. Parmar, C.V., Prajapati, D.L., Chavda, V.V., Gokhale, P.A., Mehta, H.B.: A Study of Cardiac Parameters using Impedance Plethysmography (IPG) in Healthy Volunteers. J. Phys. Pharm. Adv. 2(11), 365–379 (2012)
15. Pranali, C.: Choudhari Denoising of Radial Bioimpedance Signals using Adaptive Wavelet Packet Transform and Kalman Filter, 01-08. IOSR Journal of VLSI and Signal Processing (IOSR-JVSP) 5(1), 01–08 (2015). Ver. II (Jan–Feb. 2015)
16. Piskorski, J., Guzik, P.: Filtering Poincare´ plots. Computational Methods in Science and Technology 11, 39–48 (2005)

Detection of Parkinson's Disease Using Fuzzy Inference System

Atanu Chakraborty, Aruna Chakraborty and Bhaskar Mukherjee

Abstract Parkinson's disease is a degenerative neurological disorder which severely affects the ability to move, speak and behave. Parkinsonism is clinically diagnosed but the laboratory confirmation is hard to obtain and needs very sophisticated investigations with high economic burden. In this work a fuzzy based approach is adapted for efficient and proper detection of Parkinson's disease using biomedical measurements of voice which is cheap and cost effective. The proposed system can also be utilised as a supplementary test to confirm the clinical diagnosis of Parkinson's disease in a remote place using the telephony system to track the voice signal. Moreover, it can be used as a guide for the treatment of Parkinson's disease as it gives a quantitative measure to signify the extent of the disease. The model is constructed using a Sugeno-Takagi Fuzzy Inference System (FIS) based on Fuzzy C-Means clustering. The result shows that the detection accuracy of the system is up to 96-97 percent with a reasonable efficiency. The study also compares the results with the Subtractive Clustering based Fuzzy Inference System. The accuracy of the Fuzzy C-Means based Fuzzy Inference System is found to be higher than the other one.

Keywords Parkinson's disease · Fuzzy C-Means · Sugeno-Takagi Fuzzy Inference System

---

A. Chakraborty(✉) · A. Chakraborty
Department of Computer Science and Engineering,
St. Thomas' College of Engineering & Technology, Kolkata, India
e-mail: {atanu.here2011,aruna.stcet}@gmail.com

B. Mukherjee
Department of Psychiatry, Malda Medical College & Hospital, Malda, India
e-mail: dr.bhaskar.mukherjee78@gmail.com

B. Mukherjee
Senior Consultant Psychiatric, Antara Psychiatric Hospital, Kolkata, India

© Springer International Publishing Switzerland 2016
S. Berretti et al. (eds.), *Intelligent Systems Technologies and Applications*,
Advances in Intelligent Systems and Computing 384,
DOI: 10.1007/978-3-319-23036-8_7

79

1   Introduction

Parkinsonism is one of the prominent neurological disorders of old age and its prevalence is rising as the geriatric population is on the rise. It is estimated that approximately one percent of US population is being affected and over 40,000 new cases are reported to be positive for Parkinson's disease in every year. Previously the disease was prevalent among the people of age around 75 to 84 years but at present it is largely detected in early 60s and is even noted in 50s. Early and proper detection can be very helpful in efficient treatment and disability limitation of Parkinson's disease.

To achieve the above goal an effective fuzzy based method is proposed in this work. The detection is based on the biomedical measurement of different parameters of voice and speech.

The study by Murdoch *et al.*, shows different effects of Parkinson's disease in speech and communication [1]. Fox *et al.*, has illustrated how different levels Parkinson's disease cause a weakening of muscles that regulate the voice and speech signals. It also gives the results of improvement of voice signals during Lee Silverman Voice Treatment (LSVT), the most promising treatment of Parkinson's induced dysarthria [2].This forms the basis that voice parameters measurement can be used to signify the progress and extent of Parkinson's disease during treatment. Kris Tjaden also presents a study on dysarthria and dysphagia often caused by Parkinson's disease [3].Shimon Sapir in his work has detailed about the motor abnormalities caused by Parkinson's disease [4].

Most of the natural data includes certain amount of uncertainty and vagueness. The conventional data clustering algorithms is inefficient in dealing with this problem. The fuzzy sets and fuzzy logic based clustering can be used as an effective tool for handling natural data [5]. Fuzzy c-means and Subtractive clustering is two fuzzy based clustering algorithm that has proved to be efficient in many FIS based studies [6], [7].In the survey of R.Suganya and R.Shanthi fuzzy c-means is seen to give better results than other hard and soft clustering algorithms [8].Stephen L.Chiu has proposed a fuzzy model based on the estimation made by fuzzy c–means clustering [9]. Chopra *et al.*, has also represented a fuzzy rule structure based on subtractive clustering [10].The fuzzy inference rule structure can be constructed mainly in two ways Mamdani and Sugeno-Takagi. A comparative study made by Arshdeep Kaur and Amrit Kaur shows Sugeno-Takagi models gives better efficiency than Mamdani model in some cases [11].In this study also Sugeno type model gives better efficiency than Mamdani. Popescu M. and Khalilia M., has shown in their study that in case of disease prediction fuzzy based system outperforms the classifier based methods [12].Voice disorders are common phenomena associated with Parkinson's disease. It can cause hypophonia (reduced volume of voice) and dysphonia (hoarseness or creakiness in the voice).Acoustic tools and few other techniques (like Recurrence and Fractal Scaling) can be used to detect various parameters of voice signal [13]. Athanasios *et al.*, has proposed a noble speech signal processing algorithm for high accuracy classification of Parkinson's disease using ten dysphonia features [14]. Shahbakhi *et al.*, implemented a similar kind of system using genetic algorithm and SVM

[15]. Sharma *et. al.*, proposed ANN and SVM based model for detection of Parkinson's disease [16].

This paper is organized as follows. In Section 2, preliminaries are discussed. The proposed method is explained in Section 3. Experimental results and conclusion are listed in Section 4 and Section 5 respectively.

2   Preliminaries

Parkinson's disease: Parkinson's disease is a degenerative neurological disorder that affects central nervous system. It is also known as idiopathic or primary Parkinsonism, hypokinetic rigid syndrome (HRS) or paralysis agitans. It primarily affects the control of movement and also causes behavioural, psychiatric, speech related disorders. Though the actual cause is unknown yet it is seen that it results from the loss of dopamine producing neurons located in areas of midbrain such as substantia nigra, basal ganglia and brainstem.

It is estimated that 89% of Parkinson's disease patients suffer from voice and speech related dysfunctions [2]. This is medically known as dysarthria, which is present in all forms of Parkinsonism. Parkinson's disease results into impaired movement of speech muscle including vocal cords, diaphragm etc. This in turn weakens the voice and changes its normal frequency. So, voice signal analysis and measurements provide efficient means for detection and treatment of Parkinson's disease. It can also be used as confirmatory test if the symptoms of Parkinsonism are observed without resorting to costly tests like fluoride labelled Single Photon emission Computed tomography (SPECT).

Inputs: In this fuzzy based model 16 attributes from biomedical measurement of voice is used for creating the Sugeno-Takagi Fuzzy Inference System (FIS).It is a rule structure based on the property of fuzzy sets obtained as a result of clustering.16 input attributes are selected on the basis of experimental results obtained. Along with this attributes the original diagnosis (i.e. positive for Parkinson's disease or negative) is used to train the FIS.

Vocal frequency: fundamental frequency of the voice signal being tested.
  1. Average vocal fundamental frequency (F_o (Hz))
  2. Maximum vocal fundamental frequency (F_{hi} (Hz))
  3. Minimum vocal fundamental frequency (F_{lo} (Hz))

Jitter: Variability of fundamental frequency, gives the perception of rough or harsh voice quality.
  4. Jitter percentage
  5. Jitter absolute
  6. Relative average perturbation (RAP %)
  7. Period perturbation quotient (PPQ %)
  8. Difference of differences of Periods (DDP)

Shimmer: Variability in amplitude, gives the strength of voice signal.
  9. Shimmer percentage
  10. Shimmer in decibels

11. Three point Amplitude perturbation quotient (APQ3 %)
12. Five point Amplitude perturbation quotient (APQ5 %)
13. Eleven point Amplitude perturbation quotient (APQ11 %)
14. Difference of difference in amplitude(DDA)

Noise: Measure of noise and tonal component present in voice signal.

15. Noise to harmonic ratio (NHR)
16. Harmonic to noise ratio (HNR)

3   Proposed Method

Clustering is used to identify different sets of input parameters related with the output i.e. the detection of Parkinson's disease. The fuzzy inference system is based on the *If-Then* rule structure constructed using these sets formed during clustering. The schematic diagram of the proposed method is shown in Fig. 1. The entire model comprises of three main phases. These are as follows,

3.1 Clusterification of data.
3.2 Construction of fuzzy inference structure (FIS).
3.3 Decision making.

3.1   Clusterification

Clustering is done to extract the rules from the input and output data set which properly models the behaviour of the input output data mapping .The training data set is partitioned into several homogeneous groups based on the similarity of their properties. The system is mainly based on the fuzzy c-means clustering. However, this study also compares the result with subtractive clustering.

3.1.1   Fuzzy C-Means Clustering

Fuzzy c-means is one of the most popular fuzzy based clustering algorithms. It was first proposed by Dunn in 1973[6] and later improved by Bezdek in 1981[7].Each of data item (n) is assigned to each of the homogeneous groups with a certain membership coefficient (u). u_{ij} denotes the degree of membership of ith input data x_i into j^{th} cluster A_i^j. The distance of ith input to j^{th} cluster centre is calculated as $| x_i - c_j |^2$ which follows simple Euclidian method. The basic objective is to minimize the function J_m (1).

$$J_m = \sum_{i=1}^{n}\sum_{j=1}^{c} u_{ij}^{m}| x_i - c_j |^2 \tag{1}$$

Where m denotes the fuzziness index and is always $m \geq 1$.Maintaining the following constraints (2), (3) and (4).

$$0 \leq u_{ij} \leq 1, \forall i, j \tag{2}$$

$$\sum_{j=1}^{c} u_{ij} = 1, \forall i \tag{3}$$

$$0 < \sum_{i=1}^{n} u_{ij} < n \, , \forall n \tag{4}$$

In each iteration the objective function J_m is optimized and the membership coefficient and cluster centre is updated using the relations (5) and (6) respectively,

$$u_{ij} = \frac{1}{\sum_{k=1}^{c} \left(\frac{|x_i - c_j|}{|x_i - c_k|} \right)^{\frac{2}{m-1}}} \tag{5}$$

$$c_j = \frac{\sum_{i=1}^{n} u_{ij}^m \cdot x_i}{\sum_{i=1}^{n} u_{ij}^m} \tag{6}$$

The algorithm terminates when the change in consecutive updates of the membership coefficient falls below the pre-set threshold.

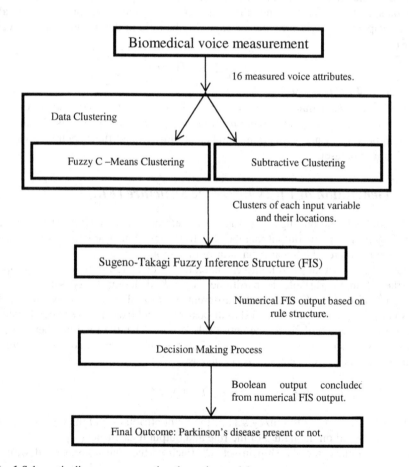

Fig. 1 Schematic diagram representing the entire model

3.1.2 Subtractive Clustering

This algorithm was proposed by Chiu [9].Subtractive clustering is another type of fuzzy based clustering which automatically detects the number of clusters and the cluster centres. In this algorithm each data point x_i is associated with a potential p_i. The potential is measured as the distance of the data point from all other data points. The initial potential of a data point is calculated using equation (7).

$$p_i = \sum_{j=1}^{n} e^{-\alpha|x_i - x_j|^2} \tag{7}$$

where $\alpha = 1/r_a^2$ and r_a^2 is the radius that effectively measures the neighbourhood distance of a datapoint.

Among all the points the data point having the maximum potential will be chosen as the first cluster center. p_1^* is the maximum potential corrosponding to the datapoint x_1^*.For finding the next cluster center the potential of all other datapoints is revised by decreasing it in the ratio of the distance of x_1^* given in (8).

$$p_i = p_i - p_1^* e^{-\beta|x_i - x_j^*|^2} \tag{8}$$

where $\beta = 4/r_b^2$, $r_b = 1.5r_a$.

r_b should be greater than r_a to get delocalized cluster centres. The second cluster centre will be the data point having maximum of the revised potentials. At the end of the procedure it gives a certain number of clusters and their centres based on the nature of dataset.

3.2 Generation of Fuzzy Inference Structure (FIS)

This model is based on Sugeno-Takagi fuzzy inference system (FIS).The speciali-ty of Sugeno FIS is the output membership function is linear or constant in nature. This case facilitates the linear behaviour of the output.

Each voice attributes taken into consideration is clustered into different groups. Let the ith input variable is partitioned into K different fuzzy sets A_i^1, A_i^2, A_i^3....,A_i^K. If K_1, K_2, K_3..,K_n is the different number of clusters associated with $x_1, x_2, x_3...,x_n$ input variables, then the all possible combination of the membership in these fuzzy sets will create the entire rule structure i.e. the total number of rule in the system will be $K_1 * K_2 * K_3 * ....Kn$. For a multiple input and single output system like this the rule can be represented as

If x_1 is A_1^i and x_2 is A_2^i ...... x_n is A_n^i then y is C_i .

where A_1^i, A_2^i,...., A_n^i is the fuzzy sets of the input attributes and the conclusion of the rule C_i based on the nature of the rule and output data clusters. x_j is A_j^i means that x_j has highest membership in set A_j^i. Each of the rule is associated with a firing strength or weight w_i,

$$w_i = \mu A_1^i(x_1) . \mu A_2^i(x_2).... \mu A_n^i(x_n) \tag{9}$$

It is calculated by the arithmetic product of the coefficient of membership of each input variable into the corresponding fuzzy sets (9), the coefficient of membership ($\mu\ A_j^i(x_j)$) denotes the degree of membership of input x_j to the fuzzy set A_j^i. The calculation of membership value of each input to the predetermined sets is known as fuzzification. The degree of membership is obtained using a mathematical function called membership function. The membership function used here is Gaussian membership function. After the evaluation of all rule outcomes along with the respective firing strengths the final output for j^{th} input O_j is calculated using (10).

$$O_j = \frac{\sum_{i=1}^{n} w_i . C_i}{\sum_{i=1}^{n} w_i} \tag{10}$$

This is the defuzzification phase which is computed as the weighted average of the firing strengths and the rule outcomes. Fig. 2 shows the schematic representation of the fuzzy inference system.

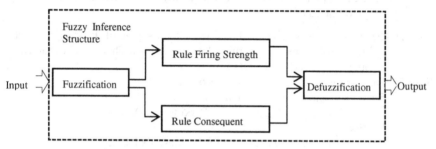

Fig. 2 Schematic diagram of FIS

3.3 Decision Making

The output of the FIS system is not boolean (i.e. Parkinson present or positive is 1 and absent or negative is 0) rather a continuous value ranging from 0.1 to 1.5. By repeated experiment with different case studies it is noted that desired negative outputs (0) gets a small value (-∞ to 0.29) as output where as positive results (1) are higher (0.3 to 1.5) than the others. So it can be concluded that, if outcome is less than certain threshold (0.3) then result goes negative else it will become positive. This decision logic is implemented to get the final boolean output i.e. the Parkinson's disease present or absent. It is done by checking the output of FIS system. If it is higher than 0.3 then the case can be concluded as positive (for Parkinson's disease) else it is negative.

3.4 Proposed Algorithm

The algorithm represents the entire process undertaken to detect the Parkinson's disease form the biomedical measurements of voice signal. The input data (16 biometrically measured voice parameters) is clustered to find the different fuzzy sets associated with the output i.e. detection of Parkinson's disease. Based on these sets the fuzzy rule structure of the fuzzy inference system is created. This fuzzy inference structure is evaluated with a set of testing inputs. The fuzzy infe-rence structure (FIS) output is fed into the decision logic for the final detection outcome.

Step 1: The collected and processed dataset is clustered into several homogenous groups using *fuzzy c- means* or *subtractive clustering*.

Step 2: Based on obtained fuzzy sets (clusters) *Sugeno-Takagi rule structure* is generated.

    2.1 All possible combination *fuzzy if-then* rules are created based on each of the clusters formed.

    2.2 A conclusion of each rule is represented as the linear function of inputs and their membership grades.

Step 3: For an input based on its degree of membership in different fuzzy sets (fuzzification) each of the rule consequent is calculated along with firing strength of the rule.

Step 4: The final output is calculated as a weighted average of the firing strength and the consequent of each rule (defuzzification). This becomes the final output of FIS structure (say O_i).

Step 5: Decision making based on the FIS output O_i.

    5.1 If $O_i <$ Threshold limit then

        i^{th} *Output = 1* (positive for Parkinson's Disease)

    5.2 Else

        i^{th} *Output = 0* (negative for Parkinson's Disease)

4   Experimental Results

The source of dataset (training and testing) is created by Max Little of the Univer-sity of Oxford, in collaboration with the National Centre for Voice and Speech, Denver, Colorado. It contains biomedical voice measurements of 31 people out of 23 are detected positive for Parkinson's disease. 5 to 6 voice instances of each person are collected.

The experiment is conducted using MATLAB (2008), in a 32-bit Windows 7 Professional machine. The processor used is Intel Pentium, 2.20 GHz and 2.00GB RAM. The threshold value in the decision making phase is calculated to be 0.3 by the experimental analysis of output of the fuzzy inference system.

Training set consists of 16 parameters obtained from biomedical measurement voice and the diagnostic result (1 if positive for Parkinson else 0).The training set comprises of 100 data items and testing is done with 28 sets. Out of 28 testing samples 27 samples turned to be same with actual diagnostic result only in 1 case it differed (Table 1). The actual diagnosis refers to the clinical diagnosis pertain-ing to the test cases by the physicians using laboratory examinations.

Table 1 Fuzzy C- Means based FIS results

Number of Cases	Actual diagnosis	Testing Results	Number of Cases	Actual diagnoses	Testing Results
1	Positive	Positive	15	Positive	Positive
2	Positive	Positive	16	Positive	Positive
3	Positive	Positive	16	Positive	Positive
4	Positive	Positive	18	Positive	Positive
5	Negative	Negative	19	Positive	Positive
6	Negative	Negative	20	Positive	Positive
7	Negative	Negative	21	Positive	Positive
8	Negative	Negative	22	Positive	Positive
9	Positive	Positive	23	Negative	Negative
10	Positive	Positive	24	Negative	Negative
11	Positive	Positive	25	Negative	Negative
12	Positive	Positive	26	Positive	Positive
13	Positive	Positive	27	Negative	Positive
14	Positive	Positive	28	Negative	Negative

Table 2 Subtractive Clustering based FIS results

Number of Cases	Actual diagnosis	Testing Results	Number of Cases	Actual diagnoses	Testing Results
1	Positive	Positive	15	Positive	Positive
2	Positive	Positive	16	Positive	Positive
3	Positive	Positive	16	Positive	Positive
4	Positive	Positive	18	Positive	Positive
5	Negative	Negative	19	Positive	Positive
6	Negative	Negative	20	Positive	Positive
7	Negative	Negative	21	Positive	Positive
8	Negative	Positive	22	Positive	Positive
9	Positive	Negative	23	Negative	Negative
10	Positive	Negative	24	Negative	Negative
11	Positive	Positive	25	Negative	Negative
12	Positive	Positive	26	Positive	Positive
13	Positive	Negative	27	Negative	Negative
14	Positive	Positive	28	Negative	Negative

Thus this model has accuracy up to 96.4 % with quite satisfactory efficiency. It clearly gives better results as compared with existing methods like approximately 85% of ANN & SVM [16] or 93-94 % of genetic algorithm & SVM [15] based systems. As a comparative study the model based on subtractive clustering is created and tested using the same dataset. The accuracy of this system is 85.71 % i.e. 24 outputs are correct (Table 2).

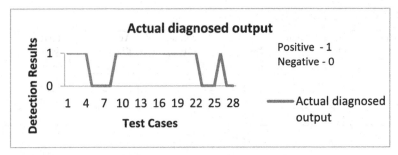

Fig. 3 The plot of actual diagnosis made clinically

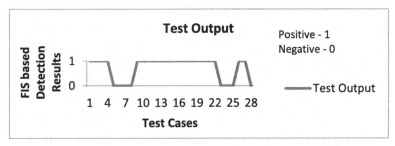

Fig. 4 The plot of testing results of the detection system

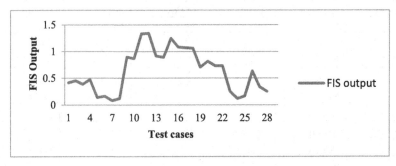

Fig. 5 The output of fuzzy inference system for the testing input

The actual diagnosed outcome line in Fig. 3. almost matches the predicted output of the proposed system in Fig. 4.The outcome of fuzzy inference system (before decision making) is shown in Fig. 5, which clearly interprets that FIS model is tuned to the behaviour of actual diagnosis. All the graphs correspond to fuzzy c-means based system.

5 Conclusion

The diagnosis of complicated neurological disorder like Parkinson's disease needs careful and expertise observation of the neurologist. Often the detection depends on several diagnostic examination procedures. It is expensive and may not be available everywhere specially in remote and distant places. Some neurologists also prefer to go for medication therapy before final decision to observe the improvements based on the symptoms, which have been seen to be fatal in many cases. The purpose of this work is to facilitate the effective detection of Parkinson's disease in a computerised environment using least cost and time. Biomedical measurement of voice signal does not involve very complicated or expensive method. It can be done over telephony system for a remote place. This system can also be used to aid the treatment of Parkinson's disease. The output obtained from the fuzzy Inference system can act as measure of degree or level of Parkinson's disease in an affected patient. A fuzzy logic approach based on clustering is undertaken to deal with the complicacy in detection of Parkinson's disease. The comparative study shows that using fuzzy c-means clustering the fuzzy inference system can be trained better than subtractive clustering with an accuracy of 96-97 percentages.

References

1. Murdoch, B., Whitehill, T., de Letter, M., Jones, H.: Communication Impairments in Parkinson's Disease. SAGE-Hindawi Access to Research, Parkinson's disease **2011**, 1–2 (2011)
2. Fox, C.M., Morrison, C.E., Ramig, L.O., Sapir, S.: Current Perspectives on the Lee Silverman Voice Treatment (LSVT) for Individuals With Idiopathic Parkinson Disease. American Journal of Speech-Language Pathology **11**, 111–123 (2002)
3. Tjaden, K.: Speech and Swallowing in Parkinson's Disease. Topics in Geriatric Rehabilitation, 115–126 (2008)
4. Sapir, S.: Multiple Factors Are Involved in the Dysarthria Associated with Parkinson's Disease: A Review With Implications for Clinical Practice and Research. Journal of Speech, Language, and Hearing Research **57**, 1330–1343 (2014)
5. Raju, G., Thomas, B., Tobgay, S., Kumar, T.S.: Fuzzy clustering methods in data mining: a comparative case analysis. In: ICACTE, pp. 489–493. IEEE publication (2008)
6. Dunn, C.: A Fuzzy Relative of the ISODATA Process and Its Use in Detecting Compact Well-Separated Clusters. Journal of Cybernetics **3**, 32–57 (1973)
7. Bezdek, J.C.: Pattern Recognition with Fuzzy Objective Function Algorithms. Plenum Press, New York (1981)
8. Suganya, R., Shanthi, R.: Fuzzy C- Means Algorithm- A Review. International Journal of Scientific and Research Publications **2**, 1–3 (2012)
9. Chiu, S.L.: Fuzzy Model Identification Based on Cluster Estimation. Journal of Intelligent and Fuzzy Systems **2**, 267–278 (1994)
10. Mitra, R., Kumar, V.: Identification of rules using subtractive clustering with application to fuzzy controllers, vol. 7, pp. 4125–4130. IEEE publications (2004)
11. Kaur, A., Kaur, A.: Comparison of Mamdani-Type and Sugeno-Type Fuzzy Inference Systems for Air Conditioning System. International Journal of Soft Computing and Engineering (IJSCE) **2**, 323–325 (2012)

12. Popescu, M., Khalilia, M.: Improving disease prediction using ICD-9 ontological features, pp. 1805–1809. IEEE publications (2011)
13. Little, M.A., McSharry, P.E., Roberts, S.J., Costello, D.A.E., Moroz, I.M.: Exploiting Nonlinear Recurrence and Fractal Scaling Properties for Voice Disorder Detection. BioMedical Engineering OnLine 6 (2007)
14. Tsanas, A., Little, M.A., Mc Sharry, P.E., Spielman, J., Ramig, L.O.: Novel Speech Processing Algorithms for High-Accuracy Classification of Parkinson's Disease. IEEE Transactions on Biomedical Engineering 59, 1264–1271 (2012)
15. Shahbakhi, M., Far, D.T., Tahami, E.: Speech Analysis for Diagnosis of Parkinson's Disease Using Genetic Algorithm and Support Vector Machine. J. Biomedical Science and Engineering 7, 147–156 (2014)
16. Sharma, A., Giri, R.N.: Automatic Recognition of Parkinson's disease via Artificial Neural Network and Support Vector Machine. International Journal of Innovative Technology and Exploring Engineering 4, 35–41 (2014)

Feature Selection for Heart Rate Variability Based Biometric Recognition Using Genetic Algorithm

Nazneen Akhter, Siddharth Dabhade, Nagsen Bansod and Karbhari Kale

Abstract Heart Rate Variability (HRV) is a prominent property of heart, so far utilized by medical community for diagnostic and prognostic purpose. There was an early attempt to employ HRV for biometric recognition purpose however due to lack of information, the methodologies applied, features used, and results obtained are not available for reference and comparison. In this article we attempt to utilize HRV for biometric purpose, and subsequently obtained 101 most commonly used HRV features. These features have been identified in the guidelines framed by the especially constituted taskforce of European Society of Cardiology and North American Society of Pacing and Electrophysiology for standardization of HRV related studies. Biometric recognition system depends basically on some strongly discriminative elements in a feature vector for accurately distinguishing individuals. The large feature vector of 101 features in addition to the useful ones, may definitely have irrelevant and redundant features. Therefore features selection becomes a crucial step before classification is attempted and feature selection from a large feature sets, cannot be done arbitrarily. The main intention of this article is to identify prominent features of HRV data that can be employed in biometric recognition. For this purpose we applied Genetic Algorithm (GA) which utilizes adaptive search techniques and have documented significant improvement on variety of search problems. GA proposed 15 prominent features out of 101. Performance analysis with the identified features is presented along with the recognition rate.

1 Introduction

In an ECG strip the tallest spike is known as R-peak. The time duration between two adjacent R-R peaks is known as R-R Interval as seen in Fig. 1. This R-R Interval varies in every adjacent pair of beat. The variance in RR-Interval is popularly known

N. Akhter(✉) · S. Dabhade · N. Bansod · K. Kale
Department of Computer Science & Information Technology, Dr. Babasaheb Ambedkar Marathwada University, Aurangabad 431001, Maharashtra, India
e-mail: {getnazneen,dabhade.siddharth,nagsenbansod,kvkale91}@gmail.com

© Springer International Publishing Switzerland 2016
S. Berretti et al. (eds.), *Intelligent Systems Technologies and Applications*,
Advances in Intelligent Systems and Computing 384,
DOI: 10.1007/978-3-319-23036-8_8

91

as Heart rate variability (HRV). This duration represents the variability property of heart rate and is an indicator of the functioning of the Autonomic Nervous System (ANS). HRV analysis is a reliable indicator of the state of the internal physiology hence it is under investigation of the medical science research community since last two decades. Since HRV is such an intrinsic property of heart governed by ANS and indicator of an individual's state of internal physiology, we were tempted to explore the same for biometric recognition. Only RR-Interval is required for HRV analysis and it were traditionally measured from ECG signals. Whenever heart beats, the resultant is a pulse. There are two different ways to measure these events one is electrical action (heart beat) measured by ECG and the other one is mechanical movement (resultant Pulse) measured by PPG. Researchers have documented evidences in favor of Photoplethysmography (PPG) to surrogate ECG for HRV analysis [1-3]. ECG from chest is the clearest but rarely used outside hospital [4] and if it has to be employed in biometric applications it faces the challenge of poor user cooperation. If heart signals are to be used in biometric recognition systems then other methods need to be explored. PPG sensors being low cost and comfortable in data collection are one of the instant choices as ECG alternative. While Y. Y. GU et al. [5] proposed a novel biometric approach for human verification using PPG signals, A. Resit et al. [6] proposed a novel feature ranking algorithm for time domain features from first and second order derivatives of PPG signals for biometric recognition. There have been several classification attempts for disease pattern identification in HRV data [7-10]. But only two early attempts of recognition using HRV are documented in literature one by A. Milliani et al. [11] attempted to recognize two different postures i.e. upright and supine of each individual using HRV, but basically their focus was more on identification of posture individual by individual and not specifically biometric recognition, while J. Irvine et al. [12] proposed HRV based human identification which is the only reported attempt specifically aimed at biometric recognition but its techniques and results are unknown due to lack of information. S. A. Israel et al. [13] gave extensive performance analysis of three different sensing methods of heart i.e. ECG, pulse oximetry and blood pressure, which documented the latter two methods to be on the lower side. Da Silva et al. [14] has presented the usability and performance study of heart signals from fingertips and also in [15] Da Silva et al. proposed a new off the person dataset of ECG data collected at fingertips which strengthens the reliability of heart signals collected at fingertips.

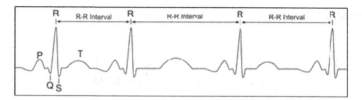

Fig. 1 ECG Strip Indicating the R-R Interval

2   Methodology

The proposed system workflow can be seen in Fig. 2, it is a six step process, and the initial three steps are executed in parallel with the data acquisition process. The acquisition software not just acquires the RR-Intervals from the connected hardware, but is also designed to simultaneously identify ectopy (noise/false beats) and suitably eliminate them from RR data. The hardware and acquisition software are in house designed and developed.

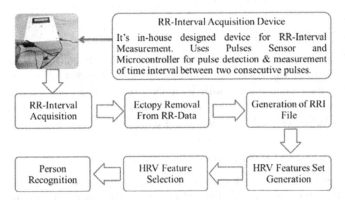

Fig. 2 Proposed System workflow

2.1   Data Collection

Using an IR based pulse detection sensor, the pulses are detected at the fingertips of the subjects as seen in Fig. 3. The connected microcontroller unit measures the time duration between two consecutive pulses and transmits it through the RS232 serial communication via serial/USB bridge adaptor. The acquisition software is designed in Visual Basic 6, which is responsible for not only acquiring data but also preprocessing of RR time series data.

Fig. 3 IR Based Pulse detection Sensor with the measuring Unit

2.2   Preprocessing

Preprocessing of the RR time series data is done simultaneously with the data acquisition. Which includes the identification of Ectopy beats (false beats) that adds lot of noise to the RR time series data, therefore its detection and elimination becomes necessarily desired. These Ectopy or noise may originate due to several reasons like: inherent variability of the heartbeats, improper threshold detection of the beats, amplitude variation, missing beat, double triggering due to beat shape, and electronic noise including pickup.

2.3   Database Specification

At present our database consists of 2430 sequences of RR-Intervals of 81 subjects (47 males and 34 Females) whose 10 samples each of 64 RR-Intervals were measured continuously for 1 minute approximately, in three different sessions spread over nine months with time interval of three months between each session. The subjects were informed of the purpose and their consent was taken before data collection.

2.4   Feature Set Generation

In an attempt to use HRV in biometric recognition system we generated the HRV parameters as suggested in [16, 17] and few more additional ones identified from literature survey. HRV parameters are obtained as a result of applying various standard linear methods like statistical and spectral techniques and nonlinear methods like Poincare and sample entropy. These HRV parameters are used by researchers for classification of diseases. We call these parameters as HRV features that can also serve as feature vector for biometric classification. In all we obtained 101 features as shown in Table 1, each of which has some significance or the other in HRV analysis for diagnostic or prognostic purpose but which of these would really prove suitable for uniquely identifying an individual is yet to be established. The names and descriptions of all the 101 features are in Appendix I. As all the features so obtained are homogeneous in nature we generated the feature set by simple concatenation.

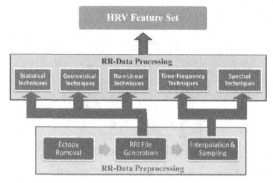

Fig. 4 HRV Feature Set Generation

Table 1 HRV Feature Set of 101 Features from five Different Techniques using Ten Algorithms

Sr. No	Technique Name	Algorithm	No. of Features
1	Statistical	Standard Methods	12
		Welch	13
2	Spectral	Burg	13
		Lomb-Scargle	13
		Wavelet	14
3	Time-Frequency	Burg	14
		Lomb	14
4	Geometrical	Histogram	2
		Poincare	2
5	Non-Linear	Sample Entropy	4
		Total	101

2.5   Feature Selection

In addition to useful ones, the feature set of 101 HRV features naturally must also be containing irrelevant and redundant features which would degrade the recognition rate. As a general understanding, for any given classification problem, the researchers have always tried to keep the number of features as low as possible to reduce the cost of computation as well as the complexity of the algorithm design[18]. Looking at the size of our feature set, feature selection through a reliable algorithm becomes necessary. Feature selection algorithms can be classified into filters and wrappers [19]. Wrapper methods are widely recognized as superior alternative in supervised learning problems [20]. Genetic Algorithm (GA) comes under the category of wrapper feature selection algorithm.

3   Genetic Algorithm (GA)

A common an efficient use of GA is proposed for function minimization [21]. It is basically a search technique which works on the theory of evolution proposed by Darwin. According to his theory the fittest species survived the evolution process and the weak perished. Talking in terms of feature selection, the strongest feature survives all the iterations (mutation phase) and others are eliminated. The parameters used in this algorithm resemble the theory of evolution, like mutation, population size, number of generations, chromosome, etc. In addition to these parameters, one fitness function has to design to evaluate the goodness of the solution provided in every mutation (iteration). Its working principle is illustrated in Fig. 5. For every mutation initial population (subset of features from original set) is created the fitness function evaluates them using KNN classifier.

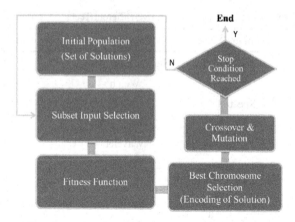

Fig. 5 Overview of Genetic Algorithm Routine with Iterative Steps

For utilizing the GA for feature selection we used genetic algorithm toolbox of Matlab 2012a. The parameters used in the GA along with their values are shown in Table 2. Table 3 shows the feature set selected by GA as potentially good performer in its classification result which consists of only 15 features out of 101 features of original feature set. This feature set is used in our recognition system and it classification results are discussed in the next section.

Table 2 Parameters of Genetic Algorithm and their values used for the problem in hand

Parameter Name	Value
Population Size	50
Number of Generations	100
Probability of Crossover	0.8
Probability of Mutation	0.001
Elite Count	2
Type of Mutation	Uniform
Type of Selection	Rank-Based
Stall Generations Limit	10
Stall Time Limit	Infinite

Table 3 Feature Vector obtained after Feature Selection by Genetic Algorithm. (For description of each feature please refer Appendix-I)

Sr. no.	Feature(s) Name	Technique Name
1	Max, min, mean, median, RMSSD, meanHR, sdHR	Statistical Technique
2	SD1, SD2	Poincare chart
3	aVLF, PLF, PLF	Spectral (Welch) (Lomb) (Burg)
4	aLF, PLF, PeakHF	Time-Frequency (Burg) (Burg) (Lomb)

4   Results and Discussions

Using the feature vector obtained from Genetic Algorithm, we implemented two layered Multilayer perceptron classifier for classification, and we randomly chose 40 subjects with 10 samples each from our HRV database. While choosing the sample we selected four samples from session one and three each from session two and three. Our neural network had 15 input nodes, 10 neurons and 40 hidden nodes and 40 output nodes. The training process was evaluated by sum square error method. Best training performance reached at 182nd epoch shown in Fig. 6.

The verification performance of the network can be seen in ROC which indicates a reasonably good system with maximum true positive cases and limited number of false positive cases. From the Fig. 7, it is quite visible that the system not only has maximum true positive cases but the true positive cases are at a much farer distance from the 45^0 diagonal indicating a highly reliable and efficient system.

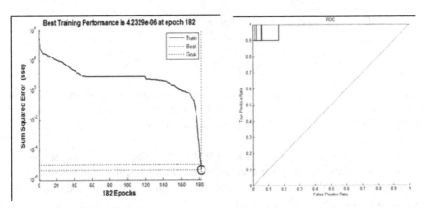

Fig. 6 Training Performance   **Fig. 7** ROC Curve indicating System Performance

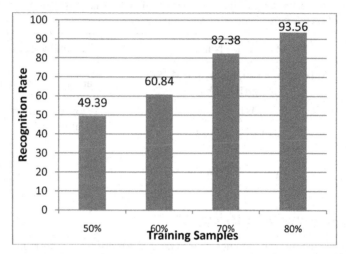

Fig. 8 Training Samples with corresponding Recognition Rates

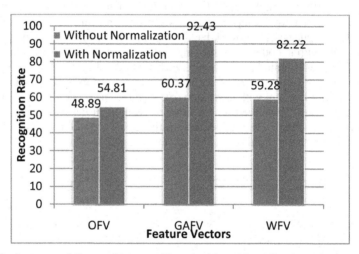

Fig. 9 Performance of Feature Vectors with and without Normalization. (OFV is Original Feature Vector of 101 features, and GAFV is the proposed Feature Vector by Genetic Algorithm, and WFV is the feature Vector proposed by other Wrapper Algorithms [22])

We trained and tested the network with different number of samples as to ascertain sample limit. Initially we used 50% sample for training and the rest 50% for testing which resulted in 49.39% recognition rate, further we increased the sample in training set to 60% and rest for testing which improved the results giving 60.84% recognition rate. With 70-30 we got 82.38% and with 80-20 we got much improved results i.e. 93.56% which indicates a significant improvement in the classification results. Fig. 8 indicates the same graphically. While Fig. 9 shows the performance of the Original feature set (OFS) with 101 features if used directly for classification, gives poor performance owning to the irrelevant and redundant features, whereas the feature vector proposed by GA showed much better performance with normalized feature values, This indicates a significant improvisation in results The performance of GAFV feature vector both with and without normalized values is significantly better i.e. 93.43% and 60.37 respectively. This is clearly evident in Fig. 9. It also shows the better performance of GA Feature vector in comparison with the feature vector proposed by five different wrapper algorithms in one our previous experiments [22]. From Fig. 9 one more thing is evident that features need to normalize on similarly scale to achieve improved results.

5   Conclusion

We have successfully used Genetic Algorithm (GA) for feature selection for Heart Rate Variability feature set of 101 features, redundant and irrelevant features were discarded and 15 promising features including seven from time domain and two from Poincare map were identified as capable of subject discrimination and to be effective in human identification, while spectral analysis gave three useful features but each one from different algorithm and something similar with time-frequency features. The significantly improved results are clear indicators of effectiveness GA suggested Feature Vector in HRV based biometric recognition. Further normalization of feature values also contributed in improving the results. We have proposed the RR-interval collection duration to be approximately one minute, can further reduction of duration be implemented for HRV based biometric recognition can be attempted.

Acknowledgement This work was carried out in Multimodal System Development laboratory established under UGC's SAP scheme SAP (II) DRS Phase-I F. No.-3-42/2009 & SAP (II) DRS Phase-II F. No.4-15/2015. This work was also supported by UGC under One Time Research Grant F. No. 4-10/2010 (BSR) & 19-132/2014 (BSR).The authors acknowledge UGC for the same. The authors also acknowledge UGC for providing BSR and RGNF fellowships.

References

1. Lin, W.-H., Wu, D., Li, C., Zhang, H., Zhang, Y.-T.: Comparison of heart rate variability from PPG with that from ECG. In: Zhang, Y.-T. (ed.) The International Conference on Health Informatics. IFMBE Proceedings, vol. 42, pp. 213–215. Springer, Heidelberg (2013)
2. Selvaraj, N., Jaryal, A., Santhosh, J., Deepak, K.K., Anand, S.: Assessment of heart rate variability derived from finger-tip photoplethysmography as compared to electrocardiography. J. Med. Eng. Technol. 32(6), 479–484 (2008)
3. Gil, E., Orini, M., Bailón, R., Vergara, J., Mainardi, L., Laguna, P.: Photoplethysmography pulse rate variability as a surrogate measurement of heart rate variability during non-stationary conditions. Physiol. Meas. 31(9), 127–1290 (2010)

4. Park, B.: Psychophysiology as a tool for HCI research: promises and pitfalls. In: Jacko, J.A. (ed.) HCI International 2009, Part I. LNCS, vol. 5610, pp. 141–148. Springer, Heidelberg (2009)
5. Gu, Y.Y., Zhang, Y., Zhang, Y.T.: A novel biometric approach in human verification by photoplethysmographic signals. In: 4th International IEEE EMBS Special Topic Conference on Information Technology Applications in Biomedicine, 2003, pp. 13–14, April 24–26, 2003 doi:10.1109/ITAB.2003.1222403
6. Kavsaoğlu, R., Polat, K., Recep Bozkurt, M.: A novel feature ranking algorithm for biometric recognition with PPG signals. Comput. Biol. Med. **49**, 1–14 (2014)
7. Lin, C.-W., Wang, J.-S., Chung, P.: Mining Physiological Conditions from Heart Rate Variability Analysis. IEEE Computational Intelligence Magazine **5**(1), 50–58 (2010)
8. Melillo, P.: Classification Tree for Risk Assessment in Patients Suffering From Congestive Heart Failure via Long-Term Heart Rate Variability. IEEE J. Biomed. Heal. Informatics **17**(3), 727–733 (2013)
9. Nizami, S., Green, J.R., Eklund, J.M., McGregor, C.: Heart disease classification through HRV analysis using parallel cascade identification and fast orthogonal search. In: 2010 IEEE International Workshop on Medical Measurements and Applications, MeMeA 2010–Proceedings, pp. 134–139 (2010)
10. Szypulska, M., Piotrowski, Z.: Prediction of fatigue and sleep onset using HRV analysis. In: Proceedings of the 19th International Conference Mixed Design of Integrated Circuits and Systems (MIXDES), pp. 543–546 (2012)
11. Malliani, A., Pagani, M., Furlan, R., Guzzetti, S., Lucini, D., Montano, N., Cerutti, S., Mela, G.S.: Individual recognition by heart rate variability of two different autonomic profiles related to posture. Circulation American Heart Association **96**, 4143–4145 (1997)
12. Irvine, J.M., Wiederhold, B.K., Gavshon, L.W., Israel, S.A., McGehee, S.B., Meyer, R., Wiederhold, M.D.: Heart rate variability:a new biometric for human identification. In: International Conference on Artificial Intelligence (IC-AI 2001), Las Vegas, Nevada, pp. 1106–1111 (2001)
13. Israel,S.A., Irvine, J.M., Wiederhold, B.K., Wiederhold, M.D.: The Heartbeat: The Living Biometric. Biometrics Theory, Methods, Appl., 429–459 (2009)
14. da Silva, H.P., Fred, A., Lourenco, A., Jain, A.K.: Finger ECG signal for user authentication: usability and performance. 2013 IEEE Sixth International Conference on Biometrics: Theory, Applications and Systems (BTAS), pp. 1–8, September 29, 2013–October 2, 2013. doi:10.1109/BTAS.2013.6712689
15. Plácido da Silva, H., Lourenço, A., Fred, A., Raposo, N., Aires-de-Sousa, M.: Check Your Biosignals Here: A new dataset for off-the-person ECG biometrics. Computer Methods and Programs in Biomedicine **113**(2), 503–514 (2014)
16. Task Force of the European Society of Cardiology and North American Society of Pacing and Electrophysiology: Heart rate variability: standards of measurement, physiological interpretation and clinical use. Eur. Heart J. **17**, 354–381 (1996)
17. Rajendra Acharya, U., Paul Joseph, K., Kannathal, N., Lim, C.M., Suri, J.S.: Heart rate variability: A Review. Med. Bio. Eng. Comput. **44**, 1031–1051 (2006). doi:10.1007/s11517-006-0119-0
18. Estevez, P.A., Tesmer, M., Perez, C.A., Zurada, J.M.: Normalized Mutual Information Feature Selection. IEEE Transactions on Neural Networks **20**(2), 189–201 (2009). doi:10.1109/TNN.2008.2005601

19. Kohavi, J., Pfleger, K.: Irrelevant features and the subset selection problem. In: Proc. 11th Int. Conf. Mach. Learn., pp. 121–129 (1994)
20. Talavera, L.: An evaluation of filter and wrapper methods for feature selection in categorical clustering. In: Famili, A., Kok, J.N., Peña, J.M., Siebes, A., Feelders, A. (eds.) IDA 2005. LNCS, vol. 3646, pp. 440–451. Springer, Heidelberg (2005)
21. Ladha, L., Deepa, T.: Feature Selection methods and algorithms. International Journal on Computer Science and Engineering (IJCSE) 3(5), May 2011
22. Akhter, N., Tahrewal, S., Kale, V., Bhalerao, A., Kale, K.V.: Heart based biometrics and use of heart rate variability in human identification systems. In: 2nd International Doctoral Symposium on applied computation and security systems 23–25 May 2015, kolkata, India

Appendix-I

Statistical Features		
No.	Name	Description
1	SDNN	Standard deviation of all Normal-Normal Intervals
2	RMSSD	Root mean square of successive differences
3	NN50	it's a count of number of adjacent pairs differing by more than 50ms
4	pNN50	(%) NN50 count divided by total intervals
5	MeanRRI	Mean of Normal-Normal Interval
6	MeanHR	Mean Heart Rate
7	Max	Maximum Interval duration in a particular RRI
8	Min	Minimum Interval duration
9	Mean	Mean of the whole RRI sequence
10	Median	Median of the RRI sequence
11	SDHR	Standard Deviation of Heart Rate

Spectral Features		
No.	Name	Description
1	aVLF	absolute value in Very low frequency Spectrum
2	aLF	absolute value in low frequency Spectrum
3	aHF	absolute value in high frequency Spectrum
4	aTotal	Total absolute value
5	pVLF	Power % of Very Low Frequency in PSD
6	pLF	Power % of Low Frequency in PSD
7	pHF	Power % of high Frequency in PSD
8	nLF	low frequency in normalized Unit
9	nHF	high frequency in normalized Unit
10	LFHF	LF to HF Ratio
11	peakVLF	Peak Value in Very low frequency
12	peakLF	Peak Value in low frequency
13	peakHF	Peak Value in high frequency

Driver Eye State Detection Based on Minimum Intensity Projection Using Tree Based Classifiers

A. Punitha and M. Kalaiselvi Geetha

Abstract Eye state identification has wide potential applications in the design of human-computer interface, recognition of facial expression, driver dowsiness detection, and so on. A novel approach to deal with the problem of detecting whether the eyes in a given face image are closed or open is presented in this paper. The projection of row wise minimum intensity and column wise minimum intensity pixel of the histogram equalized eye image is used as a feature in this work. The various tree-based classifiers such as Random Forest, Naive Bayes Tree, Random Tree, and REPTree are exploited to classify the eye as open or closed. The experimental results show that random forest classifier yields the best performance with an overall accuracy rate of 96.3% than the other classifiers.

Keywords Face and eye localization · Row-wise and Column-wise intensity projection · Tree based classifiers

1 Introduction

As one of the most prominent facial features, eyes, which reveal the individual's emotional states, have become one of the most vital source of information for face analysis. Efficiently and precisely understanding the states of the eyes in a given face image is therefore important to a wide range of face-related research work such as human-computer interface design, facial expression analysis, live-ness detection and so on.

Also,the growing number of traffic accident due to a reduced driver's attention level has become a serious problem for society. Hence there is a need to address this problem to avoid accidents by monitoring driver fatigue and alerting the driver so that road safety can be improved. Eye behavior present momentous information about a driver's tiredness and if such behavior can be detected then it will be possible to predict a driver's state of drowsiness. By observing the eyes, the symptoms of driver fatigue can be detected early enough to avoid a vehicle accident. Hence the

A. Punitha(✉) · M.K. Geetha
Speech and Vision Lab, Deparment of Computer Science and Engineering,
Annamalai University, Annamalainagar, Chidambaram, India
e-mail: {12charuka17,geesiv}@gmail.com

© Springer International Publishing Switzerland 2016
S. Berretti et al. (eds.), *Intelligent Systems Technologies and Applications*,
Advances in Intelligent Systems and Computing 384,
DOI: 10.1007/978-3-319-23036-8_9

focus of this paper is to design a system that will precisely discriminate the open or closed state of the person's eyes that can be applied to detect driver's drowsiness.

The rest of the paper is organised as follows : A review of related work is presented in Section 2. Section 3 describes the Eye Localisation and eye state detection procedure.Section 4 describes about the various tree based classifier.Section 5 shows the Experimental results of the approach and Section 6 concludes this work.

2   Related Work

An automatic system to detect eye-state action units (AU) based on Facial Action Coding System (FACS) by use of Gabor wavelets in a nearly frontal-viewed image sequence is discussed in [1].A method using synthesized gray projection in an image is used to detect closed eye,is presented in [2]. In [3], eyes closeness detection from still images with multi-scale histograms of principal oriented gradients is presented. A robust real-time computer vision-based system to detect the eye state for driver alertness monitoring is presented in [4]. In paper [5] a robust and efficient eye state detection method based on an improved algorithm called LBP+SVM mode is proposed. A distinctive local feature named Local Binary Increasing Intensity Patterns (LBIIP), which uses one decimal label to represent the intensity increasing tendency of the local region around each pixel is used for open/closed eye recognition is presented in [6]. Detection of open or closed eye state based on the complexity of the eye contours is discussed in [7].

3   Overall Architecture of the Proposed System

The overall architecture of the proposed eye state detection is given in Fig. 1 The approach uses Viola-Jones Face Cascade of classifiers for the detection of Face. Further the eye region is determined with respect to the width and height of the detected face. The row wise minimum intensity and column wise minimum intensity pixel of the histogram equalized eye image is used as a feature and is fed to the various tree based classifiers to discriminate the open and closed state of the eye.

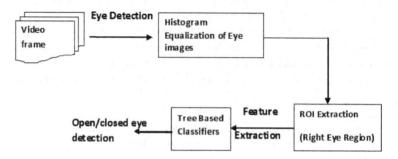

Fig. 1 WorkFlow of the Eye State Detection Approach

3.1 Face and Eye Localization

Localizing and detecting human face is often the first step in applications such as video surveillance, human computer interface, face recognition and facial expressions analysis. Face detection approach considered in this work makes use of the rapid object detection method presented in [8]. Viola- Jones object detection procedure based on cascade of classifier is used to locate the face within each frame of the video. After identifying the location of the face region, successive processing takes place within that region of interest and in particular, the eyes are estimated with respect to the width and height of the detected face. The following [Eqn. 1 - Eqn. 4] are used to draw the eye rectangle:

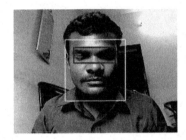

Fig. 2 Face and Eye Localisation

$$left.x = (r.x + r.width/6) \tag{1}$$

$$left.y = (r.y + r.height/4) \tag{2}$$

$$right.x = (r.x + r.width * 5/6) \tag{3}$$

$$right.y = (r.y + r.height/2) \tag{4}$$

where (1) and (2) are the top left corner co-ordinates and (3) and (4) are the bottom right co-ordinates of the eye rectangle and r.x, r.y are the top left co-ordinates and r.width, r.height are the bottom right co-ordinates of the face rectangle respectively. Fig. 2 shows the sample image with face and eye localized.

3.2 Feature Extraction from the Eye Region

Feature is an informative portion extracted from an image or a video stream. Visual data exhibit numerous types of features that could be used to identify or portray the information it reveals.

Since the left eye and the right eye open or close synchronously, their states are the same. So it is reasonable to consider only one eye for feature extraction as shown in Fig. 8 and hence redundant computation can be eliminated. The size of the eye image is 100 x 40 and the portion of right eye used for feature extraction is of size 36 x 28(width x height) respectively. Histogram Equalization is a useful pre processing technique for image analysis.

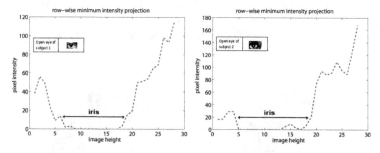

Fig. 3 Rowwise minimum intensity projection of open eye

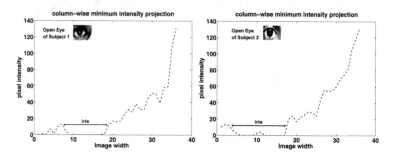

Fig. 4 Columnwise minimum intensity projection of open eye

The eye images are histogram equalized to improve the visual important features and the enhanced eye images show details better for further analysis. Each eye image is scanned horizontally to find the pixel with minimum intensity value. Since the eye image height is 28, a total of 28 values are obtained which is the minor pixel feature vector. Similarly the eye image is scanned vertically for the pixel with minimum intensity and a total of 36 values are obtained which is the column minor pixel feature vector. These row minor and column minor feature vector are combined and is of size 64 is fed as input various tree based classifier to detect the open/closed state of the eye.

Since the open eye has iris region, horizontal and vertical projection of open eye has lower intensity values than the closed eye. This forms the basis for choosing the row and column minor pixel as feature in this work and the chosen features are well able to discriminate the open eye from the closed one. Fig. 3 and Fig. 4 shows the projection of row wise and column wise minimum intensity feature vector of open eye of two different subjects and Fig. 5 and Fig. 6 shows the projection of row wise and column wise minimum intensity feature vector of closed eye of two different subjects and it is observed that the plot of the feature vector of open eye has more zero gray values that corresponds to the iris region in the open eye.

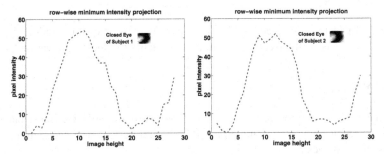

Fig. 5 Rowwise minimum intensity projection of closed eye

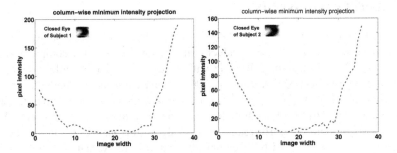

Fig. 6 Columnwise minimum intensity projection of closed eye

4   Tree Based Classifier

4.1   Random Forest

A Random Forests is a collection of decision trees trained with random features. Random Forests work as follows. Given a set of training examples, a set of random trees H is created such that for the k-th tree in the forest, a random vector ϕ_k is generated independently of the past random vectors $\phi_1..\phi_{k-1}$. This vector k is then used to grow the tree resulting in a classifier $h_k(x, \phi_k)$ where x is a feature vector. For each tree, a decision function splits the training data that reach a node at a given level in the tree. The resulting forest classifier H is used to classify a given feature vector x by taking the mode of all the classifications made by the tree classifiers $h \in H$ in the forest. Random Forests[9] are found to be the fast and robust classification techniques that can handle multi-class problems and in [10]Random forests is used for object recognition.

4.2   Random Tree

RandomTree[11]is an algorithm for constructing a tree that considers K randomly chosen attributes at each node and no pruning is performed.

4.3   RepTree

Reduced Error Pruning Tree ('REPT') is basically a fast decision tree learning and a decision tree is built based on the information gain or reducing the variance [12].

4.4   NB Tree

NBTree is a hybrid algorithm that represents a cross between Naive Bayes classifier and C4.5 Decision Tree classification and it's best described as a decision tree with nodes and branches [13].

5   Experimental Results

In this section, the experimental results obtained on Eye State detection system are presented. The experiments are conducted in Matlab 2013a on a computer with Intel Xeon X3430 Processor 2.40 GHz with 4 GB RAM. The sample videos are captured using Logitech Quick cam Pro5000. The experiments are conducted using Weka application. Weka is a complete collection of Java class libraries that is used to carry out many advanced machine learning and data mining algorithms[14]. We analyze and compare the performance of tree algorithms namely Random Forest, Random Tree, REPTree and NBTree and the features are evaluated using tenfold cross validation. In 10-fold cross validation, the data set is divided into 10 subsets. Every instance, one of the 10 subsets is used as the test set and the other left over 9 subsets are used as the training set. Performance statistics are calculated across all 10 trials.

5.1   Dataset

The experiments are conducted with 12 subjects and the subject being captured by the camera in real-time at 15 frames per second with a 640×480 resolution. From the captured images, eye region is extracted for state detection. Each subject has 200 images showing open and closed eye states. The standard experimental procedure, namely the cross-validation approach is chosen to verify the eye state detection performance. The Fig. 7 shows the sample histogram equalized eye images and Fig. 8 shows the portion of eye region used for feature extraction.

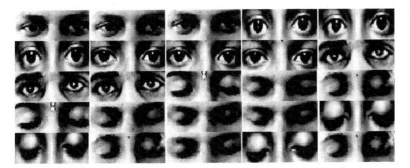

Fig. 7 Histogram Equalised Eye Images

Fig. 8 Portion of Eye Image used for Feature Extraction

5.2 Evaluation Metric

Precision (P) and Recall (R) are the commonly used evaluation metrics and these measures are used to evaluate the performance of the proposed system. These measures provide the best perspective on a classifier's performance for classification. The measures are defined as follows:

$$Precision = \frac{no.\,of\ true\ positives}{no.\,of\ true\ positives\ +\ false\ positives}$$

$$Recall = \frac{no.\,of\ true\ positives}{no.\,of\ true\ positives\ +\ false\ negatives}$$

The work used F1-measure(F1) as the combined measure of Precision (P) and recall (R) for calculating accuracy which is defined as follows:

$$F1 - measure(F1) = \frac{2PR}{P+R}$$

5.3 Accuracy and Time Taken to Generate the Classifier Model

The accuracy designates the percentage of correctly classified dataset instances by the classifier model. The results obtained with each classifier are shown in Table 1.and Fig. 9 shows the overall recognition accuracy. On the whole the accuracy generated by using Random Forest classifier is higher compared to other tree algorithms. The algorithm which gave the highest accuracy with 96.3%was Random Forest. Naive

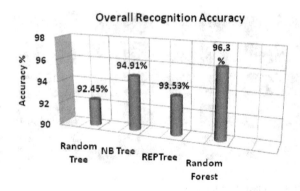

Fig. 9 Overall Recognition Accuracy

Bayes Tree was closely behind with an accuracy of 94.9%. Random tree and REPTree were other algorithms which yields an accuracy of 92.4% and 93.5% respectively. The time taken to generate the model is also important. The algorithm which takes more time to generate models would not be effective in real time applications. In this work, Random Forest algorithm which yields highest recognition accuracy builds the model in 1.2secs.

Table 1 Results obtained with various Tree based classifiers

Classifier	Class	Precision (%)	Recall (%)	F-Score (%)	Time
Random Forest	open eye	96.3	99.0	97.6	1.2 secs
	closed eye	96.3	87.3	91.6	
Random Tree	open eye	94.8	95.4	95.1	0.8 secs
	closed eye	84.4	82.7	83.5	
Rep Tree	open eye	94.7	97.0	95.8	2.4 secs
	closed eye	89.1	82.0	85.4	
NB Tree	open eye	95.9	97.6	96.7	30.4 secs
	closed eye	91.5	86.0	88.7	

6   Conclusion

This paper presents a novel approach to recognize open/closed eye states. The basic idea is based on feature extraction and using a classifier. The algorithm comprises of face and eye localization and eye state identification using minimum intensity projection. The performance of the proposed feature is evaluated with various tree based algorithms such as Random Forest, Random Tree, Naive Bayes Tree and REPTree. The random forest classifier yields the best performance with an overall accuracy rate of 96.3% than the other classifiers. In future, the proposed method can be used to effectively detect the drowsiness of driver by calculating the duration of eye closeness and if the eyes are found to be closed over some consecutive frames then it is concluded that the person is falling asleep or having condition of drowsiness and hence an alarm can be generated.

References

1. Tian, Y.L., Kanade, T., Cohn, J.: Eye-state action unit detection by Gabor wavelets. In: Proceedings of International Conference on Advances in Multi-Modal Interfaces, pp. 143–150 (2000)
2. Lu, L., Ning, X., Qian, M., Zhao, Y.: Close eye detected based on synthesized gray projection. Advances in Multimedia, Software Engineering and Computing **2**, 345–351 (2012)
3. Song, F., et al.: Eyes closeness detection from still images with multi-scale histograms of principal oriented gradients. Pattern Recognition (2014)
4. González-Ortega, D., Díaz-Pernas, F.J., Antón-Rodríguez, M., Martínez-Zarzuela, M., Dez-Higuera, J.F.: Real-time vision-based eye state detection for driver alertness Monitoring. Pattern Anal Applications **16**, 285–306 (2013)
5. Sun, R., Ma, Z.: Robust and Efficient Eye Location and Its State Detection. In: Cai, Z., Li, Z., Kang, Z., Liu, Y. (eds.) ISICA 2009. LNCS, vol. 5821, pp. 318–326. Springer, Heidelberg (2009)
6. Zhou, L., Wang, H.: Open/closed eye recognition by local binary increasing intensity patterns. In: Proceedings of IEEE Conference on Robotics, Automation and Mechatronics, pp. 7–11 (2011)
7. Tian, Z., Qin, H.: Real-time driver's eye state detection. In: Proceedings of the IEEE International Conference on Vehicular Electronics and Safety, pp 285–289 (2005)
8. Viola, P., Jones, M.: Rapid object detection using a boosted cascade of simple features. IEEE Conference on Computer Vision and Pattern Recognition, pp. 511–518 (2001)
9. Breiman, L.: Random forests. Mach. Learn. **45**(1), 5–32 (2001)
10. Bosch, A., Zisserman, A., Munoz, X.: Image classification using random forests and ferns. In: ICCV (2007)
11. Lepetit, V., Fua, P.: Keypoint recognition using randomized trees. IEEE Trans. PAMI **28**(9), 1465–1479 (2006)
12. Nor Haizan, W., Mohamed, W., Salleh, M.N.M., Omar, A.H.: A Comparative Study of Reduced Error Pruning Method in Decision Tree Algorithms. In: IEEE International Conference on Control System, Computing and Engineering, November 23–25 (2012)
13. Pumpuang P., Srivihok A., Praneetpolgrang P.: Comparisons of Classifier Algorithms: Bayesian Network, C4.5, Decision Forest and NBTree for Course Registration Planning Model of Undergraduate Students. In: IEEE International Conference, SMC (2008)
14. Witten, I.H., Frank, E., Trigg, L., Hall, M., Holmes, G., Cunningham, S.J.: Weka: Practical Machine Leraning Tools and Techniques with Java Implementations, New Zealand (1999)

Authorship Attribution Using Stylometry and Machine Learning Techniques

Hoshiladevi Ramnial, Shireen Panchoo and Sameerchand Pudaruth

Abstract Plagiarism is considered to be a highly unethical activity in the academic world. Text-alignment is currently the preferred technique for estimating the degree of similarity with existing written works. Due to its dependency on other documents it becomes increasingly tedious and time-consuming to scale up to the growing number of online and offline documents. Thus, this paper aims at studying the use of stylometric features present in a document in order to verify its authorship. Two machine learning algorithms, namely k-NN and SMO, were used to predict the authenticity of the writings. A computer program consisting of 446 features was implemented. Ten PhD theses, split into different segments of 1000, 5000 and 10000 words, were used, totaling 520 documents as our corpus. Our results show that authorship attribution using stylometry method has generated an accuracy of above 90 %, except for 7-NN with 1000 words. We also showed how authorship attribution can be used to identify potential cases of plagiarism in formal writings.

Keywords Plagiarism · Authorship verification and attribution · Stylometry · K-NN · SMO · Content analysis

1   Introduction

Authorship attribution is also known as author recognition, author verification or simply AA. It is used to identify authors or writers from books, emails, blogs and research papers. AA makes use of stylometry or textometry. The main concern of

H. Ramnial · S. Panchoo
School of Innovative Technologies and Engineering,
University of Technology, Port Louis, Mauritius
e-mail: hoshila@education.mu, s.panchoo@umail.utm.ac.mu

S. Pudaruth(✉)
Department of Ocean Engineering and ICT, Faculty of Ocean Studies,
University of Mauritius, Moka, Mauritius
e-mail: s.pudaruth@uom.ac.mu

© Springer International Publishing Switzerland 2016
S. Berretti et al. (eds.), *Intelligent Systems Technologies and Applications*,
Advances in Intelligent Systems and Computing 384,
DOI: 10.1007/978-3-319-23036-8_10

AA is to define an appropriate characterisation of documents that captures the writing style of authors [1]. Authorship attribution is often confused with author profiling or linguistics profiling. Author profiling is concerned with identifying features of the author such as her gender, name, geographical location, personality traits, etc., but it does not usually attempt to identify a specific person. Its aim is to reduce the search space so that it does not become a-needle-in-a-haystack problem. Stylometry is the statistical analysis of written texts. Some examples stylometric features are word-length frequency distribution, sentence length, word n-grams, character n-grams, PoS (parts of speech) tags, function words and content words. Authorship attribution has successfully been used to identify the works of Shakespeare and the authors of the famous Federalist Papers [2]. Authorship attribution has also become an important instrument in the identification of cybercrimes [3][4]. AA has also found applications in fraud detection.

Plagiarism is the act of copying another person's works either by altering it to a small extent or copying her original ideas without properly referencing it [5]. Plagiarism is becoming extremely difficult to identify because we need to compare a document with billions of other documents on the Internet. This takes much time and it is becoming increasingly difficult since the number of documents on the Internet is rising at a very high rate [6]. The International Center for Academic Integrity reported that 86% of high school students engage in plagiarism [7]. Furthermore, according to latest estimates, even the best search engines have only indexed only about 25% of the Internet and thus much of the Internet's content is still hidden in what is known as the deep web [8][9].

Most studies in stylometry are limited either in terms of the features used or by using unrealistic datasets. Furthermore, most of the well-known plagiarism detection engines like Turnitin [10], Viper [11], Plagium [12], PlagTracker [13] and Paper Rater [14] do not have any functionality for analysing and comparing the writing styles present in the documents that are submitted to them. Only Grammarly [15] analyses the text to some extent but then the focus is only on the identification of grammatical mistakes and not on stylometrics. Horovitz [16] said that the use of Turnitin is unethical and it is illegal to some extent as it is using the students' works in its database for commercial purposes. Turnitin is unable to trace plagiarised materials from documents that have not been submitted in its database or are not available on the visible web. It does not have the ability to detect cleverly paraphrased textds. Furthermore, websites that provide facilities for paid papers or reports make use of a file that prevents the TurnitinBot [17] from crawling and indexing their content [18]. Thus, this paper aims at the identification of plagiarism through authorship attribution in formal writings using an increased number of stylometric features.

This paper proceeds as follows: In section 2, we give a detailed description of existing works on authorship attribution and author profiling. The detailed methodology is described in section 3. Section 4 describes and evaluates the results from the large amount of experiments that we conducted. Section 5 concluded the paper with a look on future works.

2   Literature Review

With the billions of websites and documents found on the internet, it has become very time consuming, unreliable and difficult to compare a particular document with all others in order to find the potential cases of plagiarism. The 'invisible web' has pushed plagiarism systems to their limit because protected resources and undigitised materials like books appear as being invisible to the plagiarism scanner [8]. An alternate solution to the text-alignment is hence required.

2.1   Corpus Size

The predictions from any machine learning algorithm can only be as good as the dataset. Thus, in this paper, we ensured that we had an adequate number of documents and each document contained a reasonable amount of words. This is a crucial requirement for authorship attribution as the results are very dependent on the corpus size. The studies by Argamon and Levitan [19] and Hoover [20] were to predict the writer and the nationality of novels out of a corpus of 20 books written by 8 authors with an average of 10 thousand words per book. They achieved 99% classification accuracy on author and 93.5% on nationality using the SMO machine learning algorithm. In another study by Nirkhi et al., [21], they made used of the Reuters_50_50 dataset which contains 50 documents written by 50 authors. The average number of words per document is 500 words. They were able to achieve 80% and 90% accuracy for authorship identification using the k-NN and SMO algorithm, respectively. The unrealistic sizes of training data are often overestimated when compared to realistic situations [22][23].Therefore, finding the minimum size of text which is suitable for author attribution has also been investigated. Iqbal et al., [24] proposed a dynamic approach for extracting write-prints from emails which are then used to identify the author. His method does not rely on a set of predefined features instead these are dynamically extracted based on the available data.

2.2   Stylometric Features

For the proper performance of author attribution and detection of potential suspects of plagiarism, it is important to know the stylometric features used in order to perform unbiased authorship attribution. Abbassi and Chen [25] used an approach based on writeprints to identify an author of a given document. They made use of online texts which comprised of Enron mail, Ebay comments, Java forum and cyber-watch chats. Their experiments were based on 25, 50 and 100 authors respectively. They made use of content words, PoS, word length, vocabulary richness and sentence length for the prediction. Their best prediction was 94% using the sliding window algorithm. Several algorithms like SVM, Ensemble SVM, PCA and Karhunen-Loeve (KL) transforms were used as baseline in this study. The previous study by the same authors in 2006 made use of

300 messages per forum in English and Arabic. They used lexical (character based and word-based features), syntactic (function words and PoS), structural (paragraphs and greetings) and content based features (content words). They claimed that writeprints outperformed SVM where there are at least 5 instances in the training set. Writeprints failed when there is a single instance (document) of an author [26]. Moreover, Pavelec et al., [27] made use of a corpus of 150 Portuguese news articles in which 10 authors wrote 15 texts. The average number of words used was 600. They made use of conjunctions only and SVM for author prediction and they achieved an accuracy of 78%. Again Stańczyk and Cyran [28] made use of 168 novels from two famous Polish writers, Henryk Sienkiewicz and Boleslaw Prus. They used a neural network (ANN) based on function words and punctuations for author recognition. They achieved 95.8% accuracy using both features together. Furthermore, Iqbal et al., [29] used stylometric features (word-length, sentence length, punctuation, vocabulary richness, function words, structural-based and content-based) to predict the author of a given document. They made use of the Enron Email dataset where analysis was done on 10, 20, 40, 80 and 100 emails per author. An accuracy of 90% was obtained in the scenario where there were 5 authors and the k-means clustering algorithm was used.

Lopez-Monroy et al. [30] developed a new document author representation (DAR) technique for authorship attribution. It was tested on the c50 corpus from the Reuters Corpus Volume 1 [23]. Their approach produced better results than the method based on Support Tensor Machines (STM). In addition to, Koppel et al., [31] made a study to determine among who of the 10,000 bloggers is the author of a given text. The minimum number of words used was 200. They used content words, function words, strings of non-alphabetic characters and strings of non-numeric characters. They were able to achieve an accuracy of 42% using SVM. In 2012, the same dataset but this time with blogs of 2,000 words were used by Koppel et al., [32]. They used the character 4-grams feature and 46% of the authors were correctly assigned using the cosine similarity method. The objectives of Halteren et al., [33] were similar to that of Koppel et al., [34] but the corpus they made use of was different. The corpus consisted of 72 Dutch essays from 8 different authors (students) with a minimum number of 628 words and maximum number of 1342 words. They used function words, lexical (PoS) and syntactic features and they claimed an accuracy of 99%. The study by Stamatatos [35] was to make use of an Arabic corpus which enclosed 100 documents per author. They were able to achieve an accuracy of 69.1% among 10 authors. Allison and Guthrie [36] made a study on the identification of an author using the Enron Email corpus which contains 4071 emails of 160 different authors with an average number of 75 words per document. They used word-ngrams for author prediction. They were able to achieve an accuracy of 87.1%.

3   Methodology

Our corpus consists of 10 PhD theses from 10 different authors. All the 10 PhD theses were from the 'ICT in Education' field. All the authors were from UK Universities. Only the main part (introduction to conclusion) of the thesis was kept for analysis. While segmenting the documents into 1,000 words, 5,000 and 10,000 words using our document segmentor software, they undergo another cleaning process in order to remove document noise (citations, weird symbols and numbers). The documents were segmented in order to determine the optimal size of a document for authorship attribution and for the detection of plagiarism.

Based partly on the literature, a range of stylometric features has been complied. These features consist of word length, character n-grams, function words, sentence length, punctuation, unique words, vocabulary richness, PoS tags and many more. There are also some new features like symbols, combined-words (e.g. web-based), word endings ('ll, `ll, ed, ing, ion, ly, n't, s), sentences starting with the word "the" and sentences containing the following words (and, etc, e.g, what, which).

We made use of our own software to extract the 446 stylometric features to perform authorship attribution. Another subset of 153 features was selected for comparison. Two machine learning techniques k-NN and SMO were then used to do authorship attribution. Confusion matrices obtained from the machine learning algorithms help to detect plagiarism.

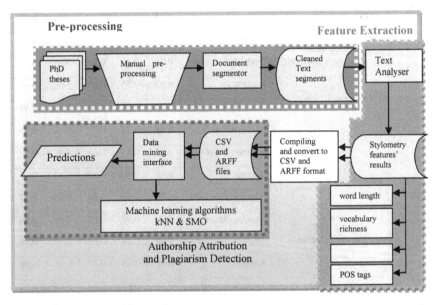

Fig. 1 System Architecture of the Authorship Attribution System and Plagiarism Detection using Stylometry

Figure 1 shows our architecture for authorship attribution and plagiarism detection. It consists of three main phases: pre-processing, feature extraction and prediction using machine learning classifiers. There are several procedures involved in each phase.

4 Results and Evaluations

4.1 Results of Prediction Based on Authors

In this section, we show that it is possible to identify different authors by deriving relevant statistics from their writing style (stylometrics). The number of features used was 446 and 153 (subset of the 446 features).

Table 1 Prediction results based on authors

				1-NN	3-NN	5-NN	7-NN	SMO
10,000 words	40 instances	446 features	Average Precision	**0.97**	0.94	0.93	N/A	**0.98**
			Average Recall	**0.95**	0.90	0.90	N/A	**0.98**
			Average F-Score	**0.95**	0.89	0.90	N/A	**0.98**
			Accuracy	**0.95**	0.90	0.90	N/A	**0.98**
		153 features	Average Precision	**0.93**	0.89	0.89	N/A	**0.98**
			Average Recall	**0.93**	0.85	0.85	N/A	**0.98**
			Average F-Score	**0.93**	0.86	0.86	N/A	**0.98**
			Accuracy	**0.93**	0.85	0.85	N/A	**0.98**
5,000 words	80 instances	446 features	Average Precision	**0.92**	0.91	0.92	0.84	**0.98**
			Average Recall	**0.91**	0.88	0.89	0.81	**0.98**
			Average F-Score	**0.91**	0.87	0.88	0.81	**0.97**
			Accuracy	**0.91**	0.88	0.89	0.81	**0.98**
		153 features	Average Precision	**0.90**	0.84	0.84	0.83	**0.97**
			Average Recall	**0.90**	0.82	0.83	0.80	**0.96**
			Average F-Score	**0.90**	0.81	0.81	0.79	**0.96**
			Accuracy	**0.90**	0.83	0.83	0.80	**0.98**
1000 words	400 instances	446 features	Average Precision	0.78	0.84	0.84	**0.86**	**0.94**
			Average Recall	0.76	0.81	0.82	**0.84**	**0.94**
			Average F-Score	0.77	0.82	0.82	**0.85**	**0.94**
			Accuracy	0.76	0.81	0.82	**0.84**	**0.94**
		153 features	Average Precision	0.74	0.78	**0.80**	**0.80**	**0.91**
			Average Recall	0.73	0.73	**0.78**	**0.78**	**0.91**
			Average F-Score	0.73	0.74	**0.78**	**0.78**	**0.91**
			Accuracy	0.73	0.73	**0.78**	**0.78**	**0.91**

Table 1 shows the detailed prediction results for author attribution when the segments of different sizes and different number of features were tested with the nearest neighbour (kNN) and support vector machines (SMO) algorithm from Weka. A cross-validation technique with 10 folds was used for all the thirty experiments. SMO outperformed the kNN algorithm in all the above experiments; the differences between these two methods become more accentuated as the document segments became smaller.

For segments of 10,000 and 5,000 words, 1-NN offered higher accuracy values than 3-NN, 5-NN or 7-NN. In general, for kNN, using 446 features produced higher classification accuracies than using 153 features. The number of features used did not have any significant effect on SMO as the performance measures are almost the same in both cases. However, for the smaller segments of 1,000 words, 7-NN performed better with an accuracy value of 86%. We also noted that the accuracy continued to increase (by small amounts) until 13-NN, after which it started to decrease (by small amounts). SMO with 446 features performed slightly better than SMO with 153 features.

Overall, the accuracy values for all the experiments were very high showing that it is indeed possible to distinguish between different authors via their writing style, if appropriate features are used. The best confusion matrix (98% accuracy) for author attribution is shown in Figure 2.

```
=== Confusion Matrix ===              === Confusion Matrix ===
    Predicted                             Predicted
 a b c d e f g h i j  <-- classified as   a b c d e f g h i j  <-- classified as
 4 0 0 0 0 0 0 0 0 0 | a = A              4 0 0 0 0 0 0 0 0 0 | a = A
 0 4 0 0 0 0 0 0 0 0 | b = B              0 3 0 0 1 0 0 0 0 0 | b = B
 0 0 4 0 0 0 0 0 0 0 | c = C              0 0 4 0 0 0 0 0 0 0 | c = C
 0 0 0 4 0 0 0 0 0 0 | d = D              0 0 0 4 0 0 0 0 0 0 | d = D
 0 0 0 0 4 0 0 0 0 0 | e = E   Actual     0 2 0 0 2 0 0 0 0 0 | e = E   Actual
 0 1 0 0 0 3 0 0 0 0 | f = F              0 1 0 0 0 3 0 0 0 0 | f = F
 0 0 0 0 0 0 4 0 0 0 | g = G              0 0 0 0 0 0 4 0 0 0 | g = G
 0 0 0 0 0 0 0 4 0 0 | h = H              0 0 0 1 0 0 3 0 0 0 | h = H
 0 0 0 0 0 0 0 0 4 0 | i = I              0 0 0 0 0 0 0 0 4 0 | i = I
 0 0 0 0 0 0 0 0 0 4 | j = J              0 1 0 0 0 0 0 0 3 | j = J
```

Fig. 2 SMO with 446 features and 10,000 words

Fig. 3 3-NN with 153 features and 10,000 words

Among the 40 instances, only 1 was wrongly classified. One document segment belonging to author F was predicted as belonging to B. The worst accuracy value of 85% author was obtained in two situations: using 3-NN and 5-NN with 153 features and 10,000 words. The confusion matrix using 3-NN with 153 features and 10,000 words is shown in Figure 3. Among the 40 instances, 6 of them were wrongly classified. Authors A, C, D, G and I had a 100% recognition rate while authors B, E, F, H and J had a mediocre recognition rate of 75%. For example, for author E, 2 segments were classified as belonging to B. We notice that Author B was an embroiling author as many document segments belonging to other authors were predicted as belonging to him. For the excellent performance of author attribution, it was found that 10,000 words document is most suitable and also SMO with the 446 features are best in these situations.

4.2   Corpus Size

According to our methodology, we need to know which size of document is suitable for AA and to detect plagiarism. Therefore each document was split into three document sizes, i.e., 1,000 words, 5,000 words and 10,000 words segments. In general, we found that the classification accuracy reduces with decreasing size of the document segments. With 10,000 words, the accuracy was as high as 98% while it got down to 73% for 1-NN with 1000 words. To our knowledge, this is the first work which splits a document into various equal-sized segments and compares each segment with each other.

4.3   Stylometric Features

Many new features have been used in this paper. For example, we have the combined-words feature which counts the total number of combined-words in each segment of a thesis. The number of unique combined-words was also considered. Earlier works have used a single count for all types of punctuation symbols. We have used a separate total for the 9 different punctuation symbols. We also kept track of the number of special symbols (such as *, %, #, etc). The sum, mean and standard deviation of some author specific writing patterns (such as ", and", "which ," and "e.g.") per sentence were also noted. In this study, we have used 125 style words (such as am, but, by, not, only, so on, very, while, will, etc). Nine different word endings (such as 'll, `ll, ed, ing, ion, ly, n't, not and s) have been considered. This is the highest number of features that have been used so far. Content words, word n-grams and character n-grams have not been used in this project. They are too tightly bound with the topic on which an author is writing and we do not regard them as being proper stylometric features, although they have been used in many works in order to dope the performance measures.

4.4   Results of Prediction Based on Authorship Attribution When Adding More Authors to the Dataset

We also analysed the accuracy of prediction of authorship attribution when adding more authors to the dataset using 153 features. Figure 4 shows how the accuracy of author prediction varies when the number of authors was varied from 2 to 10.

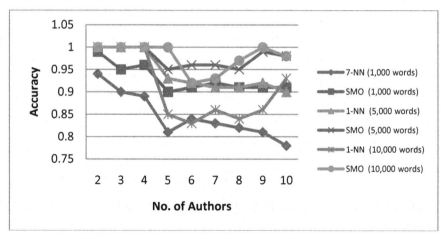

Fig. 4 Effect of author attribution among different no. of authors

From Figure 4, it can be seen that the overall performance of the classifiers show a decreasing trend when there are more authors involved. The sudden drop of accuracy, when there are 5 authors, is caused by author E whose segments are being predicted as belonging to B. From this finding, we can deduce that author B, is acting as an embroiling author i.e., B can be taken as an author who does not have her own writing style. In other words, it is very likely that B has copied texts in her work.

4.5   Plagiarism Detection

Figure 5 shows the confusion matrix when the prediction is done using SMO with 153 features and 1000 words. Based on the confusion matrix, the following recall, precision and f-measure percentage were calculated and displayed in Figure 6. In particular, we note from Figure 6, that the precision, recall and f-measure are well below 80% for only one author, i.e., author B. From Figure 5, only 31 segments (the minimum among all the ten authors) for author B has been correctly classified. The other nine segments belonging to B has been wrongly predicted into six different authors. Furthermore, thirteen document segments belonging to six different authors have been predicted as belonging to B. These results clearly show that the overall writing style of B is not very consistent. Thus, B could be considered as a potential suspect for plagiarism.

In order to prove the correctness of our proposed approach to detect potential cases of plagiarism, all the 40 document segments from the 10 authors were fed to the Turnitin plagiarism detection software [10] and then analysed to see if they contained plagiarised materials. The similarity index was noted in each case. The mean similarity index for each author was also computed.

```
=== Confusion Matrix ===
                Predicted
  a  b  c  d  e  f  g  h  i  j  <-- classified as
 39  0  0  0  1  0  0  0  0  0 |  a = A
  1 31  1  0  3  1  1  2  0  0 |  b = B
  0  3 35  0  0  0  0  1  1  0 |  c = C
  0  0  1 39  0  0  0  0  0  0 |  d = D
  0  3  2  1 33  0  0  0  0  1 |  e = E    Actual
  0  1  0  0  1 36  1  0  1  0 |  f = F
  0  2  0  0  1  1 36  0  0  0 |  g = G
  0  1  1  0  0  0  0 38  0  0 |  h = H
  0  0  0  0  0  0  0  0 40  0 |  i = I
  0  3  0  0  1  0  0  0  1 35 |  j = J
```

Fig. 5 Confusion matrix using SMO with 153 features and 1,000 words

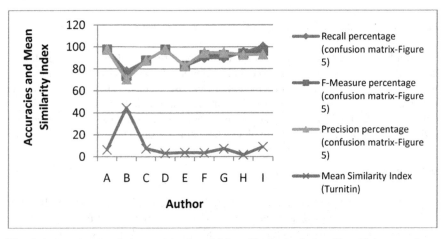

Fig. 6 Accuracies (confusion matrices) and Mean Similarity Index (Turnitin) comparison for each author

Figure 6 shows that the mean similarity index of author B is much higher compared to the other authors while the recall, precision and f-measure is much lower compared to others. Therefore, using Turnitin, we arrived at the same conclusion that Author B is a strong suspect for plagiarism. These results confirm that our proposed approach of using stylometric features to detect plagiarism is a viable one and has very high practicability in catching plagiarism in academic reports, dissertations and theses.

5   Conclusion

Plagiarism is considered a serious offence within the academic community and most universities of the world have strong policies to curb the propagation of plagiarised works. Unfortunately, due to the advancement in technology, new sources of plagiarism, such as translated texts, paper mills and the invisible web are making it more and more difficult for current text-alignment plagiarism software to offer reliable solutions. The objectives of this paper were multi-fold. Firstly, we showed that the writing style of a specific author is highly consistent. Next, we demonstrated that the writing patterns of different authors are sufficiently dissimilar and therefore these two techniques together can be used for authorship attribution with very high accuracies of over 90% in most cases. The effect of different sizes of text was also investigated. In general, the accuracies go down slowly from 10,000 words segments to 1,000 words segments. Moreover, as can be expected, the accuracies also decrease when more authors are added to the dataset. Finally, we demonstrated how we can use confusion matrices to detect prospective cases of academic dishonesty. A new dataset with new stylometric features and a novel way of processing documents have been described. Nevertheless, this approach is only suitable when large amount of texts are available and hence these techniques would not be suitable to classify shorter texts originating from student essays, emails and posts from social networks. In the future, we intend to conduct the same study with a much larger set of authors and with less than 1000 words in each segment.

References

1. Coyotl-Morales, R.M., Villaseñor-Pineda, L., Montes-y-Gómez, M., Rosso, P.: Authorship Attribution Using Word Sequences. In: Martínez-Trinidad, J.F., Carrasco Ochoa, J.A., Kittler, J. (eds.) CIARP 2006. LNCS, vol. 4225, pp. 844–853. Springer, Heidelberg (2006)
2. Kim, S., Kim, H., Weninger, T., Han, J. and Kim, H. D.: Authorship Classification: A Discriminative Syntactic Tree Mining Approach. In: Proceedings of the ACM SIGIR, July 24–28, Beijing, China (2011)
3. Nirkhi, S.M., Dharaskar, R.V.: Comparative Study of Authorship Identification Techniques for Cyber Forensics Analysis. International Journal of Advanced Computer Science and Applications 4(5), 32–35 (2013)
4. Khan, S.R., Nirkhi, S.M., Dharaskar, R..V.: E-mail Data Analysis for Application to Cyber Forensic Investigation using Data Mining. In: Proceedings of the 2nd National Conference on Innovative Paradigms in Engineering & Technology (NCIPET 2013), New York, USA (2013)
5. Maurer, H., Zaka, B.: Plagiarism–A Problem and How to Fight It. In: Proceedings of World Conference on Education Multimedia, Hypermedia and Telecommunications, AACE, pp. 4451–4458 (2007)
6. Mozgovoy, M., Kakkonen, T., Cosma, G.: Automatic student plagiarism detection: future perspectives. Journal Educational Computing Research 43(4), 511–531 (2010)
7. ICAI, Current Cheating Statistics. http://www.academicintegrity.org/icai/integrity-3.php. (accessed April 3, 2015)
8. Mechti, S., Jaoua, M. Belguith, L H.: A framework for Plagiarism Detection based on Author Profiling. In: Notebook for PAN at CLEF 2013 (2013). http://www.clef-initiative.eu/documents/71612/c7a0e432-dd82-46b1-ab9e-5d0dd98c3a8d (accessed March 3, 2015)

9. Smith, I.: The Invisible Web: Where Search Engines Fear to Go (2015). http://www.powerhomebiz.com/vol25/invisible.htm (accessed April 1, 2015)
10. Turnitin, iParadigms (2015). http://turnitin.com/ (accessed March 22, 2015)
11. Viper, Viper the Anti-plagiarism Scanner, Viper's features (2015). http://www.scanmyessay.com/features.php (accessed April 2, 2015)
12. Plagium, Plagium (2015). http://www.plagium.com/ (accessed April 2, 2015)
13. PlagTracker, PlagTracker (2015). http://www.plagtracker.com/ (accessed April 2, 2015)
14. Paper Rater, About Paper Rater (2015). http://www.paperrater.com/about (accessed April 2, 2015)
15. Grammarly, Grammarly (2015). http://www.grammarly.com (accessed April 2, 2015)
16. Horovitz, S.J.: Two Wrong Don't Negate a Copyright: Don't Make Students Turnitin if You Won't Give it Back. Florida Law Review 60(1), 229–268 (2008)
17. TurnitinBot, TurnitinBot General Information Page (2015). https://turnitin.com/robot/crawlerinfo.html (accessed: March 15, 2015)
18. Cheat For Turnitin, Limitations to Turnitin. Tips For How To Cheat Turnitin? (2015). http://cheatturnitin.blogspot.com/ (accessed March 15, 2015)
19. Argamon, S., Levitan, S.: Measuring the usefulness of function words for authorship attribution. In: Proceedings of the 2005 ACH/ALLC Conference (2005)
20. Hoover, D.L.: Frequent collocations and authorial style. Literary and Linguistic Computing 19(3), 261(28) (2004)
21. Nirkhi, S.M., Dharaskar, R.V., Thakare, V.M.: Authorship Attribution of online messages using Stylometry: An Exploratory Study. In: International Conference on Advances in Engineering and Technology (ICAET'2014) (2014)
22. Luyckx, K., Daelemans, W.: Authorship attribution and verification with many authors and limited data. In: Proceeding of the 22nd International Conference on Computational Linguistics, Vol. 1, pp. 513–520 (2008)
23. Lewis, D., Yang, Y., Rose, T., Li, F.: RCV1: A New Benchmark Collection for Text Categorisation Research. Journal of Machine Learning Research 5, 361–397 (2004)
24. Iqbal, F., Hadjidj, R., Fung, B.C.M., Debbadi, M.: A Novel Approach of Mining Write-Prints for Authorship Attribution in E-mail Forensics. Proceedings of the Digital Forensic Research Workshop, pp. 42–51. Elsevier Ltd., Quebec (2008)
25. Abbasi, A., Chen, H.: Writeprints: A stylometric approach to identity-level identification and similarity detection in cyberspace. ACM Transactions on Information Systems 2(2), Article 7 (2008)
26. Abbasi, A., Chen, H.: Visualizing Authorship for Identification. In: Mehrotra, S., Zeng, D.D., Chen, H., Thuraisingham, B., Wang, F.-Y. (eds.) ISI 2006. LNCS, vol. 3975, pp. 60–71. Springer, Heidelberg (2006)
27. Pavelec, D., Justino, E., Oliveira, L.S.: Author Identification using Stylometric Features. Inteligencia Artificial, Revista Iberoamericana de Inteligencia Artificial 11(36), 59–65 (2007)
28. Stańczyk, U., Cyran, K.A.: Machine learning approach to authorship attribution of literary texts. International Journal of Applied Mathematics & Informatics 1(4), 151–158 (2007)
29. Iqbal, F., Binsalleeh, H., Fung, B.C.M., Debbabi, M.: Mining writeprints from anonymous e-mails for forensic investigation. Digital Investigation, Science Direct 7(1), 56–64 (2010)

30. López-Monroy, A.P., Montes-y-Gómez, M., Villaseñor-Pineda, L., Carrasco-Ochoa, J.A., Martínez-Trinidad, J.F.: A New Document Author Representation for Authorship Attribution. In: Carrasco-Ochoa, J.A., Martínez-Trinidad, J.F., Olvera López, J.A., Boyer, K.L. (eds.) MCPR 2012. LNCS, vol. 7329, pp. 283–292. Springer, Heidelberg (2012)
31. Koppel, M., Schler J., Argamon, S., Winter, Y.: The Fundamental Problem of Authorship Attribution. English Studies 93(3), 284–291 (2012). Taylor & Francis
32. Koppel, M., Argamon, S., Shimoni, A.R.: Automatically Categorizing Written Texts by Author Gender. Literary and Linguistic Computing 17(4), 401–412 (2002)
33. Halteren, H.V.: Linguistic Profiling for Author Recognition and Verification. In Proceedings: 42nd Annual Meeting on Association for Computational Linguistics (ACL04), Barcelona, Spain, pp. 199–206 (2004)
34. Koppel, M., Schler, J., Argamon, S., Messeri, E.: Authorship attribution with thousands of candidate authors. In: Proceedings of the ACM SIGIR, New York, USA, pp. 659–660 (2006)
35. Stamatatos, E.: Author identification: Using text sampling to handle the class imbalance problem. ECAI, IOS Press, Vol. 44, pp. 790–799 (2008)
36. Allison, B., Guthrie, L.: Authorship Attribution of E-Mail: Comparing Classifiers over a New Corpus for Evaluation. In: International Conference on Language Resources and Evaluation, Marrakech, Morocco (2008)

Exploration of Fuzzy C Means Clustering Algorithm in External Plagiarism Detection System

N. Riya Ravi, K. Vani and Deepa Gupta

Abstract With the advent of World Wide Web, plagiarism has become a prime issue in field of academia. A plagiarized document may contain content from a number of sources available on the web and it is beyond any individual to detect such plagiarism manually. This paper focuses on the exploration of soft clustering, via, Fuzzy C Means algorithm in the candidate retrieval stage of external plagiarism detection task. Partial data sets from PAN 2013 corpus is used for the evaluation of the system and the results are compared with existing approaches, via, N-gram and K Means Clustering. The performance of the systems is measured using the standard measures, precision and recall and comparison is done.

1 Introduction

"Plagiarism is the reuse of someone else's prior ideas, processes, results, or words without explicitly acknowledging the original author and source" [1]. Plagiarism corresponds to copying the work or idea of another author and presenting it as own without acknowledging the original work. It can be a literal copy of the material or part of it from another source. However, the author may modify some portions of it by inserting or deleting some parts of text. It also includes cases where the author just copies the idea from another, writing it in his own words by paraphrasing or summarizing the original work, which is termed as intelligent plagiarism [2]. Plagiarism is of serious concern as the works of many authors are readily available to anyone through the high availability of the World Wide Web (WWW). Many tools and techniques have been built and analyzed over the years

N.R. Ravi(✉) · K. Vani · D. Gupta
Department of Computer Science, Amrita School of Engineering,
Amrita Vishwa Vidyapeetham, Bangalore, India
e-mail: riya.sanjesh@gmail.com, {k_vani,g_deepa}@blr.amrita.edu

© Springer International Publishing Switzerland 2016
S. Berretti et al. (eds.), *Intelligent Systems Technologies and Applications*,
Advances in Intelligent Systems and Computing 384,
DOI: 10.1007/978-3-319-23036-8_11

but still lacks proper detection efficiency. Plagiarism detection techniques can be Intrinsic or Extrinsic. Former detection techniques involve detecting plagiarized passages within a document using the author's style of writing. Latter detection techniques on the other hand compare multiple existing documents with suspected documents and try to figure out the plagiarized passages [2]. Some of the External detection techniques are String matching, Vector Space Model (VSM), Finger printing etc. In general, the major steps followed in an Extrinsic plagiarism detection system (PDS) are pre-processing, candidate document identification, detailed analysis and postprocessing. Pre-processing is the initial step and its purpose is to discard all irrelevant information from the source and suspicious documents .Candidate document identification is an important step , where the candidate source documents for each suspicious document is picked out and this further reduces the complexity of the final processing. Taking the whole set of the source documents for a detailed plagiarism detection is too cumbersome and a time consuming process [3]. Most of the currently available PDS depends on N-gram based approaches which are inherently slow. Thus finding naïve and efficient futuristic method or technology is highly in need. Machine learning (ML) approaches are explored in detail in the text classification domain with great success. Such ML techniques can be experimented with in various stages of the plagiarism detection systems as well.

In this paper the main focus is given to the candidate retrieval stage of extrinsic PDS. Usually the complete PDS evaluation is done and efficiency is measured rather than evaluating candidate retrieval task separately. But in practical scenario, the incorporation of this stage is important to aid the final exhaustive analysis of an extrinsic PDS. Here ML approaches, via, clustering techniques are used for candidate retrieval task. A Fuzzy C-Means (FCM) based clustering approach is used and compared with the existing candidate retrieval approaches, via, N-gram based and K Means clustering approach.

2   Literature Survey

The concept of Plagiarism was first introduced in the field of Arts in the first century but the word was adopted in English language in around 1620. In the twentieth century plagiarism was getting noticed as a big menace in Academia and Journalism [4]. Serious actions were taken against the people involved in the act. As pointed out by Alzahrani et.al [2], the first code plagiarism detection tools came in 1970s but it was not before 1990s we saw first tool to detect extrinsic plagiarism detection in natural language documents. Alzahrani et.al [2] describes the taxonomy of Plagiarism and plagiarism detection types, via, Extrinsic and Intrinsic. Some of the popular tools available for plagiarism detection are COPS, SCAM, CHECK and Turnitin which are discussed by Zdenek Ceaska [3]. These tools mainly focuses on copy & paste detection in the extrinsic plagiarism. Documents were represented as set of words/sentences and compared. This made the detection slow. To circumvent this problem, feature reduction was introduced to reduce the detection time. The features used to represent the document are reduced with one of the Feature reduction techniques like TF-IDF, Information Gain, Chi^2 [2]. The other techniques used to reduce the complexity are the Stop words removal, lemma etc. To reduce the number of the sources against which a given suspicious document is compared could be further reduced using one of the Candidate retrieval technique like Fingerprint, Hash based, Vector Space Model, Latent Semantic Indexing [2] etc. Other document representations and comparison techniques were developed to improve the detection. One of the ways was to represent the document as a graph [5]. Alzahrani and Salim introduced a semantic based plagiarism detection technique which used fuzzy semantic-based string similarity. Their process can be subdivided into four stages. In the first stage text is tokenized, stop words removed and stemmed. Second stage involved identifying a set of candidate documents for each suspicious document using Jaccard coefficient and shingling algorithm. Detailed comparison is carried out next between the suspicious document and the corresponding candidate documents. In this stage a fuzzy similarity is calculated. If the similarity is more than a threshold then the sentences are marked as similar. In the end a post processing is carried out where the consecutive sentences are combined to form a paragraph. Asif Ekbal et. al. [7] proposed a Vector Space Model (VSM) to overcome the problems which arises due to matching of strings. In this case the document is represented using vectors and similarity is calculated based on the cosine similarity measure. The downside of VSM model is failure to detect disguised plagiarism. It also involves lot of effort in computation.RasiaNaseemet.al. [8] utilizes VSM approach for candidate retrieval task and uses a fuzzy semantic similarity measure for detailed analysis. A further improvement on this technique was done by Ahmed Hamza et.al. [9] where Semantic Role Labeling (SRL) is used. Although this technique gives good detection, it takes lot of time and have scalability issues. Alberto and Paolo [10] proposes a basic N-gram approach for locating the plagiarized fragments. Word n-gram documents are compared with each other to detect the similarity. This approach is time consuming and did not consider any intelligent aspects, via, synonyms. N-gram method is the base for most of the detection techniques like character based, syntactic based, vector based, semantic based etc. These techniques have been traditionally slow and thus the plagiarism detection tools that utilizes these techniques are also slow in nature. Most of the PDS employ string matching approach only mainly in candidate retrieval stage, which means that such systems cannot detect intelligent plagiarism effectively.

Computational Intelligence or ML techniques are not much explored in plagiarism detection domain but are popular in document classification domains. Document classification deals with classification of text documents into multiple categories. In the field of text classification, ML techniques such as Naïve Bayes, Support Vector Machine (SVM) etc. along with clustering techniques have been explored in detail. Chanzing et. al. [11] focus on feature selection using Information Gain along with SVM for classification. K Nearest Neighbour (KNN), Naïve Bayes and Term Graph techniques are well compared by Vishwanath et. al. [12]. Even

Genetic algorithm technique have been employed in [13] for document classification but with not much of success. KNN gives good accuracy but with high time complexity. K Means, KNN and the combination and variations of them are used for text classification applications widely [14,15,16,17]. But these ML techniques are less explored in plagiarism domain. Most of the papers in plagiarism detection domain talks about the exhaustive analysis stage giving less importance to the pioneer candidate retrieval stage. A recent paper by Vani and Gupta [18] explores the ML technique, via, K Means clustering based technique specifically for the candidate retrieval stage. K Means clustering creates K non-overlapping clusters of the document features and thus produce hard boundaries.

To overcome this problem, the proposed method uses a soft clustering approach, via, Fuzzy C Means (FCM) clustering which creates K clusters which can be overlapping. Thus the objective of this paper is the exploration of FCM clustering algorithm in the candidate retrieval step of Extrinsic Plagiarism Detection and the comparative study with existing candidate retrieval approaches.

Further the method is explored using NLP techniques, via, stemming, lemmatization and chunking. These algorithms are evaluated on PAN 2013 data set and compared.

3 Existing Candidate Retrieval Approaches

Traditional N-gram based approach [10] and K Means [18] based clustering approach, the two existing algorithms for the candidate retrieval stage are explained in detail in Sub-Sections 3.1 and 3.2 respectively.

3.1 N-gram Approach

N-gram is a contiguous sequence of N items from a given sequence of text or speech. This contiguous sequence of N items can be words or character. Given a suspicious document and a set of source documents, the objective is to find out the relevant candidate source documents using N-gram method. Here, N-grams are generated from both suspicious and source. Similarity between each pair of document is calculated based on the Dice coefficient measure as in (1).

$$C(d_{sus}, d_{src}) = \frac{2|N(d_{sus}) \cap N(d_{src})|}{|N(d_{sus}) \cup N(d_{src})|} \tag{1}$$

Here N(*) is the set of N-grams in (*), C is the similarity measure, d_{sus} and d_{src} are the suspicious and source document respectively. Now for any pair of suspicious and source documents, if the computed similarity value is more than a specified threshold, then the source document is added to the candidate set of the particular suspicious document.

3.2   K Means Approach

In the basic K Means algorithm, the main problem is to decide the value of K and the initial centroids, which is important. This is well tackled by K Vani and Gupta D [18], where the value of K is fixed as the number of suspicious documents, and the initial centroids are considered as the individual suspicious documents itself. Here each centroid is fixed as a suspicious document, as the clusters/candidate set with respect to each suspicious document has to be formed. Initially Stop Words, which are the semantically irrelevant words, via, pre-position, conjunction etc are removed from the documents. Then the documents, d_{src} and d_{sus} is represented using VSM model with TF-IDF representation using Equation (2) to form vectors $V(d_{sus})$ and $V(d_{src})$.

$$wt, d = tft, d * \log \frac{|D|}{\left|\left\{d' \in D; t \in d'\right\}\right|} \tag{2}$$

Here $tf_{t, d}$ is term frequency of term t in document d and $\log \dfrac{|D|}{\left|\left\{d' \in D; t \in d'\right\}\right|}$ is inverse document frequency where numerator defines the total number of documents in the document set and denominator is the number of documents containing the term t. Then the similarity between document vectors is computed using cosine measure using Equation (3).

$$Cos(d_{sus}, d_{src}) = \frac{V(d_{sus}) \cdot V(d_{src})}{||V(d_{sus})|| \, ||V(d_{src})||} \tag{3}$$

In Equation (2), Cos (*) is the cosine similarity and the dot product of document vectors are represented in the numerator and the denominator computes the Euclidean norms of these document vectors

4   Proposed Approach

The proposed Fuzzy C Means algorithm is explained in Algorithm 1 and its variations with NLP techniques such as lemmatization, chunking and stemming are discussed in Sub-Section 4.1. Here initially, fuzzy membership is calculated for each d_{src} with all d_{sus} using Equation (4). In Equation (4), μ_{ij} is the degree of membership of x_i in the cluster j, x_i is the ith data point and c_j is the centroid.

Here m is the fuzziness factor and C is cluster centre. d_{src} with a μ_{ij} greater than a particular threshold is considered as candidates for the given suspicious document, d_{sus}.

$$\mu_{ij} = \frac{1}{\sum_{k=1}^{C} \left(\frac{\left\| x_i - c_j \right\|}{\left\| x_i - c_k \right\|} \right)^{\frac{2}{m-1}}} \tag{4}$$

4.1   Proposed Algorithm

Algorithm 1. Algorithm based on Fuzzy C Means
Input:
D_{sus} = Set of suspicious documents
D_{src} = Set of source documents
Output:
Candidate sets (*Cand*)
Begin
Convert all $D_{sus}$and D_{src} into TF-IDF form using Equation(2)
Set K = # D_{sus}
Initialize each centroid c_j as each xj in D_{sus}
Repeat for each x_j in D_{sus}
Compute μ_{ij} for each $x_i$in D_{src} using Equation (4)
Normalize μ_{ij} between 0 and 1
If $\mu_{ij} > \alpha$, then Add x_i to the $Cand(x_j)$
Repeat for each x_i in $Cand(x_j)$: Compute
$dis = 1- Cos\ (x_i, x_j)$
If $dis > \theta$,then Remove xi from $Cand(x_j)$
End

Here, α is the threshold for selecting the initial candidate documents with respect to a particular suspicious document. Further a cut off θ is applied for each candidate set formed, which helps in pruning out the highly dissimilar documents from the candidate sets.

4.2   Fuzzy C Means Algorithm with Different NLP Techniques

This Section describes the NLP techniques which are incorporated with the proposed FCM algorithm. The NLP techniques, via, Stemming (**FCM-stem**), lemmatization (**FCM-lem**) and Chunking (**FCM-chk**) are used to explore its impact on the proposed algorithm. Stemming refers to a heuristic process that removes the ends of words that often includes the derivational affixes. Lemmatization is the process of converting the word tokens into their dictionary base forms. In chunking, the document is segmented into sub constituents, such as noun phrases, verb phrases, prepositional phrases etc. With chunking, stop words are not removed. Two other combinations of the above NLP techniques are also used, via, **Chunking with Lemma** (FCM-chklem) and **Chunking with Stem** (FCM-chk stem).The examples demonstrating the NLP techniques used are given in Fig.1.

English Sent: "This Document describes strategies carried out by companies for their agricultural chemicals."
Tokenization :['This', 'Document', 'describes', 'strategies', 'carried', 'out', 'by', 'U.S.', 'companies', 'for', 'their', 'agricultural', 'chemicals', '.']
After Stop Word Removal: ['Document', 'describes', 'strategies', 'carried', 'U.S.', 'companies', 'their', 'agricultural', 'chemicals']
Stemming: ['docu', 'describ', 'strateg', 'carr', 'compan', 'agricultur', 'chemic']
Lemmatization: ['Document', 'describes', 'strategy', 'carried', 'company', 'agricultural', 'chemical']
Chunking:['This Document', 'strategies', 'describes', 'carried', 'companies', 'their agricultural chemicals']

Fig. 1 Example of NLP techniques used with proposed algorithm

5 Experimental Settings and Result Evaluation

The existing approaches and the proposed algorithm discussed in Section 3 and 4 respectively are evaluated on PAN-13 partial data set. Then the algorithm efficiency is measured using the standard IR measures, via, Recall and Precision.

5.1 Data Set

The data set in PAN 2013 is divided based on their complexity and the statistics used is given in Table.1

Table 1 Data Statistics

	Suspicious Documents	Source Documents
Set 1	39	205
Set 2	31	213
Set 3	35	209

Here, No Obfuscation set (Set 1) consists of document pairs where the suspicious document contains exact copies of passages in the source document. In random Obfuscation (Set2) the document is manipulated via word shuffling, adding, deleting and replacing words or short phrases, synonym replacements etc. In translation obfuscation (Set 3), the given text is run through a series of translations, so that the output of one translation forms the input of next one while the last language in the sequence is the original language of the text [19].

5.2 Evaluation Metrics

The performance of each algorithm is measured using the metrics, via, Recall and precision as given in Equation (4) and (5) and comparison is done.

$$prec = \frac{|D_{ret} \cap D_{exp}|}{|D_{ret}|} \qquad (5)$$

$$rec = \frac{|D_{ret} \cap D_{exp}|}{|D_{exp}|}$$

(6)

Here D_{ret}denote the set of documents that are retrieved by the system and D_{exp} is the actual expected documents. Precision measures the ratio of correctly retrieved cases to the total documents retrieved by the system while recall measures the ratio of correctly retrieved cases to the actual expected documents.

5.3   Results Analysis

The proposed and existing candidate retrieval algorithms are evaluated on the data sets mentioned in Table 1 and performance is evaluated. The proposed algorithm results are then compared with the existing K Means and N-grams methods and performance is measured using Equation (4) and (5). The proposed FCM algorithm efficiency is determined mainly by two parameters, via, α and θ as discussed in Sub-Section 4.1. The algorithm is tuned with different values of these two parameters to obtain the optimal recall and precision. Initially, the proposed FCM algorithm is evaluated with various α values on the three data sets discussed in SubSection 5.1. The recall and precision based on different αvalues is plotted in Fig.2 (a) and (b) respectively. Due to the variation in complexity of sets, the performance exhibited by different α values on these sets also differs. After conducting many evaluations, the α value for each set is determined based on the precision and recall presented using the specific α value. From Fig.2 (a) it can be observed that with Set-1, a 100 % recall is obtained with α= 0.5 and for Set-2 and Set-3 from 0.6 to 0.9 the recall is high. Now considering both precision and recall, the α value selected for each set is given in Table. 2. In a similar way, the

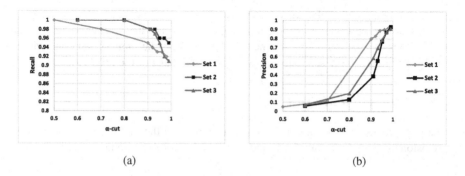

(a)   (b)

Fig. 2 (a) Recall results with different α values of FCM-basic; (b) Precision results with different α values of FCM-basic

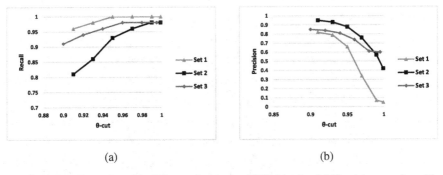

(a)   (b)

Fig. 3 (a).Recall results with different θ values of FCM-basic; (b) Precision results with different θ values of FCM-basic

Table 2 θ and α values for each set

	α	θ
Set 1	0.5	0.95
Set 2	0.91	0.98
Set 3	0.91	0.98

algorithm is evaluated with multiple θ values and the performance of the system is plotted in Fig.3 (a) and (b). The θ value that presents maximum recall and a high precision is then considered for each Set. The θ value finally selected for each set is given in Table.2.

Fig.4 and 5 plots the proposed FCM algorithm and its variations with different NLP techniques using the α and θ values as given in Table.2. Further these algorithms are compared with the existing candidate retrieval approaches, via, N-gram and K Means approach. In N-gram evaluation, N = 3 is employed with a threshold of 0.05 and the basic K Means candidate retrieval algorithm is also evaluated with the data sets. It can be noted that for Set-1 with no obfuscations, the basic FCM algorithm exhibits high recall but precision reduces. Proposed algorithm incorporated with lemmatization and stemming exhibits good performance with respect to recall. With respect to precision, FCM-chk and FCM-chklem outperforms the other NLP variations incorporated with proposed FCM. Compared to the existing approaches, it can be noted that the proposed algorithm and its variations presents a considerable improvement in recall which is the main focus of candidate stage. With Set-2 the performance variation is almost the same as in Set-1, while in Set-3 with respect to both measures FCM-chk outperforms the other FCM based variations.

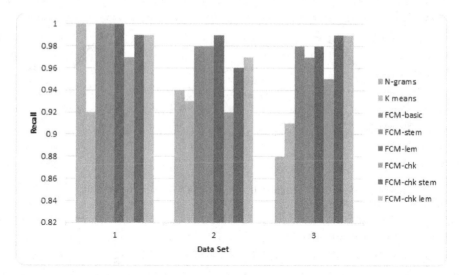

Fig. 4 Fuzzy C Means Recall values with various NLP techniques

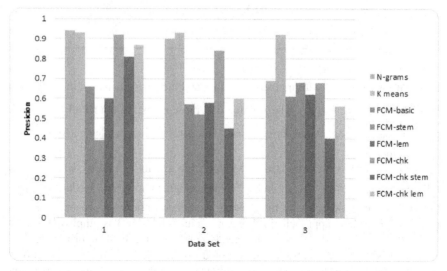

Fig. 5 Fuzzy C Means Precision values with various NLP techniques

The proposed method presents an improvement of 5% in Set-2 and 11% in Set-3 with respect to the traditional N-gram approach and compared to K Means approach an improvement of 8% in Set-1, 6% in Set-2 and 8% in Set-3. Thus compared to N-gram and K Means candidate retrieval approaches, the proposed FCM algorithm and its variations exhibit an increased performance with respect to recall, which is the main focus in candidate retrieval task.

6 Conclusions and Future Work

The proposed work focuses on exploration of clustering techniques in candidate retrieval task of extrinsic PDS. The soft clustering approach with Fuzzy C Means algorithm is employed here and FCM algorithm is tuned for the candidate retrieval task. The proposed approach exhibits a recall value of 100% in Set 1 and around 98% in Sets 2 and 3 which is considerably higher than the existing methods. Within FCM based approaches, in Set 2, FCM-lem presents about 99% recall and with Set-3 FCM-chk stem and FCM-chklem provides a recall of about 98%. K Means exhibits a good precision but with lower recall values compared to other approaches because it utilize 'hard' clustering approach. This is where Fuzzy C Means approach gains ground because of its 'soft' clustering technique. With the main focus in candidate retrieval being higher recall, it can be concluded from the analysis and discussions that the proposed algorithm and its variations exhibits a considerable improvement in recall compared to existing approaches discussed. In future, more efficient filtering techniques can be employed with the focus on improvement of precision values. For efficient comparison of performance, the algorithms will be evaluated on larger data sets.

References

1. 'A Plagiarism FAQ' IEEE.
   http://www.ieee.org/publications_standards/publications/rights/plagiarism_FAQ.html
2. Alzahrani, S.M., Salim, N., Abraham, A.: Understanding Plagiarism Linguistic Patterns, Textual Features, and Detection Methods. IEEE Transactions on Systems, Man, and Cybernetics—Part C: Applications and Reviews 42(2) (2012)
3. Ceska, Z.: The future of copy detection techniques. In: Proceedings of the First Young Researchers Conference on Applied Sciences, pp. 5–10 (2007)
4. Freedman, J.: The Ombudsman as Go-Between. In: The Fourth Annual Report of the Office of the Ombudsman (1974–1975)
5. Osman,A H., Salim, N., Bin Wahlan, S., Hentabli, H., Ali, A.M.: Conceptual similarity and graph-based method for plagiarism detection. Journal of Theoretical and Applied Information Technology 32(2), 135–145 (2011)
6. Alzahrani, S., Salim, N.: Fuzzy Semantic-based String Similarity for Extrinsic Plagiarism Detection. CLEF (Notebook Papers/LABs/Workshops) (2010)
7. Ekbal, A., Saha, S., Choudhary, G.: Plagiarism Detection in Text using Vector Space Model. IEEE Transactions on Systems, Man, and Cybernetics—Part C: Applications and Reviews 42(2) (2012)
8. Naseem, R., Kurian, S.: Extrinsic Plagiarism Detection in Text Combining Vector Space Model and Fuzzy Semantic Similarity Scheme. International Journal of Advanced Computing, Engineering and Application (IJACEA) 2(6) (2013) ISSN: 2319–281X
9. Osmana, A.H., Salima, N., Binwahlan, M.S., Alteeb, R., Abuobiedaa, A.: An improved plagiarism detection scheme based on semantic role labeling. In: Applied Soft Computing 12 (2012)

10. Barrón-Cedeño, A., Rosso, P.: On Automatic Plagiarism Detection Based on n-Grams Comparison. In: ECIR Proceedings of the 31st European Conference on IR Research on Advances in Information Retrieval, pp. 696–700 (2009)
11. Shang, C., Li, M., Feng, S., Jiang, Q., Fan, J.: Feature selection via maximizing global information gain for text classification. KnowledgeBased Systems **54**, 298–309 (2013)
12. Bijalwan, V., Kumar, V., Kumari, P., Pascual, J.: KNN based Machine Learning Approach for Text and Document Mining. International Journal of Database Theory and Application **7**(1), 61–70 (2014)
13. Uysal, A.K., Gunal, S.: Text classification using genetic algorithm oriented latent semantic features. Expert Syst. Appl. **41**, 5938–5947 (2014)
14. Buana, P.W., Jannet, S., Putra, I.: Combination of K-Nearest Neighbor and K-Means based on Term Re-weighting for Classify Indonesian News. International Journal of Computer Applications 50(11) (2012)
15. Šilić, A., Moens, M.-F., Žmak, L., Bašić, B.D.: Comparing Document Classification Schemes Using K-Means Clustering. In: Lovrek, I., Howlett, R.J., Jain, L.C. (eds.) KES 2008, Part I. LNCS (LNAI), vol. 5177, pp. 615–624. Springer, Heidelberg (2008)
16. Pappuswamy, U., Bhembe, D., Jordan, P.W., VanLehn, K.: A Supervised Clustering Method for Text Classification. In: Gelbukh, A. (ed.) CICLing 2005. LNCS, vol. 3406, pp. 704–714. Springer, Heidelberg (2005)
17. Miao, Y., Kešelj, V., Milios, E.: Document Clustering using Character N-gram: A Comparitive Evaluation with Term-based and Word-based Clustering. In: ACM International Conference on Information and Knowledge Management, pp. 357–358 (2005)
18. Vani, K., Gupta, D.: Using K-means Cluster based Techniques in External Plagiarism Detection. In: Proceedings of International Conference on Contemporary Computing and Informatics (IC3I), pp. 27–29 (2014)
19. Potthast, M., Hagen, M., Gollub, T., Tippmann, M.: Overview of 5th International Competition on Plagiarism Detection. In: CLEF 2013 Evaluation Labs and Workshop–Working Notes Papers, pp. 23–26 (2014) , ISBN 978-88-904810-3-1, ISSN 2038-4963

Issues in Formant Analysis of Emotive Speech Using Vowel-Like Region Onset Points

R. Surya, R. Ashwini, D. Pravena and D. Govind

Abstract The emotions carry crucial extra linguistic information in speech. A preliminary study on the significance and issues in processing the emotive speech anchored around the vowel-like region onset points (VLROP) is presented in this paper. The onset of each vowel-like region (VLR) in speech signals is termed as the VLROP. VLROPs are estimated by exploiting the impulse like characteristics in excitation components of speech signals. Also the work presented in the paper identifies the issue of falsified estimation of VLROPs in emotional speech. Despite the falsely estimated VLROPs, the formant based vocaltract characteristics are analyzed around the correctly estimated VLROPs from the emotional speech. The VLROPs retained for the emotion analysis are selected from those syllables which have uniquely estimated VLROPs without false detection from each emotion of same text and speaker. Based on the formant analysis performed around the VLROPs, there are significant variations in the location of the formant frequencies for the emotion utterances with respect to neutral speech utterances. This paper presents a formant frequency analysis performed from 20 syllables selected from 10 texts, 10 speakers across 4 emotions (Anger, Happy, Fear and Boredom) and neutral speech signals of German emotion speech database. The experiments presented in this paper suggest, firstly, the need for devising a new robust VLROP estimation for emotional speech. Secondly, the need for further exploring the formant characteristics for emotion speech analysis.

1 Introduction

Emotions carry extra linguistic information which add more intelligence to the spoken waveform [1, 2]. As the emotion content is spread across the whole utterance,

R. Surya(✉) · R. Ashwini · D. Pravena · D. Govind
Center for Excellence in Computational Engineering and Networking,
Amrita Vishwa Vidyapeetham, Coimbatore 641112, Tamilnadu, India
e-mail: {surya.radhakrishnan91,ashrajeev.991,d.pravena,govinddmenon}@gmail.com

© Springer International Publishing Switzerland 2016   139
S. Berretti et al. (eds.), *Intelligent Systems Technologies and Applications*,
Advances in Intelligent Systems and Computing 384,
DOI: 10.1007/978-3-319-23036-8_12

it is challenging to find out which component carries more information with regard to emotions. Also, the knowledge about emotive content in an utterance is critical for the speech processing tasks like speech recognition and speech synthesis. Hence these tasks are named as stressed speech recognition and emotional/expressive speech synthesis, respectively in the context of emotions [3, 4, 5, 6, 7]. There are many studies conducted in the literature to explore the speech parameters that are mostly affected by the emotional information [8]. Majority of the works conclude that the prosodic parameters like pitch contour and segmental duration parameters of the speech signal carry significant emotion information [9, 10, 11, 12]. For instance, the work by Ganagamohan et al. experimentally verified the significance of processing instantaneous pitch contours for extracting relevant emotional information from speech [9]. In [1], Prasanna et al. provided analysis of variations in strength of excitation measure obtained across different emotions. Also there are a few works which deal with the analysis of formant characteristics across various emotions [13, 14]. This paper attempts to analyze the variations in formant frequencies across different emotions in speech.

As the vocaltract has steady state formant configuration during the production of vowels, the work presented in the paper tries to analyze the formant parameters around each vowel across different emotions. The onsets of vowels or vowel-like regions in speech can be automatically estimated by using vowel-like region onset point (VLROP) detection algorithms. The regions of vowels, semivowels and diphthongs are generally termed as vowel like regions (VLRs) [15]. There are different algorithms available in the literature which provide the accurate vowel onset point (VOP) or vowel-like region onset point (VLROP) detection for clean or neutral speech signals [15, 16, 17].

All these algorithms exploit the impulse like excitation characteristics during production of vowels in speech. With the knowledge of VOPs or VLROPs obtained, the formant analysis can be performed around these estimated anchor points across different emotions. Eventhough, the available VLROP detection algorithms work well for neutral or clean speech signals, we have to ensure the performance of VLROP detection in emotions speech utterances where excitation source and prosodic parameters vary rapidly. Once the VLROPs are correctly and accurately estimated from emotions speech utterances, the vocaltract parameters of corresponding VLROP speech frames of each emotion can be accurately compared. Hence the novelty of the work presented in the paper are the following:

• Issues of VLROP estimation in emotional speech
• Analysis of formant frequencies in vowels of different emotions

The organization of the paper is as follows: Section 2 describes the VLROP estimation issues in emotion speech. Formant analysis in emotion speech is discussed in Section 3. Section 4 provides the experimental set up and results. Section 5 summarizes the work with conclusion and future scope for the present work.

2   Issues in VLROP Detection in Emotional Speech

2.1   VLROP Detection Using Hilbert Envelope (HE) of Linear Prediction Residual

The work presented in this paper uses Hilbert envelope of linear prediction (LP) residual based VLROP detection algorithm proposed by Prasanna et al. [16] . The speech signal is subjected to LP analysis to obtain the LP residual signal. A complex analytic signal is derived from the Hilbert transform of LP residual and the HE of the LP residual computed as the magnitude of the analytic signal. An excitation contour is constructed using the maximum value for every HE frames. The excitation contour is then convolved with first order gaussian differentiator to obtain the VLROP evidence from HE of LP residual. Finally, the peaks in the evidence signal is selected as the hypothesized VLROPs of the given speech signals. The spurious speaks in the evidence plot is removed by setting an appropriate energy based threshold [16].

2.2   VLROP Detection in Emotional Speech

Even though the LP residual HE based VLROP detection gives almost correct VL-ROPs from neutral speech, the performance of the same algorithm has to be verified

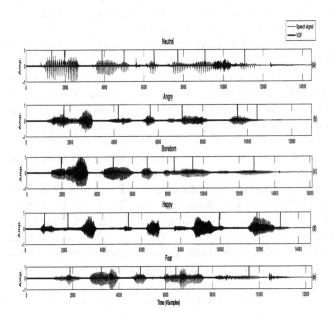

Fig. 1 Falsified and missed VLROPs in various emotions. The speech waveform and VLROPs (showed as bold stem plots) in (a) neutral, (b) anger, (c) boredom (d) happy and (f) fear.

for the speech with emotions under the conditions of the same text and the same speaker. Figure 2.2 plots different emotional speech utterances with estimated VLROP locations. Figure 2.2(a) is the neutral speech utterance with estimated VLROPs which are plotted as the bold stem plot. The number of VLROPs detected in Figure 2.2(a) is equal to the number of syllables present in the utterance. There are varied number of VLROP detections occurred in the plots (b), (c), (d) and (e) of Figure 2.2. The variations in the number of estimated VLROPs are either due to miss or spurious detection of VLROPs in emotion utterances as compared to the neutral case. The VLROPs estimated from different emotions plotted in Figure 2.2 are obtained by keeping the same threshold in the VLROP evidence plot as used for the neutral speech. Figure 2.2 gives the indications that the emotional speech utterances of the same speaker and the same text give degraded VLROP estimation as compared to the neutral speech utterances. The issues in correct VLROP estimation from emotion utterances serve as the bottleneck in the automatic independent comparison of vocaltract and excitation components of each VLR with neutral utterances.

3   Analysis of Vocaltract Characteristics in Emotional Speech

Figure 2 plots the variations in the excitation and vocaltract characteristics in various emotion utterances. The excitation characteristics are analysed for each emotion by plotting the glottal waveform obtained from the ground truth electro-glottogram recordings. Extreme right panel of the subplots shows the glottal waveforms corresponding to each emotion under consideration. A significant variations in the number of glottal cycles and the amplitudes are the emotion dependent features reflected in glottal waveforms. These information is already investigated and analysed in the emotion analysis literature by different researchers [1, 8, 10]. For instance, the segment of glottal wave plotted in Figure 2(f) for angry emotion has more glottal cycles as compared to that of the neutral glottal wave in Figure 2(c). Similarly, the amplitude of the glottal pulses in the Figure 2(f) also significantly lower as compared to that of the neutral case. The focus of the current work is to analyze the vocaltract characteristics of different emotions with respect to the neutral emotions. The vocaltract characteristics are analysed by plotting the LP spectrum of the speech frames taken around VLROPs of the emotional syllables. The locations of the formants are represented by frequencies corresponding to the spectral peaks in the LP spectrum. From the Figure 2((b), (e), (h), (k) & (n)) formants show a significant variations in the LP spectrums of different emotions. For instance, all formants in the spectrum of angry emotions show a right side shift as compared to the formants in the neutral LP spectrum shown in Figure Figure 2(b). Similarly, the LP spectrums obtained from VLRs of each emotion show different formant variations as compared to the neutral case.

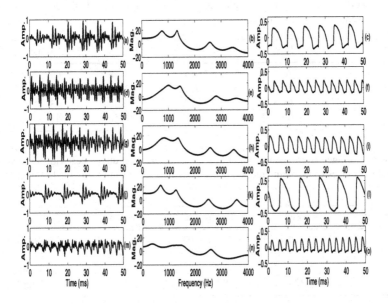

Fig. 2 Plot showing the variations in the vocaltract and excitation characteristics for different emotions. The speech waveform, LP spectrum showing the vocaltract characteristics and glottal pulse characteristics correspond to ((a)-(c)) neutral , ((d)-(f)) anger, ((g)-(i)) happy, ((j)-(l)) boredom and ((m)-(o)) fear speech segments. The speech emotion segments are of the same speaker, same text and the same syllable obtained from the German emotional speech database.

4   Experimental Results

The German emotion speech database is used for the experimental study presented in this paper. Apart from neutral utterances of the database, the emotions like Anger, Happy, Boredom and Fear are selected for the emotion analysis. Each utterance in the database has simultaneous speech and EGG recordings. The entire database consists of 7 emotions of 10 texts and 10 speakers. Also, the label files which contain the syllable boundary markings are present in the database. Each utterance which is originally sampled at 16 kHz is resampled to 8 kHz for the experiments presented in this paper. For the LP residual computation, each speech signal is divided into frames at frame size of 20 ms and a frame shift of 10 ms. The VLROPs in the neutral utterances are estimated initially for performance evaluation of the LP residual-HE based VLROP estimation for emotion speech utterances. The performance of the VLROP estimation is evaluated based on the measures like VLROP identification rate (IDR), miss rate (IR) and false alarm rate (FAR). A genuine VLROP is identified only when the syllable, in the emotion utterance corresponding to the neutral speech reference VLROP of the same neutral syllable, has an estimated VLROP present. More than one estimated VLROPs, in the emotional syllable corresponding to the

Table 1 Performance Comparison of VLROP detection with respect to the neutral utterances of German emotion speech database

Emotion	IDR	MR	FAR	No. Utt. pairs
Anger	76.22%	17.68%	6.29%	69
Happy	74.09%	18.97%	7.37%	48
Boredom	68.57 %	17.01%	15%	53
Fear	69.33%	20.38%	10.3%	41

Table 2 Average difference in formant locations of emotion utterances with respect to the neutral utterances of German emotion speech database. The values provided in the table are in Hz.

Formant	Anger	Happy	Boredom	Fear
F_1	39.64±71.23	71.70±103.73	-44 ± 51.36	-20.78±76.35
F_2	81.28±108.94	39.75±74.33	-62.9±72.35	-24.42±84.85
F_3	54.76±89.25	43.75±67.75	-33.8±119.50	19.21±91.35
F_4	44.32±191.83	48.58± 179.60	-30.05±73.43	36.78±50.98

reference VLROPs of neutral syllables, add to the false estimation. If no estimated VLROPs in the emotion syllables are found in the corresponding syllable then missing count will be incremented. Since, the manual syllable boundary labels available in the database can never be accurate, a tolerance of 10 ms on either sides of the syllable boundaries of emotion utterances are considered for searching the estimated VLROPs for the performance evaluation. Table 1 presents the average performance of VLROP estimation obtained by the pair-wise comparison of neutral utterances with each of the emotion utterances in the German emotion speech database. As per the Table 1, the VLROPs estimation performance in each emotion is far from the performance obtained from that of the neutral speech utterances. As many of the emotion utterances in the database have no corresponding neutral utterances, a varied number of neutral-emotion utterance pairs are used for performance evaluation in each emotion case.

Table 2 provides the average differences and standard deviations of the formant locations in emotion speech utterances with respect to that of the neutral utterances. The formants are computed using the LP analysis of speech frames taken around VLROPs of the neutral and emotion speech utterances. Even though, there are many missed and spurious VLROPs present in each emotion utterance, those VLROPs in the neutral syllables which has the genuine correctly identified VLROPs in the corresponding emotive syllables are only selected for the comparative formant analysis. From the Table 2, happy emotion showed a maximum right shift in F_1 and F_4 formants. Similarly, angry emotion showed highest right shift in F_2 and F_4 formants as compared to neutral speech utterances. Apart from right shift in formant locations, left shift is brought in all four formants (F_1, F_2, F_3, F_4) by boredom emotion and in case of fear, left shift is showed in F_1 and F_2. The formants for generating the Table 2 are estimated by picking the peaks of the negative difference of the LP phase spectrum as proposed by Yegnanarayana in [19].

5 Summary and Conclusions

The work in this paper presents preliminary experiments on the analysis of formants in the vowel-like regions in various emotions. As formants have steady structure in the vowel and other vowel like regions, it is essential to detect the VLRs in speech using existing VLROP detection algorithm. The paper addressed issues in the detection VLROPs from emotion speech utterances as compared to that of the neutral or clean speech cases. From the preliminary analysis of LP spectrum of different emotion syllables, it is indeed convincing that the gross level formant characteristics vary in accordance with emotions even under the conditions of the same speaker, same text and the same syllable. However, the degradation in the VLROP detection in emotion speech utterances forms bottleneck in the quantitative analysis of formants in a large database of emotion speech utterances. The reason is missed or falsified VLROPs of emotion speech signals prevent the one on one analysis of VLRS of emotion utterances with that of the neutral speech utterances. The formant analysis performed in this work is by manually comparing the VLRs in the neutral speech with the corresponding correctly estimated VLRs of each emotion speech cases.

The future work should focus on devising a new robust VLROP detection algorithm for emotion speech cases which helps us to quantitatively analyze a large database of emotion speech with least manual effort. The effect other formant parameters also have to be explored as future work as an extension to the present work.

Acknowledgments The work presented in this paper is supported by the DST fast track sponsored project titled, "Analysis, Processing and Synthesis of Emotions in Speech".

References

1. Prasanna, S.R.M., Govind, D., Rao, K.S., Yenanarayana, B.: Fast prosody modification using instants of significant excitation. In: Proc. Speech Prosody, May 2010
2. Govind, D., Prasanna, S.R.M., Yegnanarayana, B.: Neutral to target emotion conversion using source and suprasegmental information. In: Proc. INTERSPEECH 2011, August 2011
3. Cabral, J.P., Oliveira, L.C.: Pitch-synchronous time-scaling for prosodic and voice quality transformations. In: Proc. INTERSPEECH (2006)
4. Barra-Chicote, R., Yamagishi, J., King, S., Montero, J.M., Macias-Guarasa, J.: Analysis of statistical parametric and unit selection speech synthesis systems applied to emotional speech. Speech Commun. **52**(5), 394–404 (2010)
5. Fernandez, R., Ramabhadran, B.: Automatic exploration of corpus specificproperties for expressive text-to-speech: a case study in emphasis. In: Proc. ISCA Workshop on Speech Synthesis, pp. 34–39 (2007)

6. Pitrelli, J.F., Bakis, R., Eide, E.M., Fernandez, R., Hamza, W., Picheny, M.A.: The IBM expressive text to speech synthesis system for American English. IEEE Trans. Audio, Speech and Language Process. **14**, 1099–1109 (2006)
7. Hofer, G., Richmond, K., Clark, R.: Informed blending of databases for emotional speech synthesis. In: Proc. INTERSPEECH (2005)
8. Cabral, J.P., Oliveira, L.C.: Emo voice: a system to generate emotions in speech. In: Proc. INTERSPEECH, pp. 1798–1801 (2006)
9. Gangamohan, P., Mittal, V.K., Yegnanarayana, B.B.: Relative importance of different components of speech contributing to perception of emotion. In: Proc. Speech Prosody, pp. 557–660 (2012)
10. Whiteside, S.P.: Simulated emotions: an acoustic study of voice and perturbation measures. In: Proc. ICSLP, Sydney, Australia, pp. 699–703 (1998)
11. Murray, I.R., Arnott, J.L.: Towards the simulation of emotion in synthetic speech: A review of the literature on human vocal emotion. J. Acoust. Soc. Am. **93**, 1097–1108 (1993)
12. Hashizawa, Y., Hamzah, S.T.M.D., Ohyama, G.: On the differences in prosodic features of emotional expressions in japanese speech according to the degree of the emotion. In: Proc. Speech Prosody, pp. 655–658 (2004)
13. Scherer, K.R.: Vocal affect expressions: A review and a model for future research. Psychol. Bull. **99**, 143–165 (1986)
14. Erickson, D., Schochi, T., Menezes, C., Kawahara, H., Sakakibara, K.-I.: Some non-f0 cues to emotional speech: an experiment with morphing. In: Proc. Speech Prosody, pp. 677–680 (2008)
15. Prasanna, S.R.M., Pradhan, G.: Significance of vowel-like regions for speaker verification under degraded conditions. IEEE Trans. on Speech, Audio and Lang. Process. **19**(8), 2552–2565 (2011)
16. Prasanna, S.R.M., Yegananarayana, B.: Detection of vowel onset point events using excitation information. In: Proc. Interspeech 2005 (2005)
17. Prasanna, S.R.M., Sandeepreddy, B.V., Krishnamoorthy, P.: Vowel onset point detection using source, spectral peaks and modulation spectrum energies. IEEE Trans. Audio, Speech and Lang. Process. **17**(4), 556–565 (2009)
18. Burkhardt, F., Paeschke, A., Rolfes, M., Sendlemeier, W., Weiss, B.: A database of german emotional speech. In: Proc. INTERSPEECH, pp. 1517–1520 (2005)
19. Yegnanarayana, B.: Formant extraction from linear-predictions pectra. J. Acost. Soc. Am. **63**(5), 1638–1641 (1978)

Improved Phone Recognition Using Excitation Source Features

P.M. Hisham, D. Pravena, Y. Pardhu, V. Gokul, B. Abhitej and D. Govind

Abstract Phone recognizers serve as the preprocessing unit for speech recognition systems and phonetic engines. Even though, most of the state of the art speech recognition achieve relatively better accuracy at the sentence level, the phone level recognition performance falls way below the sentence level performance. The increased recognition rates at the sentence levels are achieved with help of refined language models used for the language under consideration. Therefore, the objective of the present work is to improve the phoneme level accuracy of the hidden markov model(HMM) based acoustic phone models by combining excitation source features with the conventional mel frequency cepstral coefficients (MFCC) for American English. TIMIT and CMU Arctic database, is used for the experiments in the present work. The average spectral energy around the zero-frequency region of each frame is used as the excitation source feature to combine with the 13 MFCC features. The effectiveness of the phoneme recognition is confirmed by a 0.5% increase in the phone recognition accuracy against the state of the art HMM-GMM acoustic models with MFCC features.

1 Introduction

The process of converting spoken waveform into text is known as speech recognition [1, 2]. The speech recognition at the sentence level is obtained from the

P.M. Hisham(✉) · D. Pravena · D. Govind
Center for Excellence in Computational Engineering and Networking,
Amrita Vishwa Vidyapeetham(University), Coimbatore 641112, Tamilnadu, India
e-mail: {hishamthangalpms,d.pravena,govinddmenon}@gmail.com

Y. Pardhu · V. Gokul · B. Abhitej
Department of Computer Science and Engineering, Amrita Vishwa Vidyapeetham,
Amritapuri, Kollam 690525, Kerala, India
e-mail: {abhipardhu,vutukuri.gokul,abhitej1446}@gmail.com

recognized words which in turn obtained by recognizing basic sound units called phonemes. The words are obtained by combining the recognized phonemes using a pronunciation dictionary. A final sentence level speech recognition preceded by word and phoneme recognitions. A lower sentence level accuracy is normally expected against the underlying phoneme and word recognitions. However, the recognition errors at the phoneme level is overridden by using an appropriate language model (bi-gram or tri-gram models). For instance, the state of the art recognition rate at the sentence level and phoneme level of a popular acoustic-phonetic database called TIMIT, is about 99.08% and 58.3%, respectively [3]. It is always challenging to improve the performance of the phone recognition. The phoneme recognition rate can be improved in two ways. One way is use robust statistical models and there by get the improved recognition of data. The hybrid models, by combining different statistical models, found to give better recognition rates than obtained using a single type of statistical model. For instance , the hybrid HMM-SVM and HMM-DNN (deep neural networks) models give better phoneme recognition accuracies as compared to that obtained from the conventional HMM model [4, 5, 6]. The second way is to improve the phoneme level recognition rates by finding better features and combining them with the convention MFCC features. Different works in the literature are reported by combing the vocaltract and excitation features independently towards improving the recognition rates at the phoneme level [2, 7].

The present work aims to improve the phoneme level recognition rate at the feature extraction level. The average spectral energy around the zero frequency regions of each speech frame is used as the excitation feature which is combined with conventional 13 MFCC coefficients to improve the performance of the state of the art HMM based phoneme recognizers. As the spectral information within 0-300Hz of the speech spectrum either entirely or mostly due to excitation source, the average spectral energy within the pitch frequency is computed as additional feature which represents the excitation source part of the speech. The performance of the phoneme recognizer is evaluated by building HMM phone models for each of the phonemes in English using the 14 dimensional feature vectors and their velocity ($\triangle$) and acceleration coefficients ($\triangle\triangle$). The present work is restricted only to the performance evaluation of context independent HMM based mono phone models.

The organization of the work is as follows: Section 2, describes the development of conventional phone recognizer. The proposed zero frequency filtering based excitation source spectral energy measurement is given in Section 3. Section 4 provide the experimental set up and results. Section 5 summarize the work with conclusion and future scope for the present work.

2   Development of HMM Based Phone Recognizers

A three state HMM model(with one initial, three active and one final states) is built for each phoneme. Since the work presented in this paper is for American English, there are a total of 40 phones are used for the language. The speech signal is divided into frames of size 20 ms with a shift of 10 ms. After multiplying with a hamming window, 13 MFCC features are extracted from each speech frame. The velocity (Δ) and acceleration coefficients ($\Delta\Delta$) are derived from each 13 dimensional feature vector to get a 39 dimensional feature vector. The embedded training is performed using Baum-Welsch re-estimation algorithm to estimate the HMM parameters for each phoneme. During the training, the distribution of the 39 dimensional observation symbols is modeled as 39 dimensional continuous density gaussian mixture model (GMM) with 32 component weights for each state of the HMM. The recognition of phonemes are performed by evaluating the 39 dimensional frames against each of the HMM phone models. The frame which give maximum probability value will be assigned with corresponding phone class. Figure 1 shows the schematic diagram of HMM acoustic model building and recognition of phonemes.

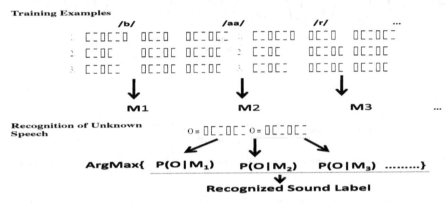

Fig. 1 The schematic diagram of HMM based phoneme recognition

3 Zero Frequency Filtering Based Excitation Source Spectral Energy Measurement

As the effect of vocaltract responses starts from 300Hz onwards of the speech spectrum, to get the the low frequency impulse like characteristics related to excitation source the zero frequency filtering of speech is performed [8, 9] [10]. Here, the speech signal is subjected to two times filtering through zero frequency resonators[1] connected in cascade and subsequent local mean subtraction of the resultant filtered output. The resulting signal is called as zero frequency filtered signal. The important excitation source information like strength excitation, instantaneous F_0, etc., are estimated from zero frequency filtered signal. The zero frequency filtering is performed by the following steps in the spectral domain:

- The spectrum of differenced speech signal, $X(z)$ is obtained; $X(z) = S(z).(1 - z^{-1})$

---

[1] The resonators whose central frequency is located at 0 Hz is referred to as zero frequency resonators

- The zero frequency resonator output spectrum, $Y(z)$ is obtained zero frequency filtering using a cascade of two second order resonators; $Y(z) = \frac{X(z)}{1-4z^{-1}+6z^{-2}-4z^{-3}+z^{-4}}$

- The trend in the output, is removed by the local mean subtraction of $Y(z)$;

$$\hat{Y}(z) = Y(z).(1 - \sum_{m-N}^{N} kz^{-m})$$ Where $k = 1/(2N + 1)$ and N is length of the

window used for local mean subtraction. The $\hat{Y}(z)$ is referred to as zero frequency filtered signal.

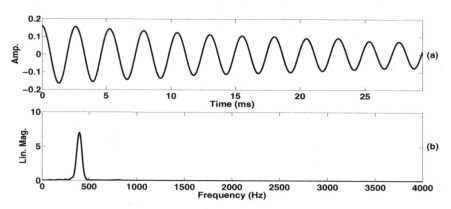

Fig. 2   (a) A segment of zero frequency filtered signal and (b) its linear magnitude spectrum.

The Figure 2(a) plots a segment zero frequency filtered signal corresponding to the voiced speech and Figure 2(b) plots the corresponding linear magnitude spectrum. The frequency corresponding the peak spectral magnitude represent the average F_0 of the voiced segment. In the present work, the average spectral magnitude from 0-F_0 Hz is computed as the additional feature value. To match the dynamic range of the MFCC coefficients, the log of the measure average spectral magnitude is taken. The measured excitation feature is combined along with other 13 MFCC features obtained for the same speech frame prior to the computation of $\triangle$ and $\triangle$ $\triangle$ coefficients. Finally, 42 dimensional feature vector is obtained for training the HMM phone models.

4   Experimental Results

The acoustic-phonetic TIMIT database is used for the performance evaluation of the proposed features for phoneme recognition [11]. The database is developed for American English speech recognition for 8 different US accents. The train set of the database contains 462 speakers and 10 utterances spoken by each speaker. The evaluation is performed with the test set of database which contains 168 speakers with 10 utterances per speaker. All speech files are down sampled to 8 kHz before the feature extraction. The speech signals are divided into 20 ms frames with a frame shift of 10 ms for MFCC feature extraction. Table 1 shows the performance comparison

Table 1 Performance evaluation of phoneme recognition for proposed features in TIMIT database

Database	Conventional Phoneme Recognition	Proposed
TIMIT	58.3%	58.8%

Table 2 Performance evaluation of phoneme recognition for proposed features in CMU Arctic database

CMU Arctic	Conventional Phoneme Recognition	Proposed
AWB	75.32%	76.24%
RMS	78.36%	78.91%
BDL	77.02%	77.06%
CLB	79.60%	79.88%
JMK	72.16%	72.87%
KSP	72.39%	73.25%
SLT	75.06%	75.36%

of traditional phoneme recognizer with HMM-GMM acoustic models with MFCC features and the proposed system with acoustic models built using the feature vectors having average spectral energy around the zero frequency band combined with MFCC features. Table 1 shows a 0.5% improvement in the phone recognition rate for the HMM-GMM acoustic models built using the proposed features. The improvement in the recognition rate is clearly due to the presence of additional phonemic information present in the excitation source component of the speech signals.

The other database we used for evaluation is CMU Arctic database which is mainly developed for Speech synthesis. Here, we used for the purpose of phoneme recognition. CMU Arctic database consists of 7 voice corpus with 1132 phonetically balanced utterances [12]. The voices in the database are recorded at the sampling rate of 32KHz(BDL, SLT, JMK) and 16KHz(KSP, AWB,CLB,RMS). The samples which are recorded at a rate of 32KHz are downsampled to 16KHz before the feature extraction . Phonetic Acoustic models are built individually by taking 1000 utterances from each speaker and remaining 132 utterances are used for testing. Table 2 shows the performance comparison of conventional and proposed method of phonetic acoustic models generated by CMU Arctic database.

Table 2 shows the improved performace of the phonetic acoustic models which reinforces the result of Table 1.

5   Summary and Conclusions

The present work explores the significance of excitation source information for phoneme recognition. The paper proposed the average spectral energy around the zero frequency band in each speech frame as a new feature for phoneme recognition. The improved phoneme recognition rate is obtained when the proposed excitation source feature is combined with the traditional MFCC features. The direct conclusion from the paper is that the excitation source of speech signals carries significant phonemic information which can be used to improve the speech recognition performance. The experiments conducted in this paper open up the need for further exploring new features of excitation source which best represents the phonemic information present in the speech signals.

Acknowledgments The works done in this paper are part of the DST sponsored fast track project titled, "Analysis, processing and synthesis of emotions in speech".

References

1. Sreejith, A., Mary, L., Riyas, K.S., Joseph, A., Augustine, A.: Automatic prosodic labeling and broad class phonetic engine for malayalam. In: Proc. Int. Conf. Control, Communication and Computing (ICCC) (2013)
2. Ghahremani, P., BabaAli, B., Povey, D., Reidhammer, K., Trmal, J., Khudanpur, S.: A pitch extraction algorithm tuned for automatic speech recognition. In: Proc. ICASSP 2014 (2014)
3. Hidden Markov Model Toolkit (HTK) Book, University of Cambridge (2003)
4. Kruger, S.E., Schaffoner, M., Katz, M., Andelic, E., Wendemuth, A.: Using support vector machines in a hmm based speech recognition system. In: Proc. SPECOM (2005)
5. Stadermann, J., Rigoll, G.: A hybrid svm/hmm acoustic modeling approach to automatic speech recognition. In: INTERSPEECH (2004)
6. Dahl, G.E., Yu, D., Dend, L., Acero, A.: Context-dependent pre-trained deep neural networks for large-vocabulary speech recognition. IEEE Trans. Audio, Speech and Lang. Process. **20**(1), 31–41 (2012)
7. Deekshitha, G., Mary, L.: Prosodically guided phonetic engine. In: Proc. IEEE International Conference on Signal Process., Informatics Commun. and Energy Sys. (2015)
8. Murty, K.S.R., Yegnanarayana, B.: Epoch extraction from speech signals. IEEE Trans. Audio, Speech and Language Process. **16**(8), 1602–1614 (2008)
9. Govind, D., Prasana, S.R.M., Yegnanarayana, B.: Significance of glottal activity detection for duration modification. In: Proc. Speech Prosody (2012)
10. Murty, K.S.R., Yegnanarayana, B.: Characterization of glottal activity from speech signals. IEEE Signal Processing Letters **16**(6), 469–472 (2009)
11. Garafolo, J., et al.: TIMIT: Acoustic-Phonetic Continuous Speech Corpus LDC93S1. Linguistic Data Consortium (1993)
12. Kominek, J., Black, A.: CMU-Arctic speech databases. In: 5th ISCA Speech Synthesis Workshop, Pittsburgh, PA, pp. 223–224 (2004)

Online Character Recognition in Multi-lingual Framework

V. Vidya, T.R. Indhu and V.K. Bhadran

Abstract Online character recognition research area becomes very prominent due to widespread popularity of handheld devices. Common people like to interact with each other on their own native language through their handheld device. Multi-lingual country like India requires a common framework to incorporate all languages. Here we propose a unified approach to recognize characters in the different languages in a single multi-lingual framework. Stroke rule generation is the first step in the character recognition. Statistical stroke model approach in which probabilities from stroke rule along with confidence measurement obtained from the PSFAM (Probabilistic Simplified Fuzzy ARTMAP) classifier is used for recognizing character. We had tested with Hindi, Malayalam, Tamil and Urdu. Accuracy varies from 85-95% for different languages depending on the number of training samples used.

1   Introduction

Development of touch screen technology has allowed the drastic growth in many different kinds of acquisition devices such as PDA, electronic tablets etc. These gadgets record pen-tip movements which result in an on-line signal which is represented as stroke trajectory in x-y coordinates with time. These technological improvements accelerated extensive activities on the online handwriting recognition problem. Emerging country like India requires these devices to reach to the hands of common people. Common people are interested in communicating with their own language. Multi-lingual country like India has several non-official languages in addition to 22 official languages. In this paper we propose a system to recognize the characters in a multi-lingual framework. Currently we have incorporated Hindi, Malayalam, Tamil and Urdu language. Other languages can also be easily integrated into this framework.

V. Vidya(✉) · T.R. Indhu · V.K. Bhadran
Center for Development and Advance Computing, Trivandrum, Kerala, India
e-mail: {vidyav,indhu,bhadran}@cdac.in

© Springer International Publishing Switzerland 2016
S. Berretti et al. (eds.), *Intelligent Systems Technologies and Applications*,
Advances in Intelligent Systems and Computing 384,
DOI: 10.1007/978-3-319-23036-8_14

Various researches are going on in Indian scripts also. H. Swethalakshmi et al. [1] proposes a character recognition system in Devanagari and Telugu using support vector machine. HMM based lexicon driven and lexicon free word recognition for Devanagari and Tamil is described in [2]. Neural network based framework to classify online Devanagari characters using discrete cosine transform features were proposed in [3]. Two dimensional Principal Component Analysis (2DPCA) method proposed by Sundaram S and A G Ramakrishnan [4], in which polynomial fits and quartiles features in addition to conventional features derived for each sample point of the Tamil character to solve the problem of online character recognition. Rituraj K.A. and G. Ramakrishnan [5] presented a fractal coding method to recognize online handwritten Tamil characters. In Malayalam script Primekumar K.P Sumam [6] presented wavelet transform and SFAM based recognition system. Freeman code based online handwritten character recognition for Malayalam using back propagation neural networks were proposed by Amritha et al. [7].Two novel approaches for Urdu script were described in [8,9].

In this paper first section explains the system architecture of proposed system. Functionalities of each module are elaborated in section 3. Multi-stroke character recognition is most challenging task due to variability in number of stroke, stroke order and direction. Strokes can vary across the characters and within the characters [2]. Here we use stroke probability based model for recognizing the multi-stroke characters. Results are discussed in section 4. This application can be used in government offices for form filling purposes such censes health survey etc instead of usual paper pen method. Novelist, poets, content writers etc will be benefited with these applications.

2   System Architecture

Data acquisition and multilingual framework are the main two components in the online character recognition system which is depicted in Fig. 1. Data acquisition is concentrated on handling user input signals and responses and standardize the input and pass them to recognizer. Multilingual framework allows users of different languages to share the common interface to write the text without keyboard. Visual characteristics of each script are analyzed in this framework. It consists of recognizer and a knowledge base for each language. The knowledge base provides information to the recognizer for identifying characters.

Application provides a user friendly GUI where user can write the text in the language supported. Recognized text should be displayed on it. The data capture module picks up pen tip movements in the form of x y coordinates as well as pen up/pen down switching. Preprocessing should remove various irregularities and writer specific variation present in the input strokes. A stroke is defined as the trajectory traversed by the pen tip from the instant when it makes contact with the writing surface to the earliest moment when the contact is broken. Feature extraction extracts characteristics representation of features from the preprocessed stroke.

The framework will allow the user to select target language and write on the user interface. With the aid of knowledge base recognizer will recognizes the characters and display it on the application GUI. The stroke classifier identifies the strokes with the help of the trained database for which language is selected. Most of the script contains multi-stroke character which is formed by two or more strokes. Stroke concatenation rules from knowledge base groups the strokes to form a valid character and then it is mapped to character code. In post processing, language dependent lexicon and rules are used to obtain the N-best list of the words for suggestions.

3 Functionalities of Character Recognition System

3.1 Preprocessing

Pre-processing step is to remove the variability in the strokes due to the writing styles as well as due to the noise in the collected data. Dehooking, duplicate point removal, normalization, interpolation, resampling and smoothing are performed over data points to reduce the distortion in the handwritten data without significant loss of information from the samples [2, 10].

Identify the location of each stroke with respect to base stroke by analyzing the bounding box of each stroke. Positions are labeled as left, right, top, bottom and middle. Exception position is also included to indicate its overlap with two positions as shown in Fig. 2. Identification of delayed strokes is the important factor to recognize the word properly. Delayed strokes are identified at the same time user write the strokes on the application and reordered it accordingly [11]. In case of Hindi and Punjabi the headline or shirorekha is identified and removed as per [2].

3.2 Feature Extraction

Global as well as local features are extracted from preprocessed data for efficient data representation. Normalized x, y coordinate, direction, curvature, vicinity aspect, curliness and slope, loop, cusp and stroke length features are extracted as mentioned in NPen++ recognizer [10]. Each handwritten stroke is recognized based on these features which are almost unique to that particular stroke/character.

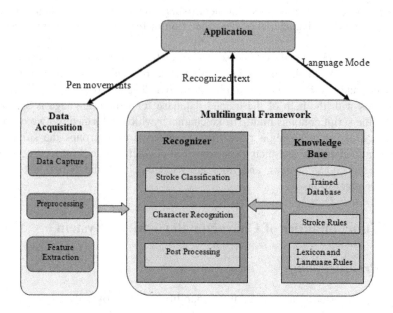

Fig. 1 System Architecture

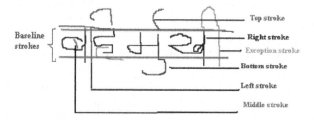

Fig. 2 Stroke positions

3.3   Stroke Classification

Feature vectors are given to Simplified Fuzzy ARTMAP for training. ARTMAP is a class of Neural Network architectures that perform incremental supervised learning of recognition categories and multidimensional maps in response to input vectors presented in arbitrary order [12, 13, 14]. Fuzzy ARTMAP is a combination of fuzzy logic and adaptive resonance theory (ART) network by incorporating computations of fuzzy subsethood to ART category choice, resonance and learning.

In the classification phase, the test vector is given to trained SFAM network. According to the sorted activation values and match function generate SFAM predications.   Determine the confidence in the classification decision by estimating the posterior probability of predicated class by Bayes classifier as given in [15, 16]. Thus the probabilistic SFAM (PSFAM) is used in the predication phase. SFAM suggestions with this confidence measure are used in next section to recognize characters.

3.4 Character Recognition

Most of Indic script contains multi-stroke characters. Stroke rule probability based character recognition is proposed here. All probable sequence of a character is called a stroke concatenation rule, simply stroke rule of that character. Data collection and data analysis is mandatory prerequisite for recognizing characters. At data collection phase, collect the handwritten ink data from different categories of informants who vary in their age, gender, profession etc. These handwritten data is annotated using our data analysis tool and stored. This analysis tool helps to identify the stroke sequence to form a character. From the analyzed data it retrieves the stroke rules for every character in the particular language. Strokes sequence obtained from different writers for the Hindi character shra(श्र) and the stroke rule is shown in Fig. 3.

Some characters share same strokes like in Hindi stroke उ is common for following characters उ ऊ अ आ ओ औ. Some of the strokes does not occur in initial position, like stroke - , which is part of character अ,झ, स. To handle such cases calculate the transition probability between strokes. Each stroke rule is prefixed and suffixed with 301 and 302 which indicate the start and end of the rule. Transitional probability is a conditional probability statistic that measures the prophecy of neighboring strokes and is defined by

$$P(S_n / S_{n-1}) = \frac{C(S_{n-1}, S_n)}{C(S_{n-1})} \tag{1}$$

Where, $C(S_{n-1}, S_n)$ represents the count of stroke S_{n-1} followed by stroke S_n. $C(S_{n-1})$ represents the count of stroke S_{n-1} in the stroke rule.

Using stroke rule probability and output from the classifier produce the multi-stage graph and compute the optimal path among them by dynamic programming. Multi-stage graph is a weighted directed graph G =<V,E> in which vertices are partitioned into k >= 2 disjoint set. K is number strokes written at time t for a character i.e. 0 <= k <= K+1. V_0 and V_{K+1} are source and sink node, its cost is assigned to 1. So each stage represents strokes written in time sequence. The node V_j^k is the jth node in k^{th} stage. j is the number of suggestions from classifier. The weight of the node Vw_j^k is confidence measure. The weight of edge (u, v),Ew_{ij}^k represents the transition probability from stroke u to stroke v obtained from stroke rule. Cost function is defined by product of node and edge weight on the path. Using forward reasoning technique we find most probable path for multi-stage graph. Also keep subset path for generating suggestions for that characters.

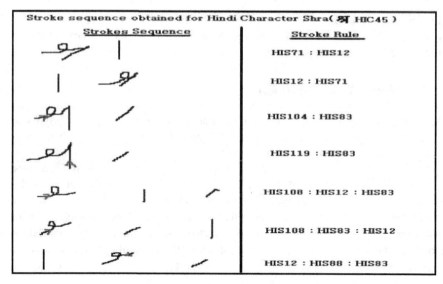

Fig. 3 Strokes sequence and stroke rule obtained for Hindi character श्र after analyzing the handwritten strokes from different writes. Stroke number, order and direction of this multi-stroke character vary from user to user.

3.5 Post Processing

Post processing helps to increase the accuracy of word recognition by incorporating language specific rules and lexicon; language rules such as two vowel signs must not be adjacent. In case of Malayalam, vowels should come only at first position of a word. These types of rules help to improve the system if extended to word recognition level.

4   Results and Discussion

143 characters are selected from Malayalam new script which includes 15 vowels, 36 consonants, 16 diacritics, 71 conjunct consonants and 5 chillus. Tamil character set includes 12 vowels, 19 consonants, 4 conjuncts and 9 vowel modifiers, total of 44 characters. Also 39 single stroke consonant vowel (CV) characters are also added like தூ,வி,பு etc. Hindi character set consist of 11 vowels, 22 consonants, 17 diacritics, 4 conjuncts and 31 half consonants. 36 basic characters are included in Urdu.

101 unique strokes are identified in case of Malayalam characters. 1321 stroke samples are trained using SFAM algorithm. For incorporating Tamil into OLCR multilingual framework, trained 1059 samples from 79 unique strokes. After the analysis of Hindi, 73 unique strokes are identified. Minimum of 12 samples of each strokes were arbitrarily chosen for training. 974 samples were given to

SFAM training module after extracting the features. In Urdu 36 basic characters are represented in 27 distinctive strokes and trained with 324 samples. One important characteristics of Urdu is that number of dots and its positions varies from character to character. Compared to other three languages multi-stroke character ratio is more in Hindi. Number of characters and strokes used in four languages is given in Table 1. Stroke classification and character recognition accuracy are listed in Table 2 and Table 3 respectively.

Table 1 No of characters and strokes used in four Indian languages.

Language	Characters	Unique Strokes
Hindi	85	73
Malayalam	143	101
Tamil	127	79
Urdu	36	27

Table 2 Stroke classification accuracy

Language	No. of strokes trained	No. of strokes tested	Accuracy (%)
Hindi	974	705	89.21
Malayalam	1321	1200	96.4
Tamil	1059	954	92.64
Urdu	324	246	90.11

Table 3 Character Recognition accuracy

Language	No. of character x No of writers	Accuracy (%)
Hindi	85 x 8	85.28
Malayalam	143 x 14	94.63
Tamil	127x7	89.81
Urdu	36x10	86.79

Small set of stroke samples are used for testing as listed in Table 2. Stroke classification accuracy is increased by training more samples. Character accuracy decreased due to occurrence of same strokes at different positions. For example in Hindi, stroke \ can be part of the character के,ट,ष in different positions such as top, bottom and middle respectively. Also in Urdu the characters vary in their position and number of dots such as ح'خ'ج'چ. Thus character accuracy can be increased by adding positional information of stroke with respect to base character in stroke rule. For example first two stroke sequence of stroke rule of Hindi character shra (श्र) mentioned in Fig. 3 is changed to HIS71R: HIS12R and HIS12R:HIS71L where R and L indicate right and left position. This help to remove some of the edges in multi-stage graph by setting value to 0 in transition probability. Thereby character recognition accuracy is increased by 2- 3 %. Recognition output of Hindi is shown in Fig. 4 and suggestion generation shown in Fig. 5.

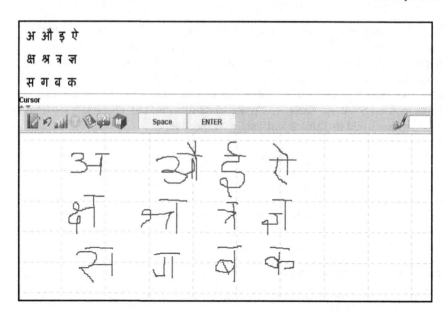

Fig. 4 Character recognition output of Hindi

Fig. 5 Suggestion generation for a character

5   Conclusions

Here a stroke rule probability and PSFAM based online character recognition method in a multilingual framework is proposed. Stroke rule and trained network vary in architecture according to switching of language to another. We have tested this approach in 4 Indian languages such as Malayalam, Hindi, Tamil and Urdu and accuracy obtained are 95.2, 87.3, 90.6 and 88.8 respectively. The positional information helps to increase accuracy by 2 to 3% compared to basic stroke rule method. Accuracy can be increased more if confusion pairs or similar looking strokes in language are handled properly.

In future we want to incorporate all Indian language into this framework. Only requirement is collection and analysis of handwritten data and generate stroke rule and train the strokes in SFAM. This method should be extended to word recognition by incorporating language rules and lexicons of corresponding language.

Acknowledgments Authors gratefully acknowledge the help received in collecting and annotating the handwritten data. Special thanks for all who were willing to spend their quality time in the data analysis and testing.

References

1. Swethalakshmi, H., Jayaram, A., Chakraborty, V.S., Sekhar, C.C.: Online handwritten character recognition of devanagari and telugu characters using support vector machines. In: Proc. 10th IWFHR, pp. 367–372
2. Bharath, A., Madhvanath, S.: HMM-Based Lexicon-Driven and Lexicon-Free Word Recognition for Online Handwritten Indic Scripts. IEEE Transactions on Pattern Analysis and Machine Intelligence **34**(4) (2012)
3. Kubatur, S., Sid-Ahmed, M., Ahmadi, M.: A neural network approach to online devanagari handwritten character recognition. In: International Conference on High Performance Computing and Simulation (HPCS) (2012)
4. Sundaram, S., Ramakrishnan, A.G.: Two dimensional principal component analysis for online tamil character recognition. In: Proc. 11th International Conference on Frontiers in Handwriting Recognition (ICFHR), pp. 88–93 (2008)
5. Kunwar, R., Ramakrishnan, A.G.: Online handwriting recognition of tamil script using fractal geometry. In: International Conference on Document Analysis and Recognition (ICDAR) (2011)
6. Primekumar, K.P., Idiculla, S.M.: On-line malayalam handwritten character recognition using wavelet transform and SFAM. In: 3rd International Conference on Electronics Computer Technology (ICECT) (2011)
7. Sampath, A., Tripti, C., Govindaru, V.: Freeman Code Based Online Handwritten Character Recognition For Malayalam Using Back propagation Neural Networks. Advanced Computing: an International Journal (ACIJ) **3**(4) (2012)
8. Husain, S.A., Sajjad, A., Anwar, F.: Online urdu character recognition system. In: IAPR Conference on Machine Vision Applications (2007)

9. Ullah Khan, K., Haider, I.: Online recognition of multi-stroke handwritten urdu characters. In: International Conference on Image Analysis and Signal Processing (IASP) (2010)
10. Jaeger, S., et al.: Online handwriting recognition: the NPen ++ recognizer. International Journal on Document Analysis and Recognition (IJDAR) 3(3), 169–180 (2001)
11. Vidya, V., Indhu, T.R., Bhadran, V.K.: Reordering of delayed strokes for online hindi word recognition. In: International Conference on Contemporary Computing and Informatics (IC3I) (2014)
12. Carpenter, G.A., et al.: Fuzzy ARTMAP: A Neural Network Architecture for Incremental Supervised Learning of Analog Multidimensional Maps. IEEE Transaction on Neural Networks 3(5) (1992)
13. Vakil-Baghmisheh, M.-T., Paves, N.: A Fast Simplified Fuzzy ARTMAP Network. Neural Processing Letters 17(3), 273–316 (2003)
14. Keyarsalan, M., Montazer, G.H.A., Kazemi, K.: Font-based persian character recognition using simplified fuzzy ARTMAP neural network improved by fuzzy sets and particle swarm optimization. In: IEEE Congress on Evolutionary Computation (2009)
15. Lim, C.P., Harrison, R.F.: Probabilistic fuzzy artmap : an autonomous neural network architecture for bayesian probability estimation. In: Fourth International Conference on Artificial Neural Networks (1995)
16. Jervis, B.W., Garcia, T., Giahnakis, E.P.: Probabilistic simplified fuzzy ARTMAP (PSFAM). In: IEE Proceeding on Science Measurement and Technology, vol. 146(4) (1999)

Personalized Multi-relational Matrix Factorization Model for Predicting Student Performance

Prema Nedungadi and T.K. Smruthy

Abstract Matrix factorization is the most popular approach to solving prediction problems. However, in the recent years multiple relationships amongst the entities have been exploited in order to improvise the state-of-the-art systems leading to a multi relational matrix factorization (MRMF) model. MRMF deals with factorization of multiple relationships existing between the main entities of the target relation and their metadata. A further improvement to MRMF is the Weighted Multi Relational Matrix Factorization (WMRMF) which treats the main relation for the prediction with more importance than the other relations. In this paper, we propose to enhance the prediction accuracy of the existing models by personalizing it based on student knowledge and task difficulty. We enhance the WMRMF model by incorporating the student and task bias for prediction in multi-relational models. Empirically we have shown using over five hundred thousand records from Knowledge Discovery dataset provided by Data Mining and Knowledge Discovery competition that the proposed approach attains a much higher accuracy and lower error(Root Mean Square Error and Mean Absolute Error) compared to the existing models.

1 Introduction

Predicting student performance plays an important role in helping students learn better [7]. By this prediction, teachers may better understand the student knowledge in each domain and thus help the students in deep learning. The problem of predicting performance of the students has been efficiently solved using different matrix

P. Nedungadi
Amrita CREATE, Amrita Vishwa Vidyapeetham, Kollam, Kerala, India
e-mail: prema@amrita.edu

T.K. Smruthy(✉)
Dept. of Computer Science, Amrita Vishwa Vidyapeetham, Kollam, Kerala, India
e-mail: smruthykrishnan@gmail.com

© Springer International Publishing Switzerland 2016
S. Berretti et al. (eds.), *Intelligent Systems Technologies and Applications*,
Advances in Intelligent Systems and Computing 384,
DOI: 10.1007/978-3-319-23036-8_15

factorization models in the past. A simple matrix factorization deals with only one main target relation (such as Student performs task). This relation is represented by a matrix where each cell represents the association between a particular student and a task. The matrix so constructed is factorized to yield two smaller matrices each representing factors of students and tasks[14]. However such a prediction depends on many other relations as well. Thus, this results in a multi relational matrix factorization model where the factors of the target entities are determined by collectively factorizing both the target relation and the dependent relations [6]. Further, weights are assigned to each relation based on their importance in the prediction problem. This results in Weighted Multi Relational Matrix Factorization (WMRMF) [16]. In this paper, we propose a biased WMRMF model which enhances the prediction accuracy of the state-of-the-art systems. For a student performance prediction problem, there are two main bias involved namely the student bias which explains a generic probability of a student in performing tasks and the task bias which explains the degree of difficulty of the task [16]. By considering the biases for prediction in multi relational model, we improve the student performance prediction accuracy drastically. Also, the state-of-the-art systems considers sum of product of latent factors, bias and global average for performance prediction. Empirically we show that considering only the bias and neglecting the global average provides more accurate results than considering both together as done in simple matrix factorization models. Hence global average has been used only for the cold start problems while for the rest, the sum of product of latent factors and biases are used.

2　Related Work

For the student performance prediction problem, many works have been done using classification and regression modeling mainly using Bayesian networks, decision tree [2][7]. There are also many models using knowledge tracing (KT) which works to find four basic parameters namely prior knowledge, learn rate, slip and guess factors using several techniques like Expectation Minimization approach, brute-force approach. [3][8]. Later, models using matrix factorization techniques were proposed which took parameters like slip and guess factors implicitly into account. The very initial model was to consider the relation Student performs tasks as a matrix where each cell value represents the performance of a student for a particular task. The matrix so constructed was factorized into two smaller latent factor matrices and prediction was done by calculating the interaction between the factors [5]. Biased model for single matrix factorization was also developed to take into consideration the user effect and item effect [5]. Both the models prove very effective for sparse matrices but only the target relation is optimized neglecting any dependent relations. Later, prediction problems were solved using multi relational models which exploited the multiple relationships existing between entities and they prove effective in recommender systems [6][11][1]. These models were used explicitly in student performance prediction domain and further, a weighted multi relational matrix factorization model was proposed and shown to be successful in terms of prediction accuracy [14][15][16]. But none of these models considered either student bias or the task bias. Further, an enhancement to matrix factorization approach was tensor factorization [4]. Tensor factorization has been used for student performance prediction which also takes the temporal or the sequential effect into consideration [13]. One of our previous work proposes an enhanced and efficient tensor decomposition technique for student performance prediction [9]. But the relationships that can be considered while modeling as a tensor are limited when compared to exploiting multiple relationships as matrices. Also, while modeling as tensors, as the number of relationships considered increases, it leads to a high space complexity. In our paper, we implement an enhanced weighted multi relational matrix factorization model considering both the student and task bias along with the latent factors that gives promising results in terms of prediction accuracy.

3 Personalized Multi-relational Matrix Factorization Model

Let there be N entities (E1, E2.. EN) e.g. (Student, Task) and M relations (R1, R2.. RM) (performs, requires) existing in a student performance prediction problem. Our aim is to predict unobserved values between two entities from already observed values, such as in a student performance prediction problem, our aim is to predict how well a student can perform a task. For this many models have been proposed. Before we move on to multi relational models, let us review the simple matrix factorization approach. A student-performs-task matrix R is approximated as a product of two smaller matrices W_1 (student) and W_2 (task) where each row contains F latent factors describing that row. Then, w_s and w_i describes the vectors of W_1 and W_2 and their elements are denoted by w_{sf} and w_{if} . Then the performance of student s for a task i can be predicted as:

$$\hat{p} = \sum_{f=1}^{F} w_{sf} w_{if} = w_s w_i^T \tag{1}$$

where $\hat{p}$ is the predicted performance value. w_s and w_i can be learnt by optimizing the objective function :

$$O^{MF} = \sum_{(s,i)\in R} error_{si}^2 + \lambda(\| W_1 + W_2 \|_F^2) \tag{2}$$

where $\| . \|_F^2$ is the Frobenius norm and λ is the regularization term to avoid over fitting, and $error_{si}^2$ is calculated as the difference between the actual performance value and the predicted value for each student-task combination and is given as:

$$error_{si}^2 = ((R)_{si} - w_s w_i^T)^2 \tag{3}$$

where $(R)_{si}$ represents actual performance value of student s for task i, Optimization is done using stochastic gradient descent model. Let us now look at the multi relational models.

3.1 Base Model 1: Multi Relational Matrix Factorization (MRMF)

Different relationships are drawn out from the domain under consideration and the factor matrices of each entity are found by a collective matrix factorization. As discussed earlier, in a system with N entities (E1, E2..EN) and M relations (R1, R2..RM), let $W_n (n \in N)$ be the latent factor matrices of each of the entities with F latent factors. These latent factors describe the entity and are built by considering every relation that the entity is associated with [6][16]. In such a system, the model parameters are learnt using the optimization function:

$$O^{MRMF} = \sum_{r=1}^{M} \sum_{(s,i) \in R_r} ((R_r)_{si} - w_{r1s} w_{r2i}^{T})^2 + \lambda (\sum_{j=1}^{N} \| W_j \|_F^2) \qquad (4)$$

3.2 Base Model 2: Weighted Multi Relational Matrix Factorization (WMRMF)

In the previous model, every relation is given equal weightage [16]. But practically, weight of the relations change for different target relations, such as in a student performance prediction, student performs task is the main relation and hence this relation should be given more weightage. Thus a weight factor (θ) is added to the previous MRMF model and the optimization function is:

$$O^{WMRMF} = \sum_{r=1}^{M} \theta_r \sum_{(s,i) \in R_r} ((R_r)_{si} - w_{r1s} w_{r2i}^{T})^2 + \lambda (\sum_{j=1}^{N} \| W_j \|_F^2) \qquad (5)$$

θ is given a value 1 for the main relation and for the rest of the relation we assign a lower value.

3.3 Proposed Approach 1: Personalized Multi Relational Matrix Factorization Using Bias and Global Average

In this paper, we propose an enhancement to the weighted multi relational model, increasing the accuracy of prediction by considering bias terms and global average for prediction. The student bias explains the probability of a student in performing tasks and the task bias explains the degree of difficulty of the task. Global average (μ) explains the average performance of all students and tasks in the training set considered [13]. The two biases and global average are added along with the dot product of latent factors while predicting performance which is given as:

$$\hat{p} = \mu + b_s + b_t + \sum_{k=1}^{K} w_{sk} h_{ik} \tag{6}$$

Thus the optimization function now becomes:

$$O^{B-WMRMF} = \sum_{r=1}^{M} \theta_r \sum_{(s,i) \in R_r} ((R_r)_{si} - (\mu + b_s + b_i + w_{r1s} w_{r2i}^T))^2 + \lambda (\sum_{j=1}^{N} \| W_j \|_F^2 + b_s^2 + b_i^2) \tag{7}$$

where b_s and b_t are the student and task bias respectively. Bias terms have been introduced in the single matrix factorization models in the past [13]. But to the best of our knowledge, this is the first paper where biases are considered in a multi relational environment. Student bias is calculated as the average of deviation of performance of student s from global average and task bias is calculated as the average of deviation of performance of task i from global average.

3.4 Proposed Approach 2: Personalized Multi Relational Matrix Factorization Using Only Bias

Along with the bias terms, the global average has also been considered for performance prediction in the previous enhancement proposed (Proposed approach 1) and this model has proved to produce higher accuracy when compared to the base models. We also propose a further enhancement to proposed approach 1 by considering only bias along with the latent factor products. Empirically we have shown that neglecting the global average and considering just the bias and the latent factors lead to better prediction accuracy in multi relational factorization. Global average has been considered for cold start problems only. Thus the optimization function for this approach is:

$$O^{B-WMRMF} = \sum_{r=1}^{M} \theta_r \sum_{(s,i) \in R_r} ((R_r)_{si} - (b_s + b_i + w_{r1s} w_{r2i}^T))^2 + \lambda (\sum_{j=1}^{N} \| W_j \|_F^2 + b_s^2 + b_i^2) \tag{8}$$

and the student performance prediction is made as:

$$\hat{p} = b_s + b_t + \sum_{k=1}^{K} w_{sk} h_{ik} \tag{9}$$

A detailed algorithm for this personalized multi relational model (Proposed approach 2) has been given in the coming section. Procedure ENHANCED-WMRMF starts with weight and latent factor initialization. The next set of steps is the iterative stochastic gradient descent process. In each iteration, for a random row of a relation, predicted value is calculated using the dot product of latent factors and the biases. Error is calculated as the difference between the predicted value and the known target value. Using this error, the latent factors and the biases are updated. The iterative process stops when the error between two consecutive iterations reaches below a threshold value(stopping criterion). Once the latent factors and the biases are updated using this algorithm, for any new combination of student-task test data, performance prediction can be made easily using equation 9.

Assuming the algorithm takes n iterations to converge, and as specified in the input section of the algorithm, for a model with N entities, M relations with R as the size of each relation, f number of factors in each latent factor matrix, the time complexity of both the base models(WMRMF and MRMF) and the proposed personalized approach is given as $O(n.M.R.f)$.

Algorithm 1. Personalized Multi relational matrix factorization model.

Input

N : $enitites$

M : $relations$

F : $number\ of\ factors$

θ : $weight$

β : $regularization\ term$

λ : $learn\ rate$

Output

b_s : $student\ bias$

b_t : $task\ bias$

w_j : $latent\ factors\ of\ each\ entity\ j$

procedure ENHANCED- WMRMF($N, M, F, \theta, \beta, \lambda$)

    Initialize weight value θ for each relation.

    Initialize latent factors w_j , ($j \in N$) for each of the N entities.

    while $stopping\ criterion\ met$ **do**

        for $i = 1\ to\ M$ **do**

            for $j = 1\ to\ size(m)$, $(m \in M)$ **do**

                $Pick\ a\ row\ (p, q)\ from\ relation\ m\ in\ random\ with\ target\ value\ p$

                $\hat{p} = b_s + b_t + w_{r1p}w_{r2q}$

                $error_{pq} = p - \hat{p}$

                $w_{r1p} \leftarrow w_{r1p} + \beta * (\theta_m * error_{pq} * w_{r2q} - \lambda w_{r1p})$

                $w_{r2q} \leftarrow w_{r2q} + \beta * (\theta_m * error_{pq} * w_{r1p} - \lambda w_{r2q})$

                $b_s \leftarrow b_s + \beta * (\theta_m * error_{pq} - \lambda b_s)$

                $b_i \leftarrow b_i + \beta * (\theta_m * error_{pq} - \lambda b_i)$

            end for

        end for

    end while

end procedure

4 Experimental Analysis

We implemented the personalized multi relational matrix factorization model on a 4GB RAM, 64 bit Operating System, Intel Core i3 machine. The model implementation is done using Java Version 8. Knowledge Discovery Data Challenge Algebra 2005-2006 dataset has been used. This data set is a click-stream log describing interaction between students and intelligent tutoring system [12]. Initially a preprocessing of the dataset, a one time activity is done using MATLAB 2012b (Version 8.0). In the preprocessing stage, data is read and unique id is assigned to each set of student, task and skill. From the dataset, we take three relations into consideration namely Student-performs-task, task-Requires-Skill and Student-haslearnt-Skill. Average performance of a student for a task is calculated from the dataset and assigned as the target value for Student-performs-task. Target value for Task requires Skill relation is either 1 (requires) or 0 (not requires). Opportunity count which represents the number of chances the student got to learn the skill forms the target value of Student-haslearnt-Skill. This value is implicitly learnt from the dataset. Thus in the pre-processing stage, the relation between entities is represented with id value assigned and corresponding target value is decided. Hence this is the most time consuming phase as compared to the actual training and prediction phase in case of multi relational models and can take hours to complete. Next, we start the training procedure - Iterative updating of the factors and bias term in order to minimize the error between the predicted and the target values as discussed in the algorithm. For experimentation, the error measure is taken as RMSE and MAE. Once we obtain the optimized latent factors and bias term, prediction can be made for any data. The global average value has been used for dealing with cold start problems. Since stochastic gradient descent is used for training, we observe a fast convergence of latent factors and bias term. Further, prediction can be done in real time.

5 Results

Dataset was divided into chunks of different sizes and experimentation was performed for the base models and the proposed approach. As explained in the previous section, Root Mean Square Error, RMSE (fig 1), Mean absolute error, MAE (fig 2) and accuracy (fig 3) were calculated for each dataset chunks considered. Graph plots giving comparison of different models are given below. Below three figures gives comparison of base models (MRMF and WMRMF) and the proposed approaches (WMRMF_bias_avg and WMRMF_bias). We notice a drastic change in accuracy while the bias term was also considered for multi relational model. Also, accuracy of model increase slightly while neglecting the global average term for performance prediction of existing students and tasks.

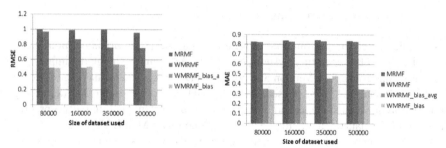

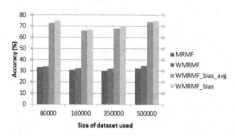

Fig. 1 Student Performance Prediction RMSE

Fig. 2 Student Performance Prediction MAE

Fig. 3 Student Performance Prediction Accuracy

6 Conclusion and Future Work

Matrix factorization models prove very effective for student performance prediction. In the past multiple relationships amongst the entities have been exploited in order to improvise the state-of-the-art systems leading to Multi-Relational Matrix Factorization (MRMF) and Weighted MRMF. In this paper, we propose to enhance the prediction accuracy by developing a personalized Weighted MRMF model. For this two approaches have been proposed. First approach considers both global average and bias terms along with the factors in predicting student performance. Second approach considers only bias with the latent factor products for performance prediction. Through experimental analysis with different dataset sizes from KDD Cup Challenge, we show that both the models achieve higher prediction accuracy and lower RMSE when compared to the base models, although considering only biases and neglecting global average prove to achieve the best results.

Training data is a snapshot taken at a particular time. As new data for prediction comes in, it could be used for online updating of the system to give out better prediction results [10]. Also, considering any other relation other than the ones considered in this paper may lead to better accuracy.

Acknowledgments This work derives direction and inspiration from the Chancellor of Amrita University, Sri Mata Amritanandamayi Devi.We would like to thank Dr.M Ramachandra Kaimal, Head, Department of Computer Science and Dr. Bhadrachalam Chitturi, Associate Professor, Department of Computer Science, Amrita University for their valuable feedback.

References

1. Krohn-Grimberghe, A., Drumond, L., Freudenthaler, C., Schmidt-Thieme, L.: Multi-relational matrix factorization using bayesian personalized ranking for social network data. In: Proceedings of thefifth ACM international conference on Web search and data mining-WSDM 2012

2. Bekele, R., Menzel, W.: A bayesian approach to predict performance of a student (bapps): a case with ethiopian students. In: Proceedings of the International Conference on Artificial Intelligence and Applications, Vienna, Austria, vol. 27, pp. 189–194 (2005)

3. Corbett, A.T., Anderson, J.R.: Knowledge tracing: Modeling the acquisition of procedural knowledge. User Modeling and User-Adapted Interaction **4**, 253–278 (1995)

4. Kolda, T.G., Bader, B.W.: Tensor decompositions and applications. SIAM Review **51**(3), 455–500 (2009)

5. Koren, Y., Bell, R., Volinsky, C.: Matrix factorization techniques for recommender systems. In: IEEE Computer Society Press, vol. 42, pp. 30–37 (2009). 38, 40, 41, 45

6. Lippert, C., Weber, S.H., Huang, Y., Tresp, V., Schubert, M., Kriegel, H.P.: Relation prediction in multi-relational domains using matrix factorization. In: NIPS 2008 Workshop: Structured Input - Structured Output (2008). 56, 57, 60, 61, 64

7. Minaei-Bidgoli, B., Kashy, D.A., Kortemeyer, G., Punch, W.F.: Predicting student performance: an application of data mining methods with an educational web-based system. In: The 33rd IEEE Conference on Frontiers in Education (FIE 2003), pp. 13–18 (2003). 27

8. Nedungadi, P., Remya, M.S.: Predicting students' performance on intelligent tutoring system personalized clustered BKT (PC-BKT) model. In: Frontiers in Education Conference (FIE). IEEE (2014)

9. Nedungadi, P., Smruthy, T.K.: Enhanced higher order orthogonal iteration algorithm for student performance prediction. In: International Conference on Computer and Communication Technologies (2015). AISC Springer

10. Rendle, S., Schmidt-Thieme, L.: Online-updating regularized kernel matrix factorization models for large-scale recommender systems. In: Proceedings of the ACM conference on Recommender Systems (RecSys 2008), pp. 251–258. ACM, New York (2008). 40

11. Singh, A.P., Gordon, G.J.: Relational learning via collective matrix factorization. In: Proceeding of the 14th ACM SIGKDD International Conference on Knowledge Discovery and Data Mining (KDD 2008), KDD 2008, pp. 650–658. ACM, NewYork (2008). 56, 58, 60

12. Stamper, J., Niculescu-Mizil, A., Ritter, S., Gordon, G.J., Koedinger, K.R.: Algebra 2005–06 Challenge dataset from KDD Cup 2010 Educational Data Mining Challenge (2010)

13. Thai-Nghe, N., Drumond, L., Horvath, T., Nanopoulos, A., Schmidt-Thieme, L.: Matrix and tensor factorization for predicting student performance. In: Proceedings of the 3rd International Conference on Computer Supported Education (CSEDU 2011) (2011a)

14. Thai-Nghe, N., Drumond, L., Krohn-Grimberghe, A., Schmidt-Thieme, L.: Recommender system for predicting student performance. In: Proceedings of the ACM RecSys 2010 Workshop on Recommender Systems for Technology Enhanced Learning (RecSys-TEL 2010), vol. 1, pp. 2811–2819. Elsevier's Procedia Computer Science (2010c)

15. Thai-Nghe, N., Drumond, L., Horvath, T., Krohn-Grimberghe, A., Nanopoulos, A., Schmidt-Thieme, L.: Factorization techniques for predicting student performance. In: Santos, O.C., Boticario, J.G. (eds.) Educational Recommender Systems and Technologies: Practices and Challenges (ERSAT 2011). IGI Global (2011)
16. Thai-Nghe, N., Drumond, L., Horvath, T., Schmidt-Thieme, L.: Multi-relational factorization models for predicting student performance. In: Proceedings of the KDD 2011 Workshop on Knowledge Discovery in Educational Data (KDDinED 2011) (2011c)

An Analysis of Best Player Selection Key Performance Indicator: The Case of Indian Premier League (IPL)

Mayank Khandelwal, Jayant Prakash and Tribikram Pradhan

Abstract IPL is the most celebrated T20 cricket festival in the world in which 8 teams give their best to reach the top team in the tournament. In such a contest there are various players from different nationalities playing for different teams. As we know that only a certain amount of players can play one match, so there is a problem for team management to choose the best combination of players for the match. In this paper, we are calculating the *Most Valuable Player (MVP)* by using a novel approach, decision tree is used to classify the players into various classes. Further, bipartite cover is used for selection of bowlers, variance analysis is used to find the similarity among players. Finally, genetic algorithm is used to select the best playing eleven. After selecting the best players, we are predicting individual strike rates with total team scores. This paper is going to give them a solution to eliminate non performing players using customized method of their performance analysis in earlier matches, assembling a decent playing eleven for any match using revolutionary methods and deciding batting order in an efficient manner.

Keywords MVP · Decision tree · Bipartite cover · Co-variance · Genetic algorithm · Regression

1 Introduction

Indian Premier League (IPL) is a Twenty20 (T20) cricket extravaganza started by BCCI (Board of Control for Cricket in India) in 2008, which is held annually in the month of April - June. The first season of IPL was sponsored by DLF, which is a leading real estate company in India. The inaugural season of the tournament took place from 18 April - 1 June 2008. The challenge is that during an IPL auction only

M. Khandelwal · J. Prakash · T. Pradhan
Department of I&CT, Manipal University, Manipal 576104, India
e-mail: {mayankkhandelwal123,jayant.11snh}@gmail.com,
     tribikram.pradhan@manipal.edu

© Springer International Publishing Switzerland 2016
S. Berretti et al. (eds.), *Intelligent Systems Technologies and Applications*,
Advances in Intelligent Systems and Computing 384,
DOI: 10.1007/978-3-319-23036-8_16

selected players can play the cricket match, hence the team owner must select the optimal combination of players. As of now, there is no fool-proof solution to this challenge nor is there any solution ranging from selection of bowlers to selection of batsmen maintaining experience within the team. A possible methodology consists of attempting to choose the best payer to buy among all the participants in an IPL auction for the team using a measure called *MVP*. MVP is dynamic in nature, which implies that comparison criteria changes over the proceedings of the auction. We then classify players according to their complete performance measured points, called *TCP*. Then, we attempt to choose the best set of bowlers using bipartite cover concept and batsmen using genetic algorithm. We attempt to decide the best batting order using genetic algorithm and using some interesting measures, we predict the results for the game.

2 Literature Survey

S. Singh, S. Gupta and V. Gupta in [1] proposed an integer programming real-time model for optimal strategy for binding processes. Spreadsheets were used to document & calculate the results since it was the optimal choice considering the flexibility of incorporation for more weight-age based on recent performance of a player to evaluate the final outcome.

S. Singh in [2] uses Data Envelopment Analysis to measure how effective teams are in IPL. The author calculates awarded points, total run rate, profit and returns by determining that total expenses including the wage price of players and staff as well as other expenses. Efficiency score is usually directly related to the performance of the player in the league. On decomposing the inefficiencies into technical and scale inefficiency, it is realize that the inefficiency is primarily due to unoptimized scale of production & unoptimized transformation of the results and the considered data.

P. Kalgotra, R. Sharda and G. Chakraborty in [3] develop predictive models which aid managers to select players for a talented team in the least possible price. This is calculated on the basis of the player's past performance. The optimal model is selected on the basis of the rate of validation data misclassification. This model helps in the selection of players by aiding in the author's bidding equation. This research also facilitates the managers to set the salaries for players.

F. Ahmed, K. Deb and A. Jindal [4] use NSGA-II algorithm to propose a new representation scheme & a multi-objective approach for selecting players in a limited budget considering the batting & bowling strengths along with the team formation. Factors such as fielding further optimize the results. The dataset to define performance is taken from IPL - 4th Edition. The author shows analysis in real-time auction events, selecting players one-by-one. The author argues that the methodology can be implemented across other fields of sports such as soccer etc.

Sonali B. and Shubhasheesh B. [5] focus on how teams strategically decide on the final bid amount based on past player & team performance in IPL and formats similar to IPL. The authors also shed light on how personalities of players can affect team performance. They analyze the possible factors based on which bidders decide and build a predictive model for pricing in the auction. The analysis is done individually for all the teams.

H. Saikia and D. Bhattacharjee [6] classify performances of all-rounders into 'Performer', 'Batting-All Rounder', 'Bowling-All Rounder' and 'Under-performer'. Further, they suggest and consider independent variables that influence an all-rounder's performance by using Step-wise Multinomial Logistic Regression (SMLR). The independent variables are used to predict the class of an all-rounder player using Naive Bayes Classification concept.

F. Ahmed, A. Jindal and K. Deb [4] suggest a multi-objective approach which optimizes and identifies the batting and bowling strengths of a team using NSGA-II algorithm. Information from the trade-off front enables decision making for final team selection. The study uses data from IPL - 4th Edition and player's statistical data as performance parameters. The authors argue that the methodology is generic is extend-able to other sports as well.

3 Methodology

In order to obtain *MVP*, here we have proposed a novel approach consisting of MVP calculation, classification using decision tree, selection of players using bipartite cover & genetic algorithm. Finally, similarity measure among players can be analyzed by using regression analysis to make a comparison between various teams. We propose the following steps in our methodology :

- Analysis of *Most Valuable Player* (MVP) by using a Set of Rules
- Dynamic Nature of MVP Calculation After Each Player's Selection Process
- Calculation of *Total Credit Point* (TCP) by using a Set of Rules
- Classification Among Selected Players of Individual Teams using Decision Tree
- Application of Bipartite Cover for the Selection of Best Bowlers & All-Rounders
- Finding similarity & dis-similarity among players using co-variance analysis
- Selection of Batsmen using Genetic Algorithm
- Deciding the Batting Order using Genetic Algorithm
- Predicting team scores with individual score & strike rate of a player using multiple linear regression analysis.

3.1 MVP Calculation

In this section, we need to find out the player's batting points (PBT), player's bowling points(PBW) and player's experience (PEX). In order to find out the above three formula's, we need to consider the following parameters : Player's Batting Average, Player's Batting Strike Rate, Number of centuries & half-centuries, Bowling Average, Bowling Strike Rate, Economy, Number of 4-wicket & 5-wicket haul and Number of Matches Played. We define the 'Most Valuable Player' (MVP) as the single parameter that can be used to compare any type of player in the auction. MVP is decided on the basis of requirement of type of player selected by the owner. For this, we need the 'Requirement Points' (minimum required in the team) for batting(BARP), bowling(BORP) and experience(ERP). 'Total Requirement' (TRP) is the sum of all requirement points (Batting + Bowling + Experience) i.e TRP = BARP + BORP + ERP

1. PBT = $(((BattingAverage * 0.3) + (BattingStrikeRate * 0.4) + \lceil NumberofHundreds \rceil + (NumberofFifties * 0.2))/10)$
2. **if that the bowler must have bowled minimum 100 bowls in his IPL career, then,**
   $PBW = (((300/BowlingAverage) + (200/BowlingStrikeRate) + (300/Economy) + \lceil Numberof4 - wicketshaul \rceil * 0.1 + \lceil Numberof5 - wicketshaul \rceil * 0.1)/10)$
3. PEX = (Number of Matches Played/Total Number of Matches in IPL so far)

Algorithm 1. Pseudo-code for MVP Calculation

```
1: procedure MVP(Set A, Set B)
2:     mvp = 0
3:     if PBW = 0) then
4:         mvp = (8*PBT*(BARP/TRP)+(ERP/TRP)*PEX)/10
5:     else
6:         if PBT/PBW>=2 then
7:             mvp = ((7*PBT*(BARP/TRP))+(2*PBW*(BORP/TRP))+(PEX*(ERP/TRP))/10)
8:         else
9:             if PBW/PBT>=2 then
10:                 mvp = ((7*PBW*(BORP/TRP))+(2*PBT*(BARP/TRP))+(PEX*(ERP/TRP))/10)
11:             else
12:                 mvp = ((9*PBW*(BORP/TRP))+(9*PBT*(BARP/TRP))+(2*PEX*(ERP/TRP))/20)
13:             end if
14:         end if
15:     end if
16: end procedure
```

3.2 Decision Tree

Decision Tree is powerful decisive tool used for Classification and Prediction. Every node is bonded with rules that help the data to be classified according to the nature defined by the rules. It is basically used in Data Warehouse for Knowledge Discovery. Following are the features of a Decision Tree:

- There must be finite number of distinct attributes for classification.
- Target values of data used for classification should be discrete.
- There should not be any missing data which are important for classification.

Algorithm 2. TCP Calculation

```
1: procedure TCP(PBT, PBW, PEX)
2:    if PBW = 0 then
3:        TCP = (8*PBT + 1*PBW + PEX)/10
4:    else if PBT/PBW>=2 then
5:        TCP = (7*PBT + 2*PBW + PEX)/10
6:    else if PBW/PBT>=2 then
7:        TCP = (2*PBT + 7*PBW + PEX)/10
8:    else
9:        TCP = (9*PBT + 9*PBW + 2*PEX)/20
10:   end if
11: end procedure
```

Following are the components of a Decision Tree:

- **Decision Node :** It is a non leaf node used to make a decision according to the relevant data taken into consideration for the classification.
- **Leaf Node :** Represents the final classification container that holds data post operations occurred at the Decision Node.
- **Path :** It represent the result used for classification of the data from the decision node.

In Decision Tree Data is classified starting from the root node using top down approach till the leaf node is encountered.

Algorithm 3 Pseudocode for Classification of Players using Decision Tree

```
1:  procedure DETERMINE_CLASS(TCP)
2:      if TCP>=6.5 then
3:          return A+
4:      else if TCP>=6.0 then
5:          return A
6:      else if TCP>=5.5 then
7:          return B
8:      else if TCP>=5 then
9:          return C
10:     else if TCP>=4.5 then
11:         return D
12:     else if TCP>=6.5 then
13:         return E
14:     else
15:         return F
16:     end if
17: end procedure
18: procedure CLASSIFICATION(PBT, PBW, TCP)
19:     if (PBW = 0) or ( PBT / PBW ) >= 4 ) then
20:         if TCP>=6.5 then
21:             player_type = Batsman
22:             determine_class(TCP)
23:         end if
24:     else
25:         if (PBW/PBT)>=1.25 then
26:             player_type = Bowler
27:             determine_class(TCP)
28:         else
29:             if TCP>=6.5 then
30:                 player_type = All-Rounder
31:                 determine_class(TCP)
32:             end if
33:         end if
34:     end if
35: end procedure
```

3.3 Bipartite Cover

An Undirected Graph is a Bipartite Graph if it has n vertices's, partitioned into two sets A and B such that no two edges of the same set are connected. This means that all edges exist only between vertices's from Set A to vertices's from Set B. If vertices's of $A1$ which is a subset of A, connects all vertices's of B, then $A1$ covers B. The size of this cover is determined by the number of vertices's in $A1$. $A1$ is said to be minimum cover if no smaller subset of A covers B. Hence, Set C gives minimum cover with size x. Conclusively, the bipartite cover problem aims at getting minimum cover for a set of vertices in a bipartite graph. We have shown the pseudo-code for bipartite cover in *Algorithm 4*.

Algorithm 4. Pseudo-code for Bipartite Cover

```
1: procedure BIPARTITE(Set A, Set B)
2:     x = 0                                    ▷ initial number of selected vertices for cover
3:     C = new_cover_set                        ▷ Cover Set
4:     for all i in A do
5:         a_set[i]=Degree[i]                   ▷ Populating degrees of vertices in A
6:     end for
7:     for all i in B do
8:         b_set[i]=false                       ▷ marking all vertices in set B unreached
9:     end for
10:    while all a_set[i] > 0 for all nodes in A do
11:        r = vertex i of A with max value in a_set[i]
12:        append r to C
13:        increment x by 1
14:        for all vertex s adjacent from r do
15:            if r is not reached then
16:                mark r is reached in b_set[r]
17:                for all vertex t adjacent from r do
18:                    decrement a_set[t] by 1
19:                end for
20:            end if
21:        end for
22:    end while
23: end procedure
```

3.4 Covariance of Numeric Data

Correlation & covariance are measures to find how much attributes vary in accordance to each other. Consider A and B, and a set of n observations $\{(a_1, b_1), ..., (a_n, b_n)\}$ The mean or expected values of A and B:

$$E(A) = \bar{A} = \frac{\sum_{i=1}^{n} a_i}{n} \quad and \quad E(B) = \bar{B} = \frac{\sum_{i=1}^{n} b_i}{n} \tag{1}$$

The covariance between A and B is defined as

$$Cov(A,B) = E((A - \bar{A})(B - \bar{B})) = \frac{\sum_{i=1}^{n}(a_i - \bar{A})(b_i - \bar{B})}{n} \tag{2}$$

We observe that, $r_{A,B}$ (Correlation Coefficient)

$$r_{(A,B)} = \frac{Cov(A,B)}{\sigma_A \sigma_B} \tag{3}$$

where $\sigma_A \sigma_B$ are the standard deviations of A and B, respectively. We show that

$$Cov(A,B) = E(A.B) - \bar{A}\bar{B} \tag{4}$$

For A and B which are likely to change together, if A is greater than $\bar{A}$ (the expected value of A), then B will tend to be greater than $\bar{B}$ (the expected value of B). Hence, the covariance between A and B is *positive*. On the contrary, if one of the attributes is likely to be greater than its expected value when the other attribute is lower than its expected value, then the covariance of A and B is *negative*.

If A and B are independent, then E(A.B) = E(A).E(B). Hence, the covariance is Cov(A,B) = E(A.B) - $\bar{A}B$ = E(A).E(B) - $\bar{A}B$ = 0. Converse isn't true. Some pairs of random variables (attributes) may have a covariance of 0 but are dependent.

3.5 Genetic Algorithm

Genetic algorithms have proved to be particularly efficient for searching, optimizing, machine learning and the like. Genetic Algorithms have proved to be very robust when searching in complex spaces both experimentally as well as theoretically. The iterative process of Genetic Algorithm creates new outcomes using the following steps:

- Selection: Select individuals randomly for reproduction with a probability which depends on fitness of others, relatively. This helps in achieving a good chance in selecting the best required individuals.
- Cross-Over: The selected individuals fuse to generate an offspring. Recombination and mutation techniques can be used to generate new chromosomes.
- Evaluation: The fitness of the generated chromosomes is computed.
- Mutation:The initially selected individuals are discarded by the new ones. Thus, the iteration begins to generate and evolve more off-springs.

We randomly select a cross site and swap the genes (bits in our case) of two parent chromosomes to generate two new offspring chromosomes. For example:

Offspring1=100110111010011   Chromosome1=100110110010010
Offspring2=011010101010010   Chromosome2=011010101010011

Cross-over is followed by Mutation.

We randomly select a gene and comple-
ment on that position. For instance, in a
binary string 0 will be converted to 1 and
1 will be converted to 0. Mutation is required to generate even more possibilities
in the Genetic Algorithm. Let us assume
that we mutate the 5th & 6th position of
Offspring 1 and Offspring 2 respectively.
We obtain the following:

Mutated Offspring1= 1001001111010011
Mutated Offspring2= 0110111010010010

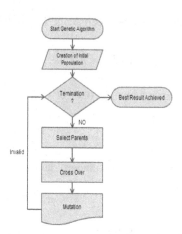

Fig. 1 Life Cycle of Genetic Algorithm

3.6 Regression

Regression Analysis involves a single predictor variable, x and a response variable, y. It is the form of:

$$y = b + wx \tag{5}$$

where we assume constant variance of y and b (Y-intercept) & w (Slope of the Line) are regression coefficients.

The regression coefficients, w and b can be analogized as weights :

$$y = w_0 + w_1 x \tag{6}$$

We can solve for the coefficients by estimating the best-fitting straight line as the one which minimizes error between estimate of the line & actual data. Let D be a training set. It will consist of predictor variable values, x, for some population and corresponding values of response variable,y. The training set contains $|D|$ data points of the form $\{(x_1, y_1), (x_2, y_2), ..., (x_{|D|}, y_{|D|})\}$. The regression coefficients are estimated using:

$$w_1 = \frac{\sum_{i=1}^{|D|} (x_i - \bar{x})(y_i - \bar{y})}{\sum_{i=1}^{|D|} (x_i - \bar{x})} \tag{7}$$

$$w_0 = \bar{y} - w_1 \bar{x} \tag{8}$$

where $\bar{x}$ and $\bar{y}$ are averages of x and y respectively. w_0 and w_1 offer good estimation of complex regression equations.

Multiple Linear Regression is an extension of straight-line regression. Here, we involve multiple predictor variables. Thus we can model response variable y as a linear function of n predictor variables $\{A_1, A_2, ..., A_n\}$, which describe a tuple, X i.e X $= \{x_1, x_2, ..., x_n\}$. The training data set, D. consists of data of the form $\{(X_1, y_1), (X_2, y_2), ..., (X_{|D|}, y_{|D|})\}$ where X_i are n - dimension tuples. Example of two predictor attributes, A_1 and A_2, is:

$$y = w_0 + w_1 x_1 + w_2 x_2 \tag{9}$$

where x_1 and x_2 attribute values A_1 and A_2, respectively in **X**.

4   Case Study

4.1   MVP Calculation and It's Dynamic Characteristics

We introduce a parameter of comparison among players which may help the team management to purchase the best set of players during an auction, called MVP value. It takes into account attributes relating to player statistics(PBT, PBW & PEX) and team management requirements (BARP, BORP & ERP). The method of calculation of MVP value is given in Algorithm 1. For fair comparison, MVP value is calculated differently for Batsman, Bowlers and All-Rounders. It is done in such a way that for a batsman, batting points(PBT) is given more preference than bowling points(PBW). Similarly MVP value calculation is done for Bowlers and All-Rounders. Also there may be different requirement of skills in different teams, so the method consider these requirement in form of points(BARP, BORP & ERP). The method compares players taking account of the these points. Hence, every player has different MVP values for different teams.

The method adapts to changing scenario in the auction. For instance, when Aaron Finch is purchased by Mumbai Indians, the method deducts the performance attributes points (PBT, PBW & PEX) of Aaron Finch from the require-

Table 1 MVP Points of Players Before Auction

Team Name				CSK	MI	SRH	DD	RCB	RR	KKR	KXIP
Fund				50,000,000	100,000,000	208,500,000	400,000,000	210,000,000	130,000,000	130,000,000	118,000,000
Batting Requirement				13	22	46	72	24	17	14	9
Bowling Requirement				10	19	42	69	34	23	21	20
Experience Requirement				9	16	43	59	21	20	13	15
Name	PBT	PBW	PEX	MVP	MVP	MVP	MVP	MVP	MVP	MVP	MVP
Yuvraj Singh	6.22	6.20	7.06	2.2067	2.2080	2.1088	2.1783	2.2398	2.0987	2.2295	2.0835
Dinesh Kartik	5.90	0.00	8.91	2.1667	2.0705	1.9486	1.9608	1.6697	1.6333	1.6170	1.2685
Kevin Pieterson	6.74	5.83	2.69	2.0888	2.0986	1.9814	2.0697	2.1732	1.9950	2.1643	2.0062
Hashim Amla	6.03	0.00	6.22	2.1344	2.0363	1.8979	1.9199	1.6307	1.5739	1.5753	1.1986
Mike Hussey	6.52	0.00	4.62	2.2493	2.1432	1.9836	2.0144	1.7077	1.6322	1.6467	1.2246
Aaron Finch	5.86	0.00	3.03	1.9886	1.8934	1.7446	1.7760	1.5039	1.4284	1.4485	1.0615
Chris Morris	6.57	6.53	6.22	2.2933	2.2950	2.1845	2.2620	2.3311	2.1738	2.3196	2.1574
Kane Williamson	6.50	5.99	6.22	2.0569	2.0543	1.9580	2.0226	2.0681	1.9364	2.0564	1.9082
Irfan Pathan	5.55	5.61	8.24	2.0388	2.0392	1.9586	2.0145	2.0618	1.9481	2.0530	1.9332

Table 2 Dynamic MVP Value of Players After Finch's Selection

Team Name		CSK	MI	SRH	DD	RCB	RR	KKR	KXIP
Fund		50,000,000	100,000,000	208,500,000	400,000,000	210,000,000	130,000,000	130,000,000	118,000,000
Batting Requirement		13	22	46	72	24	17	14	9
Bowling Requirement		10	19	42	69	34	23	21	20
Experience Requirement		9	16	43	59	21	20	13	15
Name	PBT PBW PEX	MVP	MVP	MVP	MVP	MVP	MVP	MVP	MVP
Yuvraj Singh	6.22 6.20 7.06	2.2067	2.2317	2.1088	2.1783	2.2398	2.0987	2.2295	2.0835
Dinesh Kartik	5.90 0.00 8.91	2.1667	1.8225	1.9486	1.9608	1.6697	1.6333	1.6170	1.2685
Kevin Pieterson	6.74 5.83 2.69	2.0888	2.1499	1.9814	2.0697	2.1732	1.9950	2.1643	2.0062
Hashim Amla	6.03 0.00 6.22	2.1344	1.7858	1.8979	1.9199	1.6307	1.5739	1.5753	1.1986
Mike Hussey	6.52 0.00 4.62	2.2493	1.8747	1.9836	2.0144	1.7077	1.6322	1.6467	1.2246
Chris Morris	6.57 6.53 6.22	2.2933	2.3218	2.1845	2.2620	2.3311	2.1738	2.3196	2.1574
Kane Williamson	6.50 5.99 6.22	2.0569	2.0665	1.9580	2.0226	2.0681	1.9364	2.0564	1.9082
Irfan Pathan	5.55 5.61 8.24	2.0388	2.0564	1.9586	2.0145	2.0618	1.9481	2.0530	1.9332

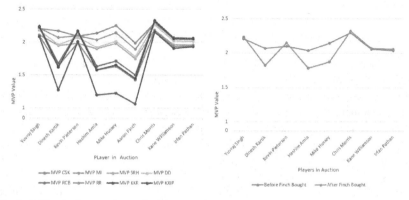

Fig. 2 Different Perspectives of MVP

ment points of Mumbai Indians. This deduction triggers the re-computation of MVP values as requirement points has changed. Since Finch was Batsman as he has PBT=5.86, PBW=0 and PEX=3.03, so deduction is more from the Batting Requirement Points(BARP). After re-computation, MVP value of Batsman such as Mike Hussey will decrease whereas Bowlers like Irfan Pathan will increase. The comparison before re-computation and after re-computation is shown in Figure 2.

4.2 Classification of Players Using Decision Tree

Every Player can be classified by the help of Decision Tree and parameters like batting points(PBT), bowling points(PBW) and MVP value. Our method classifies into batting, bowling and all-rounder based on PBT and PBW.

Table 3 Calculation of TCP and Classification of Players Using Decision Tree

Name	Mat	Runs	HS	Avg	SR	100	50	Balls	Runs	Wkts	Avg	Eco	SR	4W	5W	PBT	PBW	PEX	TCP	Type	Grade
Rohit Sharma	112	2903	109*	32.25	129.59	1	21	332	440	15	29.33	7.95	22.13	1	0	6.67	5.71	9.41	6.51	ALL ROUNDER	A+
Lasith Mailinga	83	75	17	4.68	87.2	0	0	1929	2102	119	17.66	6.53	16.21	3	1	3.63	7.57	6.97	6.72	BOWLER	A+
Kieron Pollard	77	1332	78	27.18	144.31	0	6	1076	1539	53	29.03	8.58	20.3	1	0	6.71	5.53	6.47	6.15	ALL ROUNDER	A
Ambati Rayudu	81	1710	81*	26.71	125.18	0	10	18	22	0	0	7.33	0	0	0	6.01	0.00	6.81	5.49	BATSMAN	C
Harbhajan Singh	96	574	49*	15.51	147.93	0	0	2037	2281	92	24.79	6.71	22.14	1	1	6.38	6.60	8.07	6.65	ALL ROUNDER	A+
Corey Anderson	12	265	95*	29.44	146.4	0	1	108	184	4	46	10.22	27	0	0	6.76	4.33	1.01	5.09	ALL ROUNDER	C
Aditya Tare	27	299	59*	17.58	137.15	0	1	0	0	0	0	0	0	0	0	6.03	0.00	22.27	5.05	BATSMAN	C
Jaspreet Bumbrah	13	1	1*	1	33.33	0	0	280	371	7	46.37	7.95	35	0	0	1.36	4.99	1.09	3.88	BOWLER	F
Josh Hazlewood	27	11	6*	5.5	47.82	0	0	623	787	36	21.86	7.57	17.5	1	0	2.08	6.50	2.27	5.19	BOWLER	C
Merchant de Lange	30	47	9*	11.75	134.28	0	0	638	920	38	24.21	8.65	16.7	1	0	5.72	5.91	2.52	5.49	ALL ROUNDER	C
Pawan Suyal	9	2	0	1	10	0	0	174	240	8	30	8.27	21.7	0	0	0.43	5.55	0.76	4.05	BOWLER	E
Shreyas Gopal	6	35	24	17.5	159.09	0	0	96	142	6	23.66	8.87	16	0	0	6.89	0.00	0.50	5.56	BATSMAN	B
Lendl Simmons	34	761	77	25.36	113.92	0	4	36	55	6	9.16	9.16	6	1	0	5.40	0.00	2.86	4.60	BATSMAN	D
Aaron Finch	36	888	88*	26.11	123.84	0	6	45	67	1	67	9.34	43	0	0	5.86	0.00	3.03	4.99	BATSMAN	D
Pragyan Ojha	91	17	4	1.41	37.77	0	0	1887	2309	89	25.94	7.34	21.2	0	0	1.55	6.19	7.65	5.41	BOWLER	C
McClenaghan	46	53	13*	13.25	128.25	0	0	967	1262	56	22.53	7.83	17.2	0	1	5.33	6.34	3.87	5.64	ALL ROUNDER	B
Unmukt Chand	38	731	125	20.88	114.04	2	1	0	0	0	0	0	0	0	0	5.41	0.00	3.19	4.65	BATSMAN	D
Vinay Kumar	86	291	26*	11.19	113.22	0	0	1765	2438	91	26.79	8.28	19.39	1	0	4.86	5.78	7.23	5.51	ALL ROUNDER	B
Parthiv Patel	79	1411	61	20.75	109.37	5	0	0	0	0	0	0	0	0	0	5.10	0.00	6.64	4.74	BATSMAN	D
Aiden Blizzard	81	1724	89	25.35	135.74	0	8	0	0	0	0	0	0	0	0	6.35	0.00	6.81	5.76	BATSMAN	B

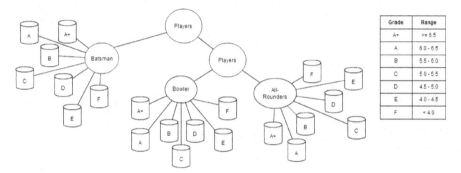

Grade	Range
A+	>= 6.5
A	6.0 - 6.5
B	5.5 - 6.0
C	5.0 - 5.5
D	4.5 - 5.0
E	4.0 - 4.5
F	< 4.0

Fig. 3 Graphical Representation of Decision Tree

Lower level of classification is assigning grades to each type of player in the dataset based on the MVP values calculated based on past performance.

4.3 Bowlers and All-Rounders Using BiPartite Cover

Out of the total number of players who can bowl, we use the concept of bipartite cover to find out the best combination of 4 bowlers + 2 all-rounders who can collectively bowl all 20 overs, as demonstrated below:

Consider Fig. 4 with two sets (Players : A to L & Overs : 1 - 20)

{Players} = { SG, LM, KP, CA, HS, JB, VK, JH, MMC, MARCHANT, PS, OJHA }
{Overs} = { 1, 2, 3, 4, 5, 6, 7, 8, 9, 10, 11, 12, 13, 14, 15, 16, 17, 18, 19, 20 }

We have to determine a subset /Players'} such that it covers the set /Overs}

Now, let each of the players contain the following values:

Shreyas Gopal = {1,3,12,19}   Lasith Malinga= {1,3,10,12,15,18,20}   Kieron Pollard = {6,7,8,14,19}

Cory Anderson = {2,4,5,18}   Harbhajan Singh = {6,9,11,13 }   Jasprit Bumrah = {7,11,14,16}

Vinay Kumar = {3,11,17}   Josh Hazlewood = {7,8,9,13}   Mitchell McClenaghan = {11,16}

Marchant de Lange = {5,13,16}   Pawan Suyal = {14,19}   Pragyan Ojha = {8,9,13,15}

We thus obtain the bipartite graph as follows:

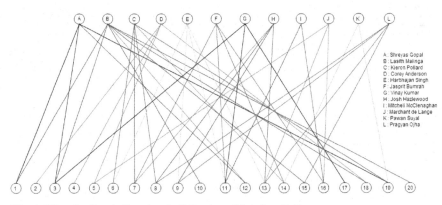

Fig. 4 Bipartite Graph Showing the Mapping of Bowlers & Overs

Initially our selected bowlers set, S = {φ} In the each iteration, we select the player covering the maximum uncovered overs. Initially, all overs are uncovered. In the first iteration, we select LM with a count of 7 uncovered overs and thus S = { 1, 3, 10, 12, 15, 18, 20}. We mark { 1, 3, 10, 12, 15, 18, 20} as covered overs. In the second iteration, we select KP with a maximum count of 5 uncovered overs and thus S = { 1, 3, 6, 7, 8, 10, 12, 14, 15, 18, 19, 20 }. We mark { 6, 7, 8, 14, 19 } as covered overs. On further calculation, we obtain the set of { LM, KP, CA, HS, JB, VK } as the most optimal solution for bowling selection which ensures possible combinations across all 20 overs. S = { 1, 2, 3, 4, 5, 6, 7, 8, 9, 10, 11, 12, 13, 14, 15, 16, 17, 18, 19, 20 }.

In Fig. 4, we have shown the relationship of players along with their best bowling performance in corresponding overs. Table 4 depicts each step during the selection and construction of solution from bipartite graph. A selected player, is marked with SELECT and the un-selected players are mentioned as the number of uncovered overs which the player can bowl.

Table 4 Stepwise Selection Process of Bowlers in Each Iteration

Player	1	2	3	4	5	6	7	8	9	10	11	12	13	14	15	16	17	18	19	20	Initially Uncovered	Iteration 1	Iteration 2	Iteration 3	Iteration 4	Iteration 5	Iteration 6
SG	Y	Y								Y										Y	4	1	0	0	0	0	0
LM	Y	Y						Y	Y			Y				Y		Y			7	SELECT	SELECT	SELECT	SELECT	SELECT	SELECT
KP				Y	Y	Y							Y							Y	5	5	SELECT	SELECT	SELECT	SELECT	SELECT
CA	Y		Y	Y														Y			4	3	3	SELECT	SELECT	SELECT	SELECT
HS						Y			Y		Y	Y									4	4	3	3	SELECT	SELECT	SELECT
JB					Y						Y			Y	Y						4	4	2	2	1	SELECT	SELECT
VK		Y									Y						Y				3	2	2	2	1	1	SELECT
JH				Y	Y	Y						Y									4	4	2	2	0	0	0
MMC								Y					Y								2	2	2	2	1	0	0
MDL			Y									Y		Y							3	3	3	2	1	0	0
PS													Y							Y	2	2	0	0	0	0	0
OJHA										Y	Y			Y	Y						4	3	2	2	0	0	0

Thus, Lasith Malina, Kieron Pollard, Corey Anderson, Harbhajan Singh, Jaspreet Bumrah, Vinay Kumar are selected.

4.4 Variance Analysis

After selecting best suitable bowlers, we are analyzing the similarity and dissimilarity among batsmen by using co-variance analysis. Here we are finding the similarity between Ambati Rayudu and Rohit Sharma. The individual scores for the last fifteen matches of IPL - 2014 are considered.

Table 5 Similarity measure among Rohit Sharma & Ambati Rayudu

Innings	1	2	3	4	5	6	7	8	9	10	11	12	13	14
A(Rayudu)	48	95	1	14	35	16	9	59	68	33	17	2	30	2
B(Rohit)	27	2	50	4	1	39	59	9	14	51	18	30	16	20
A*B	1296	190	50	56	35	624	531	531	952	1683	306	60	480	40

Average(A) = 30.6428   Average(B) = 24.2857   Sum of **A*B** = 6834
Average of A * Average of B = 744.1837   Sum/14 = 488.1429

Since (**Sum/14**) - (**Average of A * Average of B**) = -256.0408 which is less than 0, Hence Negatively Correlated

Next, we find the similarity between Aditya Tare and Parthiv Patel.

Table 6 Similarity measure among Aditya Tare & Parthiv Patel

Innings	1	2	3	4	5	6	7	8	9	10
A(Taare)	24	17	23	8	7	16	9	2	14	6
B(Parthiv)	37	57	21	2	3	26	13	29	10	4
A*B	888	969	483	16	21	416	117	58	140	24

Average(A) = 26.2   Average(B) = 20.2   Sum of **A*B** = 3132
Average of A * Average of B = 254.52   Sum/14 = 313.2

Since (**Sum/14**) - (**Average of A * Average of B**) = 58.68 which is greater than 0, Hence Positively Correlated

Similarly, we obtain a positive correlation among Lendl Simmons & Aaron Finch, and also between Aidn Blizzard & Unmukt Chand. Since Rayudu and Rohit are negatively correlated, hence both players are advised to be in the team. Similarly, Tare & Parthiv, Simmons & Finch and Blizzard & Unmukt are positively correlated, hence either one of the player's among above pairs are advised to be inducted into the team. Since, both players in a pair are similar, therefore there is no drastic effect on the team's performance.

4.5 Selection of Batsmen Using Genetic Algorithm

The selection of batsmen using Genetic Algorithm can be done with the following strategy. We have devised an encoding scheme to achieve selection of batsmen. The encoding bits are allotted in the following manner:

- First and second bits are allotted to the opening batsmen selection.
- Fourth and fifth bits are allotted to the middle order batsmen selection.
- Seventh and eighth bits are allotted to the wicketkeeper selection.
- Tenth and eleventh bits are allotted to the sloggers batsmen selection.

Table 7 Initial Encoding Scheme for Batsman Selection

0 1 0 1	1 1 0	1 0 1	1 1
Opener	Middle Order	Wicketkeeper	Slogger

Type of Batsman	Name of Batsman	Genetic Code
Opener	Simmons	10
Opener	Finch	01
Middle-Order	Rohit	11
Middle-Order	Rayudu	11
Wicketkeeper	Tare	01
Wicketkeepre	Parthiv	10
Slogger	Blizzard	11
Slogger	Unmukt	10

Now, we select twelve random chromosomes (12-bit) for the selection of batsmen as per the encoding scheme. For each 12-bit chromosome, we count the number of *1's* in the chromosome and the count is called **Fitness of the Chromosome**. After calculating the fitness for each chromosome, we determine the average fitness i.e =7 (in our case).

Table 8 Initial Chromosome Selection & Fitness Evaluation for Batsman

Bit 0	Bit 1	Bit 2	Bit 3	Bit 4	Bit 5	Bit 6	Bit 7	Bit 8	Bit 9	Bit 10	Bit 11	Fitness	Selected (Y/N)
0	1	0	1	1	1	0	1	0	1	1	1	8	Y
1	1	0	1	0	1	1	1	1	0	0	1	8	Y
0	0	0	1	0	1	1	0	0	0	1	1	5	N
1	0	1	1	0	1	1	0	1	1	1	0	8	Y
0	0	0	1	1	1	0	1	1	0	1	0	6	N
0	1	1	0	1	1	0	1	0	0	1	0	6	N
1	1	1	0	0	1	1	1	1	1	0	0	8	Y
0	1	0	1	1	1	0	1	1	1	0	1	8	Y
1	1	1	1	0	0	0	0	1	1	0	1	7	Y
0	1	1	0	1	1	0	1	1	0	0	0	6	N
1	1	0	1	1	0	0	1	1	0	1	0	7	Y
1	0	1	1	0	1	1	1	0	0	0	1	7	Y

Average Fitness = 84/12 = 7

We find those chromosomes which have a fitness value $\geq$ average fitness. So, out of the twelve chromosomes, eight are selected. Now, we take two chromosomes in a pair and randomly decide the cross-over site. we apply cross-over on the chromosome pairs. After the cross-over result, we randomly decide the mutation site for individual resultant chromosomes. After mutation, we obtain the desired sets of offsprings. We select the valid off-spring among the resultant off-spring obtain the batsmen for the team.

Table 9 Intermediate Steps of Crossover and Mutation For Batsman Selection

S.No	Selected Chromosomes	Cross-Over Sites	Result of Cross-Over	Mutation Sites	Result
1	010101101011	7	010101111001	8	010101101001
2	110101111001	7	110101101011	9	110101100011
3	101101101110	6	101101111100	2	111101111100
4	111001111100	6	111001101110	3	110001101110
5	**010111011101**	**8**	**010111011101**	**11**	**010111011111**
6	111100001101	8	111100001101	4	111000001101
7	110110011010	5	110111110001	6	110110110001
8	101101110001	5	101100011010	10	101100011110

All combinations obtained from the genetic algorithm fails to give the desired selection of players (at least one opener batsman, two middle-order batsman, one wicketkeeper and one slogger) except the fifth result, which is 010111011111. This result gives:

Table 10 Implication of Genetic Algorithm

01	11	01	11
Finch	Rohit & Rayudu	Tare	Blizzard

After the selection of batsmen, the team needs to decide the batting order for the day's play. For this, the encoding scheme is given in Table 11, the chromosome fitness selection is done in Table 12 and cross-over & mutation of the chromosome to the resultant offspring is done in Table 13. Using these genetic operations, we get the batting order shown in Table 14.

Table 11 Encoding Scheme for Selection of Batting Order

0 1 0	1	1 1	0	1
Opener	Middle Order	Slogger		Tail Enders

Type of Batsman	Name of Batsman	Genetic Code
Opener	Finch - Tare	0
Opener	Tare - Finch	1
Middle-Order	Anderson - Rohit - Rayudu	00
Middle-Order	Rohit - Anderson - Rayudu	01
Middle-Order	Rohit - Rayudu - Anderson	10
Middle-Order	Rayudu - Rohit - Anderson	11
Slogger	Pollard - Blizzard	0
Slogger	Blizzard - Polllard	1
Tail-Enders	Harbhajan - Malinga - Bumrah - Vinay	0
Tail-Enders	Harbhajan - Malinga - Vinay - Bumrah	1

Table 12 Chromosome Selection and Fitness Evaluation

Bit 0	Bit 1	Bit 2	Bit 3	Bit 4	Bit 5	Bit 6	Bit 7	Fitness	Selected (Y/N)
0	1	0	1	1	1	0	1	5	Y
1	1	0	1	0	1	1	1	6	Y
0	0	0	1	0	1	1	1	4	N
1	0	1	1	0	1	1	0	5	Y
0	0	0	1	1	1	0	1	4	N
0	1	1	0	1	1	0	1	5	Y
1	1	1	0	0	1	1	1	6	Y
0	1	0	1	1	1	0	1	5	Y

Table 13 Intermediate Steps of Crossover and Mutation of Batting Order

S.No	Selected Chromosomes	Cross-over Sites [0 - 8]	Result of Crossover	Mutation Sites	Result
1	01011101	6	01011111	5	01010111
2	11010111	6	11010101	4	11000101
3	10110110	4	10111101	4	10101101
4	01101101	4	01100110	2	00100110
5	11100111	3	11111101	3	11011101
6	01011101	3	01000111	6	01000011

Table 14 Selection of Batting Order

Batting Number	1	2	3	4	5	6	7	8	9	10	11
Player	Finch	Tare	Rohit	Anderson	Rayudu	Blizzard	Pollard	Harbhajan	Malinga	V Kumar	J Bumrah

4.6 Linear Regression

We have taken a dataset comprising of runs & strike-rate of *Rohit Sharma* in the IPL-2014 matches and the corresponding total score of *Mumbai Indians*. Refer to Table 15.

Table 15 Multiple Linear Regression Analysis for Rohit Sharma

Match	1	2	3	4	5	6	7	8	9	10	11	12	13	14	15
(X) StrikeRate (S/R)	135	40	121.95	80	20	114.7	168.57	100	233.33	113.33	210.52	90	142.85	145.45	125
(Y) Runs	27	2	50	4	1	39	59	19	14	51	40	18	30	16	20
(Z) Total Score	122	115	141	125	157	170	187	157	160	141	178	159	173	195	173
(R) S/R - S/R Avg	12.29	-82.71	-0.76	-42.71	-102.71	-8.01	45.86	-22.71	110.62	-9.38	87.81	-32.71	20.14	22.74	2.29
(S) Runs - Runs Avg	1	-24	24	-22	-25	13	33	-7	-12	25	14	-8	4	-10	-6
(T) Total - Total Avg	-34.87	-41.87	-15.87	-31.87	0.13	13.13	30.13	0.13	3.13	-15.87	21.13	2.13	16.13	38.13	16.13
$(S/R - Avg(S/R))^2$	151.04	6840.94	0.58	1824.14	10549.34	64.16	2103.14	515.74	12236.78	87.98	7710.59	1069.94	405.62	517.11	5.24
$(Runs - Avg(Runs))^2$	1	576	576	484	625	169	1089	49	144	625	196	64	16	100	36
R*T	-428.39	3462.93	12.11	1361	-13.69	-105.24	1381.81	-3.02	346.56	148.88	1855.65	-69.79	324.87	867.03	36.89
S*T	-34.87	1004.78	-380.78	701.05	-3.35	170.74	994.42	-0.94	-37.61	-396.65	295.88	-17.07	64.54	-381.34	-96.80

$$\text{Average(X)} = 122.71 \qquad \text{Average(Y)} = 26 \qquad \text{Average(Z)} = 156.86$$

$$\Sigma(R*T) = 9177.76 \qquad \Sigma(S*T) = 1882 \qquad \Sigma(S/R - Avg(S/R))^2 = 44082.38$$

$$\Sigma(Runs - Avg(Runs))^2 = 4750$$

$$w_1 = \frac{\Sigma(R*T)}{\Sigma(S/R - Avg(S/R))^2} = 0.208196 \quad \text{and}$$

$$w_2 = \frac{\Sigma(S*T)}{\Sigma(Runs - Avg(Runs))^2} = 0.39621$$

$$w_0 = \text{Average(Z)} - (w_1 * \text{Average(X)}) - (w_2 * \text{Average(Y)}) = 121.01254$$

After the analysis of the above mentioned dataset, we obtain the equation which predicts the total team score based on Rohit Sharma's runs & individual strike-rate in the match. The equation is as follows:

$$TotalScore = w_0 + (w_1 * S/R) + (w_2 * Runs) \tag{10}$$

Suppose Rohit Sharma scores 45 runs at a strike-rate of 153. The predicted score is 172.

5 Conclusion

In the paper, we demonstrate the dynamic changing requirement of a player in the duration of an auction, it results to selection of best balanced team for the team management. When a team have a full set of players, method classify them based on their role and performance using decision tree. This classification helps in choosing the playing XI. Using Bipartite Cover, method tried to find certain set of bowlers and all rounders who can bowl well in all possible overs in the match. This will give more option for the captain to utilize his bowling strength during the match. Also method utilizes the differences and similarity in performance of batsman using variance to constrain the possible accept cases for the output of genetic algorithm. After that method again applies genetic algorithm to decide the batting order of the whole team after considering the constraint from the performance of players at various batting positions. The random selection of batsmen and batting order may give a better results in the ever unpredictable game. The total score prediction's accuracy using regression shows the dependency of a team's batting on an individual player.

References

1. Singh, S., Gupta, S., Gupta, V.: Dynamic bidding strategy for players auction in IPL. International Journal of Sports Science and Engineering 5(01), 3–16 (2011)
2. Singh, S.: Measuring the performance of teams in the Indian Premier League. American Journal of Operations Research 1, 180–184 (2011)
3. Kalgotra, P., Sharda, R., Chakraborty, G.: Predictive Modeling in sports leagues: an application in Indian Premier League. In: SAS Global Forum 2013, pp. 019–2013 (2013)
4. Ahmed, F., Jindal, A., Deb, K.: Cricket team selection using evolutionary multi-objective optimization. In: Panigrahi, B.K., Suganthan, P.N., Das, S., Satapathy, S.C. (eds.) SEMCCO 2011, Part II. LNCS, vol. 7077, pp. 71–78. Springer, Heidelberg (2011)
5. Sonali, B., Shubhasheesh, B.: Auction of players in Indian premier league: the strategic perspective. International Journal of Multidisciplinary Research. ISSN 2231 5780
6. Saikia, H., Bhattacharjee, D.: On classification of all-rounders of the Indian premier league (IPL): A Bayesian approach. Vikalpa 36(4), 25–40

Ranking Student Ability and Problem Difficulty Using Learning Velocities

H.A. Ananya, I. Akhilesh Hegde, Akshay G. Joshi and Viraj Kumar

Abstract Several educational software tools allow students to hone their problem-solving skills using practice problems and feedback in the form of hints. If one can meaningfully define the "distance" between any incorrect student attempt and the correct solution, it is possible to define the student's *learning velocity* for that problem: the rate at which the student is able to decrease this distance. In this paper, we present an extension to one such educational software tool (JFLAP) that permits us to compute learning velocities for each student on practice problems involving finite automata construction. These learning velocities are helpful in at least two ways: (1) instructors can rank students (e.g., by identifying students whose learning velocities on most problems are significantly below the class average, and who may therefore require the instructor's attention), and (2) instructors can rank problems according to difficulty (e.g., while designing a question paper, a "difficult" problem might be one where only a few students have quickly converged to the correct solution).

1 Introduction and Related Work

Instructors and students in several domains have access to a wealth of educational software tools, many of which are freely available. The data generated by learners as they interact with such tools can be exploited using Educational Data Mining (EDM) techniques to help educators answer a variety of interesting questions, such as "which learning material will a particular sub-category of students benefit most from" [2], or "how do different types of student behavior impact their learning" [3]. In this paper, we are concerned with the following question: "how effectively can a particular student solve a particular problem". To answer such a

H.A. Ananya · I.A. Hegde · A.G. Joshi
PES Institute of Technology, Bangalore, India
e-mail: {ananya.h.a,akhil.d26,aksh0222}@gmail.com

V. Kumar(✉)
PES University, Bangalore, India
e-mail: viraj.kumar@pes.edu

© Springer International Publishing Switzerland 2016
S. Berretti et al. (eds.), *Intelligent Systems Technologies and Applications*,
Advances in Intelligent Systems and Computing 384,
DOI: 10.1007/978-3-319-23036-8_17

191

question on the basis of data, it is necessary to quantify "effectiveness" of the student's approach. A rudimentary measure is the time required by the student to solve the problem: an "effective" student is one who can solve the problem quickly. Unfortunately, this measure is unsuitable for weak students who may struggle to completely solve the problem, but may nevertheless demonstrate signs of progress in their approach.

In some problem domains, it is possible to define a notion of *distance* between the correct solution to a problem and any incorrect solution. Here, we can define "effectiveness" as the rate at which the student is able to decrease this distance (ideally, but not necessarily, to zero). We call this rate the *learning velocity* of the particular student for the particular problem. The novel contributions in this paper are: (1) an extension of an educational software tool (JFLAP [7]) to measure learning velocities for a specific problem domain (constructing deterministic finite automata) as an input to other EDM tools; (2) studying how this measure can answer the target question "how effectively can a particular student solve a particular problem" in this domain, thereby yielding a new way to rank students by ability; and (3) using this data as a new way to rank problems according to their difficulty.

Finite automata are used extensively to model computational processes, and their study is a core component of the undergraduate Computer Science curriculum [6]. JFLAP is an extremely popular open-source tool that is used world-wide for teaching finite automata and related concepts (see [5] for details and usage statistics). The JFLAP community encourages researchers to contribute extensions to the core software [8], and one such recent extension allows students to solve practice problems [9]. In our work, we log data generated by students as they practice these problems, and thereby compute learning velocities. There are a number of ways to define *distance* between two deterministic finite automata (DFA)[1], and a combination of three such distance functions has proved to be quite effective for grading DFA assignments [1]. These distance functions are computationally nontrivial, and add substantial bulk to the JFLAP package, so we define simpler functions that nevertheless capture distance effectively. We describe these in Section 2 and define learning velocity for DFA construction. We describe our experiments to validate this notion of learning velocity in Section 3 and Section 4 documents our results. Lastly, in Section 5 we discuss our findings and plans for future work.

2   Distance and Learning Velocity for DFA Problems

Our distance functions are motivated by observing student errors. An important kind of error (also pointed out in [1]) is misunderstanding what the question is asking. This kind of error is often observed in students unfamiliar with the

---

[1] For *non*-deterministic finite automata (NFA), the distance between two NFAs could be defined by determinizing them and computing the distance between the corresponding DFAs. This induced notion of distance seems inadequate, because small changes to an NFA can greatly alter the corresponding DFA. A distance measure for NFAs in this context remains an open problem.

language of instruction, but it can also occur because of inherent ambiguities in a natural language such as English. For example, consider constructing a DFA for *binary strings where every 1 is followed by at least two 0's*. This is a fairly typical kind of problem, but there are at least two ways of interpreting the italicized expression: (1) binary strings where every 1 is *immediately* followed by at least two 0's (solution DFA$_1$ shown in Fig. 1) and (2) binary strings where every 1 is *eventually* followed by at least two 0's (solution DFA$_2$ shown in Fig. 2). Suppose DFA$_1$ is the expected solution, and the student's incorrect solution is DFA$_2$. Note that the two DFAs are similar: both have only one final state, which is also the initial state. We capture such structural information for every state q of a DFA as a Boolean vector $\mathbf{v}_q$. Specifically, this vector encodes the following basic information: whether state q is the initial state or not, and whether state q is a final (or accepting) state or not. Also, for each letter a of the DFA's input alphabet ($a \in \{0, 1\}$ in this example), $\mathbf{v}_q$ encodes whether the transition from state q on input a is to itself or to a different state, and whether the transition's target is a final state or not. For example, the vectors associated with state p_0 in DFA$_1$ and state q_0 in DFA$_2$ are *identical*. In contrast, the vectors associated with states p_1 and q_1 are different because the transition in DFA$_1$ from state p_1 on input 1 is to another state (p_3) whereas the transition in DFA$_2$ from state q_1 on input 1 is to itself. For an input alphabet consisting of k letters, the number of distinct Boolean vectors is exactly 2^{2k+2} (for the binary alphabet, $k = 2$).

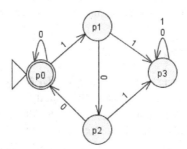

Fig. 1 DFA1 for binary strings where every 1 is *immediately* followed by at least two 0's.

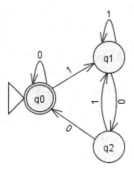

Fig. 2 DFA2 for binary strings where every 1 is *eventually* followed by at least two 0's.

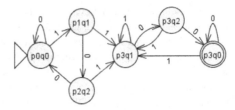

Fig. 3 Symmetric difference of DFA$_1$ and DFA$_2$.

Since any DFA can be effectively minimized into a unique representation, we define distances between minimized DFAs. This ensures that two distinct but equivalent student DFAs will be at equal distance from the solution DFA.

2.1 Normalized Vector Distance

For a given DFA, we can compute the number of states it has for each of the 2^{2k+2} types of vectors. We scale this vector by dividing by the total number of states in the DFA. Given two DFAs, we define the **normalized vector distance** between these DFAs as the Euclidean distance between the two scaled vectors. For instance, the normalized vector distance between DFA$_1$ and DFA$_2$ is 0.5, but this jumps to 0.6455 if q_2 is converted into a final state. In practice, this function seems to be the most reliable among the three distance functions we have considered.

2.2 Symmetric Distance

Let us define the symmetric difference for two DFAs to be the minimized DFA that accepts precisely the strings accepted by exactly one of the two given DFAs, and not by the other. As an example, Fig. 3 shows the symmetric difference DFA for DFA$_1$ and DFA$_2$. If the two DFAs are equivalent, then the symmetric difference DFA has a single (rejecting) state, and in this special case we define the **symmetric distance** between the two DFAs as zero. In all other cases, the symmetric difference between the two DFAs is defined as the number of states in the symmetric difference DFA. The symmetric difference between DFA$_1$ and DFA$_2$ above is 6, and this value is *unchanged* if q_2 is converted into a final state. This notion of distance appears to be quite natural, because it captures the complexity (as measured by number of DFA states) of the language representing the extent to which the two DFAs differ. However, this distance appears to be useful only in a few cases. For instance, if the two DFAs are complements of each other (i.e., they are identical except that any accepting state in one DFA is a rejecting state in the other, and vice versa), then the symmetric distance between the DFAs is 1 (suggesting that the DFAs are "close", which indeed they are), whereas all other distances suggest that the DFAs are "far" from each other.

2.3   *Structural Distance*

Using the notion of structural information defined earlier (as a vector $\mathbf{v}_q$ for each state q), we define the **structural distance** between two DFAs as the number of distinct vectors for states in the symmetric difference DFA. For instance, the structural distance between DFA_1 and DFA_2 is 5, but this increases to 6 if q_2 is converted into a final state. Once again, in the special case where the two DFAs are identical, the structural distance between the two DFAs is defined to be zero.

In this paper, we have used a linear combination of these three distance functions, using linear regression to tune the relative weights for each function so that the distance computed by the combined function closely matches the edit distance function (defined in [1]) on a set of test data.

3   Experiments

We have modified an existing open-source extension to JFLAP [9], which provides students with two kinds of feedback, shown in Fig. 4 and Fig. 5. Before the student gets started on a practice problem and at any time thereafter, she can request two types of hints that help her understand the problem better (Fig. 4): she can view (randomly chosen) strings that the DFA should accept, and she can test whether specific strings should be accepted or rejected by the DFA. Once the student is ready to tackle the problem and draws a candidate DFA, additional feedback is generated (Fig. 5). This always informs the student whether her solution is correct or incorrect. In the latter case, the student can optionally request additional hints (as shown in Fig. 5), and the tool generates strings that her DFA should accept (but doesn't), as well as strings that her DFA should reject (but doesn't). In this paper, we have modified JFLAP to record the time at which these interactions take place. Further, we have logged the actual DFA constructed by the student at each attempt, and whether the student consulted any additional hints (as shown in Fig. 4 and Fig. 5) in between the previous attempt and the current one.

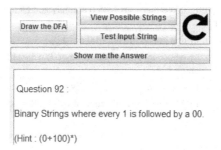

Fig. 4 Feedback available to students before/during problem solving.

Fig. 5 Feedback after testing a potential solution.

For our experiments, we designed a set of 15 unique questions involving DFA constructions, as shown in Table 1. We requested 30 student volunteers to attempt to solve as many of these problems as possible within a fixed amount of time (90 minutes). Our tool ordered questions randomly from the pool of 15 questions, and all students worked independently. After completing their work, we requested students to allow us to investigate the (anonymized) logs of their attempts recorded by JFLAP. All 30 students agreed to submit their logs, and we obtained a total of 350 attempts for all students (this includes multiple attempts for the same problem).

Table 1 The 15 DFA construction questions

Question	Description
1	Strings over $\{0, 1\}$ whose length is a multiple of 3
2	Strings over $\{a, b\}$ whose 2nd letter from the left is a
3	Strings over $\{0, 1\}$ that contain 011
4	Strings over $\{a, b\}$ with exactly two a's
5	Strings over $\{a, b\}$ with at least two consecutive a's
6	Strings over $\{a, b, c\}$ with at least three letters not equal to a
7	Strings over $\{a, b\}$ with 0 or more a's followed by odd number of b's
8	Strings over $\{a, b\}$ that start with ab and end with aa
9	Strings over $\{a, b\}$ that are palindromes of length 3
10	Strings over $\{0, 1\}$ whose integer value is divisible by 6
11	Strings over $\{0, 1\}$ that either start with 0 and have odd length OR start with 1 and have even length
12	Strings over $\{0, 1\}$ that contain neither 00 nor 11
13	Strings over $\{a, b\}$ whose length is a multiple of 3, which have 0 or more a's followed by 0 or more b's
14	Strings over $\{0, 1\}$ whose integer value is divisible by 3 but NOT divisible by 4
15	Strings over $\{0, 1\}$ where each occurrence of 01 is followed immediately by 11

4 Results

We were interested in determining whether the logs could help us rank students (according to ability) and problems (according to difficulty). While analyzing the data, we discovered that several students made multiple attempts to solve a problem, sometimes spending almost 20 minutes on a single problem. We were curious to see whether these students were making "progress" over these multiple attempts. Hence, we defined our notions of distance and learning velocity to quantify the "progress" (if any) made by students.

Our tool can compare the attempts made for a particular problem by one student against the attempts made for the same problem by one or more others. The comparison can be plotted as shown in Fig. 6, which compares the attempts made by student 9 and student 16 on a particular problem. The x-axis represents the time spent (in seconds), the y-axis represents distance (as defined in Section 2), and the slope of the line represents the learning velocity. Fig. 6 also shows the additional hints used by each student as they worked their way through the problem.

Both students were able to solve the problem, but while student 16 required 1084 seconds (~18 minutes) to solve this problem, student 9 required only 284 seconds (less than 5 minutes). Furthermore, student 9 used just one additional hint (apart from the "incorrect" feedback after each attempt except the last), whereas student 16 required several hints. (It is interesting to note that student 16 actually made more initial progress, before moving *away* from the solution.) Apart from these differences, there are interesting similarities between the two plots in Fig. 6. The initial attempts of both students is quite far from the correct solution, and both try several attempts while making little or no apparent "progress", as represented by the plateaus in the two plots. Then, both students finally "get it" and the distance falls sharply. On closer investigation, we discovered that both students were making exactly the same error before they finally recognized the issue.

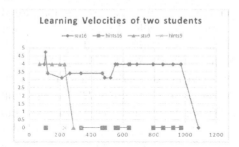

Fig. 6 Comparison of learning velocities of two students for the same problem.

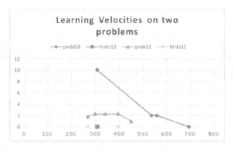

Fig. 7 Comparison of learning velocities of the same student on two different problems.

On the basis of this data, we can confidently rank student 9 higher than student 16 *for this type of problem*. However, a student's learning velocity varies (as one would expect) from problem to problem. Fig. 7 compares the learning velocities of a *single* student on two problems (10 and 12) of similar difficulty[2]. The student's initial attempt for problem 12 was quite close to being correct, but subsequent attempts moved slightly away from the solution, before finally getting very close to (but not equal to) the correct DFA. At this point, the student appears to have given up. In contrast, the same student began problem 10 from much further away, and spent considerably more time (making steady progress throughout), before obtaining the correct solution.

5   Discussion and Future Directions

Our extension to JFLAP can compute learning velocities based on any distance function. This data is presented to the instructor for visualization as explained in Section 4, and can also be fed into other EDM tools. Our findings raise several interesting questions. For instance, consider the two learning velocities curves shown in Fig. 6. What explains the long time both students spent while making no "progress"? In this case, we discovered that the question supported at least two valid interpretations (similar to the example presented in Section 2). Thus, our visualization can help the instructor identify such problematic questions by probing curious patterns further. Another question is raised by the visualization in Fig. 7: why did this student abandon problem 12 despite getting "so close" to the solution? When we investigated this, we discovered that JFLAP was giving feedback/hints that were confusing. Tools which automatically generate useful feedback for students are naturally desirable, but the possibility that students fail to understand these hints has also been noted elsewhere [4].

Our distance functions, being simpler than those in [1], can be computed very rapidly. Thus, we can add a *"How close am I?"* feedback button for students to check their progress. We are investigating the usefulness of this feature. We have also added features to JFLAP that permit instructors to identify questions for which the learning velocities are generally fast (which corresponds to easy questions) or generally slow (hard questions), based on a corpus of student logs.

Finally, we believe that administrators can use learning velocities to assess the quality of learning for a given batch of students, by determining learning velocities for all students on a pre-defined type of problem. These statistics can, for instance, be used to measure the quality of instruction given by the course instructor, relative to similar statistics gathered with previous batches of students.

---

[2] In the (subjective) opinion of the last author, who has experience teaching this material, students should find problem 10 slightly easier than problem 12.

References

1. Alur, R., D'Antoni, L., Gulwani, S., Kini, D., Viswanathan, M.: Automated grading of DFA constructions. In: Proceedings of the Twenty-Third International Joint Conference on Artificial Intelligence (IJCAI 2013), pp. 1976–1982 (2013)
2. Beck, J.E., Mostow, J.: How who should practice: using learning decomposition to evaluate the efficacy of different types of practice for different types of students. In: Woolf, B.P., Aimeur, E., Nkambou, R., Lajoie, S. (eds.) ITS 2008. LNCS, vol. 5091, pp. 353–362. Springer, Heidelberg (2008)
3. Cocea, M., Hershkovitz, A., Baker, R.S.J.D.: The impact of off-task and gaming behaviors on learning: immediate or aggregate?. In: Proceedings of the 14th International Conference on Artificial Intelligence in Education, pp. 507–514 (2009)
4. D'Antoni, L., Kini, D., Alur, R., Gulwani, S., Viswanathan, M., Hartmann, B.: How Can Automatic Feedback Help Students Construct Automata? ACM Transactions on Computer-Human Interaction **22**(2), Article 9 (2015)
5. JFLAP website. http://www.jflap.org
6. Joint Task Force on Computing Curricula, Association for Computing Machinery (ACM) and IEEE Computer Society: Computer Science Curricula 2013: Curriculum Guidelines for Undergraduate Degree Programs in Computer Science. ACM, New York, NY (2013)
7. Rodger, S.H., Finley, T.W.: JFLAP – An Interactive Formal Languages and Automata Package. Jones and Bartlett, Sudbury (2006)
8. Roger, S.H., Lim, J., Reading, S.: Increasing interaction and support in the formal languages and automata theory course. In: Innovation and Technology in Computer Science Education (ITiCSE 2007), pp. 58–62 (2007)
9. Shekhar, V.S., Agarwalla, A., Agarwal, A., Nitish B., Kumar, V.: Enhancing JFLAP with automata construction problems and automated feedback. In: Proceedings of the 7th International Conference on Contemporary Computing (IC3 2014), pp. 19–23 (2014)

A Discrete Krill Herd Method with Multilayer Coding Strategy for Flexible Job-Shop Scheduling Problem

Gai-Ge Wang, Suash Deb and Sabu M. Thampi

Abstract Krill herd (KH) algorithm is a novel swarm-based approach which mimics the herding and foraging behavior of krill species in sea. In our current work, KH method is discretized and incorporated into some heuristic strategies so as to form an effective approach, called discrete krill herd (DKH). The intention has been to use DKH towards solving the flexible job-shop scheduling problem (FJSSP). Firstly, instead of continuous code, a multilayer coding strategy is used in preprocessing stage which enables the KH method to deal with FJSSP. Subsequently, the proposed DKH method is applied to find the best scheduling sequence within the promising domain. In addition, elitism strategy is integrated to DKH with the aim of making the krill swarm move towards the better solutions all the time. The performance of the proposed discrete krill herd algorithm is verified

G.-G. Wang(✉)
School of Computer Science and Technology, Jiangsu Normal University, Xuzhou 221116, Jiangsu, China
e-mail: gaigewang@163.com

Institute of Algorithm and Big Data Analysis, Northeast Normal University, Changchun, 130117, China

School of Computer Science, Northeast Normal University, Changchun 130117, China

S. Deb
Department of Computer Science & Engineering, Cambridge Institute of Technology, Cambridge Village, Tatisilwai, Ranchi, 835103, Jharkhand, India
e-mail: suashdeb@gmail.com

S.M. Thampi
Department of Computer Engineering, Indian Institute of Information Technology & Management - Kerala (IIITM-K), Technopark Campus, Trivandrum 695581, Kerala, India
e-mail: smthampi@ieee.org

© Springer International Publishing Switzerland 2016
S. Berretti et al. (eds.), *Intelligent Systems Technologies and Applications*,
Advances in Intelligent Systems and Computing 384,
DOI: 10.1007/978-3-319-23036-8_18

by two FJSSP instances, and the results clearly demonstrate that our approach is able to find the better scheduling in most cases than some existing state-of-the-art algorithms.

1   Introduction

Scheduling is one of the most popular problems in management and computer science. Among various scheduling problems, the job-shop scheduling problem (JSSP) has turned out to be increasingly essential, not only in modern industrial field, but also in engineering optimization process so as to save resources. The flexible job-shop scheduling problem (FJSSP) is an extended version of classical job-shop scheduling problem. Compared with the JSSP, an operation can be processed on more than one machine in FJSSP. Therefore, FJSSP being a NP-hard problem [1], is more difficult to deal with since it needs to confirm the appointment of machines to operations as well as the sequencing of the operations on the assigned machines. Since the introduction of FJSS, a great number of methods have been proposed, such as biogeography-based optimization (BBO) [2], differential evolution (DE) [3], teaching-learning-based optimization (TLBO) [4], estimation of distribution algorithm (EDA) [5], artificial bee colony (ABC) [6,7], particle swarm optimization (PSO) [8-10], shuffled frog-leaping algorithm (SFLA) [11]. For all these methods, metaheuristic plays the most significant role. Recently, an array of effective algorithms have been proposed, such as particle swarm optimization (PSO) [12-15], dragonfly algorithm (DA) [16], ant colony optimization (ACO) [17-19], bat algorithm (BA) [20-24], monarch butterfly optimization (MBO) [25], differential evolution (DE) [26-29], firefly algorithm (FA) [30-32], biogeography-based optimization (BBO) [33-39], wolf search algorithm (WSA) [40], ant colony optimization (ACO) [17-19], cuckoo search (CS) [41-46], artificial bee colony (ABC) [47-49], ant lion optimizer (ALO) [50], multi-verse optimizer (MVO) [51], gravitational search algorithm (GSA) [52-54], animal migration optimization (AMO) [55], interior search algorithm (ISA) [56], grey wolf optimizer (GWO) [57,58], harmony search (HS) [59-61], flower pollination algorithm (FOA) [62], and krill herd (KH) [63,64].

KH method [63] is a relatively new swarm-based heuristic algorithm for global optimization. Since then, various improved technologies have been incorporated into the basic KH method to enhance its performance. Wang et al. have combined chaos theory with KH method [65,66]. In order to make good balance between exploration and exploitation and improve the population diversity, some new exploration and exploitation strategies have been added to the KH method. These added strategies are originated from SA [67], CS [68], HS [69], GA [70], BBO [71]. However, the basic KH method and these improved ones are mainly intended towards solving continues optimization problem. In our present work, a discrete version of KH method, called discrete krill herd (DKH) algorithm, is proposed for discrete optimization. In DKH, KH method is combined with some heuristic strategies with the aim of solving the flexible job-shop scheduling problem. Firstly, in order to make KH method deal with FJSSP, a new encoding method, named multilayer coding method, is used in the preprocessing stage. Then, the discretized version of the KH method is applied to find the best scheduling by using three movements. Further, in order to accelerate the convergence of the proposed method, an elitism strategy is added to KH. The performance of the proposed discrete krill herd algorithm is verified by two FJSSP instances, and the results show that our approach is able to find the better scheduling in most cases than some existing state-of-the-art algorithms.

The rest of paper is structured as follows. Section 2 provides the description of KH method and FJSSP, while Section 3 gives the mainframe of discrete KH method and discusses how the discrete KH method can be used to solve FJSSP. With the aim of showing the performance of DKH method, several simulation results, comparing DKH with other optimization methods for two scheduling cases, are presented in Section 4. Section 5 presents some concluding remarks and scope for further works.

2   Preliminaries

2.1   KH Algorithm

The herding behavior of krill individuals is simplified and idealized by Gandomi and Alavi, and KH [63] is put forward as a generic heuristic optimization method for the global optimization problem. In KH method, three movements are repeated until the given stopping criteria are satisfied. The directions of krill individuals are determined by three movements and they can be described as follows:

i.   foraging action;

ii.   movement influenced by other krill;

iii.   physical diffusion [63].

Therefore, the position of krill can be updated as [63]:

$$\frac{dX_i}{dt} = F_i + N_i + D_i \tag{1}$$

where F_i, N_i, and D_i are three above corresponding movements for the krill i, respectively [63].

The first movement F_i can be further divided into two parts: the present location and the knowledge about the previous one. Accordingly, this motion can be formulated as follows:

$$F_i = V_f \beta_i + \omega_f F_i^{old} \tag{2}$$

where

$$\beta_i = \beta_i^{food} + \beta_i^{best} \tag{3}$$

and V_f is the foraging speed, ω_f is the inertia weight in (0, 1), F_i^{old} is the last foraging motion.

The second one, N_i, can be estimated by the three effects: target effect, local effect and repulsive effect [63]. For the krill i, it can be formulated as shown below:

$$N_i^{new} = N^{max}\alpha_i + \omega_n N_i^{old} \tag{4}$$

and N^{max} is the maximum induced speed, ω_n is the inertia weight in (0, 1), N_i^{old} is the last motion.

In essence, the third movement is a random process. This movement is able to increase the diversity of population and make krill swarm escape from local optima. Also, it can make trade-off between exploration and exploitation by adjusting the walk steps [63]. Accordingly, this movement can be given below:

$$D_i = D^{max}\delta \tag{5}$$

where D^{max} is the maximum diffusion speed, and δ is the oriented vector [63].

More details about regular KH approach and the three movements can be referred as [63,72].

2.2 FJSSP

The task of job shop scheduling consists of allocating the processing job shop in accordance with the reasonable requirements of product manufacturing so as to enable any organization to achieve the rational use of resources and improve economic efficiency of enterprises in product manufacturing system. Job shop scheduling problem can be described mathematically as n parts to be processed on m machines, and job shop scheduling mathematical model can be given as follows:

(1) Machine set $M=\{m_1, m_2, \cdots, m_m\}$, m_j indicates the j-th machine, $j=1, 2, \cdots, m$.
(2) Part set $P=\{p_1, p_2, \cdots, p_n\}$, p_i denotes the i-th part to be process, $i=1, 2, \cdots, n$.
(3) Operation set $OP=\{op_1, op_2, \cdots, op_n\}$, $op_i=\{op_{i1}, op_{i2}, \cdots, op_{ik}\}$ denotes operations of part p_i.
(4) Potential machine set $OPM=\{op_{i1}, op_{i2}, \cdots, op_{ik}\}$, $op_{ij}=\{op_{ij1}, op_{ij2}, \cdots, op_{ijk}\}$. It shows that operation j for part p_i can be process on any machine in op_{ij}.
(5) Time matrix $\mathbf{T}$, $t_{ij}\in\mathbf{T}$ represents the processed time for part p_i on machine j.

Job shop scheduling problem has the following features: universality, complexity, dynamic fuzzy and multi constraint and it can be generally solved by optimization scheduling algorithm and heuristic methods. For the present work, a discrete version of KH method is used to solve flexible job shop scheduling problem.

3   Discrete KH Method for FJSSP

3.1   Multilayer Coding Strategy

In DKH, integer coding is used to encode each krill individual that represents processing sequence of all the parts. If the total number of parts to be processed is n and the number of operations for part n_i is m_j, then a krill individual can be represented by an integer string with the length of $2\sum_{i=1}^{k} n_i m_j$. The front part of the integer string denotes the processing sequence on the machine, while the latter part denotes processing machine number for each operation. If a krill individual is encoded as follows:

$$[5\ 3\ 1\ 4\ 1\ 3\ 2\ 4\ 5\ 2\ 1\ 2\ 2\ 3\ 4\ 3\ 2\ 1\ 4\ 3] \qquad (6)$$

The Eq. (6) represents processing sequence of five parts twice on four machines. Among them, the first 10 integers indicate the processing sequence of parts: Part 5-> Part 3-> Part 1-> Part 4-> Part 1-> Part 3-> Part 2-> Part 4-> Part 5-> Part 2. The integers from the 11th to 20th denote processing machine, and they follow the machine sequences below: Machine 1-> Machine 2-> Machine 2-> Machine 3-> Machine 4-> Machine 3-> Machine 2-> Machine 1-> Machine 4-> Machine 3.

Of course, the best krill individual should be decoded at the end of the optimization process in order to get the final scheduling. The decoding process is just the opposite of the multilayer encoding process.

3.2   Fitness Values

Fitness of each krill individual is the completion time of the whole parts, and for krill i, its fitness value can be evaluated as:

$$fitness(i) = \text{time} \qquad (7)$$

where time signifies the completion time of the whole parts. If a krill individual i can provide scheduling sequences by using shorter time than krill j, it is better than krill j.

3.3   Discrete KH for FJSSP

In order to make the population converge towards much better positions, the elitism strategy is also incorporated into the DKH method. At the end of the optimization process, *KEEP* best krill individuals are saved. And then, all the krill individuals in krill swarm are updated according to three movements. At the end of the search process at each generation, *KEEP* worst krill individuals will be replaced by *KEEP* saved best ones. This elitism strategy can make the updated population not worse than previously updated ones.

By merging multilayer coding strategy and elitism strategy with discrete KH method, the discrete KH algorithm for FJSSP has been developed, and its mainframe can be described in Algorithm 1. Here, *NP* is population size.

Algorithm 1. *Discrete KH algorithm for FJSSP*

Begin

    Step 1: Initialization. Set the function evaluations FES; initialize the population P of *NP* krill by using multilayer coding method; set the foraging speed V_f, the maximum diffusion speed D^{max}, and the maximum induced speed N^{max}; the number of elitism *KEEP*.

    Step 2: Evaluating population. Evaluate the krill population based on its position as Eq. (7).

    Step 3: While the function evaluations is less than FES **do**

        Sort all the krill according to their fitness.

        Save *KEEP* best krill individuals;

    for i=1:*NP* (all krill) **do**

        Perform foraging action.

        Perform movement influenced by other krill.

        Perform physical diffusion.

        Update position for krill i.

        Evaluate each krill based on its new position X_{i+1} as Eq. (7).

    end for i

        Sort all the krill and find the current best.

        Replace *KEEP* worst krill individuals with *KEEP* best ones.

    $t = t+1$.

    Step 4: end while

    Step 5: Decode the best krill individual to get final scheduling sequence.

End.

---

4 Simulation Results

In this section, the DKH method is verified from various aspects by using two job shop scheduling problems in comparison with ABC [47], ACO [17] and GA [73]. In order to obtain fair results, all the implementations are conducted under the same conditions as shown in [69].

In the simulations below, we will use the unchanged parameters for DKH: the foraging speed V_f =0.6, the maximum diffusion speed D^{max}=0.7 and the maximum induced speed N^{max}=0.7. For ABC, ACO and GA, we set the parameters as [33]. We ran two hundred times for each method on each scheduling instance to achieve typical performances. The optimal solution achieved by each method for each benchmark is marked in bold through 2500 and 5000 function evaluations (FES).

4.1 Instance 1

The detailed information about scheduling Instance 1 can be shown in Table 1. In Instance 1, the number of parts and operations are six. By using multilayer coding strategy, the length of each individual is $2\times(6\times6)=72$ in for method. The results of Instance 1 are shown in Table 2, which indicate the average, the worst, the best as well as Std (standard deviation) performance of each method.

Table 1 The detailed information for Instance 1

		Part 1	Part 2	Part 3	Part 4	Part 5	Part 6
	Operation 1	5	6	4	[2,9]	[3,7]	5
	Operation 2	4	[2,9]	8	[6,7]	5	[1,10]
Jm	Operation 3	3	[6,8]	7	[2,1]	[4,10]	5
	Operation 4	5	2	[4,7]	10	[2,5]	[3,6]
	Operation 5	[4,5]	5	[9,10]	6	2	[3,8]
	Operation 6	[2,6]	4	[6,9]	7	8	[3,9]
		Part 1	Part 2	Part 3	Part 4	Part 5	Part 6
	Operation 1	3	10	9	[5,4]	[3,3]	10
	Operation 2	6	[8,6]	4	[2,6]	3	[3,3]
T	Operation 3	4	[5,7]	7	[5,5]	[9,11]	1
	Operation 4	7	3	[4,6]	3	[1,7]	[3,6]
	Operation 5	[6,4]	10	[7,9]	8	5	[4,7]
	Operation 6	[3,7]	10	[8,7]	9	4	[9,4]

Table 2 Optimization results obtained via four methods over 200 runs and 2500/5000 FEs for Instance 1

		ABC	ACO	GA	DKH
FES=2500	Best	47	48	46	**44**
	Mean	53.96	51.94	51.53	**49.22**
	Worst	**60**	**60**	**60**	60
	Std	1.84	**1.01**	2.07	1.61
FES=5000	Best	47	47	46	**43**
	Mean	53.05	51.28	50.29	**49.10**
	Worst	**59**	**59**	**59**	59
	Std	1.61	**1.07**	1.85	1.50

From Table 2, for the FES=2500, DKH has the best performance among four methods for scheduling Instance 1. KH can successfully solve this problem by using 44*s*, which is far less than ABC, ACO and GA. For the worst performance, four methods have similar performance. Convergent process of DKH method for Instance 1 can be shown in Fig. 1 and the final solution is also shown in Gantt chart (see Fig. 2). For FES=5000, DKH has also the best performance among four methods. Carefully looking at Table 2 reveals that the minimum time obtained by ABC and GA when FES=2500 is equal to FES=5000, while DKH and ACO has better time than FES=2500. For Std performance, ACO has the smallest Std among four methods. In addition, Optimization process of KH and its final solution in the form of Gantt chart can be shown in Fig. 3 and Fig. 4, respectively.

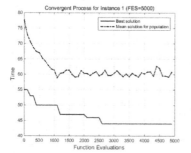

Fig. 1 Convergent process of DKH method for Instance 1 (FES=2500)

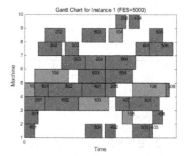

Fig. 2 Gantt chart for Instance 1 (FES=2500)

Fig. 3 Convergent process of DKH method for Instance 1 (FES=5000)

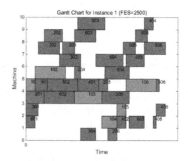

Fig. 4 Gantt chart for Instance 1 (FES=5000)

4.2   Instance 2

In addition, DKH method is tested by scheduling Instance 2 (see Table 3). Similar to Instance 1, the number of parts and machines in Instance 2 are also six. By using multilayer coding strategy, the length of each individual is 2×(6×6)=72 for each method. The optimization results of Instance 2 are shown in Table 4.

From Table 4, for the FES=2500, ACO, GA and DKH have the same best performance for scheduling Instance 2, and they can successfully solve this problem by using 40s. For average time, DKH performs better than other three methods. For the worst performance, four methods have similar performance. Convergent process of DKH method for Instance 2 can be seen in Fig. 5 and the final solution is also shown in Gantt chart (see Fig. 6). For FES=5000, ACO, GA and DKH have the same best performance for scheduling Instance 2, and they can successfully solve this problem by using 40s. For average time, DKH has the minimum time though it is little less than ACO. Carefully looking at Table 4, the minimum time obtained by four methods when FES=2500 is equal to FES=5000. This indicates that the increment of FES does not improve the performance of the method all the time. For Std performance, ACO has the smallest Std among four methods. In addition, Optimization process of DKH and its final solution in the form of Gantt chart can be shown in Fig. 7 and Fig. 8, respectively.

Table 3 The detailed information for Instance 2

		Part 1	Part 2	Part 3	Part 4	Part 5	Part 6
	Operation 1	[3,10]	2	[3,9]	4	5	2
	Operation 2	1	3	[4,7]	[1,9]	[2,7]	[4,7]
Jm	Operation 3	2	[5,8]	[6,8]	[3,7]	[3,10]	[6,9]
	Operation 4	[4,7]	[6,7]	1	[2,8]	[6,9]	1
	Operation 5	[6,8]	1	[2,10]	5	1	[5,8]
	Operation 6	5	[4,10]	5	6	[4,8]	3
		Part 1	Part 2	Part 3	Part 4	Part 5	Part 6
	Operation 1	[3,5]	6	[1,4]	7	6	2
	Operation 2	10	8	[5,7]	[4,3]	[10,12]	[4,7]
T	Operation 3	9	[1,4]	[5,6]	[4,6]	[7,9]	[6,9]
	Operation 4	[5,4]	[5,6]	5	[3,5]	[8,8]	1
	Operation 5	[3,3]	3	[9,11]	1	5	[5,8]
	Operation 6	10	[3,3]	1	3	[4,7]	3

Table 4 Optimization results obtained via four methods over 200 runs and 2500/5000 FEs for Instance 2

		ABC	ACO	GA	DKH
	Best	42	40	40	40
FES=2500	Mean	47.41	40.45	44.95	40.21
	Worst	54	54	54	54
	Std	1.77	0.75	1.63	1.13
	Best	42	40	40	40
FES=5000	Mean	47.21	40.20	43.71	40.15
	Worst	51	51	51	51
	Std	1.43	0.07	1.67	1.15

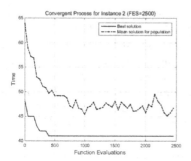

Fig. 5 Convergent process of DKH method for Instance 2 (FES=2500)

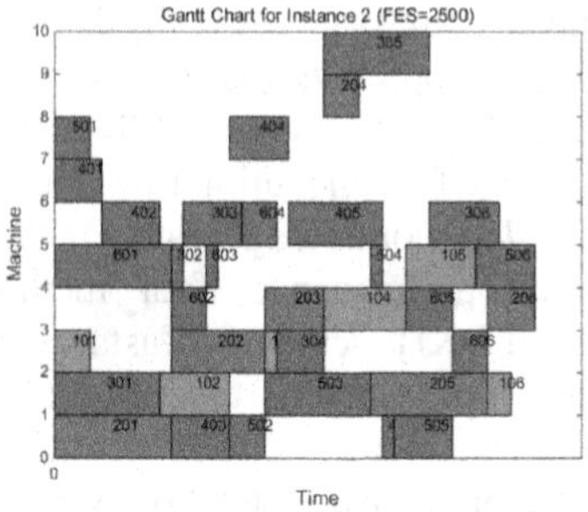

Fig. 6 Gantt chart for Instance 2 (FES=2500)

Fig. 7 Convergent process of DKH method for Instance 2 (FES=5000)

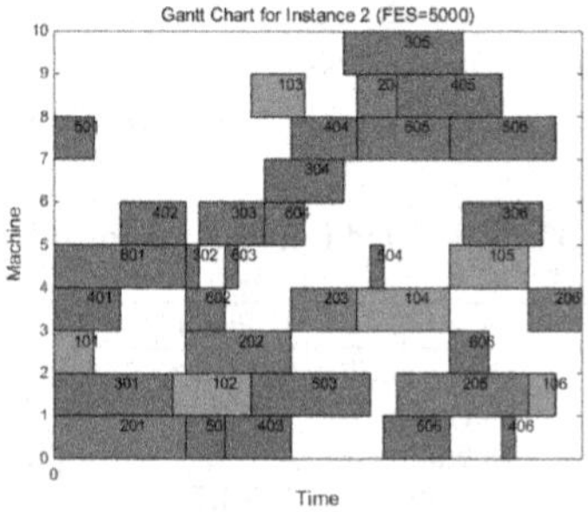

Fig. 8 Gantt chart for Instance 2 (FES=5000)The results obtained by four methods on two scheduling instances indicate that DKH is well capable of solving scheduling problem in most cases. DKH method is an effective method for discrete optimization.

5 Conclusion

In our current work, the basic KH method for continuous optimization is discretized so as to introduce, what is called, discrete krill herd (DKH), for discrete optimization. In DKH, KH method is combined with some heuristic strategies with the aim of solving the flexible job-shop scheduling problem. Firstly, in order to make KH method deal with FJSSP, multilayer coding method, is used in the preprocessing stage. Then, DKH method is applied to find the best scheduling by using three movements. Furthermore, an elitism strategy is added to DKH in order to preserve the best krill individuals. The performance of the proposed discrete krill herd algorithm is verified by two FJSSP instances, and the results obtained by four methods on two scheduling instances clearly indicate that DKH is well capable of solving scheduling problem in most cases. In other words, DKH method is clearly an effective method for discrete optimization.

Despite various advantages of the DKH method, the following points should be clarified and dealt with in the future research.

Firstly, the parameters used in a metaheuristic method largely influence its performance. In the present work, the parameters are not adjusted carefully. How to select the best parameter settings is worth further study.

Secondly, computational requirements are of vital importance for a metaheuristic method. How to make the DKH method implement faster is another area which merits investigation.

At last, despite superior performance of DKH w.r.t. ABC, ACO and GA in most cases, the Std obtained by DKH is worse than ACO. The poor Std will limit the application of DKH. How to decrease the Std range is worthy of further scrutiny. This may be done by theoretically analyzing its convergence through dynamic systems and Markov chain.

Acknowledgments This work was supported by Research Fund for the Doctoral Program of Jiangsu Normal University (No. 13XLR041).

References

1. Hu, Y., Yin, M., Li, X.: A novel objective function for job-shop scheduling problem with fuzzy processing time and fuzzy due date using differential evolution algorithm. Int. J. Adv. Manuf. Tech. **56**(9), 1125–1138 (2011). doi:10.1007/s00170-011-3244-3

2. Lin, J.: A hybrid biogeography-based optimization for the fuzzy flexible job shop scheduling problem. Knowl.-Based Syst. (2015). doi:10.1016/j.knosys.2015.01.017

3. Li, X., Yin, M.: An opposition-based differential evolution algorithm for permutation flow shop scheduling based on diversity measure. Adv. Eng. Softw. **55**, 10–31 (2013). doi:10.1016/j.advengsoft.2012.09.003

4. Xu, Y., Wang, L., S-y, W., Liu, M.: An effective teaching–learning-based optimization algorithm for the flexible job-shop scheduling problem with fuzzy processing time. Neurocomputing **148**, 260–268 (2015). doi:10.1016/j.neucom.2013.10.042

5. Wang, S., Wang, L., Xu, Y., Liu, M.: An effective estimation of distribution algorithm for the flexible job-shop scheduling problem with fuzzy processing time. Int. J. Prod. Res. **51**(12), 3778–3793 (2013). doi:10.1080/00207543.2013.765077

6. Wang, L., Zhou, G., Xu, Y., Liu, M.: A hybrid artificial bee colony algorithm for the fuzzy flexible job-shop scheduling problem. Int. J. Prod. Res. **51**(12), 3593–3608 (2013). doi:10.1080/00207543.2012.754549

7. Caniyilmaz, E., Benli, B., Ilkay, M.S.: An artificial bee colony algorithm approach for unrelated parallel machine scheduling with processing set restrictions, job sequence-dependent setup times, and due date. Int. J. Adv. Manuf. Tech. **77**(9–12), 2105–2115 (2014). doi:10.1007/s00170-014-6614-9

8. J-q, L., Y-x, P.: A hybrid discrete particle swarm optimization algorithm for solving fuzzy job shop scheduling problem. Int. J. Adv. Manuf. Tech. **66**(1–4), 583–596 (2012). doi:10.1007/s00170-012-4337-3

9. Lei, D.: Pareto archive particle swarm optimization for multi-objective fuzzy job shop scheduling problems. Int. J. Adv. Manuf. Tech. **37**(1–2), 157–165 (2007). doi:10.1007/s00170-007-0945-8

10. Chen, C.-L., Huang, S.-Y., Tzeng, Y.-R., Chen, C.-L.: A revised discrete particle swarm optimization algorithm for permutation flow-shop scheduling problem. Soft. Comput. **18**(11), 2271–2282 (2013). doi:10.1007/s00500-013-1199-z

11. Fang, C., Wang, L.: An effective shuffled frog-leaping algorithm for resource-constrained project scheduling problem. Comput. Oper. Res. **39**(5), 890–901 (2012). doi:10.1016/j.cor.2011.07.010

12. Kennedy, J., Eberhart, R.: Particle swarm optimization. Paper presented at the Proceeding of the IEEE International Conference on Neural Networks, Perth, Australia, November 27–December 1, 1995

13. Zhao, X., Liu, Z., Yang, X.: A multi-swarm cooperative multistage perturbation guiding particle swarm optimizer. Appl. Soft. Compt. **22**, 77–93 (2014). doi:10.1016/j.asoc.2014.04.042

14. Mirjalili, S., Lewis, A.: S-shaped versus V-shaped transfer functions for binary Particle Swarm Optimization. Swarm. Evol. Comput. **9**, 1–14 (2013). doi:10.1016/j.swevo.2012.09.002

15. Wang, G.-G., Gandomi, A.H., Yang, X.-S., Alavi, A.H.: A novel improved accelerated particle swarm optimization algorithm for global numerical optimization. Eng. Computation **31**(7), 1198–1220 (2014). doi:10.1108/EC-10-2012-0232

16. Mirjalili, S.: Dragonfly algorithm: a new meta-heuristic optimization technique for solving single-objective, discrete, and multi-objective problems. Neural Comput. Appl. (2015). doi:10.1007/s00521-015-1920-1

17. Dorigo, M., Maniezzo, V., Colorni, A.: Ant system: optimization by a colony of cooperating agents. IEEE Trans. Syst. Man Cybern. B Cybern. **26**(1), 29–41 (1996). doi:10.1109/3477.484436

18. Zhang, Z., Feng, Z.: Two-stage updating pheromone for invariant ant colony optimization algorithm. Expert Syst. Appl. **39**(1), 706–712 (2012). doi:10.1016/j.eswa.2011.07.062

19. Zhang, Z., Zhang, N., Feng, Z.: Multi-satellite control resource scheduling based on ant colony optimization. Expert Syst. Appl. **41**(6), 2816–2823 (2014). doi:10.1016/j.eswa.2013.10.014

20. Gandomi, A.H., Yang, X.-S., Alavi, A.H., Talatahari, S.: Bat algorithm for constrained optimization tasks. Neural Comput. Appl. **22**(6), 1239–1255 (2013). doi:10.1007/s00521-012-1028-9

21. Yang, X.S., Gandomi, A.H.: Bat algorithm: a novel approach for global engineering optimization. Eng. Computation **29**(5), 464–483 (2012). doi:10.1108/02644401211235834

22. Yang, X.-S.: A new metaheuristic bat-inspired algorithm. In: González, J.R., Pelta, D.A., Cruz, C., Terrazas, G., Krasnogor, N. (eds.) NICSO 2010. SCI, vol. 284, pp. 65–74. Springer, Heidelberg (2010)

23. Mirjalili, S., Mirjalili, S.M., Yang, X.-S.: Binary bat algorithm. Neural Comput. Appl. **25**(3–4), 663–681 (2013). doi:10.1007/s00521-013-1525-5

24. Yang, X.S.: Nature-inspired metaheuristic algorithms, 2nd edn. Luniver Press, Frome (2010)

25. Wang, G.-G., Deb, S., Cui, Z.: Monarch butterfly optimization. Neural Comput. Appl. (2015). doi:10.1007/s00521-015-1923-y

26. Storn, R., Price, K.: Differential evolution-a simple and efficient heuristic for global optimization over continuous spaces. J. Global Optim. **11**(4), 341–359 (1997). doi:10.1023/A:1008202821328

27. Gandomi, A.H., Yang, X.-S., Talatahari, S., Deb, S.: Coupled eagle strategy and differential evolution for unconstrained and constrained global optimization. Comput. Math Appl. **63**(1), 191–200 (2012). doi:10.1016/j.camwa.2011.11.010

28. Zou, D., Wu, J., Gao, L., Li, S.: A modified differential evolution algorithm for unconstrained optimization problems. Neurocomputing **120**, 469–481 (2013). doi:10.1016/j.neucom.2013.04.036

29. Wang, G.-G., Gandomi, A.H., Alavi, A.H., Hao, G.-S.: Hybrid krill herd algorithm with differential evolution for global numerical optimization. Neural Comput. Appl. **25**(2), 297–308 (2014). doi:10.1007/s00521-013-1485-9

30. Gandomi, A.H., Yang, X.-S., Alavi, A.H.: Mixed variable structural optimization using firefly algorithm. Comput. Struct. **89**(23–24), 2325–2336 (2011). doi:10.1016/j.compstruc.2011.08.002

31. Yang, X.S.: Firefly algorithm, stochastic test functions and design optimisation. Int. J. of Bio-Inspired Computation **2**(2), 78–84 (2010). doi:10.1504/IJBIC.2010.032124

32. Wang, G.-G., Guo, L., Duan, H., Wang, H.: A new improved firefly algorithm for global numerical optimization. J. Comput. Theor. Nanos. **11**(2), 477–485 (2014). doi:10.1166/jctn.2014.3383

33. Simon, D.: Biogeography-based optimization. IEEE Trans. Evolut. Comput. **12**(6), 702–713 (2008). doi:10.1109/TEVC.2008.919004

34. Mirjalili, S., Mirjalili, S.M., Lewis, A.: Let a biogeography-based optimizer train your Multi-Layer Perceptron. Inf. Sci. **269**, 188–209 (2014). doi:10.1016/j.ins.2014.01.038

35. Li, X., Yin, M.: Multi-operator based biogeography based optimization with mutation for global numerical optimization. Comput. Math Appl. **64**(9), 2833–2844 (2012). doi:10.1016/j.camwa.2012.04.015

36. Li, X., Yin, M.: Multiobjective binary biogeography based optimization for feature selection using gene expression data. IEEE Trans. Nanobiosci. **12**(4), 343–353 (2013). doi:10.1109/TNB.2013.2294716

37. Lin, J.: Parameter estimation for time-delay chaotic systems by hybrid biogeography-based optimization. Nonlinear Dynam. **77**(3), 983–992 (2014). doi:10.1007/s11071-014-1356-7

38. Lin, J., Xu, L., Zhang, H.: Hybrid biogeography based optimization for constrained optimal spot color matching. Color Research & Application **39**(6), 607–615 (2014). doi:10.1002/col.21836

39. Wang, G., Guo, L., Duan, H., Liu, L., Wang, H., Shao, M.: Path Planning for Uninhabited Combat Aerial Vehicle Using Hybrid Meta-Heuristic DE/BBO Algorithm. Adv. Sci. Eng. Med. **4**(6), 550–564 (2012). doi:10.1166/asem.2012.1223

40. Fong, S., Deb, S., Yang, X.-S.: A heuristic optimization method inspired by wolf preying behavior. Neural Comput. Appl. (2015). doi:10.1007/s00521-015-1836-9

41. Yang, X.S., Deb, S.: Cuckoo search via Lévy flights. In: Abraham, A., Carvalho, A., Herrera, F., Pai, V. (eds.) Proceeding of World Congress on Nature & Biologically Inspired Computing (NaBIC 2009), pp. 210–214. IEEE Publications, USA (2009)

42. Li, X., Wang, J., Yin, M.: Enhancing the performance of cuckoo search algorithm using orthogonal learning method. Neural Comput. Appl. **24**(6), 1233–1247 (2013). doi:10.1007/s00521-013-1354-6

43. Li, X., Yin, M.: Modified cuckoo search algorithm with self adaptive parameter method. Inf. Sci. **298**, 80–97 (2015). doi:10.1016/j.ins.2014.11.042

44. Wang, G.-G., Gandomi, A.H., Zhao, X., Chu, H.E.: Hybridizing harmony search algorithm with cuckoo search for global numerical optimization. Soft. Comput. (2014). doi:10.1007/s00500-014-1502-7

45. Li, X., Yin, M.: A particle swarm inspired cuckoo search algorithm for real parameter optimization. Soft. Comput. (2015). doi:10.1007/s00500-015-1594-8

46. Wang, G.-G., Deb, S., Gandomi, A.H., Zhang, Z., Alavi, A.H.: Chaotic cuckoo search. Soft. Comput. (2015). doi:10.1007/s00500-015-1726-1

47. Karaboga, D., Basturk, B.: A powerful and efficient algorithm for numerical function optimization: artificial bee colony (ABC) algorithm. J. Global Optim. **39**(3), 459–471 (2007). doi:10.1007/s10898-007-9149-x
48. Li, X., Yin, M.: Parameter estimation for chaotic systems by hybrid differential evolution algorithm and artificial bee colony algorithm. Nonlinear Dynam. **77**(1–2), 61–71 (2014). doi:10.1007/s11071-014-1273-9
49. Li, X., Yin, M.: Self-adaptive constrained artificial bee colony for constrained numerical optimization. Neural Comput. Appl. **24**(3–4), 723–734 (2012). doi:10.1007/s00521-012-1285-7
50. Mirjalili, S.: The ant lion optimizer. Adv. Eng. Softw. **83**, 80–98 (2015). doi:10.1016/j.advengsoft.2015.01.010
51. Mirjalili, S., Mirjalili, S.M., Hatamlou, A.: Multi-verse optimizer: a nature-inspired algorithm for global optimization. Neural Comput. Appl. (2015). doi:10.1007/s00521-015-1870-7
52. Rashedi, E., Nezamabadi-pour, H., Saryazdi, S.: GSA: a gravitational search algorithm. Inf. Sci. **179**(13), 2232–2248 (2009). doi:10.1016/j.ins.2009.03.004
53. Mirjalili, S., Wang, G.-G., Coelho, L.: Binary optimization using hybrid particle swarm optimization and gravitational search algorithm. Neural Comput. Appl. **25**(6), 1423–1435 (2014). doi:10.1007/s00521-014-1629-6
54. Mirjalili, S., Lewis, A.: Adaptive gbest-guided gravitational search algorithm. Neural Comput. Appl. **25**(7–8), 1569–1584 (2014). doi:10.1007/s00521-014-1640-y
55. Li, X., Zhang, J., Yin, M.: Animal migration optimization: an optimization algorithm inspired by animal migration behavior. Neural Comput. Appl. **24**(7–8), 1867–1877 (2014). doi:10.1007/s00521-013-1433-8
56. Gandomi, A.H.: Interior search algorithm (ISA): a novel approach for global optimization. ISA Trans. **53**(4), 1168–1183 (2014). doi:10.1016/j.isatra.2014.03.018
57. Mirjalili, S., Mirjalili, S.M., Lewis, A.: Grey wolf optimizer. Adv. Eng. Softw. **69**, 46–61 (2014). doi:10.1016/j.advengsoft.2013.12.007
58. Mirjalili, S.: How effective is the Grey Wolf optimizer in training multi-layer perceptrons. Appl. Intell. (2015). doi:10.1007/s10489-014-0645-7
59. Geem, Z.W., Kim, J.H., Loganathan, G.V.: A new heuristic optimization algorithm: harmony search. Simulation **76**(2), 60–68 (2001). doi:10.1177/003754970107600201
60. Wang, G., Guo, L., Duan, H., Wang, H., Liu, L., Shao, M.: Hybridizing harmony search with biogeography based optimization for global numerical optimization. J. Comput. Theor. Nanos. **10**(10), 2318–2328 (2013). doi:10.1166/jctn.2013.3207
61. Zou, D., Gao, L., Li, S., Wu, J.: Solving 0-1 knapsack problem by a novel global harmony search algorithm. Appl. Soft. Compt. **11**(2), 1556–1564 (2011). doi:10.1016/j.asoc.2010.07.019
62. Yang, X.-S., Karamanoglu, M., He, X.: Flower pollination algorithm: A novel approach for multiobjective optimization. Eng. Optimiz., 1–16 (2013). doi:10.1080/0305215X.2013.832237
63. Gandomi, A.H., Alavi, A.H.: Krill herd: a new bio-inspired optimization algorithm. Commun. Nonlinear Sci. Numer. Simulat. **17**(12), 4831–4845 (2012). doi:10.1016/j.cnsns.2012.05.010
64. Wang, G.-G., Gandomi, A.H., Alavi, A.H., Deb, S.: A hybrid method based on krill herd and quantum-behaved particle swarm optimization. Neural Comput. Appl. (2015). doi:10.1007/s00521-015-1914-z
65. Wang, G.-G., Guo, L., Gandomi, A.H., Hao, G.-S., Wang, H.: Chaotic krill herd algorithm. Inf. Sci. **274**, 17–34 (2014). doi:10.1016/j.ins.2014.02.123
66. Wang, G.-G., Gandomi, A.H., Alavi, A.H.: A chaotic particle-swarm krill herd algorithm for global numerical optimization. Kybernetes **42**(6), 962–978 (2013). doi:10.1108/K-11-2012-0108

67. Wang, G.-G., Guo, L., Gandomi, A.H., Alavi, A.H., Duan, H.: Simulated annealing-based krill herd algorithm for global optimization. Abstr. Appl. Anal. **2013**, 1–11 (2013). doi:10.1155/2013/213853
68. Wang, G.-G., Gandomi, A.H., Yang, X.-S., Alavi, A.H.: A new hybrid method based on krill herd and cuckoo search for global optimization tasks. Int. J. of Bio-Inspired Computation (2014)
69. Wang, G., Guo, L., Wang, H., Duan, H., Liu, L., Li, J.: Incorporating mutation scheme into krill herd algorithm for global numerical optimization. Neural Comput. Appl. **24**(3–4), 853–871 (2014). doi:10.1007/s00521-012-1304-8
70. Wang, G.-G., Gandomi, A.H., Alavi, A.H.: Stud krill herd algorithm. Neurocomputing **128**, 363–370 (2014). doi:10.1016/j.neucom.2013.08.031
71. Wang, G.-G., Gandomi, A.H., Alavi, A.H.: An effective krill herd algorithm with migration operator in biogeography-based optimization. Appl. Math. Model. **38**(9–10), 2454–2462 (2014). doi:10.1016/j.apm.2013.10.052
72. Guo, L., Wang, G.-G., Gandomi, A.H., Alavi, A.H., Duan, H.: A new improved krill herd algorithm for global numerical optimization. Neurocomputing **138**, 392–402 (2014). doi:10.1016/j.neucom.2014.01.023
73. Goldberg, D.E.: Genetic Algorithms in Search, Optimization and Machine learning. Addison-Wesley, New York (1998)

Fine Tuning Machine Fault Diagnosis System Towards Mission Critical Applications

R. Gopinath, C. Santhosh Kumar, V. Vaijeyanthi and K.I. Ramachandran

Abstract Machine condition monitoring has become increasingly important in enhancing productivity, reducing maintenance cost, and casualties in mission critical applications (for example, aerospace). It may be noted that we cannot afford to miss an alarm (fault) condition in a mission critical application as the cost of a failure can be fatal and very expensive, and this means that we need to have a high alarm accuracy, i.e., sensitivity. A better alarm accuracy is desired in mission critical applications, even at the expense of a lower no-alarm detection accuracy, i.e., specificity. A lower specificity means more false alarms. In this work, we propose a novel approach to fine tune a decision tree based machine fault identification system towards a mission critical application like aerospace by improving the sensitivity. We experiment the proposed approach with a three phase 3kVA synchronous generator to diagnose the inter-turn short circuit faults. The proposed approach outperforms the baseline system by an absolute improvement in sensitivity of 1.08%, 0.82%, and 0.53% for the R, Y, and B phase faults respectively.

1 Introduction

In recent years, condition based maintenance (CBM) has become more popular compared to other techniques such as run-to-failure, and periodical maintenance techniques [1]. CBM is performed through continuous monitoring of the condition of a machine under observation. Electrical and/or mechanical signatures are collected from the machine to diagnose the condition of the machine. CBM recommends maintenance actions only when there is an evidence of abnormal behavior in the

R. Gopinath(✉) · C. Santhosh Kumar · V. Vaijeyanthi · K.I. Ramachandran
Machine Intelligence Research Laboratory, Department of Electronics and Communication
Engineering, Amrita School of Engineering, Amrita Vishwa Vidyapeetham, Coimbatore,
Tamil Nadu, India
e-mail: r_gopinath@cb.amrita.edu

© Springer International Publishing Switzerland 2016
S. Berretti et al. (eds.), *Intelligent Systems Technologies and Applications*,
Advances in Intelligent Systems and Computing 384,
DOI: 10.1007/978-3-319-23036-8_19

signatures of the machine [1]. When a fault occurs in the machine, it affects the system characteristics, and this in turn changes the characteristics of voltage, current, vibration, and temperature signals, as the case maybe, and these changes are continuously monitored to track the health condition of the machine.

Machine learning algorithms such as Support Vector Machine (SVM) [2], Artificial Neural Network (ANN) [3], Adaptive Neuro Fuzzy Interference System (ANFIS) [4, 5], decision tree [6], and nearest neighbor rule are some of the popular approaches [7] used in machine condition monitoring. One or more of these algorithms are used in most of the state-of-the-art CBM systems. Among these algorithms, the decision tree is widely used in CBM systems for its improved performance [8]. The advantages of decision tree based CBM systems are that the rules generated through the algorithm can be easily interpreted and easily integrated with other popular approaches [9].

Tran et al. used classification and regression tree (CART) [10] for selecting the prominent features and applied to ANFIS to diagnose the faults in an induction motor. The classifier performance was measured by computing sensitivity, specificity, and overall classification accuracy. The CART based hybrid fuzzy min-max neural network model was used [11] to diagnose the faults in the induction motor using motor current signature analysis. The condition of the machine has been tracked using CART-ANFIS model and identified as a potential tool for machine condition prognosis [12]. Lomax and Vadera [13] review a class of algorithms to optimize several decision tree classifiers and improve the overall classification accuracy by studying the misclassified data samples, and optimizing the tree for improved performance. Kusiak and Verma [14] proposed an algorithm that weights each example in a manner that improves the overall classification accuracy. The weights associated with each of the samples are determined using a heuristic algorithm.

While it is important to optimize the performance of a CBM for overall classification accuracy, it would be of great interest to explore how a CBM system could be fine tuned towards a specific application, by improving the sensitivity. Brummer and Preez [15] proposed an application-independent evaluation of the speaker recognition systems which outputs soft decisions as log-likelihood-ratios, using detection error trade off (DET), as a measure for evaluation. Further, using DET, we can evaluate the performance of the system for the specific application using a predefined detection cost function (DCF). Then, we may fine tune the system by selecting suitable thresholds for the probabilities to optimize the performance at the selected operating point, and thereby minimize the corresponding DCF [15]. The DCF can be used as a measure of the expected cost that needs to be minimized. However, it is not clear how the DCF can be applied to fine tune a decision tree classifier.

In this paper, we propose a cost sensitive CART algorithm to fine tune the machine condition monitoring systems towards mission critical applications using DCF [15] as the measure to be minimized. We accordingly modify the Gini impurity index used in the CART algorithm to improve the sensitivity. To demonstrate the effectiveness of the cost sensitive CART algorithm, we experiment with the detection of interturn short circuit faults in a 3kVA synchronous generator in the R, Y, and B phases separately, considering each of them as a two class classification problem. The details of the experimental setup are discussed in section 2. Section 3 elaborates on the data collection. Section 4 and 5 describe the CART and fine tuning the CART algorithm respectively. Further, section 6 elaborates the experiments and results of this study, and finally section 7 concludes.

2   Experimental Setup

In synchronous generators, short circuit faults may happen in the stator winding or field winding coils. We designed the stator and rotor winding terminals of the synchronous generator to inject short circuit faults in the corresponding winding terminals. Each phase of the stator winding consists of six coils and each coil has three terminal leads taken to the front panel, making 18 taps per phase and 54 taps in total, to be able to inject short circuit faults at different turns of the coil. Details of a similar experimental setup of a brushless synchronous generator may be found in [16]. The synchronous generator with the fault injection capability and the experimental setup are shown in the Fig. 1(a) and 1(b) respectively. In this experimental setup, we can inject faults in the stator winding at 30, 60, and 82 per cent of the total number of turns in each coil. Fig. 1(c) shows the current signature time domain waveforms of the three phases under no-fault condition and with an inter-turn fault injected in the R-phase of the winding.

3   Data Collection and Feature Extraction

Inter-turn short circuit faults are injected in the stator winding terminals available in the front panel of the generator. Short circuit experiments are conducted in a controlled manner. The 3kVA synchronous generator is connected to a resistive load to emulate different load conditions. NI-PXI-6221[1] data acquisition system is used to interface the current sensors with the computer. Each experiment is conducted for ten seconds and the three phase output current signal is sampled at 1 kHz, ensuring Nyquist sampling rate. This process is repeated several times, and sufficient amount of data is collected to train, test and evaluate the performance of the system. Subsequently, the data is segmented into frames of 512 samples. For the identification of the fault, we only consider a frame of 512 samples as the input, that corresponds to 0.5 s data. The signals are then transformed to the frequency domain using Fast Fourier Transform (FFT) to get 512 FFT coefficients. The size of the frame was empirically selected for the best performance. Subsequently, the amplitude of the FFT coefficients, ignoring the phase information, is calculated, and the frequency domain statistical features [17] are calculated. The details of the statistical features are listed in Table 1.

4   Classification and Regression Tree (CART)

The decision tree algorithm is one among the most extensively used approaches for the classification problems in various disciplines. It consists of nodes with a rooted

---

[1] http://sine.ni.com/ds/app/doc/p/id/ds-15/lang/en

(a) Customized 3 kVA Synchronous Generator   (b) Experimental Facility

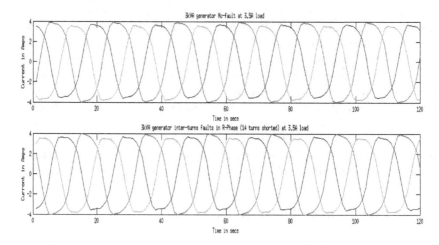

(c) Current signature captured during no fault and inter turn fault conditions for the 3kVA generator

Fig. 1 Experimental setup and Current signature

tree. A node which directs the tree is called root node and it has no incoming edges. All other nodes have one incoming edge. A node with an outgoing edge is called internal node or test node. All other nodes are leaves or terminal nodes. Decision tree inducers will construct decision trees automatically for the input data [18]. The objective is to construct an optimal decision tree by minimizing the generalized errors. There are various decision tree inducers such as, ID3, C4.5, and Classification and Regression Tree (CART) which can be used to construct the tree. In this work, we used CART for its improved performance [19]. The CART algorithm generates a set of questions, the answers to which determine the next set of questions. The result of these questions forms a tree structure whose ends are decision nodes. CART uses Gini-impurity index to split the internal nodes, which measures the divergence between the probability distributions of the target attributes value. The Gini index of a node for a two class problem with alarm and no-alarm as classes may be expressed as [20]

$$I_G = 1 - p(a|t)^2 - p(na|t)^2 \tag{1}$$

Table 1 Frequency Domain Features

Features

$$Feat_1 = \frac{\sum_{k=1}^{K} s(k)}{K}$$

$$Feat_2 = \frac{\sum_{k=1}^{K}(s(k)-Feat_1)^2}{K-1}$$

$$Feat_3 = \frac{\sum_{k=1}^{K}(s(k)-Feat_1)^3}{K(\sqrt{Feat_2})^3}$$

$$Feat_4 = \frac{\sum_{k=1}^{K}(s(k)-Feat_1)^4}{K(Feat_2)^2}$$

$$Feat_5 = \frac{\sum_{k=1}^{K} f_k s(k)}{\sum_{k=1}^{K} s(k)}$$

$$Feat_6 = \sqrt{\frac{\sum_{k=1}^{K}(f_k-Feat_5)^2 s(k)}{K}}$$

$$Feat_7 = \sqrt{\frac{\sum_{k=1}^{K} f_k^2 s(k)}{\sum_{k=1}^{K} s(k)}}$$

$$Feat_8 = \sqrt{\frac{\sum_{k=1}^{K} f_k^4 s(k)}{\sum_{k=1}^{K} f_k^2 s(k)}}$$

$$Feat_9 = \frac{\sum_{k=1}^{K} f_k^2 s(k)}{\sqrt{\sum_{k=1}^{K} s(k) \sum_{k=1}^{K} f_k^4 s(k)}}$$

$$Feat_{10} = \frac{Feat_6}{Feat_5}$$

$$Feat_{11} = \frac{\sum_{k=1}^{K}(f_k-Feat_5)^3 s(k)}{K(Feat_6)^3}$$

$$Feat_{12} = \frac{\sum_{k=1}^{K}(f_k-Feat_5)^4 s(k)}{K(Feat_6)^4}$$

$$Feat_{13} = \frac{\sum_{k=1}^{K}(f_k-Feat_5)^{\frac{1}{2}} s(k)}{K\sqrt{Feat_6}}$$

$s(k)$ is the spectrum for $k = 1, 2..K$, where K is the number of spectrum lines f_k is the frequency value of the kth spectrum.

Where $p(a|t)$ and $p(na|t)$ are the probabilities for alarm and no-alarm respectively at the target node t.

5 Fine Tuning the CART Algorithm

Sensitivity and specificity are used to analyze the performance of a binary classifier. Sensitivity measures the proportion of actual positives (faults) which are correctly classified as positive, whereas specificity measures the proportion of actual negatives (no faults) which are correctly classified as negative. Accuracy measures the degree of right predictions of a classifier [21]. In this paper, we consider systems used in mission critical applications that require high sensitivity for fault detection. In such systems, every possibility of an alarm (fault) needs to be detected, and the alarm detection accuracy has to be as high as possible. This may be even desirable at the expense of a few more false alarms. Let P_a be the prior probability for an alarm

Table 2 Performance of CART decision-tree algorithm using current features for inter-turn short circuit faults in the R phase winding

C_{miss}	C_{fa}	Sensitivity (%)	Specificity (%)	Overall classification accuracy (%)
1	1	98.02	98.47	98.27
10	1	98.24	98.47	98.37
100	1	98.32	98.47	98.41
1000	1	98.43	98.47	98.46
10000	1	**99.10**	98.47	**98.75**
100000	1	99.10	98.47	98.75

to occur. Then we may write the prior probability for the no-alarm condition as $P_{na} = 1 - P_a$. We may now write a detection cost function, C_{DCF} as [15]:

$$C_{DCF} = C_{miss} P_a P_{miss} + C_{fa} P_{na} P_{fa} \qquad (2)$$

where C_{miss} and C_{fa} are the costs for missing an alarm and a false alarm respectively, P_{miss} is the probability for missing an alarm, and P_{fa} is the probability for false alarm (falsely reporting a no-alarm condition as an alarm). If we are required to have an increased sensitivity, we need to have higher cost associated with missing an alarm, C_{miss}, while keeping the value for C_{fa} to be the same. It may be noted that Eq. (2) has many operating points depending on the value of P_{miss} and P_{fa} [15]. We chose the point at which the missing and false alarm probabilities are equal, usually referred to as equal error rate (EER) [15], to evaluate C_{DCF}. At the EER point, $P_{miss} = P_{fa}$. The Eq. (2) can be written as $C_{DCF} = C_{miss} P_a + C_{fa} P_{na}$. Now the $p(a|t)$ and $p(na|t)$ can be written as

$$\hat{p}(a|t) = \frac{2p(a|t)C_{miss}}{C_{miss} + C_{fa}} \qquad (3)$$

$$\hat{p}(na|t) = \frac{2p(na|t)C_{fa}}{C_{miss} + C_{fa}} \qquad (4)$$

Incorporating the overall detection cost function C_{DCF} into the Gini index, we obtain the modified Gini index $\hat{I}_G$ for the node t as:

$$\hat{I}_G = 1 - \frac{4p(a|t)^2 C_{miss}^2}{(C_{miss} + C_{fa})^2} - \frac{4p(na|t)^2 C_{fa}^2}{(C_{miss} + C_{fa})^2} \qquad (5)$$

Modified Gini index in Eq. (5) may be seen as the Gini index (Eq. (1)) with the probabilities for alarm and no-alarm, $p(a|t)$ and $p(na|t)$, weighted with C_{miss} and C_{fa} respectively. It may be noted that in the default case of training the decision tree

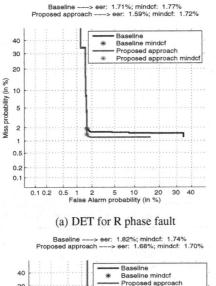

(a) DET for R phase fault

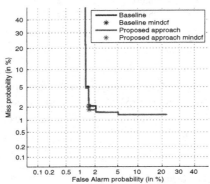

(b) DET for Y phase fault

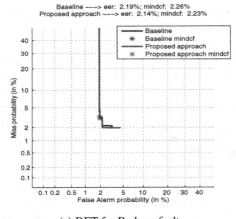

(c) DET for B phase fault

Fig. 2 Detection Error Trade-off (DET) plots for the fault diagnosis system

Table 3 Performance of CART decision-tree algorithm using current features for inter-turn short circuit faults in the Y phase winding

C_{miss}	C_{fa}	Sensitivity (%)	Specificity (%)	Overall classification accuracy (%)
1	1	97.65	98.30	98.01
10	1	98.43	98.30	98.36
100	1	**98.47**	98.30	**98.37**
1000	1	98.47	98.30	98.37

Table 4 Performance of CART decision-tree algorithm using current features for inter-turn short circuit faults in the B phase winding

C_{miss}	C_{fa}	Sensitivity (%)	Specificity (%)	Overall classification accuracy (%)
1	1	96.56	98.80	97.81
10	1	96.60	98.80	97.82
100	1	**97.09**	98.80	**98.04**
1000	1	97.09	98.80	98.04

algorithm, we set the C_{miss} and C_{fa} to unity, with the same penalty for missing an alarm and for a false alarm. Equation (5) reduces to Eq. (1) when $C_{miss} = C_{fa} = 1$. To achieve high sensitivity of the system, we should weigh the cost of missing alarm C_{miss} more than the cost of a false alarm.

6 Experiments and Results

It may be noted that an N class problem may be realized as an N two class problem. For its simplicity, we chose to experiment with a two class problem, fault or no-fault, with the R or Y or B phase of a 3kVA generator. We loaded the generator at 2 A, 2.5 A, 3 A, and 3.5 A and performed the experiments on all these loads. Training and testing data are collected separately, and no data is shared between the training and testing sets. We used 7600 samples of no fault data and 6080 samples of inter-turn fault data for training the system. For testing, we used 3344 and 2676 samples of no fault and fault data respectively. In this work, we used 13 frequency domain features that are listed in Table 1. For the three phase signal, we extracted 39 features. They are $Feat_1 - Feat_{13}$ for the R-phase, $Feat_{14} - Feat_{26}$, and $Feat_{27} - Feat_{39}$, for the Y-phase and B-phase respectively.

The baseline system has equal weights for the cost of missing an alarm, C_{miss}, and cost of a false alarm, C_{fa}. We then increase the value of C_{miss} and note that the sensitivity of the condition monitoring system is improved with the increase in C_{miss}, while keeping C_{fa} constant. From our experiments, we empirically found that, the system has learned the fault conditions at large C_{miss} values. Further increase in weights does not help the system to learn about the faulty conditions. We also note that, specificity remains unchanged while sensitivity increases for appropriate weights. Since the no fault data is homogeneous, the system has learned enough about the no fault conditions. Thus, cost modifications do not affect the system specificity while sensitivity increases.

It may be seen from the Tables 2-4 that the sensitivity is improved by 1.08% and the overall classification accuracy by 0.48% for R phase inter-turn fault detection experiment, with the selection of a suitable C_{miss}. Subsequently, for Y and B phases, the sensitivity has improved by 0.82% and 0.53% respectively, and the overall classification accuracy by 0.36 and 0.23% respectively. The effectiveness of the proposed approach is illustrated using detection error trade-off (DET) plot. From the Fig. 2 we observed that, for the proposed approach, equal error rate (EER) has been reduced by 0.12%, 0.14%, and 0.05% for the R, Y, and B phase faults respectively. Similarly, the detection cost function (DCF) has been minimized by 0.05%, 0.04%, and 0.03% for the R, Y, and B phase faults respectively.

7 Conclusions

We presented an approach to fine tune a machine fault identification system towards a mission critical application. We achieve this with the help of a detection cost function (DCF) that weighs missing an alarm and a false alarm differently. We illustrated the effectiveness of the approach using an experiment to detect the stator winding short circuit faults independently in the R, Y, and B phase winding of a 3kVA synchronous generator. Using the proposed approach, we improved the sensitivity of the fault identification system by 1.08%, 0.82%, and 0.53% for the R, Y, and B phase faults respectively.

Acknowledgments We are grateful to Aeronautical Development Agency (ADA): National Programme for Micro and Smart Systems (NPMASS) Bangalore, India for supporting this work, through a research grant. Authors would like to thank Prof. T.N.P. Nambiar, Mr. Kuruvachan K. George, and Mr. Thirugnanam for their help during the course of this work. Mr. Thomas, Bharath Electricals, Coimbatore fabricated the experimental 3kVA generator with the fault injection capability.

References

1. Jardine, A.K.S., Lin, D., Banjevic, D.: A review on machinery diagnostics and prognostics implementing condition-based maintenance. Mechanical Systems and Signal Processing **20**, 1483–1510 (2006)
2. Jack, L.B., Nandi, A.K.: Fault detection using support vector machines and artificial neural networks, augmented by genetic algorithms. Mechanical Systems and Signal Processing **16**, 373–390 (2002)
3. Samanta, B., Al-Balushi, K.R., Al-Araimi, S.A.: Artificial neural networks and genetic algorithm for bearing fault detection. Soft Computing **10**, 264–271 (2006)

4. Lei, Y., He, Z., Zi, Y., Hu, Q.: Fault diagnosis of rotating machinery based on multiple ANFIS combination with GAs. Mechanical Systems and Signal Processing **21**, 2280–2294 (2007)

5. Ye, Z., Sadeghian, A., Wu, B.: Mechanical fault diagnostics for induction motor with variable speed drives using Adaptive Neuro-fuzzy Inference System. Electric Power Systems Research **76**, 742–752 (2006)

6. Sun, W., Chen, J., Li, J.: Decision tree and PCA-based fault diagnosis of rotating machinery. Mechanical Systems and Signal Processing **21**, 1300–1317 (2007)

7. Casimir, R., Boutleux, E., Clerc, G., Yahoui, A.: The use of features selection and nearest neighbors rule for faults diagnostic in induction motors. Engineering Applications of Artificial Intelligence **19**, 69–177 (2006)

8. Karabadji, N.E.I., Khelf, I., Seridi, H., Laouar, L.: Genetic optimization of decision tree choice for fault diagnosis in an industrial ventilator. In: Condition Monitoring of Machinery in Non-Stationary Operations, pp. 277–283 (2012)

9. Saimurugan, M., Ramachandran, K.I., Sugumaran, V., Sakthivel, N.R.: Multi component fault diagnosis of rotational mechanical system based on decision tree and support vector machine. Expert Systems with Applications **38**, 3819–3826 (2011)

10. Tran, V.T., Yang, B.S., Oh, M.S., Tan, A.C.C.: Fault diagnosis of induction motor based on decision trees and adaptive neuro-fuzzy inference. Expert Systems with Applications **36**, 1840–1849 (2009)

11. Seera, M., Lim, C., Ishak, D., Singh, H.: Fault detection and diagnosis of induction motors using motor current signature analysis and a hybrid FMM-CART Model. IEEE Transactions on Neural Networks and Learning Systems **23**, 97–108 (2012)

12. Tran, V.T., Yang, B.S., Tan, A.C.C.: Multi-step ahead direct prediction for the machine condition prognosis using regression trees and neuro-fuzzy systems. Expert Systems with Applications **36**, 9378–9387 (2009)

13. Lomax, S., Vadera, S.: A survey of cost-sensitive decision tree induction algorithms. ACM Computing Surveys (CSUR) **45**, 16 (2013)

14. Kusiak, A., Verma, A.: A data-driven approach for monitoring blade pitch faults in wind turbines. IEEE Transactions on Sustainable Energy **2**, 87–96 (2011)

15. Brummer, N., du Preez, J.: Application-independent evaluation of speaker detection. Computer Speech and Language **20**, 230–275 (2006)

16. Sottile, J., Trutt, F.C., Leedy, A.W.: Condition monitoring of brushless three-phase synchronous generators with stator winding or rotor circuit deterioration. IEEE Transactions on Industry Applications **42**, 1209–1215 (2006)

17. Lei, Y., He, Z., Zi, Y.: A new approach to intelligent fault diagnosis of rotating machinery. Expert Systems with Applications **35**, 1593–1600 (2008)

18. Rokach, L.: Data mining with decision trees: theory and applications, vol. 69. World Scientific (2008)

19. Sathyadevi, G.: Application of CART algorithm in hepatitis disease diagnosis. In: 2011 IEEE International Conference on Recent Trends in Information Technology (ICRTIT), pp. 1283–1287 (2011)

20. Breiman, L., Friedman, J., Stone, C.J., Olshen, R.A.: Classification and regression trees. CRC Press (1984)

21. Ferri, C., Hernández-Orallo, J., Modroiu, R.: An experimental comparison of performance measures for classification. Pattern Recognition Letters **30**(1), 27–38 (2009)

Hardware Support for Adaptive Task Scheduler in RTOS

Dinesh G. Harkut and M.S. Ali

Abstract A real-time system is a system that reacts on event in the environment and executes functions based on these within a precise time. Traditionally, because of these strict requirements where the system must not fail in any situation, systems have been constructed from hardware and software components specifically designed for real-time. In embedded system, a real-time operating system (RTOs) is often used to structure the application code and ensure that the deadlines are met. Generally RTOs are implemented in software, which in turns increases computational overheads, jitter and memory footprint which can be reduced even if not remove completely by utilizing latest FPGA technology, which enables the implementation of a full featured and flexible hardware based RTOs. This article is a survey focusing on describing previous work in this domain and conclusion drawn from the research over the years. This paper also proposes the novel FIS based adaptive hardware task scheduler which uses fuzzy logic to model the uncertainty at first stage along with adaptive framework that uses feedback which allows processors share of task running on multiprocessor to be controlled dynamically at runtime. The increased computation overheads resulted from proposed model can be compensated by exploiting the parallelism of the hardware as being migrated to FPGA.

Keywords Hardware scheduler · Job priority · Real-time operating system · Reconfigurable computing · Scheduling algorithms · FIS · HW/SW co design

D.G. Harkut(✉)
Department of Computer Science & Engineering,
Prof. Ram Meghe College of Engineering & Management, Badnera-Amravati, India
e-mail: dg.harkut@gmail.com

M.S. Ali
Prof. Ram Meghe College of Engineering & Management, Badnera-Amravati, India
e-mail: softalis@hotmail.com

© Springer International Publishing Switzerland 2016   227
S. Berretti et al. (eds.), *Intelligent Systems Technologies and Applications*,
Advances in Intelligent Systems and Computing 384,
DOI: 10.1007/978-3-319-23036-8_20

1   Introduction

Today's consumer market is driven by technology innovations. Many technologies that were not available a few years ago are quickly being adopted into common use. Equipment for these services requires microprocessors inside and can be regarded as embedded system. Embedded devices are often designed to serve their unique purpose and are included in a variety of products within different technical areas such as industrial automation, consumer electronics, automotive industry and communications and multimedia systems. Embedded systems find application in almost all the product ranging from train and airplanes to microwave ovens and washing machines. As semiconductor prices drop and their performance improves, there is a rapid increase in the complexity of embedded applications. The increased complexity of embedded applications and the intensified market pressure to rapidly develop cheaper product have caused the industry to streamline software development. Use of embedded operating system or Real Time Operating System (RTOS) is one technique used to reduce development time of such system as it has effects on hardware abstraction, multitasking, code size, learning curve and the initial investment. Unfortunately, operating systems do introduce several forms of overheads.

Real time systems are embedded systems in which the correctness of application implementations is not only dependent upon the logical accuracy of its computations, but its ability to meet its timing constraints as well [1]. Thus the design of the RTOses have dual goal of minimizing the overheads and maximizing the determinism.

This paper is organized as follows. Section II is an overview of the Hardware/Software co-design approaches. Section III describes related work of other research projects, proposed model is discussed in section IV and section V covers summary and conclusion from mainly previous work and related work

2   Hardware Software Co-design Architecture

RTOs are often used in embedded systems to structure the application code to ensure that deadlines are met. The notions of best-effort and real-time processing have fractured into a spectrum of processing classes with different timeliness requirements including desktop multimedia, soft real-time, firm real-time, adaptive soft real-time and traditional hard real-time [2-4]. Many Real-Time systems are hard and missing deadline is catastrophic where as in soft real-time system, occasional violation of deadline may not result in useless execution of the application but decreases utilization [5].

Traditionally RTOS's are implemented in software, but major drawbacks of standard software based RTOS's is that they suffer from computational overheads, indeterminism, jitter and often a large memory footprint. RTOS computational overheads is caused mainly by tick interrupt management, which get even worse with more task and high tick frequencies, but also task scheduling , resource allocation and de-allocation, deadlock detection and various other OS/API functions take execution time from the task running on the CPU.

Embedded system always consists of software and hardware components and can no longer depend in independent hardware or software solutions to real time problem due to cost, efficiency, flexibility, upgradability, scalability and development time.

Task implemented as software programs running on microprocessor have the properties of high flexibility but poor performance. On the other hand, task implemented as hardware modules placed in Hardware have the characteristics of high performance along with low flexibility and high cost. The FPGA technology, which can be programmed virtually an n number of times (depends upon the technology), which paved the way for enhanced flexibility and made it possible to implement established software algorithms in hardware i.e. real-time kernel activity like scheduling, inter-process communications, interrupt management, resource management, synchronization and time management controls. Algorithm implemented in hardware has unique characteristics of high level parallelism and improved determinism that consequently decreases system overhead, improve predictability and increases response time.

As a tradeoffs, reconfigurable and hardware/software co-design approaches that offer real time capabilities while maintaining flexibility to support increasing complex systems become more feasible solution to allow software tasks running on a microprocessor along with hardware task running in an FPGA device (Fig.1).

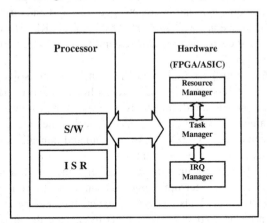

Fig. 1 Hardware Software System Architecture

This hardware/software co-design approach reach a level of maturity that are allowing system designers to perform operating systems core and housekeeping functionality such as time management and task scheduling in hardware harness the advantages of higher level program development while achieving the performance potential offered by executions of these functions in parallel hardware circuits.

3   Literature Review

The main source of indeterminism in real time systems are varying instruction cycle time caused by pipeline, caches, varying execution time of RTOs kernel functions, external asynchronous interrupts etc. By migrating real time kernel from software to hardware it is possible to remove jitter, lessen CPU overhead and improve the indeterminism due to cache and pipeline problems. Various models and systems have been proposed to overcome this problem and some of them were discussed in remaining section.

Lennart Lindh *et al.* [7-10] proposed a system FASTCHART, consist of hardware based RT kernel capable of handling 64 tasks with 8 different priorities. It is an RISC based uniprocessor system which puts ID of tasks into various queues making the implementation efficient.

POLIS – Co-Design Finite State Machine (CSFM) synthesis model, which supports globally asynchronous and locally synchronous computation, is proposed by F. Balarin, G. Berry, F. Boussinot *et al.* [11-13]. Implementation is splits between Software and ASICs and Co-simulation is provided using Ptolemy environment [14]. The output of POLIS is the C-code for the selected processor and the optimized hardware. Complexity of processors makes static estimation is difficult and does no support for large design.

Lennart Lindh *et al.* [15,16] also proposed FASTHARD which is extension to his earlier work FASTCHART, is a hardware based general purpose processors. Features like rendezvous, external interrupts, periodic start and termination of task without CPU interference are well supported by FASTHARD. However system is limited in supports for customization and scalability. Paper does not provided any benchmarks or test results.

The COSYMA system proposed by [17,18] aims speedup software executions to meet timing constraints. Profiling and symbolic [19] approaches are used to calculate timing information and thus helps in HW/SW partitioning which can be fine or coarse grained. List and path based [20] techniques are used to estimate execution time of hardware. COSYMA uses simulated annealing for partitioning and does not support burst-mode communication.

J. Adomat *et al.* [21] come up with RTU (Real Time Unit), a multi-processor system which uses single interrupt input of each CPU to control and context switching. Lindh *et al.* [22] also proposes extensible multiprocessor system - SARA, which can be used together with RTU to remove the all scheduling and tick processing overheads.

T. Samuelsson *et al.* [23] benchmark the RTU, SARA on Rhealstone. The result seems to be more deterministic. J. Lee *et al.* [24] further integrate the RTU with δ-framework for co-design. In [25] S. Nordstrom *et al.* adapt μC/OS II RTOS with uniprocessor to boost the performance. S. Nordstrom *et al.* [26] configurability is added and commercialize by Prevas AB. Lac of counting semaphore, mutex, deadline detection/preventions & no support for dynamic priorities limits the practical usefulness of RTU.

T. Nakano *et al.* [27] come up with STRON system which is based on μTRON project, wherein basic system calls and functionality has been implemented in hardware kernel which results in increasing speedup and reducing jitter. In this system a small micro kernel has been implemented to take care of the features not implemented in hardware. This system has tick frequency limitations and does not have hardware support to prevent unbounded priority inversion.

In [28,29], R. Gupta *et al.* developed VULCAN - Hardware/Software partitioning tool, which uses heuristic graph partitioning algorithm and runs in polynomial time. The goal is to minimize hardware cost while maintaining timing constraints. The original description was in Hardware-C [30], which is mapped to fine grained Control-Data Flow Graph. Test results are missing in the paper.

P. Chou *et al.* [31-35], proposes automated interface synthesis hardware software co-design framework for embedded system- CHINOOK. It supports mapping of an embedded system model to one (or more) processor and peripherals while ensuring timing constraints [36]. In [37, 38] more emphasis is put on distributed architecture and the system is inflexible and more complex.

H. De. Man *et al.* proposed heterogeneous hardware/software DSP system COWARE in [39], which is originally based on [40-43] and is basis of commercial CoWare N2C [44]. COWARE system allows co-specification using existing languages VHDL, DFL, Sliage & C [45] but imposes increased demands on generation of exhaustive library elements. This model supports the re-use and encapsulation of hardware and software by a clear separation between functional and communication behavior of a system components.

In [46-48] Bjorn B. Brandenburg *et al.* discuss the LITMUS^{RT} project is a soft real-time extension of the Linux kernel with focus on multiprocessor real-time scheduling and synchronization. The primary focus is to provide a useful experimental platform for applied real-time systems research. It supports the sporadic task model with both partitioned and global scheduling [49-51]. LITMUS^{RT} is not yet stable interfaces.

A. Parisoto *et al.* [52] suggested F-Timer framework which is FPGA based task scheduler targeted at general purpose processor, capable of managing 32 tasks with 64 different priorities. System does not have any hardware support for task synchronization and resource handling. Paper also does not discussed about scheduling algorithm employed.

J. Stankovic *et al.* [53,54] takes a radically different approach task scheduling than normal RTOS's. Spring kernel is basically designed for large and complex multiprocessor based RTOS. Task management is based on dynamic and speculative planning based scheduling implement through heuristic algorithm and tree search which makes it capable to handle fine granularity of task deadlines. However large amount of pre-calculation overheads become the major bottleneck for overall speedup of the system.

J. Hildebrandt *et al.* describes hardware implementation of dynamic scheduling coprocessor in [55,56]. It is basically hardware scheduling accelerator which can be configured for several different algorithms along with the most advanced Enhanced Least Laxity First (ELLF) [57,58]. This system has increases the overall determinism but at the cost of higher complex logic and could no address trashing of task.

In [59-63], I. Mooney *et al.* presents hardware/software co-design RTOs framework, δ-Framework which supports 30 different processors. This framework generates all HDL code which can be implemented in FPGA. The system is cost effective as far as overall speedup and hardware area (number of gates) is concerned. More work on SOC was conducted [64, 65] to integrate priority inheritance and deadlock avoidance mechanism.

V.Mooney *et al.* [66], design and developed configurable hardware scheduler to improve response time, interrupt latencies and CPU utilization and supports high tick frequency. The interrupt controller in scheduler supports 8 external interrupts each can be configured for dispatching a specific task and supports three different algorithms which can be change at run time.

Issues of extension to OS and flexibility arises out of moving entire OS to hardware can be overcome in model propose by Z. M. Wirthlin *et al.* in [67]. The nano-processor provides upgradability, flexibility and also enhancing the execution time by moving selected inefficient OS services in hardware to save on power consumption to a great extent as shown in several studies [68-70].

Leveraging the potential of hardware parallelism, Paul Kohot *et al.* in [71], developed Real-Time Manager (RTM). Routine housekeeping tasks are implemented in hardware and thus relieve the processor for critical functions which boosts the overall performance. RTM supports static priority scheduling and handles task, time and event management. The author claims RTM decreases RTOS overheads by 90% decreases response latency by 81%.

In [72] M.Vetromille *et al.* describes how low tick granularity can cause jitter and result in deadline misses. HaRTS supports high tick frequency and thus reduce jitter without lower CPU available time for task to process. It requires less chip area and uses less power than additional processor but more complex to implement.

Communication speed between RTOS and hardware overshadowed the speed gain by hardware scheduler. This problem has been overcome by intelligent design proposed by S. Chandra *et al.* in [73]. The Hardware RTOS implemented for accelerating eCos, HW-eCos is interfaced to an ARM processor requires less gates to implement and provides better speedup. Paper does not discuss the number of tasks and resources supported by this system.

Z. Murtaza, S. Khan *et al.* describe SRTOS in [74] is aim at real-time DSP application targeted on AVZ21 DSP processor. However paper doesn't provide any experimental test result but system supports additional instruction for fast resource allocation and context switching.

H-Kernel is an outcome of through use of FPGA and thoughtful HW/SW co-design for specific application as describe by M. Song *et al.* [75]. Nevertheless H-Kernel is suitable for system with small numbers of task and increases performance in the tune of 50-60%, system become more complex and bulky as number of task increases.

A. S. R. Oliveira *et al.* in [76] presented ARPA-MT system, which is multi-threading processor with five stage pipeline. It supports heterogeneous task and context switches without hampering the processor performance.

Luis Almeida *et al.* in [77,78] describe the details about OReK_CoP, Hardware implementation of OReK Real-Time Kernel. All kernel functions execute in absolute time and almost in parallel, without interfering CPU – improves determinism and improve resource utilization. Better and more direct connections between CPU and coprocessor would have removed quiet large latency introduced due to PLB bus interface in this system.

Many HW SW based solutions have been proposed to reduce the context switching time [79-81]. Xaingrong Zhou, Peter Petrov *et al.* presented model by converging compiler, micro-architecture and OS kernel to reduce the context switching cost and improve overall responsiveness. In this system context switching may be deferred until next switch point to limit the number of context registers required to hold state. This arrangement may result in more deadline miss which can be avoided by more complex and good RTOS kernel design.

N. Maruyama *et al.* in [82] proposed architecture ARTESSO, where RTOS, checksum calculation, memory copying and TCP header rearrangement are ported to hardware. It uses novel virtual queue instead of FIFO based queues used in RTU and STRON, which are logic expensive. The author claims that this system is 6-9 times faster than STRON and 7 times more energy efficient than its software counterpart.

Number of research projects have approached the task of designing OS for FPGA based reconfigurable computers (RC) [83- 87]. Hayden Kwok-Hay *et al.* [88-90] describe BORPH, an operating system designed for FPGA-based RC. By providing native kernel support for FPGA hardware, BORPH offers a homogeneous UNIX interface for both software and hardware processes. Hardware processes inherit the same level of service from the kernel. Performance of real-time wireless digital signal processing system based on BORPH will be presented [91].

HartOS proposed by Lange A.B. *et al.* is Hardware implemented Real-Time Operating System) in [92,93] is designed to be very flexible and support most of the features normally found in a standard software RTOS directly in hardware without sacrificing flexibility. HartOS has been compared to the commercial Sierra kernel, its performance outperforms that of the Sierra Kernel [94,95]. The HartOS the ability to let the kernel run at a higher clock frequency than the microprocessor, which allows more tasks to be processed serially at the same tick frequency, and speed up the part of the API functions executed in the kernel.

Comparative study of various methodologies/models reviewed in the literature is given in the Table (Table 1).

Scheduling algorithm plays as important role in the design of real-time systems which involves allocation of resources and time to jobs in such way that certain performance requirements are met.

Most of the model discussed and reviewed are mainly focused on to improve the performance by migrating some of the house keeping routine jobs from software to hardware with a aim to leverage the potential of parallel processing of hardware which can further be improved to a greater extent if more realistic scheduling algorithm is devise and migrate it on hardware to assist processor and RTos so as to increase the overall performance without increasing memory footprint and power consumptions.

Table 1 Comparative study of various methodologies/models.

Methodology/ Model	Category	Main Objectives	Architecture Used	Claims by Authors	Shortcoming
FASTCHART [7,8,9] (1991)	Hybrid	To improve deterministic & remove jitter.	RISC based processor with Load Store architecture	Migrated full kernel to Hardware	Lacks interrupts capabilities, pipelining, cache & supports single processor.
POLIS [11] (1991)	Hybrid	To design dashboard controller with a mixed implementation of SW & ASICs.	Co-design Finite State Machine (CFSM)	Flexibility to evaluate HW/SW partitioning, architecture & scheduler.	Static estimation is difficult (complexity of processors). No support for large design.
FASTHARD [15] (1992)	Hybrid	To develop H/W based RT kernel for general purpose processor.	Memory mapped address/data bus scheme.	Support external interrupts & rendezvous.	Limited scalability and customization supported. No test result available.
RTU [21] (1994-2001)	HW	To reduce system overhead and improve predictability	Memory mapped VME bus scheme.	Supports multiple task, binary semaphores, event flags, watchdogs.	Does not supports dynamic priority & bus latency time slows the response time.
Silicon TRON [27] 1995	Hybrid	To increase timing predictions & reduce/remove jitter	Memory mapped address/data bus scheme.	Supports task mgt., flags, semaphores, timers & external interrupt.	No mechanism to prevent unbounded priority inversion.
VULCAN [28] 1995	Hybrid	To reducing ASIC hardware cost.	Control-Data Flow Graph based fine grained mapping.	Hardware/software partitioning results in reducing the overall cost.	Limited scope as support bus-oriented fixed architecture only.
CHINOOK [31,32] 1996	Hybrid	To provide automated interface synthesis	Distributed architecture.	Supports mapping of processor & peripherals with strict timing constraints	Protocol of selected port incompatible & hence inflexible.
COSYMA [17,18] (1997)	Hybrid	To speeding up S/W execution to meet timing constraints	Memory mapped address/data bus scheme.	Uses novel list & path-based scheduling to estimate HW execution time	Does not support burst mode communications.
Spring Coprocessor [53,54] 1999	Hybrid	To design systems with guaranteed scheduling without blocking resources.	Memory mapped address/data bus scheme.	Supports fine granularity of task deadlines & multiprocessors.	Small overall speed as long time is used pre-calculations on EAT of all resources.
The δ-Framework [60] 2002	S/W	To provide framework targeted for HW/SW co design.	Memory mapped address/data bus scheme.	Uses less nos. of gates for equivalent HW area.	It just provides framework, design details missing.
Mooney [66] 2003	Hybrid	To design & develop configurable scheduler.	Both memory mapped & instruction set acceleration.	Supports Priority based, Rate monotonic & EDF algorithms & high tick rate.	Inflexible and bench-mark results are not available.
Nano-processor [67] 2003	Hybrid	To increasing flexibility & compatibility with range of hardware.	Memory mapped address/data bus scheme.	Provides flexibility of choosing services to perform in HW with faster execution.	Needs some optimization to reduce power consumption.
Real-Time Task Manager [71] 2003	Hybrid	To improve performance by migrating routine task to HW.	Memory mapped address/data bus scheme.	Supports static priority & handles task, time & event mgt. with same tree.	Hardware scales linearly with number of records.
HaRTS [72] 2006	Hybrid	To reducing jitter with increase in granularity.	OPB bus scheme.	Requires less power, less chip area and supports high tick frequency.	Design is very complex to implement.
LITMUSRT [46] 2006	S/W	To provide testbed to evaluate different RT scheduler on multicore platforms	Push/pull approach as std Linux RT scheduler.	Supports G-EDF based scheduling with private queue for each processor.	Maintainability & portability is major issues.

Table 1 (*continued*)

Methodology/ Model	Category	Main Objectives	Architecture Used	Claims by Authors	Shortcoming
HW- eCos [73] 2006	Hybrid	To improve system performance & reduce the code size of the RTOS.	Memory mapped address/data bus scheme.	Removes context switching overheads through interrupt line to CPU.	Chip-to Chip slow communication speed become bottleneck in overall performance.
Silicon RTOS [74] 2006	Hybrid	To design efficient real time DSP applications.	Memory mapped address/data bus scheme.	Supports external interrupt management & uses priority based scheduling	Use of register file for each task limits the usage for system with many tasks.
H-Kernel [75] 2007	Hybrid	To gain large performance through thoughtful HW/SW co design.	Memory mapped address/data bus scheme.	Supports priority based task, interrupt, event & time mgt through H-kernel.	Support small number of task. Complexity increases with increase in task numbers
OReK_CoP [77,78] 2009	Hybrid	To port OReK kernel to HW coprocessor to improve performance.	PLB bus interfacing with stack based priority ceiling.	Supports asynchronous interrupt handling which improve determinism	Comparatively higher latency time and supports only binary semaphores.
Xiangrong et al [80] 2010	S/W	To improve responsiveness by reducing the context-switching time.	Micro-architecture & OS kernel.	Uses micro-architecture to lower context switching.	Design is very complex and consumes more power.
ARTESSO [82] 2010	Hybrid	To improve throughput by moving TCP header calculation & checksum to HW.	TCP/IP Protocol.	Supports priority based FCFS scheduler & uses novel virtual queue structure.	Comparatively less reliable and scalable.
BORPH [88,89] 2010	S/W	To compatible with FPGA based reconfigurable hardware.	Unix based OS uses virtual file system.	Reduces context switching drastically by exploiting the benefits of parallelism.	Hardware is more complex to design and maintain.
ARPA-MT [76] 2011	Hybrid	To produce specialized predictable & customizable processor.	Stack based priority ceiling.	Supports heterogeneous task & schedules using RM or EDF protocol.	Comparatively difficult to achieve claimed determinism & resource efficiency.
HartOS [92,93] 2011	Hybrid	To reduce jitter, computational overheads & memory footprint.	Coprocessor/stream & FSL-AXI stream interface.	Interrupt handled as task & mutex are protected by stack based priority ceiling.	Support for deadlock preventions and flag management missing.

4 FIS Based Adaptive Hardware Task Scheduler

We proposed FIS based adaptive hardware task scheduler framework which is discussed in subsequent paragraph basically consist

- Global Fuzzy scheduler – Long term scheduler
- Local Adaptive scheduler – Short term scheduler.

Both of this scheduler work in cascade and are migrated on hardware which will work in synchronous with processor and RTos to fulfill the overall systems objectives as illustrated in figure (Fig. 2).

In this framework, incoming tasks are maintained in task queue in FCFS basis which is feed to Fuzzy Task scheduler which acts as long term global scheduler. To build a fuzzy system, inputs and output(s) to it must be first selected and partitioned into appropriate conceptual categories which actually represent a fuzzy set on a given input or output domain. Job priority, deadline and CPU time are selected as input to the Fuzzy Inference System (FIS) [96,97], which consist of five stages:

1. Fuzzifying inputs
2. Applying fuzzy operators
3. Applying implication methods
4. Aggregating outputs
5. De-fuzzifying outputs

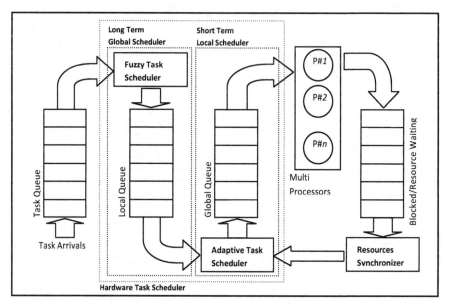

Fig. 2 Proposed FIS based Adaptive Hardware Task Scheduler

Here Madani's Fuzzy inference method of TSK or simply Sugeno method of fuzzy inference may be used [98-101]. Output is single value which is treated as Job Processing Priority (JPP) and maintained in global queue as per the newly calculated JPP which in turn feed to Adaptive Task Scheduler a second stage scheduler.

Under traditional task model like periodic, sporadic etc., the schedulability of system is based on each task's worst-case execution time (WCET), which defined the maximum amount of time each of its jobs can execute. The disadvantage of using WCETs is that system may be deemed un-schedulable even if they would function correctly most of the time when deployed. This drawback can be over-come by making our scheduler adaptive to the runtime varying conditions, to allocate per-task processors time share, instead of always using constant share allocation based on constant WCET and readjusting the priority of task. When there is variation in the WCET and the actual execution time of a particular job beyond some predetermined threshold value, adaptive task schedulers is invoked with actual execution time and reschedule the task and refresh and reorder the tasks in local queue accordingly. This results into adjusting the per task processor time share based on the runtime conditions which will effectively increases the overall schedulability and processor utilization. Overall quality-of-service (QoS) can be improved by ignoring the transient overload conditions. Dispatcher will dispatch the task from local queue to processors bank to get serve. Further resource synchronization is used to optimize scheduling of the tasks blocked on shared resource which are parked on blocked or waiting queue. Processors share allocations are adjusted using feedback and resource synchronization techniques [102]. This proposed fuzzy-adaptive feedback based scheduling algorithm based framework in capable of handling multiprocessor environment. As this entire scheduling task is proposed to be migrated to hardware (FPGA), increased computation resulted from two stage FIS scheduler can be compensated by exploiting parallelism characteristic of hardware with additional advantage of increase in determinism and lessening burdening of the processor, which results in increase in overall performance of the system.

5   Conclusion and Future Work

The conclusion from a comprehensive literature review of the publication throughout the last three decades, is that the major drawback from software based RTO's can be removed by implementing the entire/ partial kernel of a real-time operating system in hardware. It is also found that through clever design and utilization of the latest FPGA technologies, the implementation of a full featured and flexible hardware based RTOs should be possible.

Implementing a real-time kernel in hardware makes it possible to draw benefits from hardware characteristics such as parallelism and determinism. The execution time of real-time functions gets deterministic and task switch can be performed without any CPU time delay [9,52]. When real-time kernels are implemented in software, one of the disadvantages is that the execution time for the service calls will have a minimum and a maximum time. The time gap can be big and the worst-case time is one of the factors that will decide the utilization factor of the system. The scheduling time varies with the number of tasks and scheduling algorithm and must be bounded by a pessimistic worst case execution time, which decrease the determinism. In hardware, the time gap can be designed to be close to 0, which leads to predictable time behaviour, simpler timing analysis of the system and almost no overhead. It is also easier to debug tasks since different protection modes are not required [103-106].

A hardware kernel executes in parallel to the CPU, which relieves pressure from the CPU which gets almost full execution time for the application tasks. There is less software code in memory since the functionality is implemented in hardware instead [10,52].

A software OS will generate a clock tick interrupt to the CPU when either it is executed or the lists of tasks (queues) are worked at or new periodic delay times are calculated for the tasks. With the hardware kernel in the system, it checks all queues concurrently and only generates an interrupt to the CPU when there is to be a task switch [103,107]. Another advantage of having the kernel in hardware is the possibility to use complex scheduling algorithms, unlimited of different queue types without any performance loss. Also there is an improved understandability and complexity reduction when the system is divided into parts [103,104]. We have proposed FIS based hardware task scheduler which uses fuzzy logic to model the uncertainty at first stage along with adaptive framework that uses feedback which allows processors share of task running on multiprocessor to be controlled dynamically at runtime. Further, resource synchronization module is used to remove the deadlocks on resources among task from block task queue which will increase the overall performance of the RTOs.

Our future work is to map this proposed model on MicroBlaze soft processor core as MicroBlaze FPGA designs are readily available and can be implemented with little effort. The FreeRTOS port in MicroBlaze is being targeted to be modified and run tasks concurrently on multiple processors as FreeRTOS provides simple, easy-to-use and highly portable kernel. The aim to produce a version of FreeRTOS that supports multicore hardware and efficient hardware based task scheduler.

References

1. Stewart, D.: Introduction to Real Time. Embedded systems programming, CMP Media, November 2001
2. Deng, Z., Liu, J.W., Sun, S.: Dynamic scheduling of hard real-time application in open system environment. Tech. Rep., University of Illinois at Urbana-Champaign (1996)
3. Buttazzo, G., Stankovie, J.A.: RED: robust earliest deadline scheduling. In: Proceeding of 3rd International Workshop Responsive Computing Systems, Lincoln, NH, pp. 100–111 (1993)
4. Petters, S.M.: Bounding the execution time of real-time task on modern processors. In: Proceeding of 7th International Conference Real-Time Computing Systems and Applications, Cheju Island, pp. 498–502 (2000)
5. Zhu, J., Lewis, T.G., Jackson, W., Wilson, R.L.: Scheduling in hard real-time applications. IEEE Software 12, 54–63 (1995)
6. Lindh, L., Sjoholm, S.: VHDL for konsruktion, 3rd edn. (1999). ISBN: 91-44-01250-0
7. Lindh, L.: FASTCHART - a fast deterministic CPU and hardware based real-time kernel, pp. 36–40, January 1991
8. Lindh, L., Stanischewski, F.: FASTCHART - performance, benefits and disadvantages of the architecture. In: Proceeding of 5th Euromicro Workshop on Real-Time Systems (1993)
9. Lindh, L., Stanischewski, F.: FASTCHART- idea and implementation. In: IEEE International Conference on Computer Design (ICCD), Boston, USA, October 1991
10. Lindh, L.: FASTCHARD – a fast time deterministic hardware based real-time kernel. In: IEEE International Conference on Computer Design (ICCD), Cambrdge, USA, October 1991

11. Balarin, F., Chiodo, M., Giusto, P., Hsieh, H., Jurecska, A., Lavagno, L., Passerone, C., Sangiovanni-Vincentelli, A., Sentovich, E., Suzuki, K., Tabbara, B.: Hardware-Software Co-Design of Embedded Systems: The POLIS Approach. Kluwer Academic Publishers (1997)

12. Berry, G., Gonthier, G.: The Esterel Synchronous Programming Language: Design, Semantics, Implementations. Journal Science of Computer Programming archive, Elsevier North-Holland, Inc. Amsterdam, The Netherlands 19(2), 87–152, November 1992

13. Boussinot, F., De Simone, R.: The ESTEREL Language. Proceedings of the IEEE **79**(9), 1293–1303 (1991)

14. Buck, J.T., Ha, S., Lee, E.A., Messerschmitt, D.G.: Ptolemy: A Framework for Simulating and Prototyping Heterogeneous Systems. International Journal of Computer Simulation, special issue on "Simulation Software Development", 155–182, April 1994

15. Lindh, L.: FASTHARD - a fast time deterministic hardware based real-time kernel. In: 4th Euromicro Workshop on Proceedings of Real-Time Systems, pp. 21–25, June 1992

16. Lindh, L., Starner, J., Furunas, J.: From single to multiprocessor real-time kernels in hardware. pp. 42–43, May 1995

17. Ernst, R., Henkel, J., Benner, T., Ye, W., Holtmann, U., Herrman, D., Trawny, M.: The COSYMA environment for hardware software co-synthesis of small embedded systems. IEEE Micro, 159–166 (1996)

18. Osterling, A., Benner, T., Ernst, R., Herman, D., Scholz, T., Ye, W.: The COSYMA system. In: Hardware Software Co-Design: Principles and Practice. Kluwer Academic Publishers (1997)

19. Ernst, R., Ye, W.: Embedded program timing analysis based on path clustering and architecture classification. In: Proceedings of the International Conference on Computer-Aided Design, Santa Clara, CA, pp. 598–604, November 1997

20. Henkel, J., Ernst, R.: A path –based technique for estimating hardware runtime in HW SW-cosynthesis. In: Proceeding of the International Symposium on System Synthesis, pp. 116–121, September 1995

21. Adomat, J., Furunas, J., Lindh, L., Starner, J.: Real-time kernel in hardware RTU: a step towards deterministic and high-performance real-time systems. In: Proceedings of the 8th Euromicro Workshop on Real-Time Systems, L'Aquila, pp. 164–168, June 1996

22. Lindh, L., Klevin, T., Klevin, L.L.T., Furunäs, J.: Scalable architecture for real-time applications sara. In: CAD & CG 1999, pp. 208–211 (1999)

23. Samuelsson, T., Åkerholm, M., Nygren, P., Starner, J., Lindh, L.: A comparison of multiprocessor real-time operating systems implemented in hardware and software. In: International Workshop on Systems Implemented in Hardware and Software, Advanced Real-Time Operating System Services (ARTOSS) (2003)

24. J. Lee, Mooney III, V.J., Daleby, A., Ingstrom, K., Klevin, T., Lindh, L.: A comparison of the RTU hardware RTOS with a hardware/software RTOS. In: Proceeding ASP-DAC 2003 Design Automation Conference Asia and South Pacific, ACM New York, USA, pp. 683–688, January 2003

25. Nordstrom, S., Lindh, L., Johansson, L., Skoglund, T.: Application specific real-time microkernel in hardware. In: Proceedings of 14th IEEE-NPSS Real Time Conference, pp. 4–9, June 2005

26. Nordstrom, S., Asplund, L.: Configurable hardware/software support for single processor real-time kernels, pp. 1–4, November 2007
27. Nakano, T., Utama, A., Itabashi, M., Shiomi, A., Imai, M.: Hardware implementation of a real-time operating system. In: Proceeding of IEEE International Symposium of 12th TRON Project, Tokoy, Japan, pp. 34–42, November 1995
28. Gupta, R.: Co-Synthesis of Hardware and Software for Digital Embedded Systems. The Springer International Series in Engineering and Computer Science, vol. 329 (1995)
29. DeMicheli, G., Gupta, R., Ku, D.C., Mailhot, F., Truong, T.: The Olympus Synthesis System for Digital Design. Design & Test of Computers, IEEE 7(5), 37–53 (1990)
30. Ku, D.C., DeMicheli, G.: HardwareC – a language for hardware design Ver 2.0. CSL Technical Report CSL-TR-90-419, Stanford, April 1990
31. Chou, P., Borriello, G., Ortega, R.B.: Embedded system co-design: towards portability and rapid integration. In: De Micheli, G., Sami, M. (eds.) Hardware/Software Co-Design. Kluwer Acadmic Publishers, pp. 243–264 (1996)
32. Chou, P., Ortega, R., Borriello, G.: The chinook hardware software co-synthesis system. In: Proceedings of the International Symphosium on System Synthesis, pp. 22–27, September 1995
33. Chou, P., Borriello, G.: Software scheduling in the co-synthesis of reactive real-time systems. In: Proceedings of the 31st Design Automation Conference, pp. 1–4, June 1994
34. Chou, P., Walkup, E., Borriello, G.: Scheduling for Reactive Real-Time Systems. IEEE Micro archive Journal, IEEE Computer Society Press Los Alamitos, CA, USA 14(4), 37–47, August 1994
35. Chou, P., Ortega, R., Borriello, G.: Synthesis of the hardware software interface in microcontroller-based systems. In: Proceedings of the International Conference on Computer-Aided Design, Santa Clara, pp. 488–495, November 1992
36. Ku, D., De Micheli, G.: High-level Synthesis of ASICs under Timing and Synchronization Constraints. Kluwer Academic Publishers, Norwell (1992). ISBN: 0-7923-9244-2
37. Hines, K., Borriello, G.: Optimizing communication in embedded system co-simulation. In: International Workshop on Hardware Software Co-Design, pp. 121–125 (1997)
38. Ortega, R., Borriello, G.: Communication synthesis for embedded systems with global considerations. In: International Workshop on Hardware Software Co-Design, pp. 69–73 (1997)
39. De Man, H., Blosens, I., Lin, B., Van Rompaey, K., Vercauteren, S., Verkest, D.: Co-design of DSP systems. In: De Micheli, G., Sami, M. (eds.) Hardware/Software Co-Design. Kluwer Academic Publishers, pp. 75–104 (1996)
40. De Man, H., Verkest, D., Van Rompary, K., Bolsens, I.: Coware – A Design Environment for Heterogeneous Hardware Software Systems. Design Automation of Embedded Systems, 357–386, October 1996
41. De Man, H., Vercauteren, S., Lin, B.: Constructing application specific heterogeneous embedded architectures from custom HW SW applications. In: Proceedings of the Design Automation Conference, pp. 521–526, June 1996
42. De Man, H., Vercauteren, S., Lin, B.: Embedded architecture co-synthesis and system integration. In: Proceedings of 4th International Workshop on Hardware/Software Co-Design, Pittsburgh, PA, pp. 2–9, March 1996

43. De Man, H., Van Rompaey, K., Verkest, D., Bolsens, I.: Coware – a design environment for heterogeneous hardware software systems. In: Proceedings of the European Design Automation Conference, pp. 252–257, September 1996
44. Santarini, M.: CoWare revs tool for SoC platform design. Electronic Engineering Times, pp. 54–58, August 2000
45. Willekens, P., Devisch, D., Van Canneyt, M., Conflitti, P., Genin, D.: Algorithm Specification in DSP Station using Data Flow Language. DSP Applications, pp. 8–16, January 1994
46. Brandenburg, B., Calandrino, J., Leontyev, H., Block, A., Devi, U., Anderson, J.: LITMUSRT: a test-bed for empirically comparing real-time multiprocessor schedulers. In: Proceedings of the 27th IEEE Real-Time Systems Symposium, pp. 111–123, December 2006
47. Brandenburg, B., Anderson, J.: A comparison of the M-PCP, D-PCP, and FMLP on LITMUSRT. In: Proceedings of the 12th International Conference on Principles of Distributed Systems, pp. 105–124, December 2008
48. Brandenburg, B., Block, A., Calandrino, J., Devi, U., Leontyev, H., Anderson, J.: LITMUSRT: a status report. In: Proceedings of the 9th Real-Time Linux Workshop, pp. 107–123, November 2007
49. Mollison, M., Brandenburg, B., Anderson, J.: Towards unit testing real-time schedulers in LITMUSRT. In: Proceedings of the 5th International Workshop on Operating Systems Platforms for Embedded Real-Time Applications, pp. 33–39, July 2009
50. Cerqueira, F., Brandenburg, B.: A comparison of scheduling latency in linux, PREEMPT-RT, and LITMUSRT. In: Proceedings of the 9th Annual Workshop on Operating Systems Platforms for Embedded Real-Time applications, pp. 19–29, July 2013
51. Spliet, R., Vanga, M., Brandenburg, B., Dziadek, S.: Fast on average, predictable in the worst case: exploring real-time futexes in LITMUSRT. In: Proceedings of the 35th IEEE Real-Time Systems Symposium, Rome, Italy, pp. 96–105, December 2014
52. Parisoto, A., Souza Jr., A., Carro, L., Pontremoli, M., Pereira, C., Suzim, A.: F-timer: dedicated FPGS to real-time systems design support. In: Proceeding of 9th Euromicro Workshop on RTS, Toledo, Spain, pp. 35–40, June 1997
53. Stankovic, J., Ramamritham, K.: The spring kernel: a new paradigm for real-time systems. Software, IEEE **8**(3), 62–72 (1991)
54. Stankovic, J., Burleson, W., Ko, J., Niehaus, D., Ramamritham, K., Wallace, G., Weems, C.: The spring scheduling coprocessor: a scheduling accelerator. IEEE Transactions on Very Large Scale Integration Systems **7**, 38–47 (1999)
55. Hildebrandt, J., Golatowski, F., Timmermann, D.: Scheduling coprocessor for enhanced least-laxity-first scheduling in hard real-time systems. In: Proceedings of the 11th Euromicro Conference on Real-Time Systems, pp. 208–215 (1999)
56. Hildebrandt, J., Timmermann, D.: An FPGA based scheduling coprocessor for dynamic priority scheduling in hard real-time systems. In: Grünbacher, H., Hartenstein, R.W. (eds.) FPL 2000. LNCS, vol. 1896, pp. 777–780. Springer, Heidelberg (2000)
57. Morton, A., Loucks, W.: A hardware/software kernel for system on chip designs. In: Proceeding of the 2004 ACM Symposium on Applied Computing, Nicosia, Cyprus, pp. 869–875 (2004)
58. Kuacharoen, P., Shalan, M., Mooney III, V.: A configurable hardware scheduler for real-time systems. In: Proceeding of International Conference on Engineering of Reconfigurable Systems and Algorithms, Nevada, USA (2003)

59. Mooney, V., Hartenstein, R., Grünbacher, H. (eds.), vol. 1896 of Lecture Notes in Computer Science, pp. 777–780. Springer Berlin / Heidelberg (2000). doi:10.1007/3-540-44614-1_83

60. Mooney, V., Lee, J., Ryu, K.: A framework for automatic generation of configuration files for a custom hardware/software RTOS. In: Proceedings of the International Conference on Engineering of Reconfigurable Systems and Algorithms (ERSA 2002), pp. 31–37, June 2002

61. Mooney, V., Blough, D.: A Hardware/Software Real-Time Operating System Framework for SOCs. Design & Test of Computers, IEEE **19**(6), 44–51 (2002). USA

62. Mooney, V.: Hardware/software partitioning of operating systems. In: Proceeding of Conference and Exhibition on Design, Automation and Test in Europe. IEEE, Munich, pp. 339–339 (2003)

63. Mooney, V., Lee, J.: Hardware/software partitioning of operating systems: focus on deadlock detection and avoidance. In: IEEE Proceeding, Computer and Digital Techniques, UK, pp. 167–182, July 2005

64. Akgul, B., Mooney, V., Thane, H., Kuacharoen, P.: Hardware support for priority inheritance. In: Proceeding of 24th IEEE International Real-Time Systems Symposium, Cancun, Mexico, pp. 246–255 (2003)

65. Lee, J., Mooney III, V.: RTPOS: a novel deadlock avoidance algorithm and its hardware implementation. In: CODES+ISSS 2004 Proceedings of the international conference on Hardware/Software Codesign and System Synthesis, pp. 200–205. IEEE Computer Society, Washington, DC (2004)

66. Mooney III, V., Kuacharoen, P., Shalan, M.A.: A configurable hardware scheduler for real-time systems. In: Proceedings of the International Conference on Engineering of Reconfigurable Systems and Algorithms, pp. 96–101. CSREA Press (2003)

67. Wirthlin, M., Hutchings, B., Gilson, K.: The nano processor: a low resource reconfigurable processor. In: IEEE Workshop on FPGAs for Custom Computing Machines, Napa, CA, pp. 23–30, April 1994

68. Dick, R., Lakshminarayana, G., Raghunathan, A., Jha, N.: Power analysis of embedded operating systems. In: proceedings of the 37th Design Automation Conference, Los Angeles, CA, pp. 312–315, June 2000

69. Kuacharoen, P., Shalan, M.A., Mooney III, V.J.: A configurable hardware scheduler for real-time systems. In: Proceedings of the International Conference on Engineering of Reconfigurable Systems and Algorithms (ERSA 2003), Las Vegas, USA, pp. 96–101, June 2003

70. Moon, S.W., Rexford, J., Shin, K.: Scalable Hardware Priority Queue Architectures for High-Speed Packet Switches. IEEE Transaction on Computers **49**(11), 1215–1226 (2000)

71. Kohout, P., Ganesh, B., Jacob, B.: Hardware support for real-time operating systems. In: Proceeding of First IEEE/ACM/IFIP International Conference on Hardware/Software Codesign and System Synthesis (CODES+ISSS 2003), Newport Beach CA, pp. 45–51, October 2003

72. Vetromille, M., Ost, L., Marcon, C., Reif, C., Hessel, F.: RTOS scheduler implementation in hardware and software for real time applications. In: 17th IEEE International Workshop on Rapid System Prototyping, pp. 163–168, June 2006

73. Chandra, S., Regazzoni, F., Lajolo, M.: Hardware/software partitioning of operating systems: a behavioral synthesis approach. In: Proceedings of the 16th ACM Great Lakes symposium on VLSI, GLSVLSI 2006, NY, USA, pp. 324–329. ACM (2006)

74. Murtaza, Z., Khan, S., Rafique, A., Bajwa, K., Zaman, U.: Silicon real time operating system for embedded DSPs. In: ICET 2006: Proceedings of International Conference on Emerging Technologies, Peshwar, pp. 188–191. IEEE, November 2006

75. Song, M., Hong, S.H., Chung, Y.: Reducing the overhead of real-time operating system through reconfigurable hardware. In: proceedings of 10th Euromicro Conference on Digital System Design Architectures, Methods and Tools, pp. 311–316, August 2007

76. Oliveira, A.S.R., Almeida, L., Ferrari, A.B.: The ARPA-MT embedded SMT processor and its RTOS hardware accelerator. IEEE Transactions on Industrial Electronics **58**(3), 890–904 (2011)

77. Almeida, L., Oliveira, A.S.R., Ferrari, A.B.: A specialized and predictable processor for real-time systems. In: Workshop on Application Specific Processors, pp. 32–38, November 2009

78. Almeida, L., Silva, N., Oliveira, A., Santos, R.: The OReK real-time micro kernel for FPGA-based systems-on-chip. In: Proceedings of 6th Workshop on Embedded Systems for Real-time Multimedia, (ESTImedia 2008). IEEE Xplore, Atlanta Georgia, pp. 75–80, October 2008

79. Wu, Z., Li, H., Gao, Z., Sun, J., Li, J.: An improved method of task context switching in OSEK operating system. In: 20th International Conference on Advanced Information Networking and Applications, AINA 2006, p. 6, April 2006

80. Zhou, X., Petrov, P.: Rapid and low-cost context-switch through embedded processor customization for real-time and control applications, DAC San Francisco, CA, pp. 352–357, July 2006

81. So, K.-H., Brodersen, R.W.: Improving usability of FPGA-based reconfigurable computers through operating system support. In: Proceedings of 2006 International Conference on Field Programmable Logic and Applications (FPL), pp. 349–354, August 2006

82. Maruyama, N., Ishihara, T., Yasuura, H.: An RTOS in hardware for energy efficient software-based TCP/IP processing. In: IEEE Symposium on Application Specific Processors, pp. 58–63, June 2010

83. Danne, K., Muehlenbernd, R., Platzner, M.: Executing hardware tasks on dynamically reconfigurable devices under real-time conditions. In: 16th International Conference on Field Programmable Logic and Applications, pp. 541–546 (2006)

84. Gotz, M., Dittmann, F.: Reconfigurable microkernel-based RTOS: mechanisms and methods for run-time reconfiguration. In: Proceedings of IEEE International Conference on Reconfigurable Computing and FPGA's (ReConFig 2006), pp. 1–8, 2006

85. Mei, B., Vernalde, S., Verkest, D., Lauwereins, R.: Design methodology for a tightly coupled VLIW/reconfigurable matrix architecture: a case study. In: Proceedings of the Conference on Design, Automation and Test in Europe, pp. 1224–1229. IEEE Computer Society, Washington, DC, February. 2004

86. Walder, H., Platzner, M.: A runtime environment for reconfigurable hardware operating systems. In: Becker, J., Platzner, M., Vernalde, S. (eds.) FPL 2004. LNCS, vol. 3203, pp. 831–835. Springer, Heidelberg (2004)

87. Wigley, G.B., Kearney, D.A., Warren, D.S.: Introducing ReConfigME: an operating system for reconfigurable computing. In: Glesner, M., Zipf, P., Renovell, M. (eds.) FPL 2002. LNCS, vol. 2438, pp. 687–697. Springer, Heidelberg (2002)

88. So, H.K.-H., Changqing, X., Mei, W., Nan, W., Chunyuan, Z.: Extending BORPH for shared memory reconfigurable computers field programmable logic and applications (FPL). In: 22nd International Conference on IEEE Improving Usability of FPGA-Based Reconfigurable Computers Through Operating System Support, Oslo, pp. 563–566, August 2012
89. So, H.K.-H., Broderson, R.W.: BORPH: an operating system for FPGA-based reconfigurable computers. DAC University of California, Berkeley, Technical Report No. UCB/EECS, pp. 92–96, July 2007
90. So, H.K.-H., Brodersen, R.W.: A unified hardware/software runtime environment for FPGA-based reconfigurable computers using BORPH. ACM Transactions on Embedded Computing Systems (TECS) TECS Homepage archive 7(2), February 2008. Article No. 14, ACM New York, USA
91. Brodersen, W., Tkachenko, R., So, H.K.-H.: A unified hardware/software runtime environment for FPGA-based reconfigurable computers using BORPH. In: Proceedings of the 4th International Conference on Hardware/Software Co-design and System Synthesis, pp. 259–264. IEEE, October 2006
92. Lange, A.B., Andersen, K.H., Schultz, U.P., Sørensen, A.S.: HartOS - a hardware implemented RTOS for hard real-time applications. In: Proceedings of the 11th IFAC/IEEE International Conference on Programmable Devices and Embedded Systems, Brno, Czech Republic (2012)
93. Lange, A.B.: Hardware RTOS for FPGA based embedded systems. Master's thesis, University of Southern Denmark. http://www.hartos.dk/publications/thesis/hartos.pdf (accessed on November 2015)
94. Kohout, P., Ganesh, B., Jacob, B.: Hardware support for real-time operating systems. In: Proceedings of the First IEEE/ACM/IFIP International Conference on HW/SW co-design and System Synthesis, pp. 45–51 (2003)
95. Kuacharoen, P., Shalan, M.A., Mooney III, V.J.: A configurable hardware scheduler for real-time systems. In: Proceedings of the International Conference on Engineering of Reconfigurable Systems and Algorithms, pp. 96–101 (2003)
96. Sabeghi, M., Naghibzadeh, M., Taghavi, T.: Scheduling non-preemptive periodic task in soft real-time systems using fuzzy inference. In: 9th IEEE International Symposium on Object and Component-Oriented Real-Time Distributed Computing (ISORC), April 2006
97. Hamzeh, M., Fakhraie, S.M., Lucas, C.: Soft real-time fuzzy task scheduling for multiprocessor systems. International Journal of Intelligent Technology 2(4), 211–216 (2007)
98. Mamdami, E.H., Assilian, S.: An experiment in linguistic synthesis with a fuzzy logic controller. International Journal of Man-Machine Studies 7(1), 1–13 (1975)
99. Sugeno, M.: Industrial applications of fuzzy control. Elsevier Science Inc., New York (1985)
100. Siar, H., Nabavi, S.H., Shahaboddin, S.: Static task scheduling in cooperative distributed systems based on soft computing techniques. Australian Journal of Basic and Applied Sciences 4(6), 1518–1526 (2010)
101. Dubois, D., Fargier, H., Fortemps, P.: Fuzzy scheduling: modelling flexible constraints vs. coping with incomplete knowledge. European Journal of Operational Research 147(2), 231–252 (2003)
102. Jang, J.S.R.: ANFIS: Adaptive-Network-based Fuzzy Inference Systems. IEEE Transactions on Systems, Man, and Cybernetics 23(3), 665–685 (1993)

103. Lindh, L., Stärner, J., Furunäs, J.: From single to multiprocessor real-time kernels in hardware. In: IEEE Real Time Technology and Applications Symposium, Chicago, May 1995
104. Furunäs, J., Adomat, J., Lindh, L., Stärner, J., Vörös, P.: A prototype for interprocess communication support, in hardware in real-time systems. In: Proceedings of EUROMICRO Workshop, Toledo, Spain, pp. 18–24, June 1997
105. Lindh, L.: A real-time kernel implemented in one chip. In: Real-Time Workshop. IEEE Press, Oulu, June 1993
106. Adomat, J., Furunäs, J., Lindh, L., Stärner, J.: Real-time kernel in hardware RTU: a step towards deterministic and high-performance real-time systems. In: Proceedings of the Euromicro Workshop on Real-Time Systems, L'Aquila, Italy, June 1996
107. Lindh, L.: Utilization of Hardware Parallelism in Realizing Real Time Kernels. Doctoral Thesis, TRITA – TDE 1994:1, ISSN 0280-4506, ISRN KTH/TDE/FR-94/1-SE, Department of Electronics, Royal Institute of technology, Stockholm, Sweden, 1994, November 2015

Experimental Investigation of Cyclic Variation in a Diesel Engine Using Wavelets

Rakesh Kumar Maurya, Mohit Raj Saxena and Nekkanti Akhil

Abstract In this study, cyclic variations of combustion parameters (IMEP and THR) are analyzed using morlet wavelet transform in a diesel engine. Experiments were conducted at 1500 rpm for different engine load conditions and compression ratios. Combustion parameters were calculated from measured cylinder pressure trace. In-cylinder pressure data of 2500 consecutive engine cycles were acquired and processed for the analysis of cyclic variations. Results revealed that cyclic variability in THR decreases with increase in engine load and compression ratio. Cyclic variability is highest in idle load conditions at lowest compression ratio. Low frequency cyclic variations are observed at low loads conditions and with increase in engine load variations shift to high frequency range. Results can be utilized for the development of effective engine control strategies.

Keywords Diesel engine · Wavelets · Cyclic variability · IMEP · Heat release

1 Introduction

The internal combustion (IC) engines are well accepted and most significant prime movers since the last century due to their performance, economy, durability, controllability and lack of other competitive alternatives. Considering future projections, it seems reasonable to expect that the number of vehicles will rise in future, especially on account of the rapid economic development. Increasing number of vehicles will increase fuel consumption and emissions. Stringent emission legislations and depleting primary energy resources requires the development of clean combustion technologies of lower fuel consumptions. Automotive engines face two main challenges of reducing exhaust emissions and improve the efficiency of engines. These challenges motivate for research and development of progressively

R.K. Maurya(✉) · M.R. Saxena · N. Akhil
Indian Institute of Technology Ropar, Rupnagar 140001, Punjab, India
e-mail: {rakesh.maurya,mohitraj,nekkantia}@iitrpr.ac.in

© Springer International Publishing Switzerland 2016   247
S. Berretti et al. (eds.), *Intelligent Systems Technologies and Applications*,
Advances in Intelligent Systems and Computing 384,
DOI: 10.1007/978-3-319-23036-8_21

more sophisticated engine control and exhaust after-treatment strategies. IC Engine combustion processes are affected by several factors resulting in to complex dynamics and cycle to cycle variations. These factors include physical and chemical properties of air, fuel and residual gas mixtures, flow characteristics of mixture in the cylinder, amount of residual gases, ambient pressure and temperature, engine operating conditions (load and speed). Variations in these wide range of parameters leads to cyclic variations in combustion process. Combustion variability adversely affects the engine performance, efficiency, power as well as affects the noise [1-2]. Therefore there is a strong motivation for understanding, diagnosing and controlling cyclic variability in IC engines.

Combustion can be controlled by variation air fuel ratio, amount of residual gas fraction and ignition/injection timings using in-cylinder pressure information [3]. Cyclic variations in the combustion variables have been investigated in conventional spark ignition and compression ignition engines [4-8] as well as advanced premixed compression ignition engines [9-12] mainly using statistical methods. It is important to know the complex dynamics of combustion variables for designing effective engine control strategies. Several studies are conducted by using deterministic methods from nonlinear dynamical systems and chaos theory to understand the dynamics of cyclic variation [13-16].

Few researchers also used continuous wavelet transform (CWT) to understand the cyclic variability [17-18]. Sen et al. used wavelet analysis to understand the cyclic variability in spark ignition engine and found the presence of long, intermediate and short-term periodicities in the pressure signal to estimate the cyclic variations of a spark ignition engine [17]. In another study authors show that the variations in mean effective pressure may have a strongly periodic component and/or be intermittent depending on engine speed [19]. Study conducted by Shojji reveals that the variations in spray shape, spray direction and the spray cone angle are the main factors responsible for cyclic variation in diesel engine at low speed and low load [20]. In this study the cyclic variability in diesel engine is analyzed using continuous wavelet transform. Effect of engine load and compression ratio on cyclic variation of two combustion variables (total heat release and indicated mean effective pressure) is systematically analyzed and discussed.

2   Experimental Setup

A single cylinder, four stroke, water cooled, naturally aspirated, direct injection conventional diesel engine with variable compression ratio (range 15–18) was used for the present investigation. Engine displacement volume is 661cc and rated power is 3 kW at 1500 rpm at compression ratio 17.5. A tilting cylinder block arrangement is used for varying the compression ratio of engine without altering the combustion chamber geometry of the engine. Fuel injection timing was 23^0 BTDC in this study. Schematic diagram of experimental setup used in this study is shown in Figure 1.

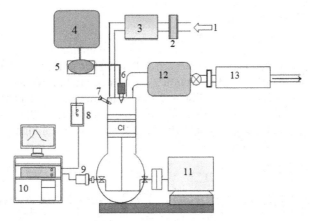

1: Intake Air, 2: Air Filter, 3: Air Flow Meter, 4: Fuel Tank 5: Fuel
Flow Meter, 6: Fuel Pump & Injector, 7: Pressure Transducer
8: Charge Amplifier, 9: Shaft Encoder 10: Data Acquisition System
11: Dynamometer 12: Exhaust Plenum 13: Exhaust Muffler

Fig. 1 Schematic diagram of experimental setup.

Optical shaft encoder with resolution of 1 crank angle degree (**CAD**) is used for
measuring angular position of crankshaft of engine. The in-cylinder pressure is
measured using piezoelectric transducer having measuring range 0-350 bars. Pressure data acquisition and analysis is done using LabVIEW based program developed at IIT Ropar. In-cylinder pressure trace was logged for 2500 consecutive
engine cycles for each operating condition. Details of the equations used for the
calculation of different combustion parameter in this study are given as follows.

Indicated mean effective pressure (IMEP) is ratio of indicated work (W_{ind}) and
displacement volume (V_d) as

$$IMEP = \frac{W_{ind}}{V_d} \tag{1}$$

$$W_{ind} = \frac{2\pi}{360} \int_{-180}^{180} \left(P(\theta) \frac{dV}{d\theta} \right) d\theta \tag{2}$$

Where, P is pressure and V is volume as function of crank angle position (θ).

Rate of heat release is calculated using equation in [21]

$$\frac{dQ(\theta)}{d\theta} = \left(\frac{1}{\gamma-1} \right) V(\theta) \frac{dP(\theta)}{d\theta} + \left(\frac{\gamma}{\gamma-1} \right) P(\theta) \frac{dV(\theta)}{d\theta} \tag{3}$$

Where Q is heat release and γ is ratio of specific heat.

Total heat release (THR) is calculated by integrating rate of heat release
between start of combustion and end of combustion.

3   Results and Discussion

In this section, cyclic variations in two calculated combustion parameters (IMEP and THR) are analyzed using wavelets. IMEP and THR time series are calculated using the equations described in section 2, from measured in-cylinder pressure of 2500 consecutive engine cycles. Figure 2 shows the time series of THR and IMEP for different load conditions at fixed compression ratio (CR) and constant engine speed (1500 rpm). Mean values of THR and IMEP are higher at higher load due to higher amount of fuel burned per cycle.

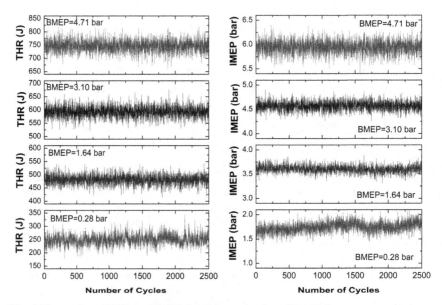

Fig. 2 Time series of THR and IMEP for different loads at CR= 17.5

Figure 3 shows the time series of THR for lowest engine load at constant engine speed (1500 rpm) for CR 17.5, 16.0 and 15.0 respectively. Coefficient of variation are higher at lower loads therefore lower load condition is selected for anlysis at different compression ratios.

Wavelet transform is capable of providing the time and frequency information simultaneously. Short-time Fourier transform (STFT) is the other approach for analysing a signal in time and frequency domain. Frequency components and their locations in time domain of signal are not known simultaneously in STFT analysis. Wavelet has merit over STFT analysis. Detailed procedure and calculation method of wavelet analysis is explained in reference [17-18].

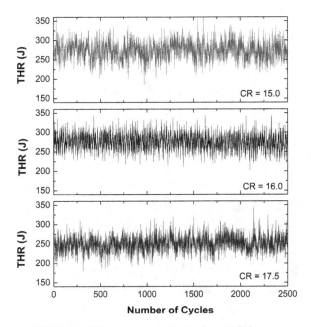

Fig. 3 Time series of THR for different compression ratios at 1500 rpm.

CWT is used to create the wavelet power spectrum and global power spectrum from the time series of THR and IMEP. Spectrogram is used for analysis of the frequency band contribution to the energy of signal and changes in periodicity (long, intermediate and short periodicity) with time. In this study spectrogram is used to find the variation of periodicity with engine load. The wavelet power spectra (WPS) and global wavelet spectra (GWS) of THR and IMEP time series for different loads are plotted by using Morlet wavelet. Matlab code is used for calculation of WPS and GPS. Base code is taken from reference [22] and modified for current study. Figure 4 and 5 shows the WPS and GWS computed using time series of the THR and IMEP (figure 2) for different engine loads at constant engine speed. Colours in the WPS represent the signal energy, which is defined as magnitude of wavelet transforms [18]. Red colour indicates that those frequency contributing maximum to the energy of signal over that interval.

It can be observed from Figure 4a that periodic cycle bands of 110-220, 66-102, 120-540 periods have the highest variability (strongest intensity) during cycle ranges 390-870, 1100-1400, 1300-2070 respectively. Few comparatively lower intensity bands of 32-64 and 11-25 periods occurs in cycles ranges 180-360 and 1580-1660 cycles respectively. The presence of the large number of strong and moderate intensity periodic bands represents high cycle to cycle variability for this load condition (BMEP = 0.28 bar).

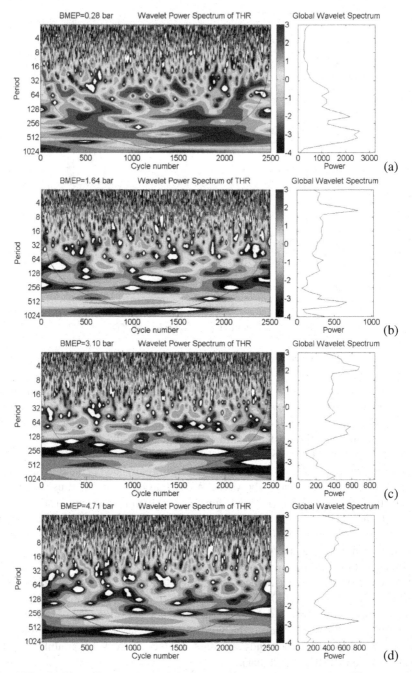

Fig. 4 Wavelet Power Spectrum and Global Wavelet Spectrum of THR for different engine loads at CR=17.5. The thick contour lines represent the 5% significance level below which denotes the cone of influence (COI).

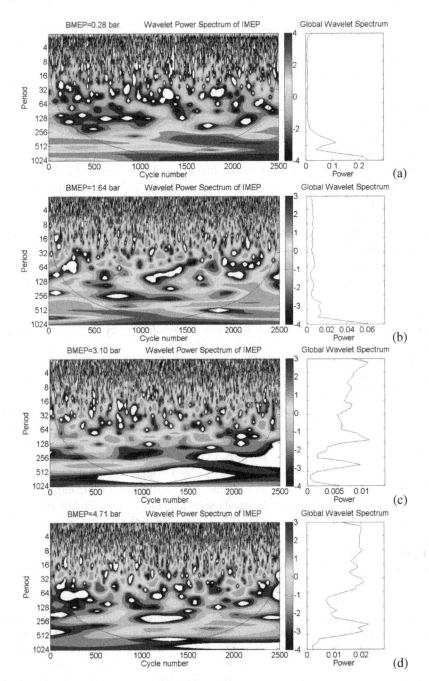

Fig. 5 Wavelet Power Spectrum and Global Wavelet Spectrum of IMEP for different engine loads at CR=17.5. The thick contour lines represent the 5% significance level below which denotes the cone of influence (COI).

Figure 4b shows the WPS and GWS for BMEP of 1.64 bar. It can be noticed from figure that a strong periodic band of 4-8 period occurs intermittently between 115-350, 730-880, 1020-1300, 1770-1940, 2040-2170, 2300-2450 cycles. Other lower intensity periodic bands of 94-131 and 440-625 cycles occurs in the range of 1350-1650 and 785-1480 cycles respectively. It can also be noticed from figures 4a and 4b that GWS power decreases with increase in engine load, which suggest decrease in variability in THR. At low engine load, smaller amount of fuel is burnt in cylinder therefore combustion temperature will also lower. Then residual gas and the cylinder wall would be at a lower temperature, which would increase the ignition delay and causes higher cyclic variations. As the load increases more amount fuel is injected and combustion chamber temperature increase leads to lower cyclic variations. It can be observed from figure 4c that the strongest intensity of 0-8 periodic band occurring in the cycle range intermittently during 110-625 and 1450-1910 cycles. Further increasing engine load (figure 4d), a strong periodic band (0-8 period) occurs in 518-597, 705-854, 1381-1446 and 1860-1890 cycle ranges intermittently. A weaker periodic band (314-445 period) is present during 740-1562 cycles (figure 4d) but GWS indicates high power for low intensity period. This is possibly due to amplification of GWS power at higher periods as reported by Shu et al. [23]. It can also be noticed from figure 4 that strongest intensity variation shift from low to high frequency with increasing engine load.

It can be noticed from figure 5a that strongest intensity periodic band (314-528 period) occurs in the cycle range of 1313-1985 adjacent to the COI. Strongest intensity IMEP variations are spread over lesser number of cycles whereas strongest intensity THR variations are spread over a large number of cycles. It can be clearly observed from figure 5b that the numbers of strong intensity periodic bands have decreased by increasing engine load. The periodic band of 46-78 periods that occurs in the range of 1698-1835 cycles has highest intensity. GWS power has decreased with increase in engine load (Figure 5b). Figure 5c illustrates the decrease in intensity of cyclic variations with further increase in engine load. It consists of only weaker periodic bands such as 314-442 and 78-110 cycles that occur in between 480-782 and 1551-1764 cycles respectively. The power of GWS in this case is lower than the previous cases (figure 5b). In figure 5d the strongest intensity periodic band (39-55 period) occur during 497-594 and 1597-1655 cycle ranges. Power of GWS slightly increases in comparison with previous load. Similar variations are observed for different fuels in reference [24].

Figure 6 shows the WPS and GWS of THR time series at CR= 17.5, 16.0 and 15.0 for constant engine speed and load. Variations are higher at lower engine loads (figure 4) therefore variations in lowest engine load are compared for different compression ratios. It can be noticed from figure 6a that periodic cycle bands of 110-220, 66-102, 120-540 periods have the highest variability (strongest intensity) during cycle ranges 390-870, 1100-1400, 1300-2070 respectively. The presence of the large number of strong and moderate periodic bands represents high cycle to cycle variability in this case (CR=17.5).

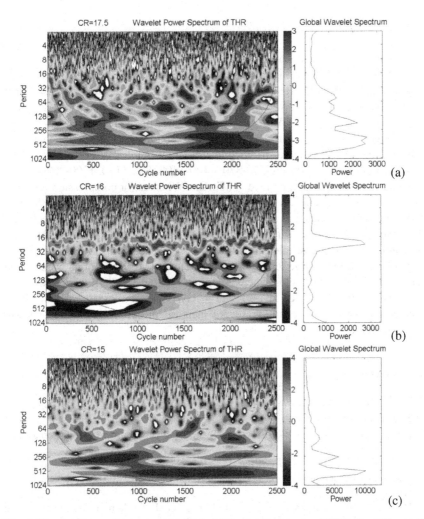

Fig. 6 Wavelet Power Spectrum and Global Wavelet Spectrum of THR for different CR (17.5, 16, 15). The thick contour lines represent the 5% significance level below which denotes the cone of influence (COI).

The strongest periodic band with period 11-33 cycles occurs in the range of 35-287, 584-885, 940-1920, 1953-2458 cycles intermittently for CR = 16.0 (figure 6b). The GWS power increases with decrease in CR, which indicates higher variations in lower compression ratio. Further decrease in compression ratio (figure 6c) strongest intensity periodic bands with periods 222-314 and 444-628 cycles occur during 448-1053 and 980-1802 respectively. The GWS power further increases heavily at this compression ratio, which indicates higher high variability in THR.

4 Conclusions

Experimental investigation of cyclic variations in combustion parameters (THR and IMEP) was conducted in a diesel engine at constant speed for different loads using continuous wavelet transform. It is found that lowest load has highest global wavelet spectrum (GWS) power indicating higher cyclic variability in both parameter (THR and IMEP). GWS power decreases with increase in engine load, which suggest decrease in variability. At no load condition, it was found that maximum cyclic variability was distributed over a larger number of cycles. Maximum cyclic variability occurs during 390-870, 1100-1400, 880-1150, 1300-2070 cycles at compression ratio 17.5 at no load conditions. Whereas for compression ratios 16 and 15, maximum cyclic variability was distributed over 35-287, 584-885, 940-1920, 1953-2458 cycles and 1890-2150, 980-1802, 310-500 cycles respectively. Large number of strong and moderate intensity periodic bands is found for lower load conditions. Strongest intensity variations shift from lower to higher frequencies with increase in engine load. Strongest intensity IMEP variations were spread over lesser number of cycles whereas strongest intensity THR variations were spread over a large number of cycles. The GWS power increases drastically at lower compression ratio resulting severe increase in cyclic variability at lower compression ratios.

References

1. Heywood, J.B.: Internal Combustion Engines Fundamentals. McGraw Hill Education, New York (1988)
2. Atkins, R.D.: An Introduction to Engine Testing and Development. SAE International, USA (2009)
3. Leonhardt, S., Muller, N., Isermann, R.: Methods for engine supervision and control based on cylinder pressure information. Mechatronics, IEEE/ASME **4**(3), 235–245 (1999)
4. Huang, B., Hu, E., Huang, Z., Zheng, J., Liu, B., et al.: Cycle-by-cycle variations in a spark ignition engine fueled with natural gas–hydrogen blends combined with EGR. Int. J. Hydrogen Energ. **34**, 8405–8414 (2009)
5. Zhang, H.G., Han, X.J., Yao, B.F., Li, G.X.: Study on the effect of engine operation parameters on cyclic combustion variations and correlation coefficient between the pressure-related parameters of a CNG engine. Appl. Energ. **104**, 992–1002 (2013)
6. Reyes, M., Tinaut, F.V., Giménez, B., Pérez, A.: Characterization of cycle-to-cycle variations in a natural gas spark ignition engine. Fuel **140**, 752–761 (2015)
7. Gürbüz, H., Akçay, I.H., Buran, D.: An investigation on effect of in-cylinder swirl flow on performance, combustion and cyclic variations in hydrogen fuelled spark ignition engine. J. Energy Inst. **87**, 1–10 (2014)
8. Rakopoulos, D.C., Rakopoulos, C.D., Giakoumis, E.G., Dimaratos, A.M.: Studying combustion and cyclic irregularity of diethyl ether as supplement fuel in diesel engine. Fuel **109**, 325–335 (2013)

9. Wang, Y., Xiao, F., Zhao, Y., Li, D., Lei, X.: Study on cycle-by-cycle variations in a diesel engine with dimethyl ether as port premixing fuel. Appl. Energ. **143**, 58–70 (2015)
10. Maurya, R.K., Agarwal, A.K.: Experimental investigation on the effect of intake air temperature and air–fuel ratio on cycle-to-cycle variations of HCCI combustion and performance parameters. Appl. Energ. **88**, 1153–1163 (2011)
11. Maurya, R.K., Agarwal, A.K.: Statistical analysis of the cyclic variations of heat release parameters in HCCI combustion of methanol and gasoline. Appl. Energ. **89**, 228–236 (2012)
12. Maurya, R.K., Agarwal, A.K.: Experimental investigation of cyclic variations in HCCI combustion parameters for gasoline like fuels using statistical methods. Appl. Energ. **111**, 310–323 (2013)
13. Daily, J.W.: Cycle-to-cycle variations: a chaotic process. Combust. Sci. Technol. **57**, 149–162 (1988)
14. Daw, C.S., Finney, C.E.A., Green, J.B., Kennel, M.B., Thomas, J.F., et al.: A simple model for cyclic variations in a spark ignition engine, SAE Paper 962086 (1996)
15. Daw, C.S., Finney, C.E.A., Kennel, M.B., Connelly, F.T.: Observing and modeling nonlinear dynamics in an internal combustion engine. Phys. Rev. E **57**, 2811–2819 (1998)
16. Foakes, A.P., Pollard, D.C.: Investigation of a chaotic mechanism for cycle-to-cycle variations. Combust. Sci. Technol. **90**, 281–287 (1993)
17. Sen, A.K., Litak, G., Taccani, R., Radu, R.: Wavelet analysis of cycle-to-cycle pressure variations in an internal combustion engine. Chaos, Solitons & Fractals **38**(3), 886–893 (2008)
18. Tily, R., Brace, C.J.: Analysis of cyclic variability in combustion in internal combustion engines using wavelets. Proceedings of the Institution of Me-chanical Engineers. Part D: Journal of Automobile Engineering **225**(3), 341–353 (2011)
19. Sen, A.K., Longwic, R., Litak, G., Górski, K.: Analysis of cycle-to-cycle pressure oscillations in a diesel engine. Mech. Syst. Signal Pr. **22**(2), 362–373 (2008)
20. Shoji, T.: Effect of Cycle-to-Cycle Variations in Spray Characteristics on Hydrocarbon Emission in DI Diesel Engines: Visualization of Sac Inner Flow, Needle Valve Motion and Cycle-to-Cycle Variations in Diesel Spray. JSME Int. J. B **40**(2), 312–319 (1997)
21. Maurya, R.K., Agarwal, A.K.: Experimental Investigation of Effect of Intake Air Temperature and Mixture Quality on Combustion of Methanol and Gasoline Fuelled HCCI Engine. Journal of Automobile Engineering, Proceedings of IMechE, Part-D **223**(11), 1445–1458 (2009)
22. http://paos.colorado.edu/research/wavelets/faq.html#bias (accessed January 11, 2014)
23. Wu, S., Liu, Q.: Some problems on the global wavelet spectrum. Journal of Ocean University of China **4**(4), 398–402 (2005)
24. Longwic, R., Sen, A.K., Górski, K., Lotko, W., Litak, G.: Cycle-to-Cycle Variation of the Combustion Process in a Diesel Engine Powered by Different Fuels. Journal of Vibroengineering **13**(1), 120–127 (2011)

Adaptive Markov Model Analysis for Improving the Design of Unmanned Aerial Vehicles Autopilot

R. Krishnaprasad, Manju Nanda and J. Jayanthi

Abstract The need for Unmanned Aerial Vehicles (UAVs) is increasing as they are being used across the world for various civil, defense and aerospace applications such as surveillance, remote sensing, rescue, geographic studies, and security applications. The functionalities provided by the system is based on the system health. Monitoring the health of the system such as healthy, degraded (partially healthy or partially unhealthy) and unhealthy accurately without any impact on safety and security is of utmost importance. Hence in order to monitor the health of the system to provides the functionality for a longer period of time system fault detection and isolation techniques should be incorporated. This paper discusses Fault Detection and Isolation (FDI approach used in Unmanned Aerial Vehicle (UAV) autopilot to make its functionality more robust and available for a longer period of time. We proposes an integrated Adaptive Markov Model Analysis (AMMA) to detect and isolate faults in critical components of the system. The effectiveness of the novel approach is demonstrated by simulating the modified system design with FDI incorporation for the UAV autopilot. The proposed FDI approach helps in identifying the gyro sensor failure and provides a degraded mode to the system functionality which did not exist earlier in the design. The simulation demonstrates the system modes such as healthy, degraded (partially healthy or partially unhealthy) and unhealthy to understand the functionality better as the current design which works in only two modes i.e. healthy and unhealthy.

Keywords UAV · AMMA · FDI · Functionality modes

R. Krishnaprasad
Jawaharlal Collage of Engineering and Technology, Palakkad, Kerala, India
e-mail: krishnaprasad.rajan90@gmail.com

M. Nanda(✉) · J. Jayanthi
CSIR- National Aerospace Laboratories, Bangalore, India
e-mail: {manjun,jayanthi}@nal.res.in

© Springer International Publishing Switzerland 2016   259
S. Berretti et al. (eds.), *Intelligent Systems Technologies and Applications*,
Advances in Intelligent Systems and Computing 384,
DOI: 10.1007/978-3-319-23036-8_22

1   Introduction

Most of the industries today are highly dependent on software for their basic and advanced functionalities. Safe and reliable software operation is significant requirement for many types systems, such as, in aircraft and air traffic control, space applications [1]. Nowadays, modern systems and control systems are often faced with unexpected changes such as sensor faults, component faults and communication faults, leading to degradation of the overall system performance.

In this paper we proposes to introduce the health monitoring concept into the critical sub-system design of autopilot (AP) of the Unmanned Aerial Vehicles (UAVs) to provides the functionality for a longer time by providing the degraded functionality of the sub-system in addition to complete functionality. UAV has a lot of advantages such as small volume, low cost, low environmental requirements, high maneuvering, provides higher efficiency, easier to handle, and easy to maintain and operate. Due to these attractive features of UAV, it has been used in vast application areas [2]. Few of the applications are : reconnaissance, surveillance, search and rescue, remote sensing, traffic monitoring missions, archaeology, forest fire detection, armed attacks, research, oil, commercial and motion picture filmmaking, maritime patrol, aerial target practice, and parceling of raw materials[3],[4] and [5].

The proposed design introduces a robust fault detection & isolation in the design to provide degraded functionality of the autopilot which was earlier not available. With the introduction of FDI logic reasoner the autopilot functionality is analyzed by its health modes such as healthy, partially healthy, partially unhealthy, and unhealthy modes. The analysis of these modes is realized by means of adaptive Markov model analysis (AMMA), hybrid mathematical modeling with Hidden Markov Model (HMM) and Bayesian inference [9], [18], [20] and [22]. In order to analyze the health of the critical component of the autopilot, FDI approach is developed for the UAV autopilot as a case study. The autopilot of the UAV is a critical sub-system and monitoring its health and identifying the fault improves the overall performance of the UAV. The behavior of the gyroscope is analyzed as it provides most of the critical parameters. These parameters are: Bias (B), Quantization (Q), Scale Factor (SF), Sensitivity (S) and Random Walk Noise (RW) shown in Table 1 [12].

The behavior analysis of the parameters is done based on the parameter data. Various parameter data for healthy, unhealthy mode is randomly generated, behavior of the gyroscope is studied and its effect on the system modes and functionality is studied. Based on the behavior of the sensors, FDI technique is introduced to modify the system modes. Simulink tool suite is used to design, implement and simulate the modified autopilot design using FDI (AMMA) approach. The system functionality modes under various sensors failure is studied for the modified design and compared with the original design to demonstrate the effectiveness of this approach. Comparison of the designs shows the improvement of system robustness and availability by incorporating the design with the FDI. The modified autopilot design demonstrate the sub-system functionalities for a longer period as the AMMA provides the continuous health of the gyros leading to slow degradation of functionality under the fault conditions and also provides indication of degradation to incorporate corrective action. With this novel approach, we can easily monitor the health of the gyros and its effect on the system functional behaviors. The data driven analysis provides the effectiveness of this approach as data helps in understanding the gyros better [21]. This approach can be utilized for other sub-system of the UAV for an overall improvement in the functionality and availability.

The paper is organized as follows: Section 2 discussed the various FDI techniques that have been used for safety critical systems, Section 3 discusses data driven behavioral analysis of gyroscope, Section 4 discusses adaptive Markov model and Section 5 describes the future work & conclusion.

2   Background Work

FDI systems classified based on their redundancies are: hardware redundancy and analytical redundancy. Large variety of methods has been proposed for designing FDI systems. Papers [6], [7], [8] and [17] discuss various approaches, methods, and techniques for FDI based systems. Some of the FDI approaches are: pattern recognition, parameter estimation, observers (Luenberger observers, extended Luenberger observers and unscented observers), filters (Kalman filters, extended Kalman filters, unscented Kalman filters), multiple model adaptive estimation and intelligent technology. Intelligent technology consists of fault diagnosis such as expert system method, graphical method, soft computing, and distributed artificial intelligence. Graphical systems include fault tree analysis and Bayesian network. Soft computing method consists of rough set, artificial neural network, support vector machine, fuzzy logic, artificial immune system and decision tree. There exist advantages and drawbacks of hybrid methods [10].

Recent FDI techniques such as linear parameter varying system [13], worst-case false alarm analysis [14] and wavelet neural network [16] are proving to be effective approach in FDI. Work in [9], [20], and [22] explain stochastic, mathematical logic relations of hidden Markov model, and Bayesian networks. In particular Bayesian network describe the relationship amongst project state impact factors, decisions taken and the satisfaction levels related to each decision [15].FDI techniques are proving effective in monitoring the health of the system and making the system more robust.

3   Data Driven Behavioral Analysis of Gyro Sensor

Data-driven techniques are often used to detect discrete changes in a process and it helps in mimicking the behavior. Only, model-based approach does not work very well when it is not feasible to create analytic relations describing all the observed data, e.g., for vibration data which is usually sampled at very high rates and requires very detailed finite element models to describe its behavior [21]. The data generated for analyzing the gyro sensor is shown in the Table 1. Gyro sensors are used in many applications such as inertial navigation, flight control, automotive testing, laboratory use, platform stabilization, pointing and robotics etc. [11]. Gyro sensors consist of various kinds of mechanical & electrical characteristics which can produce gyro bias errors at its output digital signals. The parameters of interest are accuracy, repeatability, size, cost, weight, maintainability, and reliability. Some typical faults in gyros are sensitivity error, non-linearity, bias stability error, random walk noise, and input axes alignment error [12].

Table 1 shows the state table, which indicates system functionality modes of gyro sensor based on these parameters. We choose critical & non-critical parameters of the sensors and study the effect of the parameter on the system health. Bias, scale factor, sensitivity and random walk noise are chosen as critical parameters (C), and quantization noise as Non critical (NC) parameter. With five parameters there are 2^5 possible input combinations need to be considered. Here binary symbol '0' indicate normal condition for gyro sensor and binary symbol '1' indicate gyro sensor drift away from the normal condition. Next, we analyses how the C and NC parameters effect the gyro sensor functionality. Various scenarios of the parameters in healthy, partially healthy and unhealthy as considered. For example : If all the parameters are in normal condition, one of the parameter will drift away from the normal condition and if only one critical and non- critical parameter will drift away from normal condition, we assumed as system under healthy state (H). If two critical parameters drift away, then the system under partially healthy state (PH), suppose the two critical and one non critical parameter, three critical parameters, three critical parameters and one non critical and four critical and one non critical parameter will drift away from the normal condition, so we assumed as system under unhealthy state (UH).

The health of the system is considered in different modes: healthy mode, partially healthy mode, partially unhealthy mode and the unhealthy mode.

- Healthy mode: the autopilot provides reliable & complete functionality.
- Partially healthy: autopilot provides reliable but partial functionality.
- Partially unhealthy: autopilot provides unreliable but partial functionality.
- Unhealthy: No functionality provided.

Table 1 Gyro Sensor State Table Based on Designer Assumptions.

SL. No	GYRO CRITICAL PARAMETERS					STATES		
	C B	C SF	C S	NC Q	C RW	H	PH	UH
1	0	0	0	0	0	1		
2	0	0	0	0	1	1		
3	0	0	0	1	0	1		
4	0	0	0	1	1	1		
5	0	0	1	0	0	1		
6	0	0	1	0	1		1	
7	0	0	1	1	0	1		
8	0	0	1	1	1			1
9	0	1	0	0	0	1		
10	0	1	0	0	1		1	
11	0	1	0	1	0	1		
12	0	1	0	1	1			1
13	0	1	1	0	0		1	
14	0	1	1	0	1			1
15	0	1	1	1	0			1
16	0	1	1	1	1			1
17	1	0	0	0	0	1		
18	1	0	0	0	1		1	
19	1	0	0	1	0	1		
20	1	0	0	1	1		1	
21	1	0	1	0	0		1	
22	1	0	1	0	1			1
23	1	0	1	1	0		1	
24	1	0	1	1	1			1
25	1	1	0	0	0		1	
26	1	1	0	0	1			1
27	1	1	0	1	0		1	
28	1	1	0	1	1			1
29	1	1	1	0	0			1
30	1	1	1	0	1			1
31	1	1	1	1	0			1
32	1	1	1	1	1			1

Figure 1 show the corresponding state diagram based on above shown table. Three modes are indicated as brown colored nodes, transition between one modes to another indicated as arrows. This state diagram shows clearly the 32allowed input combinations based on the 5 parameters. Based on the real time application, once the gyro sensor fails or is in unhealthy mode, sensor cannot come back to the healthy mode. AMMA has an advantage to use the sensor data and detect and identify the faults more effectively as it uses analytical approach [20] and [15].

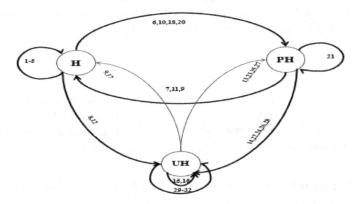

Fig. 1 Gyro Sensor Behavior State Diagram Based on State Table Shown in Table 1.

4   Proposed Fault Diagnosis Approach Based on Adaptive Markov Model Analysis.

Figure 2 shows the modified autopilot design with the adaptive Markov model analysis. The red color bordered blocks are the additional design introduced in the original system design. Novel FDI approach consists of three blocks which are gyro sensor reference model, HMM estimator, and Bayesian Network reasoning logic.

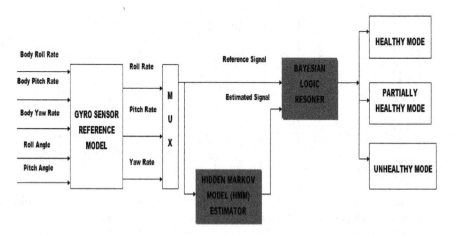

Fig. 2 Fault Detection and Isolation System for UAV Autopilot.

In the design, Bayesian Hidden Markov Model is considered, because of its unique characteristics such as doubly stochastic process, transition parameters model temporal variability, and output distribution model spatial variability [19]. Implementation of Simulink model is based on general residual generation and residual evaluation equations as given in equation (1).

$$\dot{X} = AX + BU$$
$$Y = ZX \tag{1}$$

In the equation (1), Y, $\dot{X}$ is the output of HMM, A & B are the state transition & observation symbol observation matrices respectively, X and U are the gyro output signals & gyro control input signals, Z is the sensitivity matrix and Y is the FDI approach output signals.

4.1   Using Hidden Markov Estimator for Fault Detection of Gyro Sensor

HMM is a doubly stochastic process with an underlying stochastic process that is not observable (it is hidden) but can only be observed through another set of stochastic processes that produce the sequence of observed symbols [18],[20] and [22]. Mathematically, the complete specification of an HMM consists of the following elements and represented by a compact notation $\lambda = (A, B, \pi)$ [20].Accordingly, N= 3, M= 5, so hidden modes are S = {H, PH, UH} or {S1, S2 & S3}. These are shown in the figure 1 as state diagram & table 1 as state table. We assumed Healthy mode as an initial state, its probability distribution is '1', '0.5' for partial healthy mode & '0' for Unhealthy mode.

Special Case. In real time applications, gyro sensor cannot transits to healthy mode (H) from Unhealthy mode (H) or Unhealthy (UH) to partial Healthy mode (PH). Due to the limited allowed transitions as described for real time systems. Adaptive Markov Model Analysis (AMMA) becomes effective to understand the behavior of such systems. In order to simulate the mode transition of the systems as discussed in figure 1 and table 1, the reference model is modified such that the mode index either stays in the same mode or increases by one. In other words, gyro sensor can only stay in the current health mode or move to the next one. In this case, the coefficients of the transition matrix of the model satisfy the following properties given in equation (2).

$$
\begin{aligned}
&\text{a.} \quad a_{ij} = 0, j < i \\
&\text{b.} \quad a_{ji} = 0, j > i + 1 \\
&\text{c.} \quad a_{NN} = 1, \\
&\text{d.} \quad a_{Ni} = 0, i < N
\end{aligned}
\tag{2}
$$

An unhealthy mode considered as final (failure) state, based on these properties, to construct 3×3 matrix as state transition probability distribution i.e., $A = \{a_{ij}\}$, where $a_{ij} = P (q_{t+1} = Sj| q_t = S_i)$, $1 \le i, \le 3$, $1 \le j \le 3$.H, PH & UH are the hidden modes, corresponding matrix given in equation (3).

$$
A = \begin{array}{c} \\ S_1 \\ S_2 \\ S_3 \end{array}
\begin{array}{ccc} S_1 & S_2 & S_3 \end{array}
\begin{bmatrix} 1 & 0 & 0 \\ 0 & 0.5 & 0.5 \\ 0 & 0 & 1 \end{bmatrix}
\tag{3}
$$

To find the observation symbol distribution at each of the hidden states from the corresponding state table, i.e., B = {b_j(k)} where $b_j(k) = P (Y_k| q_j = S_j)$, $1 \le j \le 3$, $1 \le k \le 5$, so we can easily construct 3×5 matrix as observation symbol distribution given in equation (4).

$$
B = \begin{array}{c} \\ S_1 \\ S_2 \\ S_3 \end{array}
\begin{array}{ccccc} B & SF & S & Q & RW \end{array}
\begin{bmatrix} 0.063 & 0.063 & 0.063 & 0.063 & 0.063 \\ 0.188 & 0.125 & 0.125 & 0.0938 & 0.125 \\ 0.25 & 0.313 & 0.344 & 0.281 & 0.281 \end{bmatrix}
\tag{4}
$$

The initial probability distribution, $\pi = \{\pi_i\}$ where $\pi_i = P\ \{q_t = S_i\}$, $1 \le i \le 3$. From table1, we assume initial probabilities of each hidden modes given in equation (5).

$$\Pi = [0.3125\ 0.28125\ 0.40625] \tag{5}$$

With computational easiness, HMM helps in identifying the occurrence of faults which are discussed in following sections.

4.2 Bayesian Inference Reasoning Logic

We use the Bayesian Inference logic to derive the sensitivity matrix of the gyroscope parameters. Bayesian inference uses conditional probabilities to represent uncertainty, here, our ultimate goal to isolate the faults by using the decision theory. For instance, the gyro sensor model that has the maximum a posterior probability, given the data, can be chosen in order to minimize the misclassification rate [19].

$$\hat{k} = \text{avg max}_k\ P\ (\lambda_k/Y) \tag{6}$$

According to the Bayesian theorem, formally expressed given in equation (7):

$$P(\lambda_k/Y) = \frac{P(Y/\lambda_k)\pi_k}{\sum\limits_{k=1}^{N} P(Y/\lambda_k)\pi_k} \tag{7}$$

Where, k= 1, 2..... M, P (Y/λ_k) is the likelihood of the model λ_i given the data Y and can be calculated thanks to the popular forward- backward analysis. Here, given the model P (Y/λ) so called forward algorithm used to determine the likelihood of the observed sequence Y. For k= 1, 2... M, i = 1, 2,... N; given in equation (8).

$$\alpha_k(i) = P\ (Y_1,\ Y_2,\dots\dots\ Y_M,\ S_k = q_{i,}/\lambda) \tag{8}$$

$\alpha_k(i)$ is the probability of the partial observation sequence up to time k, where the underlying Markov process in state q_i at time k [20]. Here, is that $\alpha_k(i)$ can be computed recursively given in equation (9).

a. Let $\alpha_0(i) = \ _ib_i(Y_1)$, for i = 1,2, ...,N

b. $\alpha_k(i) = [\sum\limits_{j=1}^{N} \alpha_{k-1}(j)a_{ji}]\ b_i(Y_k)$, for k = 1...........M-1 and i = 1,2,...N $\tag{9}$

Likelihood observation sequence value given in equation (10).

$$P\ (Y/\lambda) = \sum\limits_{i=1}^{3} \alpha_5(i) = 0.001098 \tag{10}$$

This likelihood observation sequence value setting as threshold value. On Other hand, our goal is to find the most likely state sequence. In other words we want to uncover the hidden part of the hidden Markov model. For HMMs we want to maximize the expected number of correct states. The backward algorithm addressed likelihood state sequence. For, k = 1, 2,... M, and i = 1, 2,......N; given in equation (11).

$$\beta_k(i) = P\ (Y_{k+1},\ Y_{k+2},\ \ldots\ldots\ Y_M,\ S_k = q_i/\lambda) \tag{11}$$

$\beta_k(i)$ can be computed recursively given in equation (12).

1. $\beta_M(i) = 1$; for i = 1,2,...........,M

2. $\beta_k(i) = \displaystyle\sum_{j=1}^{N} a_{ij} b_j \left(Y_{k+1}\right) \beta_{k+1}(j)$; for k= M-1, M-2....0 and i = 1, 2, ...N   (12)

After computing the forward $\alpha_k(i)$ backward $\beta_k(i)$ coefficients, these values are require to solve $\gamma_k(i)$ coefficient. For k = 1, 2, ... M-1, i = 1, 2, ...N; given in equation (13).

$$\gamma_k(i) = P(S_k = q_i/Y, \lambda) \tag{13}$$

Where, k= 1,......M-1, i= 1,......N, $\alpha_k(i)$ measures the relevant probability up to time and $\beta_k(i)$ measures the relevant probability after time k. Here, is that $\gamma_k(i)$ can be computed recursively given in equation (14).

$$\gamma_k(i) = \frac{\alpha_k(i)\beta_k(i)}{\displaystyle\sum_{k=1}^{N} \alpha_s(i)} \tag{14}$$

After computing values of $\gamma_k(i)$, we can easily construct 3×5 sensitivity matrix denoted by Z given in equation (15).

$$Z = \begin{bmatrix} 1.76\text{X}10^{-5} & 1.64\text{X}10^{-5} & 1.74\text{X}10^{-5} & 1.77\text{X}10^{-5} & 1.78\text{X}10^{-5} \\ 0.0752 & 0.4664 & 0.03930 & 0.0284 & 0.04289 \\ 0.1969 & 0.2464 & 0.2708 & 0.2215 & 0.2216 \end{bmatrix} \tag{15}$$

Matrix Z isolates the functionality modes of gyro sensor such as healthy, partially healthy and unhealthy. This matrix helps in analyzing the probability of the occurrence of the system modes which enables in fault detection and the source of fault for determining the system health mode. With the probability figures for the parameters we determine functionality modes of the gyro sensor.

5   Implementation of Model and Simulation Results

Figure 3 shows the modified design is modeled in Matlab/Simulink tool suite. Implementation of gyro sensor reference model is done by using mathematical relations of gyro sensors. In order to generate parameter data we choose the operating range roll angle, pitch angle and control signals such as body roll rate, body pitch rate and body yaw rates as -10° to 40° for the gyro sensor. Signal delays are introduced from inputs to generate random flight data outputs. Analysis of the sensor outputs (roll rate (P), pitch rate (Q), yaw rate (R), PQ, PQ and QR) with respect to the inputs for various combinations of gyro sensor parameters.

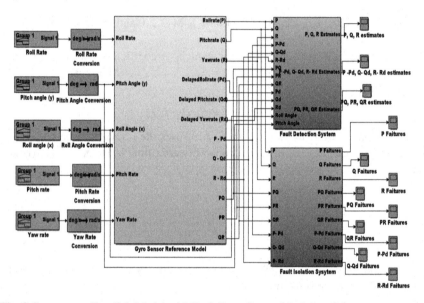

Fig. 3 Gyro sensor Simulink Model with Fault Detection and Isolation Subsystems.

The HMM matrices derived helps in detection in faulty parameter. HMM notation $\lambda = (A, B, \pi)$ is used for implementing. HMM fault detection model generates the error estimates, which depends on input signals and control signals of gyro sensor. For the Bayesian inference, we set the threshold value. Sensitivity matrix based on this value for isolating gyro sensor faults. Finally, we analyses the various combinations of gyro output, how the isolation of faults occurred. Based on the approach knowledge, we analyze by sensor output drifted away from normal condition, how slowly the sensors degraded its functionalities depend on the parameters, also we check whether the gyro sensor health monitoring which means if the sensor is healthy mode, partially healthy mode and unhealthy modes.

Simulation results shows the various failures combinations such as P failures, Q failures, R failures, PQ failures, PR failures, QR failures based on five parameters. Figures 4(a), 4(b) & 4(c) shows the P, Q, R failures and Figures 5(a), 5(b) & 5(c) shows PQ, PR and QR failures and x- axis represents time and y- axis represents parameters. Each of these figures shows three waveforms of functionality modes such as healthy, degraded and unhealthy depends on a single parameter. So, Every these figures shown below consists of fifteen waveforms. The yellow line indicate as the zero reference, Magenta colored waveform indicate as failures depend upon bias (B) errors, light blue colored waveform indicate as failures depend upon scale factor (SF) errors, red colored waveform indicate as failures depend upon sensitivity (S) errors, green colored waveform indicate as failures depend upon the quantization (Q) noise and violet colored waveform indicate as failures depend upon random walk noise (RW).

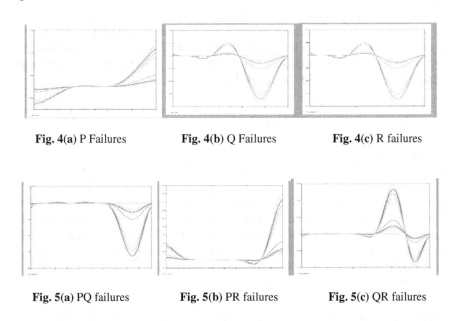

Fig. 4(a) P Failures   **Fig. 4(b)** Q Failures   **Fig. 4(c)** R failures

Fig. 5(a) PQ failures   **Fig. 5(b)** PR failures   **Fig. 5(c)** QR failures

Figures 6(a) and 6(b) show the roll rate waveforms without FDI system and with FDI system. In these figures x- axis represents time and y- axis represents functionality modes: healthy (H), partially healthy (PH) and unhealthy (UH) based on five parameters as discussed earlier. The waveforms are generated based on the gyro sensor roll rate data given shown in figure 3.

In figure 6(a), time from 0 to 4 seconds is the healthy mode and beyond 4 seconds is the unhealthy mode. This figure shows the gyro sensor health transition from healthy to unhealthy mode. In figure 6(b), there exists degraded mode in between the healthy and unhealthy modes. As seen, time from 0 to 4 seconds is healthy mode, from 4 to 6 seconds is degraded mode, and beyond 6 seconds is unhealthy mode. Comparing these waveforms, we can easily analyze how the roll rate outputs are available for a longer period of time with the introduction of the FDI approach. The FDI approach predicts transition of the roll rate from healthy to degrade to unhealthy mode and this approach can be further improved to make the roll rate available for a longer period of time.

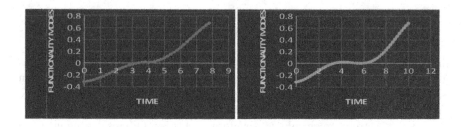

Fig. 6(a) Roll rate waveform without FDI system.   **Fig. 6(b)** Roll rate waveform with FDI system.

Similar analysis is performed for pitch and yaw rates. With the incorporation of the FDI approach in the UAV autopilot we are able to improve its functionality and performance as we are detecting the failure of the critical parameters and transit to the unhealthy mode in a smooth way.

6   Future Work and Conclusion

The research work proposes a novel approach called Adaptive Markov Model Analysis (AMMA) to incorporate a FDI approach in UAV autopilot to make its functionality robust and available for a longer period of time. This is a novel approach to detect and isolate occurrence of faults and also provides with the actual health state recognition. By using Bayesian inference based HMM models for reasoning logic, together with the a priori probabilities of each model. Future research will be focused on forecasting of faults for more complex and critical safety critical systems. The proposed approach and the extension should also be validated on the data of real systems.

Acknowledgements Authors would like to thank the Director, CSIR-NAL, for providing us the opportunity for their extensive support and acknowledge their helping hand in making them to work in this field.

References

1. Pullum, L.L.: Software Fault Tolerance Techniques and Implementation, pp. 1–358 (2001). http://www.artechhouse.com
2. Baoan, Li, Dawei, D.: Automatic test system for large unmanned aerial vehicle. IEEE, pp. 1–4 (2014)
3. Kumar, Nilesh, Jain, Sheilza: Identification Modeling and Control of Unmanned Aerial Vehicles. International Journal of Advanced Science and Technology **67**, 1–10 (2014)
4. Austin, R.: Design, UAVs Development and Deployment. Willey Publication, pp. 1–399 (2010)

5. Hoffer, N.V., Coopmans, C., Jensen, A.M., Chen, Y.Q.: International Conference on Unmanned Aircraft Systems (ICUAS), pp. 897–903 (2013)
6. Sadeghzadeh, I., Zhang, Y.: A Review on Fault-Tolerant Control for Unmanned Aerial Vehicles (UAVs). American Institute of Aeronautics and Astronautics, pp. 1–12 (2011)
7. Marzat, J, Piet-Lahanier, H., Damongeot, F., Walter, E.: Model-based fault diagnosis for aerospace systems: a survey, In: Proceedings of the Institution of Mechanical Engineers, Part G: Journal of Aerospace Engineering, 1–31 (2011)
8. Supervision, I.: Fault- Detection and Fault – Diagnosis Methods- an Introduction. Elsevier science Ltd, pp. 639–652 (1997)
9. Ghahramani.Z.: An Introduction to Hidden Markov Models and Bayesian Networks. International Journal for Pattern Recognition and Artificial Intelligence, 1–25 (2001)
10. Lv, F., Li, X., Wang, X.-Q.: A survey of intelligent network fault diagnosis technology. In: 25th Chinese Control and Decision Conference (CCDC), pp. 4874–4879, (2013)
11. Grewal, M., Andrews, A.: How Good Is Your Gyro? IEEE Control Systems Magazine, 1–4, February 2010
12. G200 Dual Axis MEMS Gyro User Guide. Gladiator Technologies (2012- 2014), pp. 1–36
13. Lopez-Estrada, F.R., Ponsart, J-C, Theillio, D., Astorga-Zaragoza, C.M., Zhang, Y.M.: Robust sensor fault diagnosis and tracking controller for a UAV modelled as LPV system. In: International Conference on Unmanned Aircraft Systems (ICUAS), pp. 1311–1316, May 2014
14. Hu, B., Seiler, P.: Worst-case false alarm analysis of fault detection systems. In: American Control Conference (ACC), pp. 654–659, June 2014
15. Ahmed Nagy, Mercy Njima and Lusine Mkrtchyan.: A bayesian based method for agile software development release planning and project health monitoring. In: International Conference on Intelligent Networking and Collaborative Systems, pp. 192–199 (2010)
16. Berenji, H.R., Wang, Y.: Wavelet neural networks for fault diagnosis and prognosis. In: IEEE International Conference on Fuzzy Systems, pp. 1334–1339 (2006)
17. Kapadia, R., Stanley, G., Walker, M.: Real World Model-based Fault Management, pp. 1–8
18. Rabiner, L.R., Juang, B.H.: An Introduction to Hidden Markov Models. IEEE ASSP Magazine, 4–16 (1986)
19. Le, T.T., Chatelain, F., Berenguer, C.: Hidden Markov Models for Diagnostics and Prognostics of Systems under Multiple Deterioration Modes, pp. 1–9, July 2014
20. Stamp, M.: A Revealing Introduction to Hidden Markov Models, pp. 1–20, September 2012
21. Narasimhan, S., Choudhury, I.R., Balaban, E., Saxena, A.: combining model-based and feature-driven diagnosis approaches – a case study on electromechanical actuators. In: 21st International Workshop on Principles of Diagnosis, pp. 1–8 (2010)
22. Shang, Y.: The limit behavior of a stochastic logistic model with individual time dependent rates. Journal of Mathematics, 1–8 (2013). http://dx.doi.org/10.1155/2013/502635

Energy Efficient Compression of Shock Data Using Compressed Sensing

Jerrin Thomas Panachakel and K.C. Finitha

Abstract This work analyses the potential of compressed sensing (CS) for compressing shock data signals of space launch vehicles. Multiple shock data signals were compressed using compressed sensing by exploiting the sparsity of the shock data signals in the time domain. Since shock data signals are sparse in wavelet domain also, thresholding based DWT compression was performed to compare the performance of compressed sensing. Three performance metrics, *viz.* Peak Root mean-square Difference (PRD), Compression Ratio (CR) and execution time were used. It is also evaluated how compression of the shock data reflects in the Shock Response Spectrum (SRS). The results clearly show that CS surpasses the DWT based compression in terms of execution time for a given CR but has slightly inferior results in terms of PRD for higher values of CR. With lower computation power requirements and dimensionality reduction, CS becomes an ideal choice for compressing shock data signals in a mobile signal processing system with constraints on processing power and for transmission over a power-hungry wireless network.

1 Introduction

International Data Corporation (IDC) forecasts that the "digital universe" will grow to an astonishing 8 ZB by the year 2015 [1]. This figure was just 2.7 ZB in the year 2012 [2][3]. This unprecedented rise in the data creation rate calls on, never like before, the need of compression techniques. The art and science of compression itself has evolved by time. The first attempt to compress electrical signals was by

J.T. Panachakel(✉)
Indian Institute of Science, Bangalore, India
e-mail: jerrin.panachakel@gmail.com

K.C. Finitha
Vikram Sarabhai Space Centre (VSSC), Trivandrum, India
e-mail: kc_finitha@vssc.gov.in

© Springer International Publishing Switzerland 2016   273
S. Berretti et al. (eds.), *Intelligent Systems Technologies and Applications*,
Advances in Intelligent Systems and Computing 384,
DOI: 10.1007/978-3-319-23036-8_23

Homer Dudley in the late 1930s where he used what is known as the "vocoder" [4][5]. Depending on the requirements, different methods are used for compressing different classes of signals.

In this paper, we propose the use of Compressed Sensing for compressing the telemetry signals transmitted from a space launch vehicle. Out of the various telemetry signals *viz.* shock data, vibration data and the acoustic data, compression of shock data is discussed in this paper. Shock data is defined as the plot of the magnitudes of shock pulses as a function of time where shock pulses are caused by the non-periodic excitations characterised by severity and suddenness and which usually causes relative displacement in the system [6]. As far as a space launch vehicle is concerned, these pulses result from rocket motor ignition, staging events, deployment events, etc. [7]. Although it is essential to compress shock data since a typical flight test requires more than ten three-axis accelerometers for generating the shock data [8], not many works have been carried out in compressing shock data mainly because [9] [5],

- Typical data compression algorithms are complex and are often difficult to maintain.
- Typical data compression algorithms may affect the inherent structure of the data making the analysis of the signal difficult.
- Computational power constraints in a mobile signal processing system.

The proposed CS based compression technique addresses these constraints efficiently. CS was our choice of compression scheme because,

- Shock data signal, as discussed in Section 2.1, is sparse in multiple domains making it an ideal candidate for compressed sensing.
- Compressed sensing has comparatively lower power requirements [10].

We believe that the reduction in the computational complexity manifests into energy efficiency.

The rest of the paper is organised as follows, Section 2.1 discusses about shock data signals, Section 3 covers, rather briefly, compressed sensing. The various results which we have obtained are given in Section 4.

Notation: Normal letters designate scalars, boldface letters designate vectors and capital boldface letters designate matrices. Also, $\langle x, y \rangle$ stands for the inner product of x & y and $||.||_p$ for the l_p norm.

2 Shock Data Signal Analysis

2.1 Saturation Analysis

Shock data signals sampled at the rate of $6400Hz$ were used in this work. First 500 samples of one of the shock data signals used in this work is shown in Fig. 1. A common problem associated with shock data signals is the saturation of the accelerometer used [7]. Tom Irvine in [7] notes that the positive and negative spectral curves in a saturated accelerometer data diverges from each other in acceleration

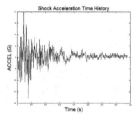

Fig. 1 Sample shock acceleration time history

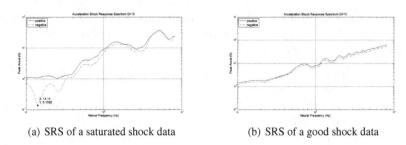

(a) SRS of a saturated shock data   (b) SRS of a good shock data

Fig. 2 Shock Response Spectra of shock data signals

Shock Response Spectrum (SRS). To analyse whether the shock data signals are obtained from saturated accelerometers, SRSs were plotted using Kelly-Richman algorithm [11], [12]. SRS of a saturated shock data is given in Fig. 2(a). It may be noted that there is wide margin between the positive and negative spectral curves at lower frequencies. This indicates accelerometer saturation. Fig. 2(b) gives the SRS of a good shock data.

The different shock data signals used in this work were analysed and it was verified that they were not the signals generated by saturated accelerometers.

2.2 Sparsity Analysis

As we will see in Section 3, one of the integral requirement on the structure of a signal for using compressed sensing is that the signal should have sparse representation in some domain. Consider a signal x which can be viewed as a $N \times 1$ column vector in $\mathbb{R}^N$. If $\boldsymbol{\Psi} = \{\boldsymbol{\psi}_1, \boldsymbol{\psi}_2, ..., \boldsymbol{\psi}_N\}$ represents an $N \times N$ dimensional basis matrix where each $\{\boldsymbol{\psi}_i\}_{i=1}^{i=N}$ is a basis vector, then x can be expressed as

$$x = \sum_{i=1}^{N} s_i \boldsymbol{\psi}_i \qquad (1)$$

or equivalently,

$$x = \Psi s \qquad (2)$$

where s contains the weighing elements, given by $s_i = \langle x, \psi_i \rangle$ [13]. x is said to be K sparse in the domain spanned by the vectors in Ψ if only K basis vectors are required for expressing x in the Ψ space [13]. Sparsity is a stringent condition and a less stringent condition is "compressibility". The signal x is said to be compressible in the domain spanned by vectors in Ψ if the sorted magnitudes in the Ψ space follows a power law decay [14].

The Shock data signal was analysed for sparsity in various domains including time domain, Discrete Cosine Transform domain, Modified DCT (MDCT) or DCT IV domain and Discrete Wavelet Transform Domain (DWT) using basis functions such as Haar, Coiets, & Daubechies wavelets. For analysing sparsity in all domains except the time domain, the following operations were performed,

- **Step 1:** The original signal in the time domain was transformed to the domain in which the analysis has to be done.
- **Step 2:** The absolute values of the coefficients or the weighing elements were computed and these values were arranged in the descending order.
- **Step 3:** The sorted values were plotted as a function of the new indices.

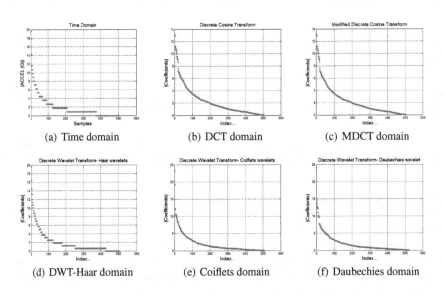

(a) Time domain	(b) DCT domain	(c) MDCT domain
(d) DWT-Haar domain	(e) Coiflets domain	(f) Daubechies domain

Fig. 3 Sparsity analysis

The results are given in Fig. 3.

Interestingly, a shock data signal is sparse in time domain as well as in the wavelet domain as evident from Fig. 3. Hence compressed sensing can indeed be applied for compressing shock data.

3   Compressed Sensing

In a conventional transform coding system, the full N-sample signal is acquired. This signal is transformed to a domain in which the signal is K-sparse. N transform coefficients or weights are computed and out of these N coefficients, (N_K) smallest coefficients are discarded and the K largest coefficients and their locations are encoded. The inefficiency of this approach is due to the following reasons [13],

1. N samples are acquired only to discard $(N - K)$ samples.
2. N coefficients have to be computed even though we need only K coefficients finally.
3. The location of the retained coefficients have to encoded. This results in an overhead.

Compressed sensing addresses these inefficiency by acquiring M samples instead of N where $M \ll N$. In the general case, M inner products between x and a collection of vectors $\{\phi_j\}_{j=1}^{M}$ each having dimension $1 \times N$ is computed. Let y_j denotes the inner product between x and ϕ_j, then arrange the measurements y_j in an $M \times 1$ vector y and the measurement vectors ϕ_j^T as rows in an $M \times N$ matrix Φ, then,

$$y = \Phi x \tag{3}$$

using (2),

$$y = \Phi \Psi s = \Theta s \tag{4}$$

The matrix Θ must satisfy what is called *Restricted Isometry Property* (RIP) so that we can efficiently reconstruct the signal back from the M measurements. RIP is given by

$$(1 - \delta_s)||s||_2 \leq ||\Theta s||_2 \leq (1 + \delta_s)||s||_2 \tag{5}$$

where δ_s is the isometric constant of the matrix Φ.

The original signal x can be approximated from y by using l_1 norm optimization in essence of compressed sensing. A much deeper discussion can be found in [16] and [13].

In our work, the sparsity of shock data signals in the time domain was exploited and the sensing matrix used was random i.i.d Gaussian matrix and partial Fourier matrix [17]. Algorithms based on standard interior point methods [18], [19] (*l1*-magic, [20]) were used at the recovery stage.

4   Results

Thresholding based DWT compression algorithm *hereafter referred to as *DWT Compression)*[10], [21] was also implemented for the purpose of comparison. Comparisons were made based on two metrics 1)PRD and 2)Execution time. These factors are explained below,

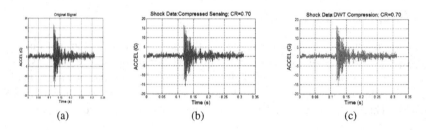

Fig. 4 Original and Recovered Signal: (a) Original Signal (b) Compressed Sensing (c) Thresholding based DWT

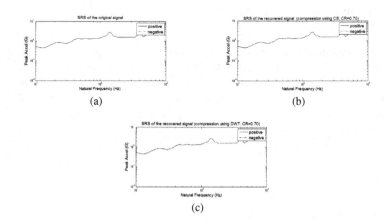

Fig. 5 Shock Response Spectra: (a) Original (b) CS (c) Thresholding based DWT

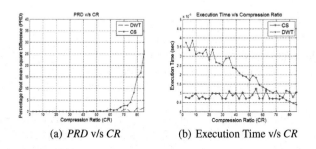

(a) *PRD* v/s *CR*   (b) Execution Time v/s *CR*

Fig. 6 Comparison of *PRD*, *CR* and execution time

- **Percentage Root mean-square Difference (PRD):** This parameter measures the quality of the recovered signal on comparison to the original signal. Mean independent form of this parameter [22], given by

$$PRD = \sqrt{\frac{\sum\limits_{n=1}^{N} (x[n] - \hat{x}[n])^2}{\sum\limits_{n=1}^{N} (x[n] - \bar{x}[n])^2}} \times 100\% \qquad (6)$$

where $x[n]$ and $\hat{x}[n]$ are respectively the original signal and the recovered signal, $\bar{x}[n]$ is the mean of $x[n]$ and N is the total number of samples.

In this work, the PRD was measured for the spectral signals in the SRS instead of the time domain signals. That is, $x[n]$ corresponds to the spectral signal of the original shock data in the respective SRS and $\hat{x}[n]$ is the spectral signal of the shock data recovered from the compressed signal.

- **Execution Time:** Execution time is the measured *stopwatch time* required for a set of operations to complete. In Matlab® *tic-toc* function can be used for measuring the execution time. In this work, the execution time was determined on a computer running on *3rd Generation Intel® CORE*™ *i3* processor running at $1.2GHz$ clock speed.

Both these metrics were plotted with respect to a third metric called *compression ratio* (CR) which gives a measure of signal compression, given by [22]

$$CR = \frac{N_{org} - N_{comp}}{N_{org}} \times 100 \qquad (7)$$

where N_{org} is the number of non-zero samples in the original signal and N_{comp} is the number of non-zero samples in the compressed signal. It can be expressed as a fraction or as percentage.

Fig. 4 shows the original signal and the signal recovered from the compressed signal when compressed using compressed sensing (Fig. 4(b)) and thresholding based DWT compression (Fig. 4(c)). It was also evaluated how compression of the shock data reflects in the respective SRS. Since there is redundancy between the positive and negative spectral curves in a non-saturated shock data signal, only the positive spectral curve is considered here.

SRS spectra of the original, the signal recovered using CS and the signal recovered using DWT compression are given in Fig. 5. The PRD v/s CR is given in Fig. 6(a) and the execution time v/s CR in Fig. 6(b). Few reasons for the reduction in the execution time are

1. The need for domain transformation of the signal is avoided.
2. If the linear transformation in the DWT compression technique is viewed as a matrix multiplication, the matrix is a full rank matrix but in case of compressed sensing, the sensing matrix is not a full rank matrix.

3. Sparsity of the signal.

Further, the overhead of storing the location of the (N_K) largest coefficients in the case of DWT compression is avoided in case of compressed sensing. Also there is a dimensionality reduction as far as compressed sensing is concerned because the sensing matrix is not full rank. This dimensionality reduction helps to decrease the power consumption when the signal is transmitted over a power hungry wireless network.

5   Conclusion

As expected, thresholding based DWT compression algorithm has better PRD for higher CR than compressed sensing but they have almost equal PRDs for moderate and low CRs. Also, compressed sensing requires lesser computational power than the DWT counterpart. This along with the dimensionality reduction achieved using compressed sensing makes this scheme suitable for a mobile signal processing system with constraints on computational power and energy. The sparsity of the shock data signal in domains other than the time domain wasn't exploited in this work and we anticipate much better results if it was also exploited. Also, the recovery makes use of a very basic algorithm; the use of other algorithms like gradient projection [23], orthogonal matching pursuit (OMP) [24] etc., may yield better results.

References

1. Vesset, D., Woo, B., Morris, H.D., Villars, R.L., Little, G., Bozman, J.S., Borovick, L., Olofson, C.W., Feldman, S., Conway, S., Eastwood, M., Yezhkova, N.: World wide big data technology and services 2012–2015 forecast. Market analysis, International Data Corporation (IDC) (2012)
2. Gens, F.: IDC predictions 2012: Competing for 2020. Report, International Data Corporation (IDC) (2011)
3. Cousins, K.: Transform your data center with cloud lifecycle managementibm cloud & smarter infrastructure product management. Presentation, IBM Cloud & Smarter Infrastructure Product Management (2013)
4. Dudley, H.: Remaking speech. The Journal of the Acoustical Society of America 11, 169–177 (1939)
5. Lynch, T.: Data compression with error-control coding for space telemetry. NASA technical report, National Aeronautics and Space Administration (1967)
6. Walter, P.L.: Selecting accelerometers for and assessing data from mechanical shock measurements. PCB Electronics Technical Note (2008)
7. Irvine, T.: An introduction to the vibration response spectrum (2012). www.vibrationdata.com
8. De-Rong, C., Xiang-Bin, L.: Airborne shock response spectra analyzer based on FPGA. In: 20th International Congress on Instrumentation in Aerospace Simulation Facilities, ICIASF 2003, pp. 224–227 (2003)
9. Honeywell: Solid-state flight data recorder 1x, 2x, 4x models. Product description, HoneyWell (2011)

10. Mamaghanian, H., Khaled, N., Atienza, D., Vandergheynst, P.: Compressed sensing for real-time energy-efficient ecg compression on wireless body sensor nodes. IEEE Transactions on Biomedical Engineering **58**, 2456–2466 (2011)
11. Kelly, R., Richman, G., Shock, Center, V.I.: Principles and techniques of shock data analysis. Number v. 5 in Shock and vibration monograph series. Shock and Vibration Information Center, U.S. Dept. of Defense (1971)
12. Martin, J.N., Sinclair, A.J., Foster, W.A.: On the shock-response-spectrum recursive algorithm of kelly and richman. Shock and Vibration **19**, 19–24 (2012)
13. Baraniuk, R.G.: Compressive sensing [lecture notes]. IEEE Signal Processing Magazine **24**, 118–121 (2007)
14. Cevher, V., Indyk, P., Carin, L., Baraniuk, R.G.: A tutorial on sparse signal acquisition and recovery with graphical models (2010)
15. Cyklucifer, nderung, L., Sonntag: Sparsity (2011)
16. Donoho, D.L.: Compressed sensing. IEEE Transactions on Information Theory **52**, 1289–1306 (2006)
17. Rudelson, M., Vershynin, R.: On sparse reconstruction from fourier and gaussian measurements **61**, 1025–1045 (2008)
18. Chen, S.S., Donoho, D.L., Saunders, M.A.: Atomic decomposition by basis pursuit. SIAM Journal on Scientific Computing **20**, 33–61 (1998)
19. Van Den Berg, E., Friedlander, M.P.: Probing the pareto frontier for basis pursuit solutions. SIAM Journal on Scientific Computing **31**, 890–912 (2008)
20. Candes, E., Romberg, J.: l1-magic: Recovery of sparse signals via convex programming, vol. 4 (2005). www.acm.caltech.edu/l1magic/downloads/l1magic.pdf
21. Benzid, R., Marir, F., Boussaad, A., Benyoucef, M., Arar, D.: Fixed percentage of wavelet coefficients to be zeroed for ecg compression. Electronics Letters **39**, 830–831 (2003)
22. Blanco-Velasco, M., Cruz-Roldán, F., Godino-Llorente, J.I., Blanco-Velasco, J., Armiens-Aparicio, C., López-Ferreras, F.: On the use of PRD and CR parameters for ECG compression. Medical Engineering & Physics **27**, 798–802 (2003)
23. Figueiredo, M.A., Nowak, R.D., Wright, S.J.: Gradient projection for sparse reconstruction: Application to compressed sensing and other inverse problems. IEEE Journal of Selected Topics in Signal Processing **1**, 586–597 (2007)
24. Pati, Y.C., Rezaiifar, R., Krishnaprasad, P.: Orthogonal matching pursuit: recursive function approximation with applications to wavelet decomposition. In: 1993 Conference Record of The Twenty-Seventh Asilomar Conference on Signals, Systems and Computers, pp. 40–44. IEEE (1993)

Design of Fourth Order GCF Compensation Filter

Mirza Amir Baig, Manjeet Kumar, Vivek Kumar Mishra
and Apoorva Aggarwal

Abstract In this paper, the design of compensated generalized comb filter (GCF) is presented using maximally flat (MF) minimization technique. The proposed filter's coefficients are obtained by MF technique. The proposed GCF compensation filter is the cascade of two second order compensation filter with GCF. Simulation result demonstrates that the passband droop of GCF is reduces to a large extent with the proposed filter.

Keywords FIR Filters · Generalized Comb Filters (GCFs) · Maximally Flat (MF) Technique

1 Introduction

The Cascaded integrator comb (CIC) filter is the simplest decimation filter proposed by E. B. Hogenauer [1]. In decimating CIC filter, the input is passed through one or more integrator section, then passes through down sampler, and finally followed by one or more comb sections. However, this filter has some of the disadvantages e.g., low stopband attenuation and high passband droop whereas Generalized Comb Filter solves this problem to some extent, it increases stopband attenuation.

A GCF is a generalized CIC decimation filter, proposed in [2], with increased stopband attenuation and having extended folding bands. These folding bands are present at each frequency points $2\pi i/D$; where D denotes decimation factor and $i = 1, 2, 3, .., D - 1$.

The word "GENERALIZED" shows that this filter has been generalized for all possible locations of their zeroes by rotating it through an angle called rotation parameter, denoted by α_p, where $p = 1, ..., N$ and N denotes order of the filter [2, 3]. Whereas traditional CIC filter has zero values of rotational parameters for

M.A. Baig(✉) · M. Kumar · V.K. Mishra · A. Aggarwal
Division of Electronics and Communication Engineering, Netaji Subhas Institute of
Technology, Dwarka, New Delhi 110078, India
e-mail: {amir.mirza002,manjeetchhillar,vivek9453,16.apoorva}@gmail.com

© Springer International Publishing Switzerland 2016   283
S. Berretti et al. (eds.), *Intelligent Systems Technologies and Applications*,
Advances in Intelligent Systems and Computing 384,
DOI: 10.1007/978-3-319-23036-8_24

all N, i.e., $\alpha_p = 0$ for all N. GCF theory provides a new viewpoint about the self fourier function [4]. GCF is used to solve the problem of recognition or chasing of quasi-periodically varying real systems [5] and further results are improved in [6]. The design of computational efficient GCF is proposed in [7]. The general method to design second order compensated GCF is shown in [8]. Special cases of GCF are proposed in [9, 10].

The transfer function of GCF is as follows [2]

$$H_{GCF}(z) = \prod_{p=1}^{N} \frac{\sin(\alpha_p/2)}{\sin(\alpha_p D/2)} \prod_{p=1}^{N} \frac{1 - z^{-D}e^{-j\alpha_p D}}{1 - z^{-1}e^{-j\alpha_p}} \tag{1}$$

where D denotes decimation factor and α_p (for $p = 1, 2, 3, .., N$) are rotation parameters.

The DTFT of eq. (1) leads to,

$$H_{GCF}(e^{j\omega}) = H(\omega)exp\left(-j\frac{(D-1)}{2}\left(\omega N + \sum_{p=1}^{N}\alpha_p\right)\right) \tag{2}$$

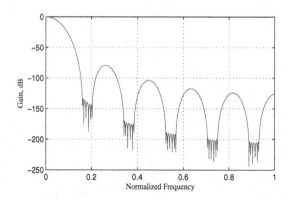

Fig. 1 Magnitude response of GCF for $N = 6$, $D = 11$, and $\nu = 4$.

where,

$$H(\omega) = \prod_{p=1}^{N} \frac{\sin(\alpha_p/2)}{\sin(\alpha_p D/2)} \prod_{p=1}^{N} \frac{\sin((\omega + \alpha_p)D/2)}{\sin((\omega + \alpha_p)/2)} \tag{3}$$

The simplified expression for α_p is given by

$$\alpha_p = \frac{q_p \pi}{\nu D}$$

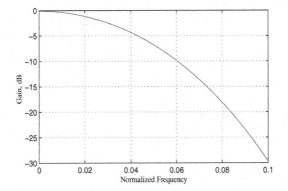

Fig. 2 Passband details of GCF for $N = 6$, $D = 11$, and $v = 4$.

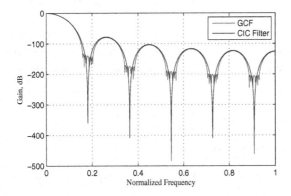

Fig. 3 Magnitude response of GCF and CIC filter for $N = 6$, $D = 11$, and $v = 4$.

where v is the positive number and $q_p \in [-1, +1]$, optimal values of q_p for different values of N are illustrated in Table I [2].

Let us design GCF using the following specifications: $N = 6$, $D = 11$, $v = 4$ and $q_p = [+0.95, +0.675, +0.25, -0.25, -0.675, -0.95]$ [2]. Fig. 1 shows the GCF's magnitude response, and its passband detail is illustrated in Fig. 2. If we put $\alpha_p = 0$ for all N in eq. (1), it leads to the transfer function of traditional CIC filter.

$$H_C(z) = \left(\frac{1}{D} \frac{1 - z^{-D}}{1 - z^{-1}} \right)^N$$

The comparison between GCF and CIC filter has been shown in Fig. 3, which depicts that GCF has better stopband attenuation than CIC filter. The increased attenuations around the folding bands can be easily seen in GCF plot. But, GCF still exhibits a high passband droop.

The motive behind the work is to reduce the passband droop of GCF. In this paper, we presents a design method for passband compensation of GCF. Two second order compensation filters are used in cascade with GCF to design the proposed compensated GCF. Simulation result demonstrates the effectiveness of proposed GCF compared with existing second order GCF.

The rest of the paper is organised as follows: In Section II, the design of proposed fourth order compensation filter is formulated using Maximally Flat technique. The simulation results and discussion obtained with proposed compensation filter are discussed in Section III. Finally, Section IV concludes the paper.

2 Proposed Compensated GCF

The transfer function of the second order compensation filter is

$$P(z^D) = a_0 + a_1 z^{-D} + a_0 z^{-2D} \tag{4}$$

where a_0 and a_1 are real valued coefficients. The GCF is cascaded with two 2nd order compensation filter as shown in Fig. 4(a). Using multirate identity [11], $P(z^D)$ moves to a lower rate which provides more efficient structure as shown in Fig. 4(b).

The overall transfer function is

$$T(z) = H_{GCF}(z)P(z^D)P(z^D) \tag{5}$$

The frequency response is given by

$$T(e^{j\omega}) = H_{GCF}(e^{j\omega})P(e^{j\omega})P(e^{j\omega}) \tag{6}$$

$P(e^{j\omega})$ can be written as

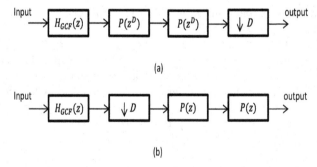

(a)

(b)

Fig. 4 Decimation block diagram. (a) Cascaded GCF and proposed Compensation filter. (b) Efficient structure for decimation.

$$P(e^{j\omega}) = [a_1 + 2a_0 \cos(D\omega)]e^{-j\omega D} \tag{7}$$

With the help of eq. (7), eq. (6) becomes

$$T(e^{j\omega}) = e^{-j\omega[((D-1)N+4D)/2]} H(\omega)[P_A(D\omega)]^2 \tag{8}$$

where $P_A(D\omega)$ is the amplitude response of $P(e^{j\omega D})$, which is given by

$$P_A(D\omega) = a_1 + 2a_0 \cos D\omega \tag{9}$$

An error function is defined in the frequency range $[0, \omega_p]$ where ω_p is the upper brink frequency of the filter.

$$E(\omega) = H(\omega)[P_A(D\omega)]^2 - 1 \tag{10}$$

To calculate the coefficients of a_0 and a_1, we apply some conditions over the error function. $E(\omega)$ must be zero at frequencies $\omega = 0$ and $\omega = \omega_0$, where $\omega_0 \leq \omega_p$, For $\omega = 0$, from eqs. (3), (9) and (10), we get

$$a_1 = 1 - 2a_0 \tag{11}$$

And for $\omega = \omega_0$, the condition results in

$$H(\omega_0)(2a_0 \cos(D\omega_0) + a_1)^2 = 1 \tag{12}$$

Now, the value of ω_0 is decided by the technique used for minimization of $E(\omega)$, as follows:
For Maximally flat compensation technique, the error function should be extremely flat at $\omega = 0$. The $E(\omega)$ has as many derivatives that vanishes at $\omega = 0$ as possible [8],[12]-[15].

$$\frac{d^2 E(\omega)}{d\omega^2} = 0 \bigg|_{\omega=0} \tag{13}$$

Using eq. (10), we get

$$E(\omega) = H(\omega)[a_1 + 2a_0 \cos D\omega]^2 - 1$$

$$E(\omega) = H(\omega)[a_1^2 + 4a_0^2 \cos^2 D\omega + 4a_0 a_1 \cos D\omega] - 1 \tag{14}$$

Taking first derivative of the error function with respect to ω, we get

$$E'(\omega) = a_1^2 H'(\omega) + 4a_0^2 [H'(\omega) \cos^2 D\omega - DH(\omega) \sin(2D\omega)]$$
$$+ 4a_0 a_1 [H'(\omega) \cos(D\omega) - DH(\omega) \sin(D\omega)] \qquad (15)$$

Using eq. (13), eq. (14) becomes

$$E''(\omega) = a_1^2 H''(\omega) + 4a_0^2 [(H''(\omega) \cos^2(D\omega) - DH'(\omega) \sin(2D\omega))$$
$$- (DH'(\omega) \sin(2D\omega) + 2D^2 H(\omega) \cos(2D\omega))] + 4a_0 a_1 [(H''(\omega) \cos(D\omega)$$
$$- DH'(\omega) \sin(D\omega)) - (DH'(\omega) \sin(D\omega) + D^2 H(\omega \cos(D\omega))] \qquad (16)$$

$$E''(\omega) = a_1^2 H(\omega) + 4a_0^2 H''(\omega) \cos^2(D\omega) - 8a_0^2 DH'(\omega) \sin(2D\omega)$$
$$- 8a_0^2 D^2 H(\omega) \cos(2D\omega) - 8a_0 a_1 DH'(\omega) \sin(D\omega) + 4a_0 a_1 H''(\omega) \cos(D\omega)$$
$$- 4a_0 a_1 D^2 H(\omega) \cos(D\omega) \qquad (17)$$

Rearranging eq. (16), we get

$$E''(\omega) = H''(\omega)[a_1^2 + 4a_0^2 \cos^2(D\omega) + 4a_0 a_1 \cos(D\omega)] - H'(\omega)[8a_0^2 D \sin(2D\omega)$$
$$+ 8a_0 a_1 D \sin(D\omega)] - H(\omega)[8a_0^2 D^2 \cos(2D\omega) + 4a_0 a_1 D^2 \cos(D\omega)] \qquad (18)$$

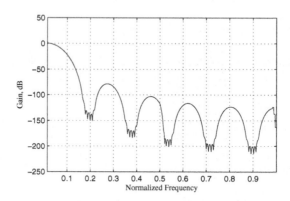

Fig. 5 Magnitude response of fourth order compensation filter.

At $\omega = 0$, all sin terms are equal to 0, all cos terms are equal to 1 and $H(0) = 1$ in eq. (3). By simplifying eqs. (13) and (18), we get

$$a_0 = \frac{H''(\omega)}{4D^2}\bigg|_{\omega=0} \qquad (19)$$

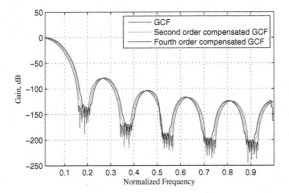

Fig. 6 Comparison of GCF, second order and fourth order compensation filter.

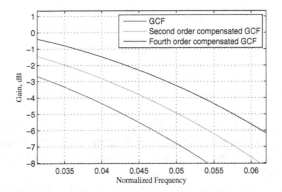

Fig. 7 Zoomed in curve of Fig. 6.

where $H''(\omega)$ is the second order derivative of $H(\omega)$. From eqs. (19) and (22), we can calculate the value of a_0.

In order to calculate $H''(\omega)$, we consider the following derivative (using division rule for derivatives):

$$\frac{d}{d\omega}\left(\frac{\sin((\omega+\alpha_p)D/2)}{\sin((\omega+\alpha_p)/2)}\right) = \frac{\sin((\omega+\alpha_p)D/2}{\sin((\omega+\alpha_p)/2}$$
$$\times\left(\frac{D}{2}\cot((\omega+\alpha_p)D/2) - \frac{1}{2}\cot((\omega+\alpha_p)/2)\right)$$

$$(20)$$

First derivative of $H(\omega)$, i.e. $H'(\omega)$ is given as

$$H'(\omega) = \frac{H(\omega)}{2} \sum_{p=1}^{N} \left(D \cot \left(D \frac{\omega + \alpha_p}{2} \right) - \cot \left(\frac{\omega + \alpha_p}{2} \right) \right) \qquad (21)$$

Using eq. (21) and product rule for derivatives, $H''(\omega)$ is expressed as

$$H''(\omega) = \frac{H(\omega)}{4} \sum_{p=1}^{N} \left(\csc^2 \left(\frac{\omega + \alpha_p}{2} \right) - D^2 \csc^2 \left(D \frac{\omega + \alpha_p}{2} \right) \right)$$

$$+ \left(\sum_{p=1}^{N} \left(D \cot \left(D \frac{\omega + \alpha_p}{2} \right) - \cot \left(\frac{\omega + \alpha_p}{2} \right) \right) \right)^2 \qquad (22)$$

Using eqs. (11) and (19), we get coefficients of the proposed compensation filter. After calculations, a_0 and a_1 comes out to be

$$a_0 = -0.123966942$$

$$a_1 = +0.752066116$$

3 Discussions and Results

In this section the simulation results has been presented for the design of fourth order compensated generalized comb filter. All the simulation has been done in MATLAB 7.13 version. The effectiveness of proposed compensation filter has been justified from the given example.

Example: The following specifications are considered to design the compensated GCF are: $N = 6$, $D = 11$, $v = 4$, $\omega_p = 0.045\pi$, $\alpha_p = q_p\pi/vD$ for $p = 1, 2, 3, 4, 5, 6$ where $q_p = [+0.95, +0.675, +0.25, -0.25, -0.675, -0.95]$ same as that of in [2]. The corresponding passband droop of GCF is $A_{PD} = -5.5$ dB. The magnitude plot of the fourth order compensation filter is shown in Fig. 5 and Fig. 6 depicts the comparison plot of GCF, second order and fourth order compensation filter.

By applying maximally flat method, using eq. (11) and (19) the values of $a_0 = -0.123966942$ and $a_1 = 0.752066116$ are obtained. The passband droop after second order compensation filter $A_{PD} = -3.8$ dB, whereas the passband droop after fourth order compensation filter $A_{PD} = -2.2$ dB. The above mentioned result can be justified from Fig. 7. From the observation it can be concluded that the fourth order compensation filter reduces passband droop more than second order compensation filter. However fourth order compensation filter increases the complexity of the filter as it increases the number of multipliers.

4 Conclusions

A novel technique for compensated GCF is presented which is based on fourth order compensation filter. The resulting compensation filter helps in effectively reducing the passband droop of Generalized Comb Filter (GCF). The simulation result proves the effectiveness of the proposed work. This work can be further extended for the design of GCF using fractional sample rate converter.

Acknowledgments The authors are highly grateful to the Director, Netaji Subhas Institute of Technology, Dwarka, New Delhi, India for providing the good research environment in the institute.

References

1. Hogenauer, E.B.: An economical class of digital filters for decimation and interpolation. IEEE Trans. Acoust., Speech, Signal Process **29**(2), 155–162 (1981)
2. Laddomada, M.: Generalized comb decimator filter for $\Sigma\Delta$ A/D converters: Analysis and design. IEEE Trans. Circuits Syst. I, Reg. Papers **54**(5), 994–1005 (2007)
3. Laddomada, M.: On the polyphase decomposition for design of generalized comb decimation filter. IEEE Trans. Circuits Syst. I, Reg. Papers **55**(8), 2287–2299 (2008)
4. Nishi, K.: Generalized Comb Function: A New Self fourier function. IEEE Trans. Circuits Syst. Analog Digit. Signal Process **46**(1), 40–50 (1999)
5. Niedźwiecki, M., Kaczmarek, P.: Generalized Adaptive Notch and Comb Filters for Identification of Quasi-Periodically Varying Systems. IEEE Trans. on Signal Processing **53**(12), 4599–4609 (2005)
6. Niedźwiecki, M., Meller, M.: Generalized adaptive comb filter with improved accuracy and robustness properties. In: 20th European Signal Processing Conference (EUSIPCO 2012), pp. 91–95 (2012)
7. Laddomada, M.: Fixed-point design of generalised comb filters: a statistical approach. IET Signal Process 4(2), 158–167 (2009, 2010)
8. Fernandez-Vazquez, A., Dolecek, G.J., General, A.: Method to Design GCF Compensation Filter. IEEE Trans. Circuits Syst. II, Express Briefs **56**(5), 409–413 (1999)
9. Presti, L.L.: Efficient modified-sinc filters for sigma-delta / converters. IEEE Trans. Circuits Syst. II, Analog Digit. Signal Process **47**(11), 1204–1213 (2000)
10. Laddomada, M.: Comb-based decimation filters for $\Sigma\Delta$ A/D converters: Novel schemes and comparisons. IEEE Trans. Signal Process **55**(5), 1769–1779 (2007)
11. Jovanovic-Dolecek, G. (ed.): Multirate Systems: Design and applications. Idea Group Publishing, Hershey (2002)
12. Selesnick, I.W.: Low-pass filter realizable as all-pass sums: Design via a new flat delay filter. IEEE Trans. Circuits Syst. II, Analog Digit. Signal Process **46**(1), 40–50 (1999)
13. Fernandez-Vazquez, A., Dolecek, G.J.: Maximally Flat CIC Compensation Filter: Design and Multiplierless Implementation. IEEE Trans. Circuits Syst. II, Express Briefs **59**(2), 113–117 (2012)
14. Aggarwal, A., Kumar, M., Rawat, T.K.: L_1 error criterion based optimal FIR filters. In: Annual IEEE India Conference (INDICON) (2014)
15. Upadhyay, K., Kumar, M., Rawat, T.K.: Optimal design of weighted least square based fractional delay FIR filter using genetic algorithm. In: IEEE Int. Conf. Signal Propag. and Comp. Tech. (ICSPCT), pp. 53–58 (2014)

Performance Analysis of Updating-QR Supported OLS Against Stochastic Gradient Descent

Remya R.K. Menon and K. Namitha

Abstract Regression model is a well-studied method for the prediction of real-valued data. Depending on the structure of the data involved, different approaches have been adopted for estimating the parameters which includes the Linear Equation solver, Gradient Descent, Least Absolute Shrinkage and Selection Operator (LASSO) and the like. The performance of each of them varies based on the data size and computation involved. Many methods have been introduced to improve their performance like QR factorization in least squares problem. Our focus is on the analysis of performance of gradient descent and QR based ordinary least squares for estimating and updating the parameters under varying data size. We have considered both tall/skinny as well as short/fat matrices. We have implemented Block Householders method of QR factorization in Compute Unified Device Architecture (CUDA) platform using Graphics Processor Unit (GPU) GTX 645 with the initial set of data. New upcoming data is updated directly to the existing Q and R factors rather than applying QR factorization from the scratch. This updating-QR platform is then utilized for performing regression analysis. This will support the regression analysis process on-the-fly. The results are then compared against the gradient descent implemented. The results prove that parallel-QR method for regression analysis achieves speed-up of up to 22x compared with the gradient descent method when the attribute size is larger than the sample size and speed-up of up to 2x when the sample size is larger than the attribute size. Our implementation results also prove that the updating-QR method achieves speed-up approaching 2x over the gradient descent method for large datasets when the sample size is less than the attribute size.

R.R.K. Menon(✉) · K. Namitha
Department of Computer Science and Engineering, Amrita Vishwa Vidhyapeetham,
Amritapuri, Kollam, Kerala, India
e-mail: ramya@am.amrita.edu, namitha.amrita@gmail.com

© Springer International Publishing Switzerland 2016   293
S. Berretti et al. (eds.), *Intelligent Systems Technologies and Applications*,
Advances in Intelligent Systems and Computing 384,
DOI: 10.1007/978-3-319-23036-8_25

1 Introduction

Enhancement and analysis of prediction model has always been an area of interest. One of the most powerful technique used in prediction modeling is Regression Analysis [2], [10], [12]. The objective of regression analysis is to discover a model that approximates $Y \in R^m$ as a function of $A \in R^{m \times n}$:

$$Y \approx f(A, \beta) , \tag{1}$$

where $A \in R^{m \times n}$ is a design matrix with values of explanatory variables, $\beta \in R^n$ is an n slope vector of regression coefficients and $Y \in R^m$ is an m vector of target variables. Ordinary Least Squares (OLS) [5] approach and iterative techniques (like gradient descent) are two well established methods for fitting regression by estimating optimal $\beta \in R^n$. From a given set of training data $(a_{11}..a_{1n}, y_1)...(a_{m1}..a_{mn}, y_m)$ least squares estimates β by minimizing the Residual Sum of Squares (RSS) between actual and predicted value. This solution can be expressed as:

$$\beta = (A^T A)^{-1} A^T Y \tag{2}$$

Gradient Descent (GD) [2] starts by guessing initial β and then updates β gradually to get least error value for cost function. Stochastic Gradient Descent (SGD) [2] is a variation of gradient decent which considers only one training example at a time for updating β. The update rule for a single training example is of the form:

$$\beta_j = \beta_j + \alpha(y^{(i)} - h(a^{(i)}))a_j^{(i)}, \quad h(a_i) = \beta_0 + \beta_1 a_{i1} + ... + \beta_n a_{in} + \varepsilon_i , \tag{3}$$

with $\varepsilon_i \sim N(0, \sigma^2)$ as the error term, n as the number of features for i^{th} training example (x_i, y_i), j takes values from $0...n$ and α is the learning rate which determines the size of each iteration step. Since it process only one example at a time, SGD can work well for on-the-fly data. For an $A \in R^{m \times n}$ matrix, training with SGD costs $O(km \bar{n})$ where k is the number of iterations and $\bar{n}$ is the number of features in each sample. When the data size is eminently large, SGD is preferred over OLS but the convergence time of SGD algorithm can be high as explained in [7].

The nature of data being dynamic, it is desirable to enhance the performance of OLS. M.Bernardelli, in his work [3] explains how QR factorization can be effectively used to improve the performance of OLS. The procedure of updating β to include a new set of data $(\hat{A}, \hat{Y})$ can also be amended by updating the existing factors Q and R [9].

The purpose of this paper is to compare the training procedure of QR supported OLS solution and stochastic gradient descent method [17]. In on-line training, new sets of examples are introduced often. We revise our learned regression model by accommodating these new examples. We also analyze the performance of updating the existing factors Q and R in OLS against the iterative convergence of SGD algorithm. Different comparisons are performed to discover which technique is most reliable for regression analysis with varying data sizes.

2 Related Work

As the focus of our paper is to bring about an improvement in the regression process especially for dynamic data, the related work that were looked into are basically that deal with least squares solution, QR factorization-based least squares, SGD and QR implementation on the Graphics Processor Unit (GPU). [3] focus on implementing the QR updating step using Givens rotation which can then be used in the least-squares problem. The authors emphasize on the fact that Q matrix is not explicitly required for the least-squares computation. The paper focuses on updating QR when a single row is added or deleted. [9] describes the usage of QR updates for solving least squares. Both Givens rotation and Householder reflection has been used. [17] focus on SGD for solving least-squares problems for on-line data. The paper performs error bounds and convergence measures for average SGD against standard SGD.

L. Bottou in his work [7] claims that large-scale learning can be improved with optimization techniques like SGD, second-order SGD and average SGD as the computational complexity becomes a limiting factor. The trade-offs of large-scale learning has been given in terms of 3 types of errors - approximation error, estimation error and optimization error. While [11] implements QR factorization on the GPU using a block Householders approach which was done on rectangular matrices for solving overdetermined least-squares problems, [1] focuses on a similar system which aims at reducing the memory traffic by making use of memory-bound kernels. Reduction-trees are implemented as kernels.

We use a parallel-QR approach on GPU using Compute Unified Device Architecture (CUDA) for the initial set of data. From then on, we update the generated Q and R using Householder reflection for data of different sizes like overdetermined as well as underdetermined system of equations.

3 Solution Approach

The focus of our paper is to analyze two well-known approaches: OLS and SGD for performing regression analysis. One of the relevance of this analysis is for the usage of on-line training data. Also, data $A \in R^{m \times n}$ comes in different forms which include $m > n$ as well as $m < n$. The sequence of computations involved in this work is summarized in Fig. 1. Parallel-QR as well as SGD is performed on the initial dataset. Later, when a new set of data is arrived, existing QR factorization is updated. β estimated in these ways are used to generate a hypothesis capable of doing prediction for any future input.

The linear regression model has the form:

$$Y = A\beta + \varepsilon \tag{4}$$

OLS is the most common analytical method for estimating β as a closed form solution. We have to search for β so that $A\beta$ is as close as possible to Y in quadratic norm by minimizing RSS as:

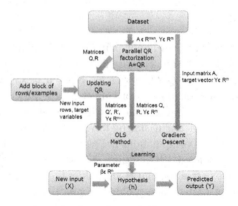

Fig. 1 Regression model using OLS with QR vs SGD

$$\text{Min. RSS} = \sum_{i=1}^{m}(y_i - \beta_0 - \sum_{j=1}^{n}(A_{ij}\beta_j))^2 \, , \tag{5}$$

with m as the total number of samples in the training set and n as the number of features for i^{th} training example. Differentiating the matrix form with respect to β and setting the first derivative to zero, we can have the solution as

$$A^T(Y - A\beta) = 0, \quad \beta = (A^T A)^{-1}A^T Y \tag{6}$$

We can do the prediction of response variable $\hat{Y}$ for a new input $\hat{A}$ as

$$\hat{Y} = \hat{A}\beta = QQ^T Y \tag{7}$$

In the geometrical representation of least squares, the residual $\hat{Y} - \hat{A}\beta$ has to be orthogonal to the approximation $\hat{A}\beta$ in the subspace spanned by $\hat{A}$. $A^T A$ is non-singular if and only if A is full rank. Through out this paper, we are assuming that A is full rank.

OLS works well when number of features in a sample is less. The major advantage of OLS over gradient descent is that it forms a solution analytically in a single step without the necessity of learning parameter α and its corresponding iterations.

3.1 Estimating Regression Coefficients using QR Factorization

QR factorization [8],[16] of an $m \times n$ matrix A is $A = QR$, Q is an $m \times m$ orthogonal matrix and R is an $m \times n$ upper triangular matrix. Householder transformation [5], [8] is one method for factorizing A into Q and R which is defined by a reflection about a plane.

The OLS (6) for solving linear regression will be numerically unstable as it may generate round-off errors while computing $A^T A$. It is relatively simple and stable to compute the regression coefficients using QR factorization. We can substitute $A = QR$ in (2) to obtain:

$$\beta = ((QR)^T QR)^{-1}(QR)^T Y = R^{-1}Q^T Y \qquad (8)$$

QR decomposition outperforms OLS method by avoiding the matrix inversion which is computationally intensive for large matrices. When $A^T A$ is nearly singular, some of the diagonal elements in its inverse will be very large leading to rounding error. QR preserves the numerical stability by avoiding the formation of $A^T A$. β can be obtained easily by back substitution with R having elements $r_{ij} = 0$ for $i > j$. The overall cost for this procedure is $2mn^2$ flops.

The QR factorization method can be accelerated by implementing it on a GPU using CUDA which is expected to outperform serial implementation. The Householder algorithm [8] for reducing A to R is rich in matrix-vector product which include lots of memory references. The block reduction method [4], [11] makes use of matrix-matrix operations which will reduce the number of memory accesses. Thus it is preferable to use block reduction method in high performance computers like GPU machines [15] having a well defined memory hierarchy.

Algorithm 1. Block Householder QR

Partition A into $\left[A_1 A_2 ... A_{n/r} \right]$, $Q \longleftarrow I$
for $k = 1$ to n/r **do**
    $s = (k - 1)r + 1$
    for $j = 1$ to r **do**
        $u = s + j - 1$
        $[v, \alpha] = householder(A(u : m; u))$
        $A_k = (I - \alpha v v^T)A_k$
        $V(:, j) = [zeros(j - 1, 1); v]$, $B(j) = \alpha$
    end for
    $Y = V(1 : end, 1)$, $W = -B(1)V(1 : end, 1)$
    for $j = 2$ to r **do**
        $v = V(:, j)$, $z = -B(j)v - B(j)WY^T v$
        $W = [W\ z]$, $Y = [Y\ v]$
    end for
    $\left[A_{k+1}..A_{n/r} \right] = (I + YW^T)\left[A_{k+1}..A_{n/r} \right]$, $Q = Q(I + WY^T)$
end for

We have implemented the QR factorization using parallel block Householder method. A generalized Householder transformation which aggregates Householder reflections to form a block representation is given in Algorithm 1. The input matrix A is partitioned into n/r blocks, each with r columns. Householder matrices $H_1...H_r$ of block A_1 is aggregated as $I + YW^T$ after reducing r columns of A_1. Then remaining blocks of A and Q are updated using this aggregation. The process is repeated until entire A is upper triangularized. The matrix-vector product, outer product of vectors in the Householder reduction and matrix-matrix products of block reduction method is parallelized on GPU.

3.2 Updating Parameters by Updating QR Factorization

Occasionally we may require to compute QR factorizations numerous times with only small changes in the underlying data. Consider the continuous flow of data which are the inputs to a linear regression model. In such instance we may need to update [9] the least squares solution (6) whenever new data arrives. But it is not always feasible to recompute factors Q and R of computationally expensive procedures like QR from scratch. Instead, we can reduce some cost by updating the factors Q and R of existing factorization. The OLS solution is updated sequentially by updating the factors of initial factorization.

Given $A = QR \in R^{m \times n}$, for any m and n, the Algorithm 2 computes $\tilde{A} = \tilde{Q}\tilde{R} \in R^{(m+p) \times n}$ where $\tilde{R}$ is upper triangular, $\tilde{Q}$ is orthogonal and $\tilde{A}$ is A with a block of rows, $U \in R^{p \times n}$, inserted at position $m + 1$. The diagonal element of existing factor matrix R and a column of U is passed into the Householder function [8] which returns Householder vector v, scalar α. Using v and α, jth row of R is updated. $j + 1 : n$ columns of U is updated using $j + 1 : n$ columns of R and v, α. The flop count for

---

Algorithm 2. Updating Q and R using Householder method

Set $\tilde{Q} = \begin{pmatrix} Q & 0 \\ 0 & I \end{pmatrix}$, $lim = min(m, n)$

for $j = 1 : lim$ **do**

    $[v, \alpha] = householder(R(j, j), U(1 : p, j))$

    Update j^{th} row of R using v and α

    Update trailing part of U

    Update j^{th} column of $\tilde{Q}$ using v, α and then update $m + 1 : p$ columns of $\tilde{Q}$

end for

    $\tilde{R} = \begin{pmatrix} R \\ 0 \end{pmatrix}$

if $m < n$ **then**

    $U(:, m + 1 : n) = Q_U R_U$

    $\tilde{R}(m + 1 : m + p, m + 1 : n) = R_U$, $\tilde{Q}(1 : m + p, m + 1 : m + p) = \tilde{Q}(1 : m + p, m + 1 : m + p)Q_U$

end if

---

updating R as $\tilde{R}$ requires $2n^2 p$ flops. $m + 1 : m + p$ portion of Q is filled with the identity matrix which now becomes $\tilde{Q}$. v_j and α_j from the jth iteration is used for updating the jth column of $\tilde{Q}$. The updated $\tilde{Q}$ as well as v_j, α_j is used for updating the $m + 1 : m + p$ columns and $1 : m + p$ rows of $\tilde{Q}$.

Table 1 Datasets used for performance analysis

Dataset	Description	Instances	Attributes	Area
D1	Predict keywords activities	64	32	Engineering
D2	Fertility Data Set	100	10	Life science
D3	Wisconsin Diagnostic Breast Cancer	569	32	Life science
D4	Parkinson Speech Data Set	1040	28	Life science
D5	Parkinsons Telemonitoring	5875	21	Life science
D6	Combined Cycle Power Plant	9568	5	Engineering
D7	Maintenance - Naval Propulsion Plants	11934	18	Engineering
D8	Musk (Version 2)	100	166	Physical science
D9A	Gas sensor flow modulation - Features	58	436	Engineering
D9B	Gas sensor flow modulation - Raw Data	110	948	Engineering
D10	Arcene Data Set	100	10000	Life science

4 Experimental Analysis

4.1 Datasets

We measure the performance of algorithms using real-life data. The datasets described in Table 1 are taken from UCI Machine Learning Repository [6] for evaluation purpose. The design matrix A and the target vector Y is initialized with data from these datasets. The new rows to be added to A and the corresponding target variables to Y are populated in the same way. Dataset D1 contains Twitter data to predict rank of each keyword in a social network using learning to rank methods. D2 provides semen samples for diagnosing sperm as normal or altered according to various environmental factors, health status, and life habits. Diagnosis of malignant or benign cancer is performed using features of each cell nucleus as presented in D3. We predict Parkinson's disease rating score from multiple types of sound recordings of diseased patients given in dataset D4. Motor and total score is predicted by data D5 collected using telemonitoring device. From D6, we predict net hourly electrical energy output from a Combined Cycle Power Plant. Data in D7 have been generated from a sophisticated simulator of Gas Turbines. Dataset D8 is to learn to predict whether new molecules will be musks or non-musks. Concentrations of acetone and ethanol is predicted from data D9A and D9B collected using a chemical sensing system based on an array of gas sensors. We distinguish cancer versus normal patterns from mass-spectrometric data presented in D10.

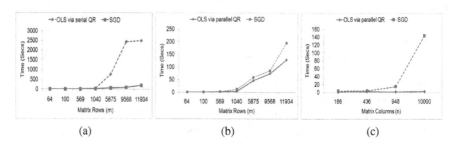

Fig. 2 Comparison of performance of SGD and (a) OLS with sequential QR (b) OLS with parallel QR for $m > n$ (c) OLS with parallel QR for $m < n$

4.2 Experimental Platform

Our platform is an Intel(R) Core(TM) i5-4590 2.6 GHz quad core CPU and an NVIDIA GeForce GTX 645 graphics processor with CUDA 6.5 [14]. All algorithms are implemented in C++ with CUDA.

4.3 Comparison of QR Supported OLS with the SGD Method

The serial code is executed completely on the CPU while the parallel part of code is executed on the GPU. We have implemented the SGD algorithm to run entirely on the CPU. All run-times are measured in seconds. The time taken for transferring the matrices from the host to the device and transferring the result back from device is included in the execution time for parallel code. The parallel block version of QR is implemented using NVIDIA's CUBLAS library [13] and CUDA kernels. The CUBLAS function *cublasSgemv()* compute the products $v^T A$ and vv^T of Algorithm 1. Matrix product operations to update A and Q is computed using CUBLAS *cublasSgemm()* function. Similarly, a custom kernel is implemented to compute the norms of the columns of each block in parallel by a single kernel invocation before the triangularization of each block which avoids the overhead of several kernel calls.

The performance analysis consists of 5 phases of comparison: performance of SGD vs OLS with sequential-QR, SGD vs OLS with parallel-QR, SGD vs OLS with update-QR, OLS with sequential-QR vs parallel-QR, Full-QR vs Update-QR.

Fig. 2a compares the time taken for prediction by SGD method and OLS using sequential QR factorization. The result clearly conveys that SGD performs faster than OLS using sequential QR.

To improve the execution time of least squares solution, we have implemented parallel QR factorization on GPU. SGD with α as 0.0001 and OLS with parallel QR performs almost similar for small m, but as the matrix becomes more tall, OLS outperforms SGD on GPU as shown in Fig. 2b. Table 2a shows the execution-time for SGD and OLS with parallel QR for an overdetermined system whereas Table 2b

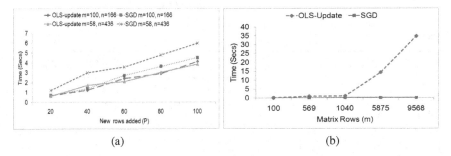

Fig. 3 Comparison of performance of SGD and (a) OLS with update-QR for different update rows, with $m < n$ (b) OLS with update-QR for varying m, with $m > n$

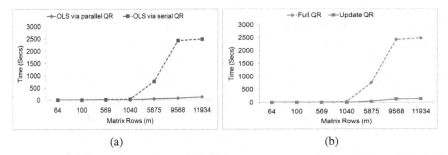

Fig. 4 Comparison of performance of (a) OLS using sequential-QR vs parallel-QR (b) full-QR vs update-QR while adding a block of rows for varying m

shows the execution-time for an underdetermined system. When the attribute size is larger than the number of data points, the time taken for updating the regression parameters will be high in the case of gradient descent method over OLS method. Thus the Fig. 2c shows the linear growth of run time of SGD with α as .001 while the time taken by OLS is almost constant. The speed-up of SGD decreases as n increases and consequently OLS with parallel QR achieves speed-up up to 22x.

The performance of SGD and OLS by updating QR while adding a block of new rows is given in Fig. 3a. From the result we can see that OLS with update-QR outperforms SGD as update size, p increases. We observe that speed-up of updating QR in OLS with update-QR over SGD reach up to 2x for large p where $m < n$. The time taken for updating $Q \in R^{m \times m}$ increases as m increases which in turn increase the overall execution time of OLS with update-QR as presented in Fig. 3b.

The comparison of sequential and parallel implementation of QR given in Fig. 4a clearly shows that parallel QR is faster than executing QR factorization serially. The time taken for serial factorization grows linearly as the number of rows m increases.

The results given in Fig. 4b compares the speed of updating algorithm for QR factorization and full QR factorization by block Householder method while adding

Table 2 Execution-time for SGD and OLS with parallel QR

Dataset	OLS	SGD
D1	0.94	0.318
D2	0.725	0.427
D3	1.336	2.346
D4	3.293	11.768
D5	46.338	58.888
D6	73.721	76.119
D7	128.21	194.789

(a) $m > n$

Dataset	OLS	SGD
D8	0.942	3.493
D9A	1.623	3.51
D9B	1.165	14.498
D10	2.013	143.646

(b) $m < n$

Table 3 Norm-wise relative errors of OLS with QR vs SGD

Dataset	OLS	SGD
D2	0.2616	0.2687
D3	0.1354	0.2250
D4	0.4415	0.6417
D5	0.3165	0.6210

(a) For different m and n

p	OLS	SGD
100	0.1554	0.1634
300	0.2816	0.2927
500	0.2786	0.3304
700	0.2831	0.3325

(b) For varying p with $m = 5875$

a block of rows with $p = 50$ for varying m. From the result we can conclude that updating algorithm is faster than performing full QR factorization and the difference between execution time of both algorithms increases as m increases.

We compute the norm-wise relative error $\frac{\|Y - A\beta\|_2}{\|Y\|_2}$ of linear regression from OLS solution with QR on GPU against solution from sequential gradient descent method. Table 3a shows that OLS solution is accurate than SGD. We also compute the relative error of solution from an updated QR factorization against solution for regression from sequential gradient descent method and the results are given in Table 3b. The errors are again higher for SGD compared to OLS with update-QR while adding a block of new rows, p to the design matrix.

5 Conclusion

The paper performs prediction based on linear regression model using an updating-QR method. Data parallelism has been introduced to the QR factorization that is initially involved by implementing it on the CUDA platform using GPU GTX 645 which speeds up the process of generating the initial Q and R factors from the existing set of data. Later, the new set of incoming data is updated directly to these QR factors. The results show that this adds up to the gain in execution time of the regression model for $m \times n$ data where $m > n$ and $n > m$ is considered. The results are compared with the gradient descent approach for prediction. Different results are provided to prove that the updating-QR based method achieves speed-up of up to 2x compared with SGD where $m < n$ and parallel QR method gives speed-up of up to 22x over gradient descent for $m < n$, speed-up of up to 2x for $m > n$.

Acknowledgments We express our fullest gratitude to our department Chairman, Dr. M.R. Kaimal, who supported us throughout with his valuable comments. We are grateful to our faculties and friends for all their support.

References

1. Anderson, M., Ballard, G., Demmel, J., Keutzer, K.: Communication-avoiding QR decomposition for GPUs. In: 2011 IEEE International Parallel & Distributed Processing Symposium (IPDPS), pp. 48–58. IEEE (2011)
2. Ng, A.Y.: Machine Learning. https://www.coursera.org/learn/machine-learning/
3. Bernardelli, M.: Usage of algorithm of fast updates of QR decomposition to solution of linear regression models, pp. 699–705. EDIS - Publishing Institution of the University of Zilina (2012)
4. Bischof, C., Van Loan, C.: The WY representation for products of Householder matrices. SIAM Journal on Scientific and Statistical Computing **8**(1), s2–s13 (1987)
5. Bjrck, K.: Numerical Methods for Least Squares Problems. Siam Philadelphia (1996)
6. Blake, C., Merz, C.J.: UCI repository of machine learning databases (1998)
7. Bottou, L.: Large-scale machine learning with stochastic gradient descent. In: Proceedings of COMPSTAT 2010, pp. 177–186. Springer (2010)
8. Golub, G.H., Van Loan, C.F.: Matrix Computations, 3rd edn. Johns Hopkins University Press, Baltimore (1996)
9. Hammarling, S., Lucas, C.: Updating the QR Factorization and the Least Squares Problem
10. Hastie, T., Tibshirani, R., Friedman, J.: The elements of statistical learning: data mining, inference and prediction, 2 edn. Springer (2009)
11. Kerr, A., Campbell, D., Richards, M.: QR decomposition on GPUs. In: Proceedings of 2nd Workshop on General Purpose Processing on Graphics Processing Units, GPGPU-2, pp. 71–78. ACM, New York (2009). doi:10.1145/1513895.1513904
12. Murphy, K.P.: Machine Learning: A Probabilistic Perspective. The MIT Press (2012)
13. Nvidia, C.: Cublas library, vol. 15. NVIDIA Corporation, Santa Clara (2008)
14. Nvidia, C.: Programming guide (2008)
15. Owens, J.D., Houston, M., Luebke, D., Green, S., Stone, J.E., Phillips, J.C.: GPU computing. Proceedings of the IEEE **96**(5), 879–899 (2008)
16. Stoilov, T., Stoilova, K.: Algorithm and software implementation of QR decomposition of rectangular matrices. In: Int. Conf. Computer Systems and Technologies-CompSysTech (2004)
17. Zhang, T.: Solving large scale linear prediction problems using stochastic gradient descent algorithms. In: Proceedings of the Twenty-first International Conference on Machine Learning, ICML 2004, p. 116. ACM, New York (2004). doi: 10.1145/1015330.1015332

Is Differential Evolution Sensitive to Pseudo Random Number Generator Quality? – An Investigation

Lekshmi Rajashekharan and C. Shunmuga Velayutham

Abstract This paper intends to investigate the sensitivity of Differential Evolution (DE) algorithm towards Pseudo Random Number Generator (PRNG) quality. Towards this, the impact of six PRNGs on the performance quality of 14 DE variants in solving nineteen 10-Dimensional benchmark functions has been studied. The results suggest that DE algorithm is insensitive to the quality of PRNG used.

1 Introduction

Evolutionary Algorithms (EAs), modeled on natural evolution, are population based search algorithms that mimic the processes of variation and selection. The population initialization, variation operation (that include recombination and mutation) and stochastic selection form the core components of an EA. In fact they are random processes and a typical implementation of an EA often employs pseudo random number generators (PRNGs) to realize these processes. Practitioners of EAs often employ built-in PRNGs (typically *rand(0,1)*) in their implementations and study the various characteristics of EAs.

However, with the multitude of PRNGs available, will the choice of PRNG impact the performance of an EA is a valid research question. In spite of scarce research efforts in this direction, results of recent studies have favored the above question [1]. It is worth pointing out that there are studies with contradictory results too! [2]

This paper intends to investigate the impact of six PRNGs on the performance of 14 variants of Differential Evolution variants (DE). DE is a recently proposed

L. Rajashekharan · C. Shunmuga Velayutham(✉)
Amrita Vishwa Vidyapeetham, Coimbatore, India
e-mail: lekshmi217@gmail.com, cs_velayutham@cb.amrita.edu

© Springer International Publishing Switzerland 2016
S. Berretti et al. (eds.), *Intelligent Systems Technologies and Applications*,
Advances in Intelligent Systems and Computing 384,
DOI: 10.1007/978-3-319-23036-8_26

continuous parameter optimization EA with robust performance [6]. The primary objective is to investigate the impact of PRNGs on the final quality of solutions obtained by the 14 DE variants.

2 Differential Evolution

Differential Evolution (DE) is a recently proposed real-parameter optimization EA. The DE algorithm has different operations as in any typical EAs viz. Initialization, *differential* mutation, crossover and selection all applied iteratively (except initialization) till a previously specified criteria for stopping is met with.

The initialization operation involves randomly choosing a population of *NP* D-dimensional vectors from the real space. After the fitness evaluation of each vector $X_{i,G}$ where i ∈ 1,....NP (called *target* or *parent* vector) in the population, the differential mutation operation creates a *mutant* vector for each of the target vector thus generating *NP* mutant vectors. Differential mutation randomly perturbs target vectors with a weighted vector difference from one or two pairs of randomly chosen target vector and creates a mutant vector.

The seven commonly employed mutation strategies are given below,

1. *DE/rand/1* $V_{i,G} = X_{r_1^i,G} + F.\left(X_{r_2^i,G} - X_{r_3^i,G}\right)$

2. *DE/best/1* $V_{i,G} = X_{best,G} + F.\left(X_{r_1^i,G} - X_{r_2^i,G}\right)$

3. *DE/rand/2* $V_{i,G} = X_{r_1^i,G} + F.\left(X_{r_2^i,G} - X_{r_3^i,G} + X_{r_4^i,G} - X_{r_5^i,G}\right)$

4. *DE/best/2* $V_{i,G} = X_{best,G} + F.\left(X_{r_1^i,G} - X_{r_2^i,G} + X_{r_3^i,G} - X_{r_4^i,G}\right)$

5. *DE/current-to-rand/1* $V_{i,G} = X_{i,G} + K.\left(X_{r_3^i,G} - X_{i,G}\right) + F.\left(X_{r_1^i,G} - X_{r_2^i,G}\right)$

6. *DE/current-to-best/1* $V_{i,G} = X_{i,G} + K.\left(X_{best,G} - X_{i,G}\right) + F.\left(X_{r_1^i,G} - X_{r_2^i,G}\right)$

7. *DE/rand-to-best/1* $V_{i,G} = X_{r_3^i,G} + K.\left(X_{best,G} - X_{r_3^i,G}\right) + F.\left(X_{r_1^i,G} - X_{r_2^i,G}\right)$

where F, K are scale factors and $X_{best,G}$ is the best target vector of current generation. The random indices $r_1^i, r_2^i, r_3^i, r_4^i,$ and r_5^i are mutually exclusive.

The crossover operation crosses *NP* mutant vectors with their respective target vectors to create *NP* trial vectors $U_{i,G}$. *Binomial* and *exponential* are two typical crossovers used in DE literature. The equation for crossover operation by way of an example is given below,

$$U_{i,G} = u_{j,i,G} = \begin{cases} v_{j,i,G} & \text{if } \left(rand_j(0,1) \le Cr \text{ or } j = j_{rand}\right) \\ x_{j,i,G} & \text{otherwise} \end{cases}$$

Finally, a one-to-one competition happens between each of the *NP* trial vectors and its respective target vector. Assuming a minimization problem, the one with the least objective value gets selected for the next generation. As can be seen from the above description of DE algorithm, it is evident that the role of PRNGs is crucial in a DE implementation and its working. This paper intends to investigate the impact of PRNGs, if there are any, on the working of DE in terms of the quality of solutions obtained by DE.

3   Related Works

Interestingly very few research attempts have been made to study the impact of PRNGs on EAs. Miguel *et al* [1] have studied four algorithms- Particle Swarm Optimization (PSO), Differential Evolution (DE), Genetic Algorithm (GA), Firefly Algorithm (FA) and their sensitivity towards two PRNGs namely Mersenne Twister and GCC RAND. They observed the sensitivity trend as DE > GA > PSO > FA with DE displaying the most sensitivity and FA the least sensitivity towards the two PRNGs.

Ville Tirronen *et al* [2] studied the sensitivity of DE/rand/1/exp on six different PRNGs.They concluded that a high quality random number generator is not required for DE as they showed least sensitivity. .However this conclusion was based on a single DE variant and through spark histograms lacking rigorous statistical analysis. It is also worth pointing out that their work considers more parameters like speed of evolution, computational cost and on the occurrence of stagnation or premature convergence.

Ivan Zelinka *et al* [3] proposed that a certain class of random number generators can be replaced by a set of deterministic process without lowering the performance of the evolutionary algorithm. The study successfully employed deterministic generators with chaos on Self Organizing Migration Algorithm and DE.

Earlier, Erick Cantu-Paz [4] studied the effect of PRNGs on Genetic Algorithms (GAs) as well as to the stochastic component operations of GAs. The results suggested that the effort of finding alternatives to uniform random initialization may prove beneficial to GA's performance.

L Skanderova *et al* [5] compared the performance of DE, in terms of speed of convergence, using typical PRNGs as well as chaotic systems. The study concluded the superiority of chaotic system in initializing DE to a faster convergence.

Motivated by the contradictory results, especially in case of DE, the current work intends to exhaustively analyze the sensitivity of 14 DE variants towards PRNGs.

4   Design of Experiments

Towards studying the impact of PRNGs on DE variants we have used the following six PRNGs:

1) GCC RAND
2) Multiply With Carry generator (MWC)
3) Complimentary Multiply With Carry generator (CMWC)
4) Linear Congruential generator (LCG)
5) XOR shift generator (XOR)
6) Mersenne Twister (MT)

GCC RAND [14] is an inbuilt PRNG available in the languages like C and C++. The rand() function has a range of about 0 to RAND_MAX and usually returns a value within this range itself. This function also conforms to the C99, C89 formats. The best thing about this PRNG is its simplicity and ease of use in different environments.

Multiply With Carry generator [7], [13] was invented by Marsagila. The underlying idea of this simple but powerful PRNG is that it uses simple integer arithmetic in the operations. And as a result, this helps to generate sequences at a very fast pace. A slight modification in it's modulo arithmetic give rise to another PRNG known as Complimentary Multiply With Carry generator or CMWC.

The principle behind the XOR PRNG [9], which is also developed by Marsagila, is that they generate random numbers using the Exclusive OR operation. The operation is performed on shifted bit forms of the current number to generate the next number in its sequence. XOR generator has succeeded in various statistical tests proving that it is reliable and is an excellent generator providing medium quality randomness in its sequence.

Linear Congruential Generator or LCG [8], [12] is one of the most famous and oldest PRNG available so far. The specialty of LCG is that these are fast and is comparatively easier to implement. The logic behind LCG is that it generates random numbers using linear equations which are not continuous in nature. One of the advantages of this PRNG is its small memory requirement when comparing to the other PRNGs.

Mersenne Twister [11] is one of the high quality generators available. The MT19937 and MT19937-64 are the standard implementations available for MT. The former uses a 32 bit word length while the latter uses 64 bit word length. One of the advantages of MT is its immensely large periods thus making it as one of the most widely used PRNG for various applications.

Considering the fact that there exists seven classical mutation strategies and two typical crossover schemes (binomial and exponential), we have employed all possible combinations i.e., fourteen different DE variants to study their sensitivity to PRNGs. The fourteen DE variants are *DE/rand/1/exp, DE/rand/1/bin, DE/rand/2/exp, DE/rand/2/bin, DE/best/1/exp, DE/best/1/bin, DE/best/2/exp, DE/best/2/bin, DE/current-to-rand/1/exp, DE/current-to-rand/1/bin, DE/current-to-best/1/exp, DE/current-to-best/1/bin, DE/rand-to-best/1/exp* and *DE/rand-to-best/1/bin.*

The above said study necessitates diverse benchmark functions with a variety of problem characteristics. In this work, we have used 19 well known benchmark functions in the evolutionary computing literature [10]. The benchmark functions are displayed in Table 4.1, 4.2 and 4.3. The benchmark functions F1 – F11 are unimodal separable [F1, F7], unimodal non-separable [F2, F8, F9, F10, F11], multimodal separable [F4, F6], and multimodal non-separable [F3, F5]. The rest of the functions i.e., F12 – F19 are the hybrid functions, which are produced by combining two functions from the set F1 – F11. They are formed by combining a non-separable function, F_{ns} [F3, F5, NS F9, NS F10] with the other functions, F' [F1, F4, NS F7]. The splitting of the functions is done using the variable m_{ns}. All the simulation experiments in this paper, has employed a dimension of 10 for all benchmark functions.

Table 1 Ranges and optimum fitness values of the benchmark functions F1 – F11 [10].

Function	Name	Range	Fitness Optimum
F1	Shifted Sphere Function	$[-100, 100]^D$	−450
F2	Shifted Schwefel's Problem 2.21	$[-100, 100]^D$	−450
F3	Shifted Rosenbrock's Function	$[-100, 100]^D$	390
F4	Shifted Rastrigin's Function	$[-5, 5]^D$	−330
F5	Shifted Griewank's Function	$[-600, 600]^D$	−180
F6	Shifted Ackley's Function	$[-32, 32]^D$	−140
F7	Shifted Schwefel's Problem 2.22	$[-10, 10]^D$	0
F8	Shifted Schwefel's Problem 1.2	$65.536, 65.536]^D$	0
F9	Shifted Extended f_{10}	$[-100, 100]^D$	0
F10	Shifted Bohachevsky	$[-15, 15]^D$	0
F11	Shifted Schaffer	$[-100, 100]^D$	0

The parameters of a typical DE are NP, the population size, F, K (scaling factor) and Cr (crossover rate). The value of NP is chosen as 30, F as 0.5, Cr as 0.8 and maximum number of generations as 3000. In case of two pair difference vectors variants, for K we have used the same value as of F. Due to the stochastic nature of the DE algorithm, 25 runs are made for each of the test function per variant per PRNG.

Table 2 Ranges and optimum fitness values of the benchmark functions F12 – F19 [10].

Name	F_{ns}	F'	m_{ns}	Range	Fitness Optimum
F12	NS-F9	F1	0.25	$[-100, 100]^D$	0
F13	NS-F9	F3	0.25	$[-100, 100]^D$	0
F14	NS-F9	F4	0.25	$[-5, 5]^D$	0
F15	NS-F10	NS-F7	0.25	$[-10, 10]^D$	0
F16*	NS-F9	F1	0.5	$[-100, 100]^D$	0
F17*	NS-F9	F3	0.75	$[-100, 100]^D$	0
F18*	NS-F9	F4	0.75	$[-5, 5]^D$	0
F19*	NS-F10	NS-F7	0.75	$[-10, 10]^D$	0

Table 3 Basic benchmark functions (F1-F11) and its mathematical formulas [10].

Function	Name	Definition
F_1	Shif. Sphere	$\sum_{i=1}^{D} z_i^2 + f\_bias$, $z = x - o$
F_2	Shif. Schwefel 2.21	$max_i\{|z_i|, 1 \leq i \leq D\} + f\_bias$, $z = x - o$
F_3	Shif. Rosenbrock	$\sum_{i=1}^{D-1} (100(z_i^2 + z_{i+1})^2 + (z_i - 1)^2) + f\_bias$, $z = x - o$
F_4	Shif. Rastrigin	$\sum_{i=1}^{D}(z_i^2 - 10\cos(2\pi z_i) + 10) + f\_bias$, $z = x - o$
F_5	Shif. Griewank	$\sum_{i=1}^{D} \frac{z_i^2}{4000} - \prod_{i=1}^{D} \cos(\frac{z_i}{\sqrt{i}}) + 1 + f\_bias$, $z = x - o$
F_6	Shif. Ackley	$-20\exp(-0.2\sqrt{\frac{1}{D}\sum_{i=1}^{D} z_i^2}) - \exp(\frac{1}{D}\sum_{i=1}^{D}\cos(2\pi z_i))$ $+20 + e + f\_bias$
F_7	Shif. Schwefel 2.22	$\sum_{i=1}^{D} |z_i| + \prod_{i=1}^{D} |z_i|$, $z = x - o$
F_8	Shif. Schwefel. 1.2	$\sum_{i=1}^{D}(\sum_{j=1}^{i} z_j)^2$, $z = x - o$
F_9	Shif. Extended f_{10}	$\left(\sum_{i=1}^{D-1} f_{10}(z_i, z_{i+1})\right) + f_{10}(z_D, z_1)$, $z = x - o$ $f_{10} = (x^2 + y^2)^{0.25} \cdot (\sin^2(50 \cdot (x^2 + y^2)^{0.1}) + 1)$
F_{10}	Shif. Bohachevsky	$\sum_{i=1}^{D} (z_i^2 + 2z_{i+1}^2 - 0.3\cos(3\pi z_i) - 0.4\cos(4\pi z_{i+1}) + 0.7)$, $z = x - o$
F_{11}	Shif. Schaffer	$\sum_{i=1}^{D-1} (z_i^2 + z_{i+1}^2)^{0.25} (\sin^2(50 \cdot (z_i^2 + z_{i+1}^2)^{0.1}) + 1)$, $z = x - o$

5   Simulation Results and Analysis

The simulation work carried out in the paper requires comparing the performance of fourteen differential evolution variants, on 19 benchmark functions, for all the six PRNGs. Consequently, pair-wise comparison between PRNGs (employed in each of the 14 DE variants) is realized by employing two-sided Wilcoxon rank sum test with the level of significance (α) chosen as 0.05. The results of Wilcoxon rank sum test are displayed in Table 5.1 and 5.2 where Table 5.1 shows the results for 'bin' variants and Table 5.2 shows the results for 'exp' variants. Columns in each table represent the pair-wise comparison between the respective two variants viz. PRNG1 vs PRNG2. Results in each cell are represented as Win-Draw-Lose with numbers corresponding to the number of functions on which the PRNG1 is superior, statistically insignificant or comparable and inferior to PRNG2.

As can be seen from both tables, the statistical insignificance of the performance between DE with compared PRNGs ranges from 7 functions to 14 functions. The superiority of PRNGs has found to be in the range of 2 functions to 7 functions. In the pair-wise comparisons between GCC vs MWC, GCC vs MT, GCC vs CMWC and XOR vs LCG, the former PRNGs (GCC mostly) displayed superiority on 4 to 6 functions on a very small number of cases. This superiority is still insignificant considering the number of cases out of 14 it has been observed.

The same arguments hold even if DE is observed in terms of the crossover employed viz. binomial and exponential. The simulation results validate the claim made in [2], that DE is not sensitive to the PRNG used in terms of search performance.

While the insensitivity of DE variants towards the choice of PRNGs is observed in case of 10 D problems in this study, similar study on very large scale optimization problem will resolve and clarify the observations made in [1] claiming that DE showed most sensitivity to PRNGs.

Table 4 Wilcoxons Rank Sum Test for DE variants with binomial crossover.

	GCC vs CMWC	GCC vs MWC	GCC vs XOR	GCC vs LCG	GCC vs MT	CMW C vs MWC	CMW C vs XOR	CMW C vs LCG	CMW C vs MT	MWC vs XOR	MWC vs LCG	MWC vs MT	XOR vs LCG	XOR vs MT	LCG vs MT
DE/rand/1/ bin	3-13-3	6-9-4	5-9-5	6-9-4	5-11-3	3-13-3	5-10-4	4-9-6	4-9-6	6-10-3	3-9-7	4-11-4	7-9-3	3-11-5	4-9-6
DE/rand/2/ bin	6-10-3	5-11-3	3-9-7	4-11-4	4-11-4	6-9-4	2-13-4	3-14-2	6-8-5	5-11-3	5-9-5	3-14-2	4-12-3	4-10-5	5-10-4
DE/best/1/ bin	7-8-4	3-11-5	6-9-4	4-9-6	7-8-4	5-8-6	2-14-3	4-10-5	5-12-2	4-10-5	2-14-3	3-13-3	6-10-3	6-9-4	4-7-8
DE/best/2/ bin	5-10-4	5-11-3	4-8-7	4-11-4	5-10-4	6-8-5	3-11-5	4-10-5	6-10-3	5-11-3	3-12-4	4-11-4	7-9-3	7-9-3	6-10-3
DE/current -to- rand/1/bin	5-12-2	7-8-4	5-8-6	5-12-2	5-11-3	4-9-6	6-7-6	7-9-3	5-12-2	6-9-4	3-15-1	2-13-4	4-10-5	4-10-5	7-8-4
DE/current -to- best/1/bin	6-9-4	4-13-2	6-10-3	6-8-5	5-11-3	5-9-5	2-14-3	5-11-3	7-9-3	3-9-7	4-9-6	5-10-4	5-11-3	5-11-3	5-11-3
DE/rand- to-best/bin	5-10-4	6-10-3	7-9-3	3-11-5	6-9-4	3-10-6	8-8-3	7-9-3	5-10-4	4-10-5	5-10-4	6-9-4	4-12-3	4-12-3	5-9-5

Table 5 Wilcoxons Rank Sum Test for DE variants with exponential crossover.

	GCC vs CMW	GCC vs MWC	GCC vs XOR	GCC vs LCG	GCC vs MT	CMW C vs MWC	CMW C vs XOR	CMW C vs LCG	CMW C vs MT	MWC vs XOR	MWC vs LCG	MWC vs MT	XOR vs LCG	XOR vs MT	LCG vs MT
DE/rand/1/ exp	5-11-3	3-10-6	3-13-3	3-11-5	6-9-4	3-12-4	4-10-5	3-14-2	4-12-3	3-13-3	3-11-5	4-11-4	4-13-2	3-11-5	4-12-3
DE/rand/2/ exp	4-9-6	5-11-3	7-9-3	6-10-3	4-12-3	5-8-6	6-11-2	6-9-4	3-10-6	5-11-3	5-8-6	3-12-4	3-9-7	4-8-7	5-10-4
DE/best/1/e xp	4-12-3	4-10-5	5-10-4	5-11-3	4-9-6	6-8-5	4-8-7	2-10-7	5-10-4	4-10-5	3-9-7	5-10-4	5-10-4	4-11-4	6-9-4
DE/best/2/e xp	3-11-5	6-8-5	6-10-3	4-9-6	6-9-4	4-8-7	7-9-3	4-12-3	4-9-6	3-10-6	4-10-5	6-10-3	5-13-1	2-11-6	3-11-5
DE/current -to- rand/1/exp	4-10-5	4-9-6	7-8-4	4-9-6	4-13-2	3-10-6	6-11-2	5-11-3	6-9-4	3-9-7	7-9-3	7-9-3	6-10-3	3-9-7	2-10-7
DE/current -to- best/1/exp	5-8-6	7-8-4	6-10-3	3-9-7	5-10-4	7-9-3	5-10-4	6-10-3	5-9-5	5-12-2	3-11-5	4-9-6	4-10-5	5-10-4	4-8-7
DE/rand- to-best/exp	4-12-3	5-9-5	3-10-6	4-12-3	3-10-6	4-11-4	6-9-4	5-9-5	7-9-3	3-10-6	5-9-5	6-10-3	5-8-6	5-8-6	4-10-5

6   Conclusion

This paper investigated the impact of the PRNGs on the performance quality of Differential Evolution (DE) algorithm. Towards this 14 variants of DE and 6 PRNGs have been considered for the empirical analysis of the sensitivity of DE towards PRNGs. A suite of 10-Dimensional 19 well known benchmark functions have been used for the study. Two-sided Wilcoxon rank sum test has been employed to pair-wise compare the input of PRNGs on 14 DE variants. The results suggest that DE is insensitive to the quality of PRNG used. As future work, a similar study involving test suite for a very large scale optimization will be carried out to further validate this claims.

References

1. Cárdenas-Montes, M., Vega-Rodríguez, M.A., Gómez-Iglesias, A.: Sensitiveness of evolutionary algorithms to the random number generator. In: Dobnikar, A., Lotrič, U., Šter, B. (eds.) ICANNGA 2011, Part I. LNCS, vol. 6593, pp. 371–380. Springer, Heidelberg (2011)
2. Tirronen, V., Äyrämö, S., Weber, M.: Study on the effects of pseudorandom generation quality on the performance of differential evolution. In: Dobnikar, A., Lotrič, U., Šter, B. (eds.) ICANNGA 2011, Part I. LNCS, vol. 6593, pp. 361–370. Springer, Heidelberg (2011)
3. Zelinka, I., Chadli, M., Davendra, D., Senkerik, R., Pluhacek, M., Lampinen, J.: Do evolutionary algorithms indeed require random numbers? extended study. In: Zelinka, I., Chen, G., Rössler, O.E., Snasel, V., Abraham, A. (eds.) Nostradamus 2013: Prediction, Modeling and Analysis of Complex Systems. AISC, vol. 210, pp. 61–75. Springer, Heidelberg (2011)
4. Cantu-Paz, E.: On random numbers and the performance of genetic algorithm. In: Proceedings of the Genetic and Evolutionary Computation Conference, GECCO 2002, pp. 311–318 (2002)
5. Skanderova, L., Rehor, A.: Comparison of pseudorandom numbers generators and chaotic numbers generators used in differential evolution. In: Zelinka, I., Suganthan, P.N., Chen, R., Snasel, V., Abraham, A., Rössler, O. (eds.) Nostradamus 2014: Prediction, Modeling and Analysis of Complex Systems. AISC, vol. 289, pp. 111–122. Springer, Heidelberg (2014)
6. Storn, R., Price, K.: Differential Evolution A Simple and Efficient Heuristic for Global Optimization over Continuous Spaces. Journal of Global Optimization, 341–359 (1997)
7. Brent, R.P.: Uniform random number generators for supercomputers. In: Proceedings Fifth Australian Supercomputer Conference, pp. 95–104 (1992)
8. Marsaglia, G.: Random numbers fall mainly in the planes. Proceedings of the National Academy of Sciences of the United States of America 61(1), 25 (1968)
9. Marsaglia, G.: Xorshift rngs. Journal of Statistical Software 8(14), 16 (2003)
10. Lozano, M., Herrera, F.: Special issue of soft computing: A fusion of foundations, methodologies and applications on scalability of evolutionary algorithms and other metaheuristics for large scale continuous optimization problems (2009)

11. Mtsumoto, M., Nishimura, T.: Mersenne twister: a 623- dimensionally equidistributed uniform pseudo random number generator. ACM transactions on modeling and Computer Simulation (TOMACS) **8**(1), 3–30 (1998)
12. Entacher, K.: A collection of selected pseudorandom number generators with linear structures. ACPC-Austrian Centre for Parallel Computation (1997)
13. Multiply-with-carry. (2014, November 24). In: Wikipedia, The Free Encyclopedia. Retrieved 04:51, April 10, 2015, from
http://en.wikipedia.org/w/index.php?title=Multiply-with-carry&oldid=635182796
14. RAND — Real pseudo-random number.
https://gcc.gnu.org/onlinedocs/gfortran/RAND.html

Control of DC–DC Converter in Presence of Uncertain Dynamics

Aniket D. Gundecha, V.V. Gohokar, Kaliprasad A. Mahapatro
and Prasheel V. Suryawanshi

Abstract The paper proposes an intelligent control of DC–DC boost converter in the presence of uncertain dynamics. The converter is desired to operate with an energy harvesting system, where the input conditions are uncertain. An extended state observer (ESO) is designed which estimates and accords with the disturbances caused by load and input uncertainties. The proposed control technique address the issues of uncertainties involved in the energy harvesting system where the system is connected to constant voltage source. The overall stability of the system is proved. The effectiveness of the proposed strategy is verified for a representative example and it assures fast voltage regulation, even in the presence of sudden load and input voltage changes. This proves that the proposed method is capable of compensating the uncertain dynamics in the system.

1 Introduction

DC–DC power converters are found in many applications including portable electronics, fuel cell technologies, wireless gadgets and hybrid power systems [1] [2]. The different types of converters; buck, boost and buck–boost converter are selected based on the application. Energy harvesting represent one of the growing applications of DC–DC converters, that provide a very small amount of power for portable

A.D. Gundecha(✉)
SSGM College of Engineering, Shegaon and MIT Academy of Engineering, Alandi (D), Pune, India
e-mail: adgundecha@entc.maepune.ac.in

V.V. Gohokar
Maharashtra Institute of Technology, Pune, India
e-mail: vvgohokar@rediffmail.com

K.A. Mahapatro · P.V. Suryawanshi
MIT Academy of Engineering, Alandi (D), Pune, India

© Springer International Publishing Switzerland 2016   315
S. Berretti et al. (eds.), *Intelligent Systems Technologies and Applications*,
Advances in Intelligent Systems and Computing 384,
DOI: 10.1007/978-3-319-23036-8_27

electronics [3]. The input to these energy harvesters is dynamic depending upon the type of source used (solar, thermoelectric, RF energy), which has to be regulated [4].

DC–DC converters play an important role in providing a constant output voltage irrespective of the input dynamics. A generalised method of modelling DC–DC converters was proposed in [5], that uses state space approach for two-state switching converters. The stability, gain and efficiency of buck, boost and buck-boost converters is dependent on switching time and duty cycle [6], [7]. Passivity based control is used to regulate the voltage for a class of state affine system [8] along with Kalman observer for estimating the state variables. An alternative source of energy is considered [9] where thermoelectric energy is harvested from a radiator which is used to power a ZigBee node. The system is designed to generate voltage from thermal gradients.

PID control is one of the most popular methods of controlling a power converter [10]. A variety of other strategies like fuzzy-logic [11], state-feedback [12] and sliding mode [13] are reported to address the problem of output regulation. A control based on constrained stabilization through bifurcation analysis [14] ensures output regulation. The control however has a bit of chatter. A control technique supplemented by the estimates of uncertainty and disturbance can be a solution. Techniques like unknown input observer (UIO) [15], perturbation observer [16], disturbance observer (DO) [17], are commonly used to estimate the effects of uncertainties and disturbances.

Extended state observer (ESO) [18] is a technique that estimates the states as well as uncertainties, with minimal information of plant. The uncertainty in plant parameters and unknown disturbances are lumped together as an additional state [19]. ESO has been applied in various applications like motion control [20], robotics [21], automotive [22], vibration [23]. The efficacy of ESO for estimating states and uncertainty is validated in [24]. It is proved that ESO performs better than high gain observer (HGO) and sliding mode observer (SMO). Additionally ESO is able to estimate even when maximum information of plant is unknown and exact calibration of sensor is not required [25].

A control strategy for regulating the output voltage of DC–DC boost converter is proposed in this paper. The algorithm is tested for uncertainties at source as well as load side. The technique gives stable, regulated output voltage even in the presence of uncertain dynamics. The main contributions of this paper are:

- The type of uncertainties considered is in line with practical application of energy harvesting
- No information of uncertainty and disturbance is required
- The estimation and tracking error are within limits
- The control strategy is tested and validated for variations in source voltage, load resistance and reference voltage

The paper is organized as follows: The mathematical model of DC–DC boost converter is introduced in Section 2. Section 3 explains the design of robust control law, by introducing the concept of extended state observer (ESO). Section 4 proves the stability of the system for uncertainties. The results along with related discussions are illustrated in Section 5 and the paper is concluded in Section 6.

2 Dynamics of DC–DC Boost Converter

A typical boost converter is shown in Fig. 1, where L is the inductance, C is the capacitance, R_L is load resistance, V_s is the external source voltage, V_o is the desired output voltage and $\mu \in [0, 1]$ is the duty cycle which is regulated with the control signal [26]. The load resistance (R_L) and external source voltage (V_s) are subject to variations on account of uncertain load and input [4] [9]. The load resistance is assumed to be completely unknown, as may be the case for variable load applications [27].

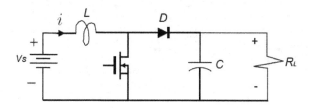

Fig. 1 Circuit diagram of boost converter

The average system model can be written as in [8],

$$\frac{dV_o}{dt} = \frac{-1}{RC}V_o + \frac{1}{C}(1-\mu)i_L$$
$$\frac{di_L}{dt} = \frac{-1}{L}(1-\mu)V_o + \frac{V_s}{L} \tag{1}$$

where i_L is the current flowing through inductor and V_o is the output voltage. Let $x_1 = C(V_o - V_{ref})$ and $x_2 = i_L$ be the state variables. The model (1) can be rewritten as,

$$\dot{x}_1 = x_2 - \frac{V_o}{R} - \mu x_2$$
$$\dot{x}_2 = (1-\mu)\frac{x_1}{LC} + \frac{1}{L}\left(V_s - V_{ref} + \mu V_{ref}\right) \tag{2}$$

Denoting $u = \left(V_s - V_{ref} + \mu V_{ref}\right)$ and $d = -\frac{V_o}{R} - \mu x_2$, the dynamics (2) can be simplified as,

$$\dot{x}_1 = x_2 + d(t) \tag{3}$$
$$\dot{x}_2 = \frac{x_1(1-\mu)}{LC} + bu$$

where $b = \frac{1}{L}$ and d is the mismatched disturbance.

3   Composite Control Design

The objective is to design an intelligent control for compensating mismatched disturbance in a DC–DC boost converter. The control uses ESO for estimating the states as well as lumped uncertainty and disturbance [18] [24] [28]. ESO can be used for compensating matched as well as mismatched uncertainties [29]. A schematic block diagram of the proposed control configuration is shown in Fig. 2.

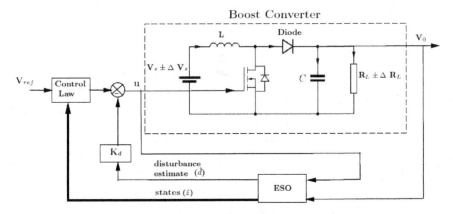

Fig. 2 Schematic of control configuration

The present work is concerned with efficient control of boost converter under varying dynamics. The estimation of states and uncertainty is based on plant output (V_o) and control input (u).

3.1   Overview of ESO

A general n^{th} order plant is mathematically represented as,

$$\begin{cases} \dot{x_1} & = x_2 \\ \dot{x_2} & = x_3 \\ \;\vdots \\ \dot{x_n} & = x_{n+1} + a_0 + b\,u \\ \dot{x_{n+1}} & = h \end{cases} \tag{4}$$

The plant in (4) is augmented with an additional state to include lumped uncertainty and disturbance.

An ESO for the augmented plant can be represented as,

$$
\begin{cases}
\dot{\hat{z}}_1 & = \hat{z}_2 + \beta_1 g_1(e) \\
\vdots \\
\dot{\hat{z}}_n & = \hat{z}_{n+1} + \beta_n g_n(e) + a_0 + bu \\
\dot{\hat{z}}_{n+1} & = \beta_{n+1} g_{n+1}(e)
\end{cases}
\tag{5}
$$

The equation (5) depicts $\hat{z}_1, \hat{z}_2 \ldots \hat{z}_n$ as the estimate of plant states and $\hat{z}_{n+1}$ as the extended state, which gives the estimate of uncertainties in plant. This estimate of uncertainty adds robustness in the control design. Additionally, $e = y - \hat{z}_1$ is the error and $g_i(.)$ is suitably constructed nonlinear gain functions.

If one chooses the nonlinear function $g_i(.)$ and their related parameters properly, the estimated state variable $\hat{z}_i$ are expected to converge to the respective state of the system x_i, i.e. $\hat{z}_i \rightarrow x_i$. The choice of g_i is heuristically given in [24]

$$
g_i(e, \alpha_i, \delta) = \begin{cases}
|e|^{\alpha_i} & , |e| > \delta \\
\dfrac{e}{\delta^{1-\alpha_i}} & , |e| \le \delta
\end{cases}
\tag{6}
$$

where δ is the small number($\delta > 0$) used to limit the gain, β is the observer gain determined by the pole-placement method. α is chosen between 0 and 1 for nonlinear ESO (NESO) and is considered unity for linear ESO (LESO). The present case is concerned with LESO. The LESO for any system is given by (5) with gains $g(e) = e$. The state-space model, of the LESO dynamics can be written as,

$$
\dot{\hat{z}} = A\hat{z} + Bu + LC(x - \hat{z})
\tag{7}
$$

where L is the observer gain vector. A control law for the second order plant in (3) is designed using ESO as in (5).

$$
\begin{cases}
\dot{\hat{z}}_1 & = \hat{z}_2 + \beta_1 e \\
\dot{\hat{z}}_2 & = \hat{z}_3 + \beta_2 e + b_0 u \\
\dot{\hat{z}}_3 & = \beta_3 e \\
y & = z_1
\end{cases}
\tag{8}
$$

The control law for 2nd order system in (8) can be stated as,

$$
u = K_z \hat{z} + K_d \hat{d}
\tag{9}
$$

where K_z is the feedback control gain determined by pole placement and K_d is the disturbance compensation gain designed as in [29]

$$
K_d = -[c_o(A + b_u K_z)^{-1} b_u)]^{-1} c_o (A + b_u K_z)^{-1} b_d
\tag{10}
$$

where $A = \begin{bmatrix} 0 & 1 \\ -\frac{1}{LC} & 0 \end{bmatrix}$, $b_u = [0 \ \ 1]^T$, $b_d = [1 \ \ 0]^T$ and $c_o = [1 \ \ 0]$

4 Closed Loop Stability

The plant model in (4) can be rewritten as

$$\begin{bmatrix} \dot{x}_1 \\ \dot{x}_2 \\ \dot{x}_3 \end{bmatrix} = \begin{bmatrix} 0 & 1 & 0 \\ 0 & 0 & 1 \\ 0 & 0 & 0 \end{bmatrix} \begin{bmatrix} x_1 \\ x_2 \\ x_3 \end{bmatrix} + \begin{bmatrix} 0 \\ b \\ 0 \end{bmatrix} u + \begin{bmatrix} 0 \\ 0 \\ h \end{bmatrix} + \begin{bmatrix} 0 \\ a_0 \\ 0 \end{bmatrix} \tag{11}$$

$$y = x_1 \begin{bmatrix} 1 & 0 & 0 \end{bmatrix} \begin{bmatrix} x_1 \\ x_2 \\ x_3 \end{bmatrix} \tag{12}$$

where h is the rate of change of uncertainty $h = \dot{d}$ and it is assumed to be a bounded function. Considering the linear extended state observer (LESO) the observer state space can be augmented as

$$\begin{bmatrix} \dot{\hat{z}}_1 \\ \dot{\hat{z}}_2 \\ \dot{\hat{z}}_3 \end{bmatrix} = \begin{bmatrix} 0 & 1 & 0 \\ 0 & 0 & 1 \\ 0 & 0 & 0 \end{bmatrix} \begin{bmatrix} \hat{z}_1 \\ \hat{z}_2 \\ \hat{z}_3 \end{bmatrix} + \begin{bmatrix} 0 \\ b \\ 0 \end{bmatrix} u + \begin{bmatrix} \beta_1 \\ \beta_2 \\ \beta_3 \end{bmatrix} e + \begin{bmatrix} 0 \\ a_0 \\ 0 \end{bmatrix} \tag{13}$$

subtracting (13) from (11) we get

$$\dot{e} = (A - LC)e + B h \tag{14}$$

where $e = x - \hat{z}$, $B = \begin{bmatrix} 0 & 0 & 1 \end{bmatrix}^T$ and $L = \begin{bmatrix} \beta_1 & \beta_2 & \beta_3 \end{bmatrix}^T$.

The stability of the system is ensured if, the observer error e should be a full rank and the eigen values of $(A - LC)$ should be negative. As the observer gains (L) is chosen by the designer, the full rank in (14) along with negative eigen values of $(A - LC)$ is guaranteed. If the disturbance is zero ($h = 0$) or slow varying ($\dot{h} \approx 0$) the error will go to zero asymptotically i.e., $\lim\limits_{t \to \infty} e(t) = 0$ which proves the stability of the system.

5 Results and Discussion

The proposed algorithm is tested for reference tracking in a boost converter for application in energy harvesting. The different cases such as variable source voltage and variable load resistance are considered. The algorithm is also tested for different reference voltages and dynamic changes in the input source. The different cases are in-line with the constraints in practical applications of energy harvesting.

The controller gains are obtained by placing the poles at $(1 + (\tau/2)s)^2$ with a time constant of $\tau = 4$ and the ESO gains $\beta's$ are obtained by placing the observer poles at $(1 + (\tau_o/3)s)^3$ with a time constant of $\tau_o = 0.03$ as stated in [30]. The switching frequency is considered as 10 KHz.

5.1 Case 1: Variable Input (V_s)

The boost converter is evaluated for tracking the reference voltage under variable input conditions. The reference voltage considered is, $V_{ref} = 3.3V$ and load resistance $R_L = 40\Omega$. The tracking performance and estimation for an input of $V_s = 1.0V$ is shown in Fig. 3. The Fig. 3a shows estimation of x_1 with control effort as shown in Fig. 3b. The duty cycle adjusted as per the requirement is shown in Fig. 3c. The reference tracking is excellent, as can be seen in Fig. 3d.

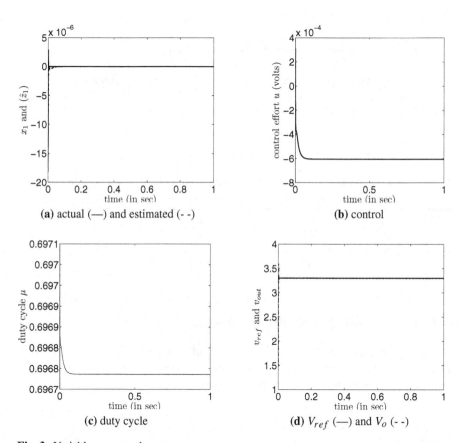

(a) actual (—) and estimated (- -)   (b) control

(c) duty cycle   (d) V_{ref} (—) and V_o (- -)

Fig. 3 Variable source voltage

The performance is tested for different values of source voltage, and results are shown in Table 1. The RMS values of estimation error ($\tilde{e}$) and tracking error ($\tilde{e}_t$) are within bounds. The variable source voltage may arise from varied sources like thermoelectric, solar, etc and the value is also uncertain. The results demonstrate the ability of ESO to effectively compensate the uncertain input.

Table 1 *Performance analysis for variable source voltage*

V_s (V)	u (mV)	Tracking error ($\tilde{e}_t$) $\tilde{e}_t = V_{ref} - V_o$	Estimation error ($\tilde{e}$) $\hat{e} = x_1 - \hat{z}_1$	Convergence time (ms)
0.5	-0.842	0.0437	2.88E-07	5.5
0.75	-1.010	0.0318	3.04E-07	2.3
1	-1.144	0.0246	2.97E-07	2.7
1.25	-1.196	0.0201	2.92E-07	3.3
1.5	-1.248	0.0170	2.86E-07	2.7

5.2 Case 2: Variable Load Resistance (R_L)

The reference voltage considered is, $V_{ref} = 3.3V$ and source voltage $V_s = 0.75V$. The performance is consistent for different values of R_L and the same is illustrated in Table 2.

Table 2 *Performance analysis for variable load resistance*

R_L (Ω)	u (mV)	Tracking error ($\tilde{e}_t$) $\tilde{e}_t = V_{ref} - V_o$	Estimation error ($\tilde{e}$) $\hat{e} = x_1 - \hat{z}_1$	Convergence time (ms)
20	-0.728	0.0287	2.17E-07	14.4
30	-0.936	0.0282	2.25E-07	9.9
40	-0.988	0.0282	2.29E-07	9.2
50	-1.092	0.0282	2.29E-07	8.5
60	-1.144	0.0282	2.25E-07	8.8

5.3 Case 3: Variable Reference Voltage (V_{ref})

The load resistance considered is, $R_L = 40\Omega$ and source voltage $V_s = 0.75V$. The results are consistent and the designed control is capable of regulating the output voltage as per the requirement. The results are shown in Table 3.

Table 3 *Performance analysis for variable reference voltage*

V_{ref} (V)	u (mV)	Tracking error ($\tilde{e}_t$) $\tilde{e}_t = V_{ref} - V_o$	Estimation error ($\tilde{e}$) $\hat{e} = x_1 - \hat{z}_1$	Convergence time (ms)
1.8	-0.15	0.0081	7.63E-08	4.5
2	-0.70	0.0018	8.7E-08	4.6
3.3	-1.10	0.0282	2.3E-07	7.2
4.5	-1.21	0.0502	3.3E-07	4.9
6	-1.42	0.0803	3.8E-07	7.9

5.4 Case 4: Dynamic Variations in Source Voltage

The results in case 1, case 2 and case 3 demonstrate the regulation of output voltage with varying dynamics of constant nature. The proposed control is used to analyse the effect of dynamic variations in source voltage. The results are shown in Fig. 4. The Fig. 4d shows the regulated output for a dynamic input as shown in Fig. 4a.

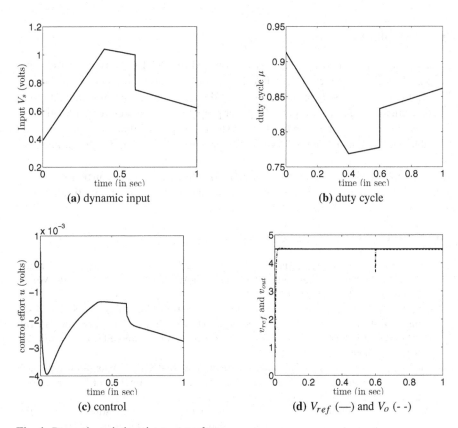

Fig. 4 Dynamic variations in source voltage

The robustness analysis in terms of estimation error ($\tilde{e}$) and tracking error ($\tilde{e}_t$) can be seen in Table 1 to Table 3. The RMS values of both; estimation error ($\tilde{e}$) and tracking error ($\tilde{e}_t$) are within acceptable bounds. The bounds can be further lowered by redesigning β. The estimation error ($\tilde{e}$) and tracking error ($\tilde{e}_t$) can be further lowered if some nominal values of plant parameters are available. However the present work is concerned with control design with no knowledge of plant parameters. The results demonstrate the efficacy of ESO for robust performance in varying types and degrees of uncertainty and disturbance.

The present work deals with robust operation of boost converter and the results are validated in simulation. The different cases considered are in tune to the conditions that may appear in an energy harvesting system. The input source can be thermo-electric or solar energy; and this is considered in case 1 and case 4. The source voltage may be variable or dynamic and the results are promising in tracking the reference voltage. The uncertain dynamics are compensated resulting in maximum power harvesting.

6 Conclusion

In this paper, a robust control strategy is proposed for DC–DC boost converter. The proposed control leverages the benefits of ESO for compensating uncertain dynamics. The ESO is able to compensate the effect of disturbances caused by load and input uncertainties on regulated output. The paper proposes an approach for application of boost converter in an energy harvesting system. It is proved that the overall system is stable and the bounds can be lowered by appropriate choice of design parameters. The efficacy of the design is tested and confirmed for multiple uncertainties that can result in a practical energy harvesting system. Results demonstrated can be applied in a straight forward manner to a solar energy or thermoelectric energy harvesting system.

References

1. Ali, M.S., Kamarudin, S.K., Masdar, M.S., Mohamed, A.: An Overview of Power Electronics Applications in Fuel Cell Systems: DC and AC Converters. The Scientific World Journal (2014). doi:10.1155/2014/103709
2. Nejabatkhah, F., Danyali, S., Hosseini, S.H., Sabahi, M., Niapour, S.M.: Modeling and Control of a New Three-Input DC-DC Boost Converter for Hybrid PV/FC/Battery Power System. IEEE Transactions on Power Electronics (2012). doi:10.1109/TPEL.2011.2172465
3. Guilar, N.J., Amirtharajah, R., Hurst, P.: Analysis of DC-DC conversion for energy harvesting systems using a mixed-signal sliding-mode controller. In: Power Electronics Specialists Conference (2007). doi:10.1109/PESC.2007.4342430
4. Weddell, A.S., et al.: A Survey of multi-source energy harvesting systems. In: Design, Automation & Test in Europe Conference & Exhibition (DATE) (2013). doi:10.7873/DATE.2013.190
5. Middlebrook, R., Ćuk, S.: A general unified approach to modelling switching-converter power stages. International Journal of Electronics Theoretical and Experimental **42**(6), 521–550 (1977)
6. Rim, C., Joung, G., Cho, G.-H.: A state-space modeling of non-ideal DC-DC converters. In: 19th Annual IEEE Power Electronics Specialists Conference (1988). doi:2523-9/88/0000-0943
7. Rim, C., Joung, G., Cho, G.H.: Practical switch based state-space modeling of DC-DC converters with all parasitics. IEEE Transactions on Power Electronics **6**, 611–617 (1991)

8. Gonzalez-Fonseca, N., Leon-Morales, J.D., Leyva-Ramos, J.: Observer-based controller for switch-mode DC-DC converters. In: 44th IEEE Conference on Decision and European Control Conference (CDC-ECC) (2005). doi:10.1109/CDC.2005.1582916

9. Lu, X., Yang, S.H.: Thermal energy harvesting for WSNs. In: IEEE International Conference on Systems Man and Cybernetics (SMC) (2010). doi:10.1109/ICSMC.2010.5641673

10. Keyser, R.D., Ionescu, C.: A comparative study of several control techniques applied to a boost converter. In: IEEE 10th Int. Conf. on Optimisation of Electrical and Electronic Equipment OPTIM, pp. 71–78 (2006)

11. Shiyas, P.R., Kumaravel, S., Ashok, S.: Fuzzy controlled dual input DC/DC converter for solar-PV/wind hybrid energy system. In: Electrical, Electronics and Computer Science (SCEECS) (2012). doi:10.1109/SCEECS.2012.6184775

12. Lakshmi, S., Raja, T.: Design and implementation of an observer controller for a buck converter. Turkish Journal of Electrical Engineering & Computer Sciences. doi:10.3906/elk-1208-41

13. Utkin, V.: Sliding mode control of DC/DC converters. Journal of the Franklin Institute **350**(8), 2146–2165 (2013). Elsevier

14. Yfoulis, C., Giaouris, D., Stergiopoulos, F., Ziogou, C., Voutetakis, S., Papadopoulou, S.: Robust constrained stabilization of boost DC-DC converters through bifurcation analysis. Control Engineering Practice (2014). doi:10.1016/j.conengprac.2014.11.004

15. Takahashi, R.H.C., Peres, P.L.D.: Unknown input observers for uncertain systems: a unifying approach and enhancements (1996). doi:10.1109/CDC.1996.572726

16. Jiang, L., Wu, Q.H.: Nonlinear adaptive control via sliding-mode state and perturbation observer. IEE Proceedings-Control Theory and Applications (2002). doi:10.1049/ip-cta:20020470

17. Chen, W.H.: Disturbance observer based control for nonlinear systems. IEEE/ASME Transactions on Mechatronics **9**(4), 706–710 (2004)

18. Han, J.: From PID to active disturbance rejection control. IEEE Trans. on Industrial Electronics **56**(3), 900–906 (2009)

19. Gao, Z., Huang, Y.I., Han, J.: An alternative paradigm for control system design. In: Proceeding of the 40th IEEE Conference on Decision and Control (2001). doi:10.1109/.2001.980926

20. Tianxu, J., Jie, C., Yongqiang, B.: A motion control design through variable structure controller based on extended state observer. In: IEEE/ASME International Conference on Mechtronic and Embedded Systems and Applications (2008). doi:10.1109/MESA.2008.4735688

21. Ma, H., Su, J.: Uncalibrated robotic 3-D hand-eye co-ordination based on the extended state observer. In: IEEE International Conference on Robotics and Automation (2003). doi:10.1109/ROBOT.2003.1242104

22. Hu, Y., Liu, Q., Gao, B., Chen, H.: ADRC based clutch slip control for automatic transmission. In: IEEE Chinese Control and Decision Conference (2011). doi:10.1109/CCDC.2011.5968672

23. Dong, L., Zheng, Q., Gao, Z.: A novel oscillation controller for vibrational MEMS gyroscopes. In: American Control Conference (2007). doi:10.1109/ACC.2007.4282883

24. Wang, W., Gao, Z.: A comparison study of advanced state observer design techniques. In: American Control Conference (2003). doi:10.1109/ACC.2003.1242474

25. Mahapatro, K.A., Chavan, A.D., Suryawanshi, P.V.: Analysis of robustness for industrial motion control using extended state observer with experimental validation. In: IEEE Conference on Industrial Instrumentation and Control (2015) (accepted)
26. Ramirez, H.S., Ortigoza, R.S.: Control Design Techniques in Power Electronics Devices. Springer-Verlag London limited (2006). ISBN-13:9781846284588
27. Sira-Ramírez, H., Silva-Navarro, G.: Generalized pid control of the average boost converter circuit model. In: Zinober, P.A., Owens, P.D. (eds.) NCN4 2001. LNCS, vol. 281, pp. 361–371. Springer, Heidelberg (2003)
28. Gao, Z.: Scaling and bandwith-parameterization based controller tuning. In: American Control Conference (2003). doi:10.1109/ACC.2003.1242516
29. Li, Shihua, Yang, Jun, Chen, Wen-Hua, Chen, Xisong: Generalized Extended State Observer Based Control for Systems With Mismatched Uncertainties. IEEE Transactions on Industrial Electronics **59**(12), 4792–4802 (2012)
30. Talole, S.E., Kolhe, J.P., Phadke, S.B.: Extended State Observer Based Control of Flexible Joint System with Experimental Validation. IEEE Transactions on Industrial Electronics **57**(4), 1411–1419 (2010)

Automated Radius Calculation of a Turn for Navigation and Safety Enhancement in Automobiles

K. Gnanasaekaran, S. Kanya, B. Suresh and Shriram K. Vasudevan

Abstract The present navigation system gives details like the route and distance of the travel, and the distance and direction of the upcoming turns. In addition, considering the safety aspect, it is important to give the radius of each turn so as to determine the maximum speed that the vehicle can take in that turn. This paper presents the calculation of the radius of the turns using LabVIEW.

1 Introduction

One accident occurs every five minutes in India, with the accident rate corresponding to 45 per 100,000 populations. Of this, 55.1% were reported due to skidding [1]. Any automobile over-speeding along a turn would cause skidding leading to many accidents. So as to avoid this, it is necessary to know the maximum speed that the vehicle can take at each turn through its course of travel. This can be done by finding the radius of each turn.

In this project, the radii of all the turns are calculated using Laboratory Virtual Instrument Engineering Workbench (LabVIEW) and Google Application Program Interface (API) server. A Graphical User Interface (GUI) using LabVIEW is created, which takes the origin and destination from the user as inputs. Using the

---

K. Gnanasaekaran · S. Kanya · B. Suresh
Department of Electrical and Electronics Engineering, Amrita School of Engineering,
Amrita Vishwa Vidyapeetham (University), Coimbatore, India

S.K. Vasudevan(✉)
Department of Computer Science and Engineering, Amrita School of Engineering, Amrita
Vishwa Vidyapeetham (University), Coimbatore, India
e-mail: Kv_shriram@cb.amrita.edu

© Springer International Publishing Switzerland 2016
S. Berretti et al. (eds.), *Intelligent Systems Technologies and Applications*,
Advances in Intelligent Systems and Computing 384,
DOI: 10.1007/978-3-319-23036-8_28

Google direction API the distance between the origin and destination is obtained along with the geometrical coordinates. The radius of all turns along that route and the distance from the origin to each turn are calculated and is provided to the user. An algorithm for determining this radius has been devised.

On entering the origin and destination in the GUI, the corresponding latitudes and longitudes are obtained from the Google API server. The geometrical coordinates of origin is varied by 2m distance in 8 different directions and the corresponding distances to the end point are obtained. Using the devised algorithm, the next geometrical point along the route is obtained. This is repeated till the destination point is reached. Using these geometrical points, the radius of the turn and the distance from the origin to each turn are calculated. The main advantage of this method is that these radii calculations can be done from any location whether the vehicle is in motion or at rest as this method does not require the Global Positioning System (GPS) data of the vehicle. This also works in bad climatic condition and does not have any problems due to blind spots.

This application can be integrated with the existing navigation system in automobiles in the future which would improve the safety aspect of it.

2   Literature Survey

As till date, there is no method to find the radius of the turns using software. So, this method was devised to overcome this difficulty which can be useful in many applications.

LabVIEW is used as the GUI as it has the option of web service with a user friendly interface [2]. The geometrical coordinates are found using the Google API server as this can be accessed with just an internet connection. This eliminates the drawbacks of bad climatic conditions. This can also be used to find the radii of the turns from any location to another irrespective of the location of the user.

For the calculation of the radii, the shortest distance (D) between the two geometrical coordinates in meters is found using the Haversines formula [3],[4].

$$X = \left(\left(\sin\left(\tfrac{lat2-lat1}{2} \right) \right)^2 + \left(\left(\sin\left(\tfrac{long2-long1}{2} \right) \right)^2 * (\cos(lat1) * \cos(lat2)) \right) \right) \tag{1}$$

$$D = 2\,\mathrm{atan2}\left(\sqrt{X}, \sqrt{1-X} \right) * 6371000 \tag{2}$$

Where, lat1 and long1 corresponds to the geometrical coordinates of the first point in radians and latitude and longitude corresponds to the second point in radians.

For the calculation of the radii (R) in meters,

$$A = \cos^{-1}\frac{(c^2+b^2-a^2)}{2bc} \tag{3}$$

$$R = \frac{a}{2\sin A} \tag{4}$$

Where a, b and c are the shortest distance between the three geometrical coordinates in meters calculated using (1) and (2).

3   Proposed Architecture

It is always easy to understand the architecture through a diagrammatic representation. The below shown figure. 1 represents the proposed architecture to solve the problem being faced. Followed by figure. 1, figure. 2 has the complete flow diagram which explains the complete flow of the process, which is very much self-explanatory.

In figure.1 LabVIEW is used for creating the GUI which request the origin and destination from the user ,once it entered the GUI contacts google API to access the maps server to get the geometrical co-ordinates of the origin and the destination and the distance between the two points then it moves to the next part where the next start points are found using the algorithm explained in figure.2,then the shortest distance is found using the next start points ,so it is used to find the geometrical co-ordinates of the turn which is used to calculate the radius of the turn ,this process is done in loop to find the radius of all the turn in the route with the distance from the origin location to the each turn and display it to the user as shown in the figure.3.

The algorithm is to find the next start points from the geometrical co-ordinates obtained from google API, this algorithm ensures that the next start is well within the route and on the road. This algorithm is devised by considering many cases so that the next start point is chosen properly. This algorithm is constructed with the help of the LabVIEW itself which is shown in figure.1.

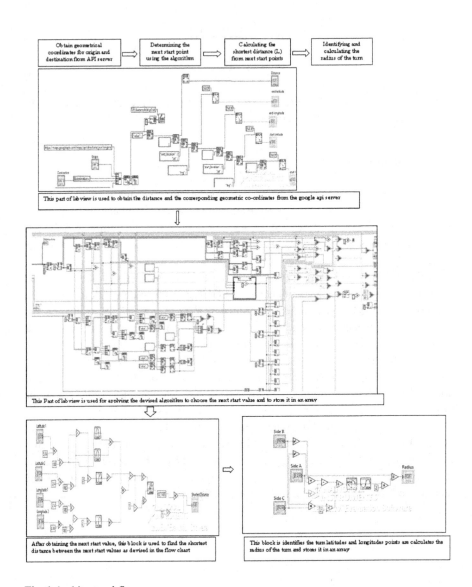

Fig. 1 Architectural flow.

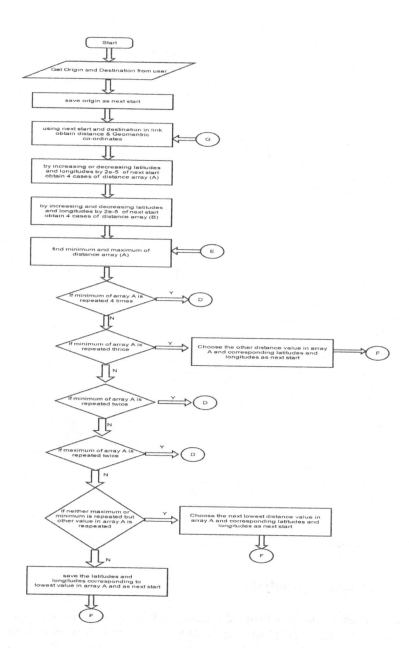

Fig. 2 Flow chart for proposed algorithm

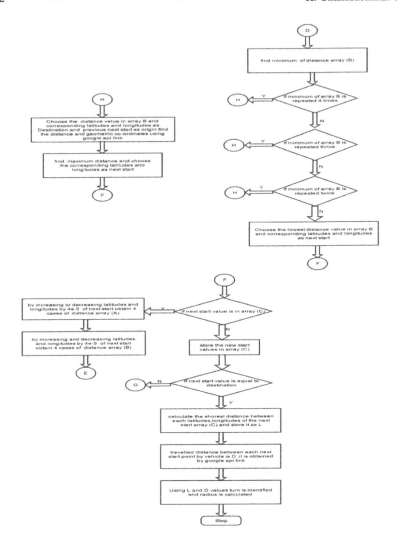

Fig. 2 *(continued)*

4   Result and Analysis

The algorithm proposed in Figure 2 is tested for various routes and the radii were found to be approximately equal to the actual radii values. A result of a sample turn radius calculation is shown in Fig. 3.

The origin entered is TANGENCO, Thirumangalam and the destination is Thirumangalam Electric Sub Station. The distance between the origin and the destination is 67 m and has 1 turn in between them. The actual distance to the turn from the origin is approximately 8m. The distance obtained from the algorithm is 7m. The actual radius of the turn is approximately 10 m, whereas, the radius obtained is 10.014 m.

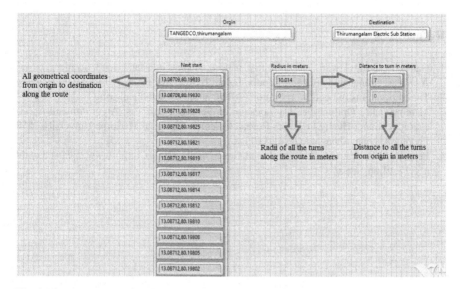

Fig. 3 Results of a case in LabVIEW

5   Hindrances Faced

The latitude and longitude of the previous start are varied by 2e-5 in eight possible ways which will give points within two meter range from the previous start. By analyzing these points it was difficult to find the next start location which is along the route required. For this, an algorithm was devised after analyzing many different cases.

It was difficult to find the starting point of a turn. So, for every 2m, it was assumed that these points were on a circle and with the shortest distance and the actual distance between these two points, the radius of the assumed circle was found. It was found to be more than 0.6 at the turns, thus identifying the turn.

The radius of the assumed circle will not give the actual radius as these points are only 2m apart. So the start and end points of the turns were found and using 3 points along the turn, the actual radius is calculated using equation (3) and (4).

6   Future Scope

Radius of turn can be included in the maps for navigation purpose. This can be used to know whether a vehicle can travel in the path or not. Radius of the turn is an essential data to determine the maximum speed with which a vehicle can turn safely. This speed can be used in automatic braking control system.

References

1. Ganveer, G.B., Tiwari, R.R.: Injury pattern among non-fatal road accident cases: A cross sectional study in central India. Indian J. Med. Sci. **59**, 9–12 (2005)
2. The National Instruments Corporation (2015). http://www.india.ni.com/
3. Mwemezi, J.J.: Optimal Facility Location on Spherical Surfaces: Algorithm and Application. Youfang Huang Logistics Research Center, Shanghai Maritime University 1550 Pudong Avenue, Shanghai, 200135 China
4. Pearson, C.H.: Latitude, Longitude and Great Circles (2009). http://www.cpearson.com/excel/LatLong.aspx/

An L_1-Method: Application to Digital Symmetric Type-II FIR Filter Design

Apoorva Aggarwal, Tarun K. Rawat, Manjeet Kumar
and Dharmendra K. Upadhyay

Abstract In this paper, the design of digital symmetric type-II linear-phase FIR low-pass (LP) and band-pass (BP) filter is formulated using the L_1 optimality criterion. In order to obtain better filter performance we compute the optimal filter coefficients using the L_1-norm based fitness function. The use of L_1 technique in digital filter design applications has the advantages of a flatter passband and high stopband attenuation over other gradient-based filter optimization methods. This technique is applied to optimally design type-II FIR filters. Simulations and statistical analysis have been performed for the 25th order LP and BP filters. It is observed, that the L_1-based filter results is an improved design in comparison with the filters obtained using the equiripple, least-square and window techniques.

Keywords Finite impulse response · L_1-error criterion · Stopband attenuation · Least-square · Window method

1   Introduction

Digital filtering is an important area of research from last few decades. Digital filters are applied in a variety of engineering applications such as, signal processing, communication, control systems and many more. They carry out the process of attenuating some band of frequencies and allow some frequencies to pass through them. The digital filter is implemented by the discrete convolution of input signal and filter coefficients. These are classified as: Finite Impulse Response (FIR) and Infinite Impulse Response (IIR) [1], [2]. In this work, we intend to design the FIR filter with optimal filter coefficients using the L_1 algorithm explained below.

A. Aggarwal(✉) · T.K. Rawat · M. Kumar · D.K. Upadhyay
Department of Electronics and Communication Engineering,
Netaji Subhas Institute of Technology, Sector-3, Dwarka, Delhi 110078, India
e-mail: {16.apoorva,tarundsp,manjeetchhillar}@gmail.com, upadhyay_d@rediffmail.com

© Springer International Publishing Switzerland 2016   335
S. Berretti et al. (eds.), *Intelligent Systems Technologies and Applications*,
Advances in Intelligent Systems and Computing 384,
DOI: 10.1007/978-3-319-23036-8_29

FIR filter design, being an approximation problem, determines the filter solution by approximating the frequency response of designed filter to the ideal response. Such approximation techniques involves, the least-square method [3, 4], equiripple design [5, 6], windowing techniques, frequency-sampling method and maximally-flat design. The most applied techniques are based on the L_2 (least-square) [7] and L_∞ (minimax) norms. In L_2-norm based filters the stopband attenuation (A_{stop}) is high with flat passsband on the cost of high overshoot at the discontinuity of ideal function. Whereas, the L_∞-norm based filters yields equal magnitude ripples in both passband and stopband.

In 2006, Grossmann and Eldar proposed a new method for the filter design, based on the L_1-norm [8]. The linear phase FIR filters are designed exploring the problems of differentiability and uniqueness of solution associated with the L_1 approach [9, 10, 11]. Considering type-II, the designed L_1 filters possess a flat passband and stopband with high A_{stop}. On its comparison with other techniques, the L_1 filter features with a higher A_{stop} than L_2 and L_∞-norms. It also eliminates the drawback of the high overshoots at the point of discontinuity. Thus, implementing the L_1 method provides a better solution in the field of filter design [12].

In this paper, the optimal LP and BP FIR filters are designed with type-II filter response using the L_1-method so as to obtain a symmetric even length filter with high A_{stop} and a flat passband response. The purpose of designing type-II filter is to obtain an even length filter which are necessary to be implemented in some applications and are not possible to be designed using the generalized type-I filter response. The obtained results are compared to the type-II equiripple filters, least-squares design and with filters designed using the kaiser window.

The rest of the paper is organized as follows: In Section 2, the framework of type-II filter design problem in mathematical formulated. The problem specific employed L_1 algorithm is described in Section 3. The simulation results and analysis are presented in Section 4. Finally, Section 5 concludes the paper.

2   Problem Formulation

In this section, the problem of designing the FIR filters with symmetric even length (N) impulse response (Type-II) is considered. The amplitude response of such filters has zero magnitude at $\omega = \pi$. Due to this, the design of high-pass and band-stop filters is not possible with type-II frequency response. In this paper,the design of type-II LP and BS FIR filters using the L_1 method is proposed. Here, the design problem is considered as an optimization problem where the frequency response of type-II filter, $H(\omega)$ is approximated to the ideal frequency response, $H_{id}(\omega)$. The ideal response for the LP and BP filters are given as

$$H_{id_{LP}}(\omega) = \begin{cases} 1, & \omega \in [0, \omega_c] \\ 0, & \omega \in (\omega_c, \pi] \end{cases} \tag{1}$$

and

$$
H_{idBP}(\omega) = \begin{cases} 1, & \omega \in [\omega_{c_1}, \omega_{c_2}] \\ 0, & \omega \in [0, \omega_{c_1}) \cup (\omega_{c_2}, \pi] \end{cases} \tag{2}
$$

The transfer function of the approximating filter, $H(z)$ is specified as

$$
H(z) = \sum_{n=0}^{N-1} h(n) z^{-n} \tag{3}
$$

The frequency response, derived from the transfer function is written as

$$
H(\omega) = \sum_{n=0}^{N-1} h(n) e^{-j\omega n} \tag{4}
$$

Eq. (4) is written as

$$
H(\omega) = \tilde{H}(\omega) e^{-j\omega \frac{N-1}{2}} \tag{5}
$$

where $\tilde{H}(\omega)$ is the amplitude response of the filter. Considering Type-II linear phase FIR filter with even length and symmetric coefficient, $\{h(n) = h(N-1-n), 0 \le n \le N-1\}$, the the amplitude response is defined as [2]

$$
\tilde{H}(\omega) = 2 \sum_{n=1}^{M} h[M-n] \cos\left[\omega\left(n - \frac{1}{2}\right)\right] \tag{6}
$$

where $M = N/2$. Assigning $b(n) = 2h[M-n], 1 \le n \le M$ and writing $\tilde{H}(\omega)$ as a function of ω and filter coefficients, $\mathbf{b}$ (where $\mathbf{b} = (b(1), b(2), \ldots, b(M))$, we get

$$
\tilde{H}(\omega) = \tilde{H}(\omega, \mathbf{b}) = \sum_{n=1}^{M} b(n) \cos\left[\omega\left(n - \frac{1}{2}\right)\right] \tag{7}
$$

Various fitness function employed for the filter design as an approximation problem are

1. Weighted Least-Squares (LS)

$$
\|E(\omega, \mathbf{b})\|_2 = \int_0^\pi W(\omega) \left|\tilde{H}(\omega, \mathbf{b}) - H_{id}(\omega)\right|^2 d\omega \tag{8}
$$

2. Weighted Chebyshev

$$\|E(\omega, \mathbf{b})\|_{\infty} = \max_{\omega \in [0,\pi]} \left\{ \int_0^{\pi} W(\omega) \left| \tilde{H}(\omega, \mathbf{b}) - H_{id}(\omega) \right| d\omega \right\} \qquad (9)$$

3. Weighted L_1-norm

$$\|E(\omega, \mathbf{b})\|_1 = \int_0^{\pi} W(\omega) \left| \tilde{H}(\omega, \mathbf{b}) - H_{id}(\omega) \right| d\omega \qquad (10)$$

where $W(\omega)$ is the weighting function and $\|.\|$ denotes the norm of the function. $E(\omega, \mathbf{b})$ is the error, measured between the approximated filter response, $\tilde{H}(\omega, \mathbf{b})$ and the ideal response, $H_{id}(\omega)$, defined as

$$E(\omega, \mathbf{b}) = \tilde{H}(\omega, \mathbf{b}) - H_{id}(\omega) d\omega \qquad (11)$$

$$= \sum_{n=1}^{M} b(n) \cos \left[\omega \left(n - \frac{1}{2} \right) \right] - H_{id}(\omega) d\omega \qquad (12)$$

The error function given in eq. (10) represents the fitness function to be minimized using the L_1 method. It evaluates the fitness function and optimize the filter coefficients. The employed algorithm for the purpose of FIR filter designing is explained in next section.

3   The L_1 Algorithm

The L_1 optimization technique remained unexplored for many years in the field of filter designing due to the above mentioned reasons. With its implementation for the optimization of filter coefficients, the error function turns out to be solvable for the case of FIR filter design. The motive behind exploring and implementing L_1 optimization is due to the smaller overshoot it yields around the discontinuity as compared with the most efficient techniques, minimax and least-squares [8]. In passband, the L_1 based filter results in a flatter response than least-square which happens to be its most desirable property. The design and optimization of linear phase FIR filters using L_1 technique and its characteristic comparison with the minimax method is being demonstrated in [13].

The linear L_1 approximation method of continuous functions defined over an interval by a finite number of basis functions was proposed in [14]. This algorithm computes the optimal coefficients of basis functions with the use of modified Newton method. This estimate was generalized and the modified Newton method was developed for the calculation of L_1-based filter coefficients in [15]. This method is described here for the design of type-II symmetric FIR filter.

The algorithm applied to formulate the L_1 problem as a linear approximation problem [15] demands for the evaluation of first and second order derivative of the error function defined in eq. (10). The nth component of gradient (first-order derivative) at $\mathbf{b}$ is given by

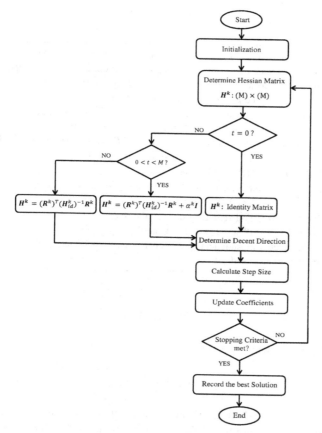

Fig. 1 Flowchart for the L_1 based FIR filter design method.

$$g_n(\mathbf{b}) = \langle \cos(n\omega), \text{sgn}(E(\omega, \mathbf{b})) \rangle \tag{13}$$

where $\text{sgn}(E(\omega, \mathbf{b}))$ gives the Signum result of the function $E(\omega, \mathbf{b})$.

The Hessian matrix (second-order derivative) is computed over the entire digital frequency, which takes one of three forms according to the number of zeros of $E(\omega, \mathbf{b})$ and is given by

$$\mathbf{H}(\mathbf{b}) = \mathbf{R}^T \mathbf{H}_{id}^{-1} \mathbf{R} \tag{14}$$

here $\mathbf{R}$ is a $t \times M$ matrix with $\mathbf{R}_{ij} = \cos((j - 0.5)z_i)$ and z_i denotes the zero of $E(\omega, \mathbf{b})$ at i^{th} position, equal to $\frac{(2i-1)\pi}{2M}$, $i = 1, 2, \ldots, M$. $\mathbf{H}_{id} = \text{diag}\{d_1, \ldots, d_t\}$ having $d_i = \frac{2\mathbf{W}(z_i)}{E'(\omega, \mathbf{a})}$.

The modified Newton's method consist of the iterations that generates a sequence of coefficients $\mathbf{b}^k$

Table 1 Algorithm for Filter Design using L_1 norm

Step 1 Design the ideal frequency response defined in (1). Set $M = (N + 1)/2$.

Step 2 Calculate initial vector $\mathbf{b}^1 \in \Re^M$, set stopping condition factor, $\epsilon > 0$, step-size selection parameters as $0 < \sigma < 1/2$, $0 < \beta < 1$ and for the control of Hessian matrix, set $\delta_1 > 0$, $\delta_2 > 0$ and $\mu > 0$. Set $k = 1$ to determine $\mathbf{b}^1$.

Step 3 Determine the Hessian matrix $\mathbf{H}^k$ of size $(M) \times (M)$, based on the value of t.

    i. If $t = 0$ or $\mathbf{H}_{id}$ is singular, then set Hessian matrix as identity matrix.

    ii. If $t \geq M$, $\mathbf{H}_{id}$ is non-singular and rank $(\mathbf{R}^k = M)$, then set $\mathbf{H}^k = (\mathbf{R}^k)^T (\mathbf{H}_{id}^k)^{-1} \mathbf{R}^k$.

    iii. If $0 < t < M$, $\mathbf{H}_{id}$ is non-singular and rank $(\mathbf{R}^k < M)$, then set $\mathbf{H}^k = (\mathbf{R}^k)^T (\mathbf{H}_{id}^k)^{-1} \mathbf{R}^k + \alpha^k \mathbf{I}$, where $\alpha^k > 0$.

Step 4 Determine the descent direction $\mathbf{d}^k$ defined in eq. (16), that obtains the unique solution.

Step 5 Stop if $|(\mathbf{d}^k)^T \mathbf{g}^k|$ is less than given threshold, ϵ.

Step 6 Calculate step-size, α^k detemined using Armijo rule.

Step 7 Set $\mathbf{b}^{k+1} = \mathbf{b}^k + \alpha^k \mathbf{d}^k$ and $k = k + 1$. Goto Step 2.

Step 8 The M coefficients are stored and the frequency response of designed Nth order FIR LP and BP filter is calculated.

$$\mathbf{b}^{k+1} = \mathbf{b}^k - \alpha^k [\mathbf{H}^k]^{-1} \mathbf{g}^k \tag{15}$$

assuming that the Newton direction

$$\mathbf{d}^k = -[\mathbf{H}^k]^{-1} \mathbf{g}^k \tag{16}$$

is a descent direction, where $\mathbf{g}^k$ is the gradient of function at $\mathbf{b}^k$, α^k is the step size, determined according to the Armijo rule [16] and $\mathbf{H^k}$ is the Hessian matrix of $\|E(\omega, \mathbf{b})\|_1$. Solving $\mathbf{d}^k$, the descent direction (also called the gradient method), involves the solution of the linear equations with M unknowns (the length of $\mathbf{d}^k$). To reduce these computations, the special structure of the matrix $\mathbf{H}^k$ in eq. (14) is exploited based on the number of zeros of $E(\omega, \mathbf{b})$. This is explained in the steps for the design of FIR LP and BP filter based on L_1 criterion, summarized in Table 1. The process flow chart is pictured in Fig. 1.

4 Simulation Results and Analysis

This section presents extensive simulations performed using the MATLAB v.7.13 platform on intel core(TM) i5 CPU, 3.20GHz with 4 GB RAM for the design of type-II 25th order FIR LP and BP filters.In order to demonstrate the superiority of the proposed design based on L_1-criterion, comparative analysis is carried out with the equiripple, least-square and windowed methods. The design parameters values used in the L_1 algorithms are as follows, $\epsilon = 10^{-6}$, $\sigma = 10^{-3}$, $\beta = 0.5$, $\delta_1 = 10^{-15}$, $\delta_2 = 10^{15}$ and $\mu = 10^{-10}$. The design examples are analyzed below.

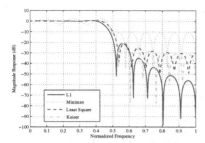

Fig. 2 Magnitude Response (dB) for 25th order FIR LPF.

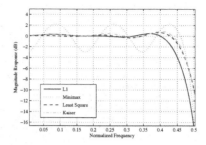

Fig. 3 Enlarged Passband Response (dB) for 25th order FIR LPF.

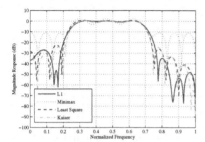

Fig. 4 Magnitude Response (dB) for 25th order FIR BPF.

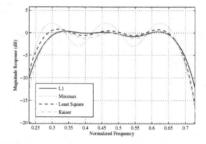

Fig. 5 Enlarged Passband Response (dB) for 25th order FIR BPF.

Filter specification specified are: For LPF, passband frequency, $\omega_p = 0.474\pi$, stopband frequency, $\omega_s = 0.493\pi$ and cut-off frequency, $\omega_c = 0.4835\pi$. For BSF, passband frequency, $\omega_{p_1} = 0.28\pi$, $\omega_{p_2} = 0.68\pi$, stopband frequency, $\omega_{s_1} = 0.23\pi$, $\omega_{s_2} = 0.28\pi$ and cut-off frequency, $\omega_{c_1} = 0.25\pi$, $\omega_{c_2} = 0.7\pi$. The magnitude response in dB of the proposed 25th order LPF and BSF is shown in Fig. 2 and 4, respectively. The response is plotted with the response obtained using already existing design methods, minimax, least-square and windowed method. Table 2 and 3 gives the optimized filter coefficients for all applied methods for the LFP and BSF design, respectively. To carry out the comparison, statistical results are analyzed and reported in Table 4. It is observed from Table 4 that the minimum stopband attenuation incurred with the L_1-method is -21.98 and -34.85, for LPF and BPF, respectively. From the stopband profile of the L_1 based filters, it can be concluded that the highest stopband attenuation is obtained with the proposed design. Furthermore, the passband ripples are 0.4672 dB and 0.3599 dB, for LPF and BPF designs, respectively. These obtained values are least among all the designs which results in the flattest passband with the L_1-based filters. The flatness in passband can be depicted from the enlarged view in Fig. 3 and 5.

Table 2 Optimized coefficients of 25th order FIR LPF filter.

Optimized Coefficients	L_1 Criterion	Minimax	Least Square	Kaiser Window
$h(0) = h(25)$	−0.004304821875159	0.015009154932070	0.003346236484176	0.003353949812098
$h(1) = h(24)$	−0.009945569482198	−0.159733218946615	−0.025832583651239	−0.026384058910212
$h(2) = h(23)$	0.002892064991698	−0.019972687908533	−0.007000222174646	−0.007087353251721
$h(3) = h(22)$	0.021425807571245	0.033344364352309	0.030978162778327	0.031656658601014
$h(4) = h(21)$	−0.004771401539193	0.013239722254575	0.012372854572430	0.012579827879823
$h(5) = h(20)$	−0.039463807888077	−0.040699493740147	−0.038297116225617	−0.039151796729370
$h(6) = h(19)$	−0.003076249234356	−0.021837804969918	−0.020951741171072	−0.021352353618087
$h(7) = h(18)$	0.057415462699534	0.051858820599784	0.050174640982908	0.051307395738706
$h(8) = h(17)$	0.019794351740752	0.037642868368773	0.036871568337463	0.037631954977186
$h(9) = h(16)$	−0.083462683176473	−0.075416760140064	−0.074476032718155	−0.076163795722620
$h(10) = h(15)$	−0.056173676201219	−0.077793990820229	−0.077508700899409	−0.079182787368566
$h(11) = h(14)$	0.179179360425898	0.161468696856027	0.161077894993653	0.164710418330001
$h(12) = h(13)$	0.424754251606035	0.438344605836526	0.440344846920456	0.448081940261748

Table 3 Optimized coefficients of 25th order FIR BPF filter.

Optimized Coefficients	L_1 Criterion	Minimax	Least Square	Kaiser Window
$h(0) = h(25)$	0.013080208237528	0.062876101398056	0.015422936597621	0.024487776955131
$h(1) = h(24)$	−0.005575502498038	−0.116269220084882	−0.015287846989148	−0.005579104536685
$h(2) = h(23)$	−0.032670100914959	−0.085345777430859	−0.051481378623170	−0.049445991844001
$h(3) = h(22)$	0.001530829610940	−0.039914129234188	0.001827636127389	−0.000997752658279
$h(4) = h(21)$	−0.011134308330060	−0.001174900404477	−0.011267776603452	−0.018418438692279
$h(5) = h(20)$	−0.018306785726551	−0.005719565444857	−0.002754812073954	−0.012640620625559
$h(6) = h(19)$	0.075542485873412	0.081156678895356	0.093059429848611	0.086412261623438
$h(7) = h(18)$	0.032384920124126	0.042306512734840	0.022096288818837	0.025221316879272
$h(8) = h(17)$	−0.007921998215940	−0.024729352404448	−0.01246705795282	−0.004696219641479
$h(9) = h(16)$	0.048396103447932	0.041439930403364	0.044780368623445	0.051389148435767
$h(10) = h(15)$	−0.195302535995375	−0.196488532016674	−0.210850541558666	−0.194403327017152
$h(11) = h(14)$	−0.221064614224406	−0.241391281601754	−0.223813361579707	−0.214944678903213
$h(12) = h(13)$	0.328305105982610	0.339512411966533	0.331916468646392	0.303651115520570

Table 4 Statistical results for the 25th order FIR LP and BP filter.

Filter	Method	Stopband Attenuation (dB)				Passband ripple
		Minimum	Mean	Variance	Standard deviation	(dB)
Low-Pass	L_1 Criterion	-21.98	-24.8977	-35.1392	-17.5556	0.4672
	Least-Square	-21.14	-23.2482	-35.9720	-17.9926	0.6806
	Kaiser Window	-21.04	-23.7017	-35.7562	-17.8626	0.8746
	Minimax	-10.80	-14.1637	-38.1315	-19.0467	2.2081
Band-Pass	L_1 Criterion	-34.85	-20.1843	-30.4865	-15.2441	0.3599
	Least-Square	-25.88	-20.7526	-31.8352	-15.9122	0.7977
	Kaiser Window	-24.62	-21.0611	-31.6042	-15.8043	0.3232
	Minimax	-10.82	-13.2948	-36.1933	-18.0896	2.1980

5   Conclusion

In this paper, the efficient design of type-II FIR LP and BP filter using the L_1-method is presented. The requirement of even length filters for specific application can be fulfilled using the proposed type-II filter design. The performance assessment for the designed filters is expressed in terms of minimum stopband attenuation and highest passband ripple. The obtained results with the L_1-method attained highest stopband attenuation and the passband with least ripples as compared to the renowned minimax, least-squares and windowed method. This method can be applied to design 2-D filters to enhance their applicability in other fields of engineering like image processing. In addition, the method can also be used to design digital differentiator and Hilbert transformer.

References

1. Mitra, S.K.: Digital Signal Processing: A Computer-based Approach. Tata Mc-Graw Hill (2008)
2. Rawat, T.K.: Digital Signal Processing, 1st edn. Oxford University Press (2014)
3. Vaidyanathan, P.P., Nquyen, T.Q.: Eigenfilters: A new approach to least squares FIR filter design and appli. IEEE Trans. Circuits Syst. **22**, 943–953 (1975)
4. Ramachandran, R.P., Sunder, S.: A unified and efficient least-squares design of linear-phase nonrecursive filters. Signal Proc. **36**, 41–53 (1994)
5. Parks, T.W., McClellan, J.H.: Chebyshev approximation for non-recursive digital filters with linear phase. IEEE Trans. Circuits Theory, 189–194 (1972)
6. Antoniou, A.: New improved method for the design of weighted chebyshev nonrecursive digital filters. IEEE Trans. Circuits Syst. **CAS–30**, 740–750 (1983)
7. Kumar, M., Rawat, T.K.: Optimal design of FIR fractional order differentiator using cuckoo search algorithm. Expert Syst. with Appli. **42**, 3433–3449 (2015)
8. Grossmann, L.D., Eldar, Y.C.: The design of optimal L_1 linear phase digital FIR filters. In: Proc. IEEE Int. Conf. Acoustics, Speech, Signal Process. (ICASSP), vol. 3, pp. 884–887 (2006)
9. Rice, J.R.: The Approximation of Functions, vol. I. Addison- Wesley, MA (1964)
10. Chen, C.K., Lee, J.H.: Design of high-order digital differentiatiors using LI error criteria. IEEE Trans. Circuits Syst., Analog Digit. Signal Process. **42**, 287–291 (1995)
11. Yu, W.S., Fong, L.K., Chang, K.C.: An L_1-Approximation based method for synthesizing FIR details. IEEE Trans. Circuits Syst. II, Analog Digit. Signal Process. **39**, 578–561 (1992)
12. Aggarwal, A., Rawat, T.K., Kumar, M., Upadhyay, D.K.: Optimal design of FIR high pass filter based on L_1 error approximation using real coded genetic algorithm. Eng. Sci. Tech., an. Int. J. (2015). doi:10.1016/j.jestch.2015.04.004
13. Aggarwal, A., Kumar, M., Rawat, T.K.: L_1 error criterion based optimal FIR filters. In: Annual IEEE India Conference (INDICON) (2014)
14. Watson, G.A.: An algorithm for linear L_1 approximation of continuous functions. IMA J. Numer. Anal. **1**, 157–167 (1981)
15. Grossmann, L.D., Eldar, Y.C.: An L_1-Method for the Design of Linear-Phase FIR Digital Filters. IEEE Trans. Signal Process. **55**(11), 5253–5266 (2007)
16. Yarlagadda, R., Bednar, J.B., Watt, T.L.: Fast algorithms for l_p deconvolution. IEEE Trans. Acoust., Speech, Signal Process. **33**, 174–184 (1985)

Application of Hybrid Neuro-Wavelet Models for Effective Prediction of Wind Speed

V. Prema, K. Uma Rao, B.S. Jnaneswar, Colathur Arvind Badarish,
Patil Shreenidhi Ashok and Siddarth Agarwal

Abstract Severe energy crisis and depletion of fossil fuels necessitates more number of installations of wind farms. Accurate wind forecast is crucial in the efficient utilization and power management of wind farms connected to a grid or in conjunction with other sources such as solar, DG, battery, etc. This paper proposes a hybrid neuro-wavelet predictive tool to predict wind speed which combines the advantages of both wavelet decomposition and neural network. Wavelet decomposition is used to filter out the high frequency outliers in the wind speed, thus making a smooth data to make the prediction accurate. The filtered data is used to train the neural network. Four different models are proposed. NAR-TS model and NAR-Wavelet models are univariate models with past values of wind speed as input. In NAR-TS model time series values are directly applied as input to neural network, whereas in NAR-Wavelet model input to the neural network is the wavelet decomposed data. In a similar way NARX-TS and NARX-Wavelet models are developed with multivariate neural network, where the inputs are air temperature, relative humidity and wind speed which is the feed back. Each of these models are used to predict 4.5 hours ahead and 18 hours ahead predictions. The Mean Average Percentage Error (MAPE) values are calculated for each model and the results are compared.

1 Introduction

Energy crisis is growing as the greatest threat for mankind. This is leading to the installation of more renewable energy sources as alternatives to fossil fuels. When it comes to renewable energy sources like wind and solar, accurate prediction is very important. Considering the fact that they are erratic in nature. When wind power and solar power are integrated with the main grid or with other sources, short-term forecasting is necessary to have an optimum power management of the system[2].

V. Prema(✉) · K. Uma Rao · B.S. Jnaneswar · C.A. Badarish · P.S. Ashok · S. Agarwal
R.V. College of Engineering, Bangalore, India
e-mail: premav@rvce.edu.in

© Springer International Publishing Switzerland 2016   345
S. Berretti et al. (eds.), *Intelligent Systems Technologies and Applications*,
Advances in Intelligent Systems and Computing 384,
DOI: 10.1007/978-3-319-23036-8_30

Literature shows various works for prediction of wind power and wind speed[1], [3], [4]. This includes physical models and statistical models. For Physical models, geographical features of the wind farm are taken as inputs [5]. These models are more suited for long term forecast than short-term forecast. One class of statistical models include time series models such as ARMA, ARIMA, exponential smoothing, etc [6][7]. In these models, models based on the historical data are developed. The past behaviour is projected onto the future using the developed time-series equations. In another class of time-series models, such as triple exponential and decomposition models, the actual data is decomposed into seasonal and trend components are these components are modelled separately [8].

The intermittency and highly erratic nature of wind speed make the above mentioned statistical methods incompetent for accurate prediction of wind speed. Thus researchers are more keen towards heuristic methods such as neural network, fuzzy logic, support vector regression techniques, etc. Researches are on with hybrid models also. Even with all these developments, it is very difficult to achieve a percentage error less than 10% in wind speed prediction. This paper proposes a hybrid mode for wind speed prediction using wavelet decomposition and Artificial Neural Network (ANN).

The data used for this work is collected from a wind farm in Bagalkote, Karnataka. The data consist of wind speed, air temperature and relative humidity recorded at every minute. This is averaged to get samples at every 10 minute interval. The wind speed data is decomposed using wavelet. The neural network is trained with this decomposed data. A data set of 1000 samples is considered for the model building and prediction. The neural network is trained with 800 samples and 200 samples are used for testing. Univariate model is developed using NAR (Non-linear Auto-regression) network by giving previous values of wind speed as the input. Multivariate prediction models have been developed using NARX (Non-linear Auto-regression with exogenous input) network by giving air temperature, relative humidity and wind speed as inputs. These proposed models are compared with simple ANN models and the MAPE (Mean Absolute Percentage Error) values are compared.

2   Wavelet Decomposition

The primary reason for choosing wavelet transformation to decompose wind speed is that wind speed is a non-stationary signal and wavelet transform is the best tool to analyse non-stationary signals. To understand the non-stationary behaviour of wind speed, let us view the wind speed plot shown in fig:1. If the pattern of wind speed is considered as a signal, it can be seen that there are number of high frequency, short duration pulses. It is not easy to figure out the time at which these high frequency variations occur. In such a situation, wavelet decomposition is the best way. In wavelet decomposition, the signal is decomposed into different sub-series. Mallat algorithm [9] is used in this paper for the wavelet decomposition. In this algorithm, the signal is decomposed into several levels of high and low frequency signals by passing the signal through different filters as show in fig:2. In each level of decomposition, an Approximation signal (C), which contains low frequency components and Detailed signal (D), which contains

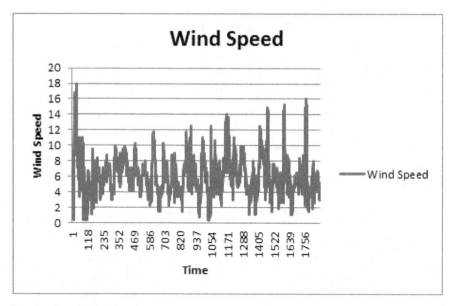

Fig. 1 Plot of Wind Speed

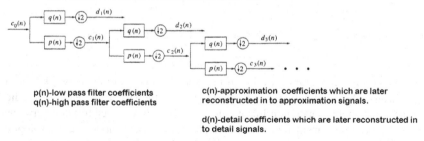

Fig. 2 Wavelet Decomposition

high frequency components are obtained. In this paper, a three level discrete wavelet transform is performed and components from each level have been fed to the ANN to train the network. It was found that the approximation signal in the 3rd level (A3) gives best result, thus only those models are discussed in the paper.

3   Neural Network Models

General time series forecasting methods are well suited when the data is linear. But for non-linear data like wind speed, techniques which can model non-linear data are needed. Neural network is one such technique where input data need not be

NAR neural network

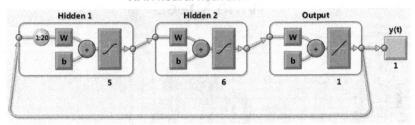

Fig. 3 Non-Linear AutoRegressive Neural Network

NARX neural network

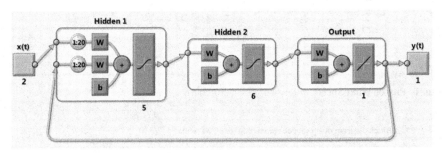

Fig. 4 Non-Linear AutoRegressive Neural Network with exogenous Input

linearly varying. In neural network, a time series data is fed to the network to train the neurons. Once the network is properly trained, the system is ready to predict future values. In this paper, non-linear auto regressive (NAR) feed forward networks shown in fig: 3are considered. Univariate and multivariate models are developed. In univariate model, only one variable in involved in modelling. The past values of wind speed are used to predict the future. In multivariate modelling three variable are fed as input to the network. The input variables chosen are air temperature, relative humidity and past wind speed. Non-linear Auto Regressive network with exogenous input (NARX) as shown if fig: 4 is implemented in MATLAB. The models are trained using Levenberg-Marquardt training algorithm.

4 Error Measures

When there are more than one model developed, a comparative analysis can be made by calculating the accuracy of each model and comparing them. The accuracy of any predictive model can be determined only by choosing appropriate error measures. In this paper, different models are compared using three error measuring parameters,

Mean Absolute Error (MAE) and Mean Absolute Percentage Error (MAPE) which can be defined as given below.

4.1 Mean Absolute Error (MAE)

MAE is defined as,

$$MAE = \frac{\Sigma \, |X_t - F_t|}{N} \times 100$$

Where Xt is the actual data at time t
Ft is the forecast at time t
N is the number of samples used in computing the error
MAE is an absolute measure. The range of MAE can be from 0 to infinity. One cannot comment on the accuracy of the model by analysing MAE. The significance of error depends on the actual value. As an example, for an actual value of 1000, an error of 10 may be negligible, whereas if the actual value is 50, an error of magnitude 10 is very huge.

4.2 Mean Absolute Percentage Error (MAPE)

MAPE can be defined as,

$$MAPE = \frac{1}{N} \times \Sigma \frac{|X_t - F_t|}{X_t} \times 100$$

MAPE is a relative measure which reflects error as the percentage of actual data. Hence the accuracy of the model can be easily judged. The MAPE is only 1% if the error is 10 for an actual value of 1000. But MAPE is 20% if the actual value is 50. Thus MAPE gives the relative size of errors. Because of this fact, MAPE is very effective in comparing various forecasting techniques. The disadvantage of MAPE is that MAPE does not give an idea whether the predicted values are more negatively deviated or positively deviated, as absolute error is considered. MAPE can go to very high value if the actual value is very small. MAPE can be greatly influenced by outliers. As an example, if the actual value is 20 and predicted value is 15, MAPE is 25%, whereas MAPE is 33.33% if actual value is 15 and predicted value is 20. Considering the above facts, MAPE is calculated for the different models developed in this paper for the comparison of the models.

5 Wind Speed Prediction Models

Two univariate models with NAR network and two multivariate models with NARX network are developed using Neural Network toolbox available in MATLAB.

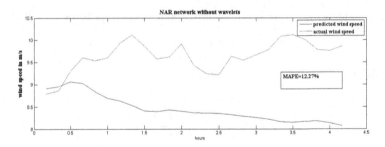

Fig. 5 NAR Network without wavelet for 25 samples ahead prediction

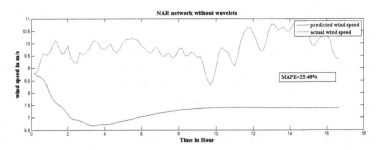

Fig. 6 NAR Network without wavelet for 100 samples ahead prediction

5.1 NAR with Time Series Input (NAR-TS) Model

In this model, the input to the neural network is the time series data which is obtained
from the wind farm. Multi-step ahead prediction is done for 25 out-of-samples and
100 out-of-samples. The results are shown in fig:5 and fig:6 respectively. The MAPE
for the models were found to be 12.27% and 25.48% respectively. It can be seen that
the MAPE reduces with increase in number of out-of-samples.

5.2 NAR with Wavelet Decomposed Input (NAR-Wavelet)

In this model, the input to the neural network is the wavelet decomposed values of
wind speed. The proposed model in this paper employs 3rd level discrete wavelet
transform and uses three daubechies wavelet db1, db2 and db3. All the approximation
and detail coefficients in the three levels were trained with neural network model.
It was found that modelling with the third level approximation signal gave the least
error. This is because at this level the high frequency outliers are removed and the
signal becomes smother and simpler for prediction. Multi-step ahead prediction is
done with 25 out-of-samples ad 100 out-of-samples. The results are shown in fig:
7 and fig: 8 respectively. The MAPE for the models were found to be 10.85% and
14.41% respectively. It can be observed that the MAPE is significantly reduced when
compared with the results without wavelet decomposition.

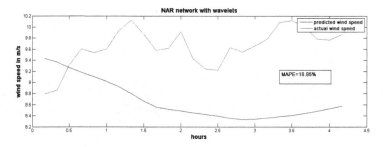

Fig. 7 NAR Network with wavelet for 25 samples ahead prediction

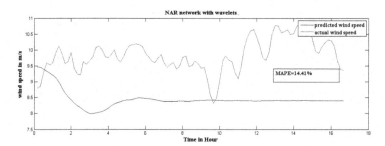

Fig. 8 NAR Network with wavelet for 100 samples ahead prediction

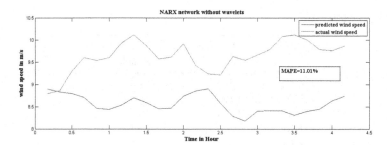

Fig. 9 NARX Network without wavelet for 25 samples ahead prediction

5.3   NARX with Time Series Input (NARX-TS)

In this model, the input to the neural network is the time series data which is obtained from the wind farm. Air temperature and Relative Humidity are considered as inputs. along with these inputs, a feed back of output, which is wind speed is also taken as an input. Multi-step ahead prediction is done for 25 out-of-samples and 100 out-of-samples. The results are shown in fig:9 and fig:10 respectively. The MAPE for the models were found to be 12.27% and 25.48% respectively. It can be seen that the MAPE increases with increase in number of out-of-samples.

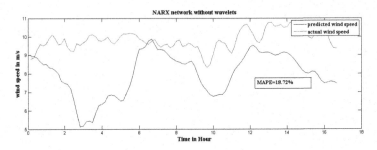

Fig. 10 NARX Network without wavelet for 100 samples ahead prediction

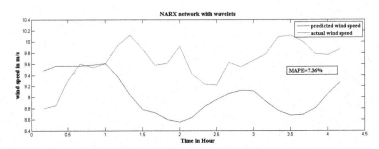

Fig. 11 NARX Network with wavelet for 25 samples ahead prediction

Table 1 Comparison of Models

Model	MAPE for 25 Samples	MAPE for 100 samples
NAR-TS	12.27	25.48
NAR-Wavelet	10.86	14.41
NARX-TS	11.01	18.72
NARX-Wavelet	7.37	11.17

5.4 NARX with Wavelet Decomposed Input (NARX-Wavelet)

In this model, the input to the neural network is the wavelet decomposed values of wind speed. Multi-step ahead prediction is done with 25 out-of-samples ad 100 out-of-samples. The results are shown in fig: 11 and fig: 12 respectively. The MAPE for the models were found to be 10.85% and 14.41% respectively. It can be observed that the MAPE is significantly reduced when compared with the results without wavelet decomposition.

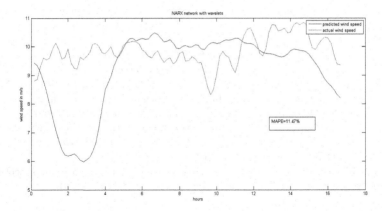

Fig. 12 NARX Network with wavelet for 100 samples ahead prediction

6 Conclusion

This paper proposes a hybrid neuro-wavelet method to predict wind speed 100 samples (18 hours) ahead. A three level wavelet decomposition is done on the available time series data. This decomposed data is fed as input to neural network. The neural network is modelled using neural network toolbox in MATLAB. The training algorithm chosen is Levenberg-Marquardt algorithm. Univariate (NAR) and multivariate (NARX) neuro wavelet models were developed. These models are compared with plain neural network models where the input is the un-filtered time series data. It can be observed from the results that NARX network with wavelet decomposed input give the least MAPE of 7.37% for 25 samples (4.5 hours) ahead prediction and 11.37% for 100 samples (18 hours) ahead prediction.

Acknowledgments The authors gratefully acknowledge the contributions of Dr. Suresh H Jangamshetti, Dept of EEE, Basaveshwar Engineering college, Bagalkot for providing the wind data without which this work would not have been possible.

References

1. Yuehui, H., Jing, L.: Comparative Study of Power Forecasting Methods for PV Stations, Master thesis, China Electric Power Research Institute (CEPRI), Beijing, China (2010)
2. Prema, V., Uma Rao, K.: Predictive models for power management of a hybrid microgrid A review. In: 2nd International Conference on Advances in Energy Conversion Technology (ICAECT 2014), pp. 185–197. Manipal Institute of Technology, Manipal (2014)
3. Saurabh, S., Hamidreza, Z.: A review of wind power and wind speed forecasting methods with different time horizons. In: 25th International Cosmic Ray Conference, Durban, South Africa (2010)
4. Foley, A.M., Leahy, P.G., Marvuglia, A., McKeogh, E.J.: Current methods and advances in forecasting of wind power generation. Renewable Energy **37**, 1–8 (2012)

5. Ma, L., Luan, S.Y., Jiang, C.W., Liu, H.L., Zhang, Y.: A review on the forecasting of wind speed and generated power. Renewable and Sustainable Energy Reviews **13**, 915–920 (2009)
6. George, S., Nikos, D.H.: An advanced Statistical Method for wind power Forecasting. IEEE Transactions on Power Systems **22**(1), 68–77 (2007)
7. Palomares-Salas, J.C., De la Rosa, J.J.G., Ramiro, J.G., Melgar, J.: ARIMA vs. Neural networks for wind speed forecasting. In: IEEE International Conference on Computational Intelligence for Measurement Systems and Applications, pp. 129–133 (2009)
8. Wan Ahmad, W.K.A., Ahmad, S.: Arima model and exponential smoothing method: a comparison. In: Proceedings of the 20th National Symposium on Mathematical Sciences (SKSM20), Malaysia, pp. 1312–1321 (2012)
9. Mallat, S.: A Theory for multiresolution signal decomposition : the wavelet representation. IEEE Pattern Analysis and Machine Intelligence **11**(7), 674–693 (1989)

Optimized Defect Prediction Model Using Statistical Process Control and Correlation-Based Feature Selection Method

J. Nanditha, K.N. Sruthi, Sreeja Ashok and M.V. Judy

Abstract Defects are the flaws in software development process that causes the software to perform in an unexpected manner and produce erroneous outputs. Detecting these defects is an important task to ensure the quality of the software product. Defect prediction models acts as quality indicators that helps in detecting the defective components in the early phases of software development cycle. These models leads to reduced rework effort, more stable products and improved customer satisfaction. It is hard to find the high risk components that are major contributors for the defects from large number of variables. Thus feature selection is a very important aspect associated with defect analysis. Here we propose a defect prediction model to control the quality of software products using statistical process control. The key contributors for building the prediction models are derived using Correlation and ANOVA based feature selection methods. The proposed model is evaluated using benchmark dataset and the results are promising when compared with standard classification models.

Keywords Feature selection · ANOVA · Correlation · Control charts · Defect analysis · Prediction models

1 Introduction

The world has moved on to a more sophisticated dimension where most of the things which required human interaction before are done by software now. Any errors in these systems can cause traumatic situation which reveals the importance of defect analysis. Defects are the errors which generate unexpected outcomes

J. Nanditha · K.N. Sruthi · S. Ashok(✉) · M.V. Judy
Department of Computer Science & I.T, Amrita School of Arts & Sciences,
Amrita Vishwa Vidyapeetham, Kochi, India
e-mail: {nandithasj,knsruthi1,sreeja.ashok,judy.nair}@gmail.com

© Springer International Publishing Switzerland 2016   355
S. Berretti et al. (eds.), *Intelligent Systems Technologies and Applications*,
Advances in Intelligent Systems and Computing 384,
DOI: 10.1007/978-3-319-23036-8_31

from the system [24, 25]. A system is prone to defects throughout its life time. The system may work perfect in the beginning since the defect involved in it may be invoked at a later point of execution in its cycle. It may then induce irrelevant outputs from then. A system evolves from smaller to larger and the largest as per the requirements change and thus the complexity. It may be developed by combining different data sources to meet the specifications requested. It is always a tedious task to analyze a huge system and interpret the errors that could have incorporated in the system [26]. But it is always important to ensure the quality of the results that are to be generated from the system. Software defect prediction ensures foreseeing the defects that would get involved and helps in reducing the cost by proper handling of the problem.

Data set contains large number of attributes. Removing irrelevant data from the data set can be a difficult and prolonged task. These subsidiary attributes may not contribute either to any of the characteristic features of the data under consideration or help in the prediction of the data. Data reduction is one of the methods in data mining task which helps to filter out the relevant features from the dataset. The relevance of attributes differs from the task for which data is analyzed and also on the quantity of the available data. If the number of attributes or instances is fewer the results obtained would tend to be less reliable or consistent. Data reduction helps to obtain reduced representation of data volume with the same or similar analytical results and contributes to the process of decision making. The hidden information from the dataset can be retrieved using classification and prediction analysis.

Here we propose a statistical process control flow where some of the statistical methods are used to reduce the data by analyzing their defect range. Prediction is done on the reduced set of data using a model that has been derived from the data set. Accuracy measures are used to analyze the efficiency of the proposed method to that of the existing algorithms.

The paper is organized as follows: Section 2 presents the summary of related work. Section 3 depicts the process flow of the defect prediction model and details the implementation steps. Section 4 illustrates the implementation results of the prediction model using benchmark dataset and the performance is analyzed and compared with existing classification models. Finally Section 5 concludes the paper with observations and future work.

2   Related Works

Numerous works in literature related to many feature selection methods and prediction models using data mining and machine learning techniques have motivated the proposed work. The defect prediction has been of great relevance since the software acquired importance in our lives. Many researchers have been working in this domain to find a proper solution for defect prediction. Data mining thus incorporated with software data paved a solution to the quality issues. The data collected for experiments appears to be highly complex and of very large size. Reduction of such complicated dataset is very important as we expect to produce a qualitative result from them. Data reduction using ANOVA has been commonly used for many similar processes. Some of the works include [3] cancer classification of bioinformatics data using ANOVA. The influence of the power of the test in theoretical aspects is discussed in [11]. Wa'el M. Mahmud, Hamdy N.Agiza, and Elsayed Radwan [13] proposed a reduction method on rough sets. PCA is a statistical technique used to obtain a set of linearly uncorrelated variables from the observations of correlated variables. Here multidimensional data are simplified to lower dimensions while retaining most of the information. However when PCA is used, all the attributes should be counted in the same unit of measurement. If any of the attribute differs in their unit of measurement, the data gives different principal components when transformed to original datasets.

The idea of clustering and classification for software prediction is mentioned in Data Mining Models for Software Defect Prediction [17].Statistical and machine learning techniques were introduced to consider the issues with software reliability in Fault Prediction Using Statistical and Machine Learning [18]. A model predicted using this method succeeded in proving the significance of the combination of statistical and machine learning methods. SVM deals with representing data as points in the space where a clear division (margin) is set within the different categories of data. This margin is set by constructing hyper planes, commonly in high dimension data. A software defect prediction model based on SVM and Naïve Bayesian is specified in [19] and [20]. The accuracy with which SVM can predict, supported the defect analysis, paved the way for more works in this area. Misuse and Anomaly detection using classification approaches are discussed by T.Subbulakshmi [14]. Classification approaches with kernel functions are used for detecting intruders.

The research paper [12] deals with the monitoring of occupational asthma using control charts. Classification is extensively used in various application domains: design of telecommunication service plans, fraud detection etc. Classification of large volume of data is maintained in the medical field .Decision tree classifiers are used extensively for diagnosis of diseases [5] [8, 9]. Support Vector Machine, Neural network, Naive Bayes, J48 etc are some of the commonly used classification algorithms. Neural network can be used to solve problems which appear to be more complex to solve using conventional technologies. They are used to solve problems that are hard to solve by computers due to the higher complexity of the algorithms to be used with them. E.g. pattern recognition and forecasting. Naive Bayes is based on Bayes' theorem and the theorem of total probability whereas J48 algorithm is based on decision tree. Naïve Bayes classifier assumes that any feature value is particularly independent of the value of another in the set [4].The characterization of Naive Bayes algorithm learns rapidly in various supervised classification problems [16].J48 predicts the missing values in the records from the values of the attributes that are already in the training set. It creates a decision tree based on a set of labeled data. Classification can be done either based on decision trees or also by generating set of rules from them [15] [7]. Accuracy of classification is usually calculated by determining the percentage of tuples that comes in a correct class using confusion matrix.

3   Proposed Method

Here we are proposing a defect prediction model using statistical process control and the performance is analyzed and compared with existing classification models like naïve Bayes and J48. The proposed system deals with extracting the relevant features using feature selection techniques like Pearson's correlation coefficient, and ANOVA which tests the dependency between various attributes. Control charts are used to analyze the performance of the key defect contributors derived from the feature reduction methods. Prediction rules are generated for each class labels. Accuracy measures helps in model evaluation. Root Cause analysis helps in defect prevention and quality control. The proposed model includes the following major steps

Step1: Data Preprocessing.
Step2: Feature/Attribute Selection using correlation (C-FS) / ANOVA (ANOVA-FS).
 Step3: Building Prediction Model using Statistical Process Control
Step4: Model Evaluation.
Step5: Quality control and Continuous Improvement

The execution flow of the proposed approach is depicted in Fig.1.

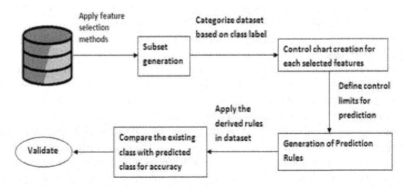

Fig. 1 Statistical Process Control Flow

3.1   Pre-processing of Data

Data preprocessing is done to remove noise or unwanted features from the dataset that is under consideration. The data collected for analysis from different sources may contain many irrelevant attributes. It can be also formed when different data has been integrated from different sources. Such unwanted attributes which may contribute to erroneous output from the data is to be removed. Data is thus transformed or reduced to overcome these defects and to produce relevant output [22]. Data reduction is done in the datasets to reduce the size of the dataset by removing the unwanted attributes from it, in such a way that the probability of obtaining the result is same as that when all the attributes remained in the set.

3.2   Feature or Attribute Selection

The major techniques for dimensionality reduction include feature selection and feature extraction where feature reduction refers to the mapping of the original high-dimensional data onto a lower-dimensional space [10]. Feature selection reduces number of features, removes irrelevant, redundant or noisy data and brings immediate results for applications [1].Feature selection selects a new set of attributes from the existing based on the extent to which the attribute or feature is relevant to the characteristics of the data of concern [21]. The features that do not contribute to the overall functionalities of the dataset, or the features which can be derived from other are removed based on such analysis and a new set of attributes are considered. Here we have used two methods for extracting relevant features from the data set. They are Correlation based feature selection and ANOVA based feature selection.

Correlation Based Feature Selection. Correlation defines the dependency of one feature to the other with its closeness to the other. The dependency can be negative or positive. Correlation coefficient points how strongly one attribute implies the other especially for numeric datasets. If an attribute or feature is more correlated to a class than it is related to another attribute in the dataset, it is regarded as non redundant attribute and can be used for further data analysis [2]. If we adopt the correlation between two variables as goodness measure, then a feature is good if it is highly correlated to the class but not highly correlated to any of the attributes [6].

Pearson's Correlation Coefficient. Equation (1) is used for finding the correlation. The closer the value of r gets to zero, the greater the variation the data points are around the line of best fit. However, the two data objects might have non-linear relationships.

$$r = \frac{\sum (X - \overline{X})(Y - \overline{Y})}{\sqrt{\sum (X - \overline{X})^2}\sqrt{\sum (Y - \overline{Y})^2}}$$

(1)

where, X and Y represent two data objects. The Pearson correlation is found for the attributes in the dataset by framing the correlation matrix. The attributes which comes in the range of boundary specified are selected from the dataset.

Anova Based Feature Selection. Anova is a technique which can be used to analyze experimental data where one or more than one response variables can be evaluated with regard to the various conditions derived from one or more classification variables. Anova checks for the variance of the attributes within the class and between the classes to find the F- ratio which specifies the significance of an attribute in a particular class. Equation (2) is used to calculate the F-ratio. Various combinations of response variables are taken and their means are analyzed. Anova can be done considering a single variable which affect the dependant variable or more than one variable which affect a dependant variable, regarded as one-way Anova and two-way Anova respectively [10]. The difference comes in the number of factors which affect the response variable. The measures sum-of-squares, degrees of freedom, and the F-ratio provides the information required to evaluate the relevance of an attribute in a class with regard to its variance from other attributes in the same class and those between the classes[3]. Anova ranks the attributes and avoids those which are of least significance which in turn produces a more efficient data set for prediction analysis [3].

$$F = \frac{MS_{Between}}{MS_{Within}} \tag{2}$$

where, MS between is Mean Square between sample groups and MS within represents mean square within sample groups.

3.3 Prediction Model – Statistical Process Control

The data that is derived from C-FS and ANOVA-FS is used for building the prediction model. Split the reduced dataset based on the class labels. For e.g.: if the class label has two values TRUE and FALSE, split the dataset into two subsets, one contain TRUE values and other set containing FALSE values. Control Charts are then used to analyze the data to extract the prediction rules. Control chart is a statistical tool which helps to analyze the variances in a process. It helps to distinguish between the common cause variation and special cause variation through plotting the process data. The graphical representation highlights the stability or instability of the data with regard to the occurrence of common or special causes of variations. It analyses the data process with regard to the lines of limit determined from the historical data. The lines of upper control limit, central line for the average and the lower control limit specifies the range in which the process data appears to be consistent and the range in which it appears to be unpredictable. These control limits are chosen so that almost all of the data points will fall within these limits as long as the process remains in-control.

An Individual Moving Chart (I-MR chart) plots individual observations and moving ranges over time for variables data. When the measurements are expensive or they have long life cycle, it would be difficult to subgroup these measurements from the observations. I-MR chart is used to identify the process centre and the variations in such scenarios. I chart or Individual chart takes each observation as a separate data point with subgroup size as 1. MR chart has a default value 2, which holds the range of value between two consecutive data points. I-MR chart plots these individual observations in one chart accompanied by the range of these observations from the consecutive data points for continuous data.

Each attribute is plotted into an I-MR chart, one for TRUE and the other for FALSE values. The graphs are analyzed to find the control limits, Upper Control limit (UCL) and lower control limit (LCL) for each attribute in each categories. Minimum value of LCL is set to 0 for positive attribute datasets. Prediction rules are generated based on the control limits of each attribute in each class label.

3.4   Evaluation using Confusion Matrix

After deriving rules from control chart limits, statistical process control prediction model (SPC) is built which is then applied to the dataset for predicting new class labels. The SPC model is then compared against the pre existing benchmark classification algorithms such as Naïve Bayes and Decision tree algorithms like J48[23].A confusion matrix is used to calculate the accuracy of these three classification algorithms. The rows in confusion matrix correspond to the known class labels in the data and the columns correspond to the predictions made by the model. This is depicted in Fig.2.

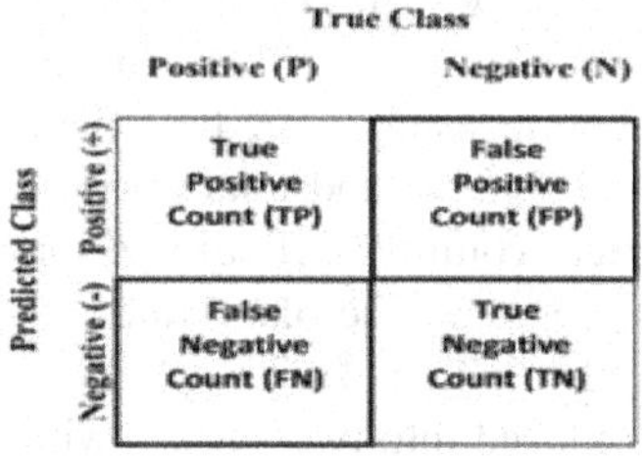

Fig. 2 Confusion Matrix

The accuracy of a prediction model denotes how many instances have been correctly classified. It is calculated using equation (3).

$$\text{Accuracy} = \frac{TP + TN}{TP + TN + FP + FN}$$

(3)

3.5   Quality Control and Continuous Optimization

Root cause analysis helps in defect prevention. It's a proactive approach and helps in quality control and continuously optimizing the performance of the system. The key defect contributors identified using the above mentioned feature selection methods can be monitored frequently using control charts to see any deviations occurred; root cause analysis with corrective and preventive actions helps in bringing back the parameters under control . Root Cause analysis can be done in many ways. It can be done based on probability, case studies and by generating rules. Probabilistic root cause analysis depicts the cause and effect relation in a graphical representation to help in the defect analysis. In case based root cause analysis different cases are formed from historic data and their solution are maintained. The data is then analysed with these case studies and those with the closest match are considered for finding the solution for the new case. Rule based root cause analysis generates a set of rules pointing to the root causes. The rule is checked with the symptom to match with the rule.

4   Experiments and Result Discussion

4.1   Dataset and Procedures

The software defect dataset (http://promise.site.uottawa.ca/SERepository) consists of 2109 records and 22 attributes with a class representing defect or not. The software metrics include McCabe, Halstead, branch-count and five different measures representing lines of code. McCabe measures consists of four metrics namely, complexity showing the extent to which the flow graph can be reduced, line of code, cyclomatic complexity depicting the number of linearly independent paths and design complexity representing the cyclomatic complexity of module's reduced flow graph. Halstead measures include the base measures, the derived measures and the line of code measures. The base measures comprise the number of unique operands, unique operators and total operator occurrences. The derived measures include intelligence, volume, difficulty, effort to write the program and time required to write the program. The blank lines, code lines and the comment lines are the measures included in line of code.

The total attributes are reduced into two subsets with 6 attributes using correlation based feature selection and feature selection using anova. The values for CFS are derived from the Pearson's correlation equation and anova results are derived from the F-ratio calculation. These attributes are ranked for selection on the basis of their class dependency. Attributes which have high correlation is ranked as high and considered as significant contributors to defect in the software program. The top 6 attributes with high rank is identified as best contributors to the defect and these are selected for the next step. These reduced attributes are enough for prediction so the rest can be removed from the dataset. Table 1 shows the ranking.

4.2   Prediction Model from Control Chart Limits

The rules for the proposed model are generated from control charts. The reduced dataset with key attributes are divided into two based on the class labels, defect = "Yes" and defect ="No". Each attribute is plotted using I-MR chart to check the process performance.   The control limits, Upper Control Limit (UCL) and Lower Control Limit (LCL) values are derived from the chart. X represents the mean value. Table 2 and Table 3 represent the control limits of the reduced attributes selected using Anova and CFS respectively.

Table 1 Attribute Prioritization Based on Anova and Cfs

CFS		ANOVA	
Attribute	Correlation Coefficient	Attribute	F-ratio
Difficulty	0.387	Design_Complexity	13.33
Unique _operand	0.387	Lo_Blank	13.23
Unique _operators	0.386	CyclomaticComplexity	12.98
Intelligence	0.363	Unique_Operators	12.9
Lo_Blank	0.355	Brach_count	10.05
Design_Complexity	0.349	Unique_Operand	9.66
Loc	0.348	Effort	8.01
Total _operands	0.343	Program_Length	7.88
IO_code	0.342	Essential_Complexity	7.84
Volume	0.34	Time	7.59
Halstead	0.339	Lo_Comment	7.27
Total_ operators	0.324	Halstead	7.03
Branch_Count	0.298	Intelligence	6.39
CyclomaticComplexity	0.296	Io_Code	5.86
Total operators and operands	0.296	Loc	5.28
Effort	0.27	Total_Operand	5.03
Time	0.27	Volume	4.89
Lo_Comment	0.233	Total_Operators	4.48
Essential Complexity	0.205	Difficulty	4.14
Lo-Code and Comment	0.005	Total operators and operands	4
Program_Length	-0.233	Io_Code_Comment	0.55

Table 2 Control Limits of Key Features Derived from Correlation for each Class Labels

Attributes	Correlation With Defects			Correlation- Without Defects		
	UCL	LCL	X	UCL	LCL	X
Difficulty	41.11	13.32	13.90	16.89	-5.95	5.47
Uniq_Operands	61.2	-20.1	20.6	25.0	-9.9	7.5
Uniq_Operators	30.35	-4.76	12.80	16.09	-2.71	6.69
Intelligence	101.2	-24.2	38.5	52.4	-16.2	18.1
Lo_Blank	18.58	-9.23	4.68	5.27	-2.82	1.23
Design Complexity	16.06	-6.30	4.88	6.07	-1.83	2.12

Table 3 Control Limits of Key Features Derived from Anova for each Class Labels

Attributes	Anova With Defects			Anova- Without Defects		
	UCL	LCL	X	UCL	LCL	X
Design Complexity	16.06	-6.30	4.88	6.07	-1.83	2.12
Lo_Blank	18.58	-9.23	4.68	5.27	-2.82	1.23
Cyclomatic Complexity	18.26	-7.19	5.53	6.99	-2.30	2.35
Uniq_Operators	30.35	-4.76	12.80	16.09	-2.71	6.69
Branch_Count	35.6	-15.4	10.1	12.89	-5.55	3.67
Uniq_Operands	61.2	-20.1	20.6	25.0	-9.9	7.5

Then prediction rules were derived out of the range obtained from Table 2 and Table 3. The rules were identified based on if-then conditions as well as the dependency or correlation of selected attributes. 'IF -THEN' rules are generated based on the upper control limits of each attribute for classifying it into a defect or not a defect .This is represented in Fig.3.An attribute with high correlation value is a significant contributor to the defect of a software. So the class attribute is marked as defect if any one of the highly correlated attribute is in defect range.

```
If (difficulty<41.11) &&If (uniq_operands <61.2) &&
If (Uniq_operators<30.35) &&If (intelligence<101.2) &&
If (Lo_blank<18.58) &&If (design complexity<16.06) then
                    Class="Defect"
Else
              Class="No Defect"
```

Fig. 3 If then Rules for Prediction

4.3   Result Analysis

The number of correctly classified attributes and the accuracy of the prediction analysis on the data set, using C-FS and ANOVA-FS are shown in Table 4. The table represents accuracy of naive Bayes and J48 on full dataset as well as the accuracy of three models on the reduced dataset. From the results it is proved that the accuracy of prediction is high when Statistical Process Control Flow (SPC) Prediction Model is applied on dataset derived from CFS (86.86 %) than the other methods evaluated.

Table 4 Classification Accuracy of Naïve Bayes , J48 and Spc-Model(with and without Feaure Reduction Using Anova and Correlation Coefficient)

Dataset	Number Of Attributes	Number of Instances	Method Used	Correctly Classified	Accuracy
Full Dataset	22	2109	J48	1788	84.77 %
			Naïve Bayes	1737	82.36 %
Anova	6	2109	J48	1784	84.58%
			Naïve Bayes	1783	84.54%
			SPC-Model	1795	85.11%
Correlation Coefficient	6	2109	J48	1791	84.93%
			Naïve Bayes	1761	83.49 %
			SPC-Model	1823	86.86%

5   Conclusion

The defect dataset has been analyzed using two different statistical methods to reduce the number of attributes based on their relevance. Experimental results showed that Feature Selection technique greatly enhances the accuracy of classification and prediction. Comparative study of the SPC prediction model against two existing standard algorithms, J48 and Naïve Bayes showed that the proposed Statistical Process Control prediction model shows more accuracy than the benchmark classification algorithms.

Our future work is to extend the analysis to find out whether the same feature selection method may lead to better prediction accuracy for various high dimension dataset in other domains like medical field, engineering, business sectors, etc.

Acknowledgement This work is supported by the DST Funded Project, (SR/CSI/81/2011) under Cognitive Science Research Initiative in the Department of Computer Science, Amrita School of Arts and Sciences, Amrita Vishwa Vidyapeetham University, Kochi.

References

1. Liu, H., Yu, L.: Toward Integrating Feature Selection Algorithms for Classification and Clustering
2. IEEE Transactions on Knowledge and Data Engineering **17**(4), 491–502 (2005)
3. International Journal of Computer Theory and Engineering. Cancer Classification of Bioinformatics data using ANOVA **2**(3),1793–8201, June 2010
4. Patil, T.R., Sherekar, S.S.: Performance Analysis of Naive Bayes and J48 Classification Algorithm for Data Classification. International Journal Of Computer Science And Applications **6**(2), April 2013. ISSN: 0974-1011
5. Dangare, C.S., Apte, S.S.: Improved Study of Heart Disease Prediction System using Data Mining Classification Techniques. International Journal of Computer Applications (0975 – 888) **47**(10), June 2012
6. Tiwari, R., Singh, M.P.: Correlation based attribute selection using Genitic Algorithm. International journal of computer Applications(0975-8887) **4**(8), August 2010
7. Shana, J., Venkatachalam, T.: Identifying Key Performance Indicators and Predicting the Result from Student Data. International Journal of Computer Applications (0975-8887) **25**(9), July 2011

8. Vlahou, A., Schorge, J.O., Gregory, B.W., Coleman, R.L.: Diagnosis of Ovarian Cancer Using Decision Tree Classification of Mass Spectral Data. Journal of Biomedicine and Biotechnology **2003**(5), 308–314 (2003)
9. Lavanya, D., Rani, K.U.: Analysis of Feature SelectionwithClassfication: BreastCancer Datasets. Indian Journal of Computer Science and Engineering (IJCSE)
10. Kalyani, P., Karnan, M.: Attribute Reduction using Forward Selection and Relative Reduct Algorithm. International Journal of Computer Applications (0975 – 8887) **11**(3), December 2010
11. Mahapoonyanont, N., Mahapoonyanont, T., Pengkaew, N., Kamhangkit, R.: Power of the test of One-Way Anova after transforming with large sample size data. International Journal Procedia Social and Behavioral Sciences **9**, 933–937 (2010)
12. Hayati, F., Maghsoodloo, S., DeVivo, M.J., Carnahan, B.J.: Control chart for monitoring occupational asthma. Journal of Safety Research **37**, 17–26 (2006)
13. Mahmud, W.M., Agiza, H.N., Radwan, E.: Intrusion detection using rough sets based parallel genetic algorithm hybrid model. In: Proceedings of the World Congress on Engineering and Computer Science, WCECS 2009, San Francisco, USA, vol. **II**, October 2009
14. Subbulakshmi, T., Ramamoorthi, A., Mercy, S.: Shalinie Ensemble design for intrusion detection systems. International Journal of Computer science & Information Technology (IJCSIT) **9**, August 2009
15. Spangler, W.E., Vargas, M.G.: Choosing Data mining Methods for Multiple Classification: Representational and performance measurement Implications for Decision Support. Journal of Management Information Sysytem **16**(1)
16. Dimitoglou, G., Adams, J.A., Jim, C.M.: Comparison of the C4.5 and a Naive Bayes-Classifier for the Prediction of Lung CancerSurvivability
17. Kaur, P.J., Pallavi: Data Mining Models for Software Defect Prediction. International Journal of Software and Web Sciences (IJSWS) (2013)
18. Malhotra, R., Jain, A: Fault Prediction Using Statistical and Machine Learning (2012)
19. Xing, F.: A Novel method for early software quality prediction based on support vector machine. In: Proc.of the 16th ISSRE, pp. 213–222 (2005)
20. Fenton, N.: Predicting Software Defects in Varying Development lifecycles Using Bayesian Nets. Information and Software Technology **49**(1), 32–43 (2007)
21. Nancy, S.G., Alias Balamurugan, S.A.: A comparative study of feature selection methods for cancer classification using gene expression dataset. Journal of Computer Applications (JCA) **VI**(3) (2013). ISSN: 0974-1925
22. Tomar, D., Agarwal, S.: A Survey on Pre-processing and Post-processing Techniques in Data Mining. International Journal of Database Theory and Application **7**(4), 99–128 (2014)
23. Jenzi, S., Priyanka, P., Alli, P.: A Reliable Classifier Model Using Data Mining Approach For Heart Disease Prediction. International Journal of Advanced Research in Computer Science and Software Engineering **3**(3), March 2013. ISSN: 2277 128X
24. Gayathri, M., Sudha, A.: Software Defect Prediction System using Multilayer Perceptron Neural Network with Data Mining. International Journal of Recent Technology and Engineering (IJRTE) **3**(2), May 2014. ISSN: 2277-3878
25. Kumaresh, S., Meenakshy Sivaguru, B.R.: Software Defect Classification using Bayesian Classification Techniques. International Journal of Computer Applications (0975 – 8887). International Conference on Communication, Computing and Information Technology (ICCCMIT-2014)
26. Azeem, N., Usmani, S.: Analysis of Data Mining Based Software Defect Prediction Techniques. Global Journal of Computer Science and Technology **11**(16) Version 1.0, September 2011

A Concept Based Graph Model for Document Representation Using Coreference Resolution

G. Veena and Sruthy Krishnan

Abstract Graph representation is an efficient way of representing text and it is used for document similarity analysis. A lot of research has been done in document similarity analysis but all of them are keyword based methods like Vector Space Model and Bag of Words. These methods do not preserve the semantics of the document. Our paper proposes a concept based graph model which follows a Triplet Representation with coreference resolution which extract the concepts in both sentence and document level. The extracted concepts are clustered using a modified DB Scan algorithm which then forms a belief network. In this paper we also propose a modified algorithm for Triplet Generation.

1 Introduction

Graph representation is one of the best ways to represent a text since it covers the disadvantages of the traditional text representation approaches like Vector Space Model and Bag of Words; where the structural and semantic information is completely ignored. In this work we are including a module called Coreference Resolution to improve the accuracy of Document Similarity analysis. In our work each concept is represented in the form of triplet <Subject, Verb, Object> [5]. Then the concepts are clustered using a modified DB scan algorithm and a mathematical model called the Belief Network is formed. This is the conceptual mining model for the document. The inclusion of Coreference module helps to resolve coreferences of Subjects and Objects in the graph and helps in improving the accuracy of clustering the concepts.

We present a brief explanation of the Related Works in Section 2 followed by the Proposed Solution in Section 3. Section 4 includes preprocessing which

G. Veena · S. Krishnan(✉)
Department of Computer Science and Application, Amrita Vishwa Vidyapeetham,
Coimbatore, India
e-mail: veenag@am.amrita.edu, krishnasruthy04@gmail.com

© Springer International Publishing Switzerland 2016   367
S. Berretti et al. (eds.), *Intelligent Systems Technologies and Applications*,
Advances in Intelligent Systems and Computing 384,
DOI: 10.1007/978-3-319-23036-8_32

includes a set of Natural Language Processing tasks and Coreference Resolution module. Section 5 explains Coreference Replacement Algorithm followed by the Graph Representation in Section 6. Section 7 shows the representation of document using conceptual mining graph model. Section 8 describes Implementation and Evaluation. Finally we lay out our conclusions in Section 9.

2   Related Works

Coreference Resolution checks whether two expressions in a text refer to the same entity. It is a very important area in text analysis. From earlier days, a lot of work has been done on resolving coreferences, but most of them does not undergo empirical evaluation and was not based on learning from an annotated corpus. Some of the earlier systems used manually designed knowledge sources. The approach in paper [1] uses a statistical model for resolving pronouns. They used the distance between the pronoun and the proposed antecedent and combine them to form a single probability to identify the coreferering antecedents. Paper [2] uses Japanese Newspaper articles as training examples for machine learning algorithm which uses C4.5 Decision Tree Classifier and compares it with the results of existing anaphora resolution approach. But their evaluation does not identify generic noun phrases; they identified only noun phrases for organizations. Paper [3] used an annotated corpus for which resolved both domain independent and domain dependent pronouns based on a large set of features. Our work uses a small set of 12 features and the Decision Tree Classifier proposed in paper [4] to solve coreferences which are domain independent and thus improves the accuracy of conceptual mining model.

3   Proposed Solution

In our work an improved triplet generation algorithm and a Coreference Resolution module is included that improves the accuracy of conceptual mining model. The proposed solution gives answers to the following questions.

1. How does Coreference Resolution solve the problems in clustering?
2. How does it improve the efficiency of similarity analysis?

The High Level Architecture includes the following steps:

- Preprocessing
- Coreference Replacement
- Triplet Generation
- Clustering
- Belief Network

First a document is passed through the preprocessing module which includes Natural Language Processing steps and Coreference Resolution. Then we extract the output of Coreference Resolution and use it to replace the pronouns. The Coreference Replacement Algorithm replaces the pronouns with the actual Subject or Object and creates a resolved document. Then the resolved document is passed into Triplet Generation followed by Clustering of triplets for the formation of Belief Network. Fig. 1 shows the High Level Architecture.

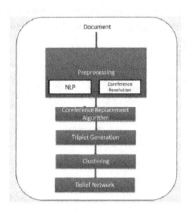
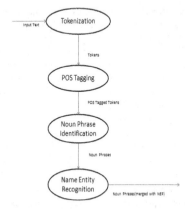

Fig. 1 High Level Architecture   **Fig. 2** Natural Language Processing

4   Preprocessing

The preprocessing includes Natural Language Processing Steps and Coreference Resolution. The Natural Language Processing steps are Tokenization, POS Tagging, Noun-Phrase identification and Name Entity Recognition. Coreference Resolution includes Feature Vector Identification, Anaphor-Antecedent Pair Selection and Coreference Resolution Algorithm.

4.1   Natural Language Processing Steps

- Tokenization
- POS Tagging
- Noun-Phrase Identification
- Name Entity Recognition

The text is input to the Tokenization module which gives tokens(smallest word units) as the output.Then the tokens are POS Tagged and are passed to Noun Phrase Identification module where the noun phrases are identified. Finally Name Entity Recognition is done on those noun phases and this output can be used for finding the corefering terms. Fig 2 shows the Natural Language Processing Steps.

4.2 Coreference Resolution

Coreference Resolution checks whether two expressions in a text refer to the same entity. Figure 3 shows a sample text and its possible coreferences. The words underlined are the entities and the words shown in italics shows its coreferences. This module is done based on the work of Wee Meng Soon[4].In his work markables are determined after preprocessing and 12 feature vectors are generated which determine whether two markables corefer or not. Markables are the union of noun phrases, name entities and nested noun phrases. Then it is passed to a Decision Tree Classifier and the corefering antecedents are found. Fig 4 shows the High Level Architecture of Coreference Resolution.

Rama was the prince of Ayodhya. *He* lived five to six thousand years or more ago. *His* wife was *Sita*. *She* was kidnapped by Ravana and was imprisoned in Lanka.

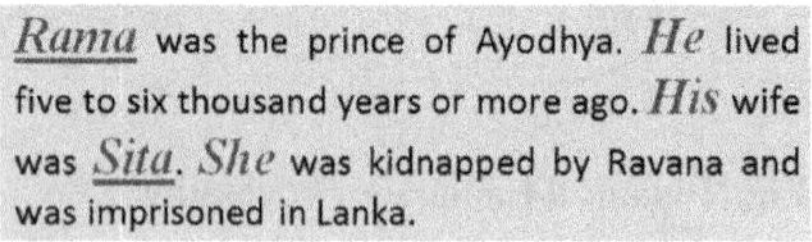

Fig. 3 Coreference Resolution Example.

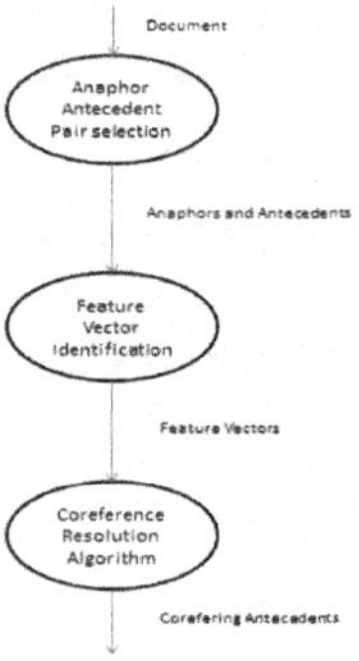

Fig. 4 High Level Architecture for Coreference Resolution

4.2.1 Anaphor-Antecedent Pair Selection

An Anaphor and an Antecedent are usually noun phrases. Every noun phrase that we get as output is an anaphor and every noun phrase that we get before the anaphor is an antecedent.

The list of anaphor-antecedent pairs used for checking coreference from the sentence in Fig 1 is shown in Table 1.

Table 1 Anaphor-Antecedent Pairs

Antecedent-Anaphor
Rama-He
He-His
His-Sita
Sita-She
She-Ravana

4.2.2 Feature Vector Identification

Feature vectors determine whether two noun phrases corefer or not. Let i be an antecedent and j be an anaphor.12 features are considered in the work of Wee Meng Soon [4]. They are Distance Feature, i-Pronoun Feature, j-Pronoun Feature, String Match Feature, Definite noun phrase feature, Demonstrative noun phrase feature, Number Agreement feature, Semantic Class Agreement feature, Gender agreement Feature, Both Proper Names Feature, Alias Feature and Appositive Feature. The list of feature for the sentence in Fig 3 is explained in Table 2.

Table 2 Feature Vectors for sentence in Fig 3

Antecedent	Anaphor	Feature Vector
Rama	He	1,0,1,0,0,0,1,-,0,0,1
He	His	1,1,1,0,0,0,1,1,0,0,1
His	Sita	0,1,0,0,0,0,1,0,0,0,0
Sita	She	1,0,1,0,0,0,0,0,1,1
She	Ravana	0,0,0,0,0,0,0,0,0,0

4.2.3 Coreference Resolution Algorithm

This algorithm is proposed by Wee Meng Soon [4] to resolve coreferences. Here i is considered as an antecedent and j as an anaphor. For each anaphor antecedent the feature vectors are found out and passed to the decision tree classifier.C4.5 which is an improvement of the ID3 algorithm[implemented in FreeLing]. If the classifier returns true a corefering antecedent is found.

Algorithm 1. Coreference Resolution Algorithm

Input: Noun Phrases
Output: Corefering Antecedents
i is an antecedent
j is an anaphor
for each i and j **do**
    generate feature vector fv and pass it to decision tree classifier
    if classifier returns true **then**
        a corefering antecedent is found
    end if
end for

Table 3 shows antecedents, anaphors and coreferences of the sentence from Fig 3.

Table 3 Coreference Resolution Module Example Output

Antecedent	Anaphor	Corefers?
Rama	He	Yes
He	His	Yes
His	Sita	No
Sita	She	Yes
She	Ravana	No

5   Coreference Replacement Algorithm

Coreference Replacement algorithm replaces ambiguous pronouns with the original Subject or Object. FreeLing [7, 8, and 9] is an opensource language analysis tool which includes a module to solve the coreferences. The output of Coreference Resolution module done in FreeLing is in XML format. From the XML file extract <mention> nodes which are childnodes of <coref> node. From <mention> extract the value of attribute 'name' and save it in an array. This array contains the coreferences. Now in the original document, check for a match with the words in the array. If a match is found the word is replaced with the original subject or the object which is most probably the first element in the coreferences array.

Algorithm 2. Coreference Replacement Algorithm

Input: XML file
Output: Resolved Text file
Let x be an XML document
Let corefarray be an empty list of coreferences
for each child node in <coreference>:
    extract <coref> node
    for each <coref> node:
        extract <mention> node
        for attribute 'name' in <mention> node:
            extract value
            add value to corefarray
        end for
    end for
for each element in corefarray:
    check for a match in document
    if match is found:
        replace original word with the coreference
    end if
end for

---

6 Graph Representation

In our work coreference resolution is included as a prerequisite for Belief Network creation. The graph is in the form of a Triplet. ie; Subject –Verb-Object form. The nodes represent either the subject or the object and the link represent the Verb (relationship) between the subject and the object. Fig 5 shows the Graph Representation of a sentence.

Fig. 5 Graph Representation of a sentence

Triplet Generation

The document is given as input. A phrase tree is generated for each sentence using the Stanford Parser. Then the phrase tree is converted to triplet form using the triplet generation algorithm. Fig 6 shows a sentence, its phrase tree and triplet.

Sita was kidnapped by Ravana and was imprisoned in Lanka.
Extract Verbs

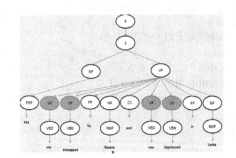

Extract Subject and Object

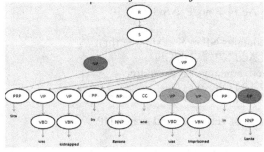

Fig. 6 Triplet Generation

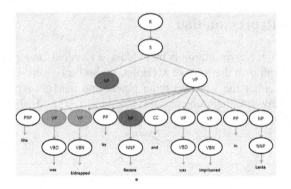

Triplets

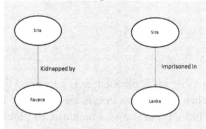

Fig. 7 (*Continued*)

Algorithm 3. Triplet Generation Algorithm
Input: A Document
Output: Concepts in the form of Triplets
S is a new sentence
Declare Lv as an empty list of Verb
Declare Sub as an empty list of Subject
Declare Obj as an empty list of Object
for each S **do**
      extract all verbs
      add verb to Lv
      for each verb in Lv **do**
            Check parent node and extract NP node
            Add NP node to Sub
      end for
      if verb contains NP or S as subtree **then**
            Add NP or S to Obj
      else if verb contains VP as subtree
            Add Object to Obj

    else
Take parent node of verb node and search NP
    end if
end for

Extract each sentence and from it extract all the verbs. Then for each verb, the subject and the object is found by checking the NP nodes.

Triplet Generation Algorithm forms a graph in the form of Subject-Verb-Object as shown in. Fig 7. Then it is clustered using the DB Scan Algorithm [5] to form a Belief Network. But there will be ambiguity in the subjects and objects while clustering. If there is no coreference it will be difficult to cluster since we do not know which entities are to be clustered. For example in Fig 7, we do not know how to cluster the nodes 'He' and 'She' which are ambiguous pronouns.

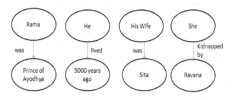

Fig. 8   Example Triplet Representation before Coreference Resolution

7   Improved Representation

As Coreference Resolution is done as a prerequisite for Triplet Graph Generation, it solves the problem of ambiguous pronouns. The Corefering pairs are found out and the anaphors are replaced by the original name.

Then the new document with coreference resolution is passed to the Triplet generation algorithm and forms the graph. Here all the corefering pairs are mapped to the original subject or object which makes the graph more accurate. Then using the DB Scan algorithm in [5,6] clustering is done and we get an accurate concept mining model. Two such models can be further used for Graph comparison. Fig 8.a shows the Triplet after Coreference Resolution is added and 8.b shows the Clustered Graph.

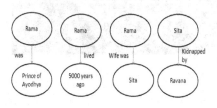

Fig. 8.a. Triplet after Coreference Resolution   **Fig. 8.b.** Clustered Graph Representation
                                                          with Coreference Resolution

8   Implementation and Evaluation

Coreference Resolution is implemented using an Opensource language analysis tool called FreeLing [7]. FreeLing support language analysis for a wide range of languages. The Coreference module in FreeLing is implemented based on the work of Wee Meng Soon [4].

Figure 9 shows a sample dataset:

Mohandas Karamchand Gandhi was the preeminent leader of Indian independence movement in British-ruled India. Employing nonviolent civil disobedience, Gandhi led India to independence and inspired movements for civil rights and freedom across the world. He is also called Bapu in India.
Born and raised in a Hindu merchant caste family in coastal Gujarat, western India, and trained in law at the Inner Temple, London, Gandhi first employed nonviolent civil disobedience as an expatriate lawyer in South Africa, in the resident Indian community's struggle for civil rights. After his return to India in 1915, he set about organising peasants, farmers, and urban labourers to protest against excessive land-tax and discrimination. Assuming leadership of the Indian National Congress in 1921, Gandhi led nationwide campaigns for easing poverty, expanding women's rights, building religious and ethnic amity, ending untouchability, but above all for achieving Swaraj or self-rule.
Gandhi famously led Indians in challenging the British-imposed salt tax with the 400 km (250 mi) Dandi Salt March in 1930, and later in calling for the British to Quit India in 1942. He was imprisoned for many years, upon many occasions, in both South Africa and India. Gandhi attempted to practise nonviolence and truth in all situations, and advocated that others do the same. He lived modestly in a self-sufficient residential community and wore the traditional Indian dhoti and shawl, woven with yarn hand spun on a charkha. He ate simple vegetarian food, and also undertook long fasts as a means to both self-purification and social protest.
Gandhi's vision of a free India based on religious pluralism, however, was challenged in the early 1940s by a new Muslim nationalism which was demanding a separate Muslim homeland carved out of India. Eventually, in August 1947, Britain granted independence, but the British Indian Empire was partitioned into two dominions, a Hindu-majority India and Muslim Pakistan. As many displaced Hindus, Muslims, and Sikhs made their way to their new lands, religious violence broke out, especially in the Punjab and Bengal. Eschewing the official celebration of independence in Delhi, Gandhi visited the affected areas, attempting to provide solace. In the months following, he undertook several fasts unto death to promote religious harmony. The last of these, undertaken on 12 January 1948 at age 78, also had the indirect goal of pressuring India to pay out some cash assets owed to Pakistan. Some Indians thought Gandhi was too accommodating. Nathuram Godse, a Hindu nationalist, assassinated Gandhi on 30 January 1948 by firing three bullets into his chest at point-blank range.

Fig. 9 Sample Dataset

Figure 10 shows the Coreferences detected.

```
- <coreferences>
  - <coref id="co98">
      <mention id="m98.1" from="t1.1" to="t1.1" words="Mohandas_Karamchand_Gandhi"/>
      <mention id="m98.2" from="t1.3" to="t1.9" words="the preeminent leader of Indian independence movement"/>
      <mention id="m98.3" from="t2.6" to="t2.6" words="Gandhi"/>
      <mention id="m98.4" from="t4.27" to="t4.27" words="Gandhi"/>
      <mention id="m98.5" from="t5.2" to="t5.2" words="his"/>
      <mention id="m98.6" from="t5.9" to="t5.9" words="he"/>
      <mention id="m98.7" from="t6.9" to="t6.9" words="Gandhi"/>
      <mention id="m98.8" from="t7.1" to="t7.1" words="Gandhi"/>
      <mention id="m98.9" from="t9.1" to="t9.1" words="Gandhi"/>
      <mention id="m98.10" from="t12.1" to="t12.1" words="Gandhi"/>
      <mention id="m98.11" from="t15.10" to="t15.10" words="Gandhi"/>
      <mention id="m98.12" from="t18.4" to="t18.4" words="Gandhi"/>
      <mention id="m98.13" from="t19.3" to="t19.3" words="Gandhi"/>
      <mention id="m98.14" from="t20.7" to="t20.7" words="Gandhi"/>
      <mention id="m98.15" from="t20.13" to="t20.13" words="his"/>
    </coref>
```

Fig. 10 XML file showing detected coreferences

The Experimental Setup considers the same datasets used in [5].
I.e.; 200 articles collected from ACM digital library on topics related to literature, hardware, computer system organization, software and data.

The runtime of Triplet Generation Algorithm is analyzed and it showed a complexity of O (n).

The result analysis is done based on two factors. They are;

- Coreference Resolution and Concept Retrieval Efficiency
- Accuracy

1) Coreference Resolution and Concept Retrieval Efficiency

This measure provides the amount of resolved coreferences and clustered concepts based on out improved model with coreference resolution.

The efficiency is calculated using three measures

- True Positives
- False Positives
- Precision

This analysis shows that the number of true positives has increased and false positives have decreased thus improving the precision from the existing concept mining model.

Table 4 shows the Performance Evaluation of extracted concepts in the existing concept mining model and Table 5 shows the Performance Evaluation of extracted concepts using the improved concept graph model with coreference resolution.

Table 4 Performance Evaluation of extracted concepts in the existing concept mining model

Performance Measure	Number of Abstract Documents		
	10	20	30
TP	10	12	20
FP	4	3	11
Precision (%)	80.4	82.1	85.7

Table 5 Performance Evaluation of extracted concepts in the improved concept mining model

Performance Measure	Number of Abstract Documents		
	10	20	30
TP	14	16	27
FP	3	2	3
Precision (%)	82.3	88.8	90

Fig 11 shows the Performance Evaluation of extracted concepts using the improved concept graph model with coreference resolution.

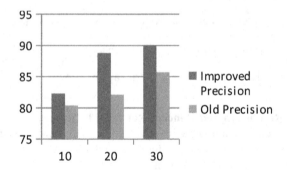

Fig. 11 Performance Evaluation comparison of Existing with the improved Concept Mining Model.

2) Accuracy

The accuracy is calculated for the comparison of three models namely Vector Space Model, Existing Concept Based Mining Model and improved Concept Graph Model.

Table 6 Result Analysis

Types	Accuracy (%)
Keyword Based Mining	0.75
Concept Based Mining	0.85
Concept Based Graph Model and mining with Coreference resolution	0.91

Table 8 shows Result Analysis and Fig 12 shows the Result Analysis Graph.

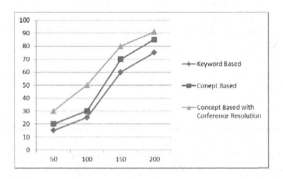

Fig. 12 Result Analysis Graph

9 Conclusion

A new concept based graph model with coreference resolution is introduced in this paper. This representation captures the concepts in a more efficient way than the existing concept mining model since the coreference of the document is resolved. The problem of ambiguous pronouns is solved with coreference resolution. The new concept based graph model can later be used for checking document similarity.

The future work includes the introduction of Semantic Role Labeling as a prerequisite for Graph Based model to improve the efficiency of Similarity analysis.

Acknowledgments We are thankful to Dr M. R. Kaimal, Chairman Department of Computer Science, Amrita University for his valuable feedback and suggestions. We are also thankful to Mr. **Lluís Padró** ,Professor Titular d'Univers tat,Departament de Llenguatges i Sistemes Informàtics,Centre de Recerca TALP,Universitat Politècnica de Catalunya for his ideas and contributions. Most of the novel ideas and solutions found in this paper are the result of our numerous stimulating discussions.

References

1. Ge, N., Hale, J., Charniak, E.: A statistical approach to anaphora resolution. In: Proceedings of the Sixth Workshop on Very Large Corpora, vol. 71 (1998)
2. Aone, C., Bennett, S.W.: Evaluating automated and manual acquisition of anaphora resolution strategies. In: Proceedings of the 33rd Annual Meeting on Association for Computational Linguistics. Association for Computational Linguistics (1995)
3. Fisher, D., et al.: Description of the UMass system as used for MUC-6. In: Proceedings of the 6th Conference on Message Understanding. Association for Computational Linguistics (1995)
4. Soon, W.M., Ng, H.T., Lim, D.C.Y.: A machine learning approach to coreference resolution of noun phrases. Computational Linguistics 27(4), 521–544 (2001)
5. Veena, G., Lekha, N.K.: A concept based clustering model for document similarity. In: 2014 International Conference on Data Science & Engineering (ICDSE). IEEE (2014)
6. Veena, G., Lekha, N.K.: An extended chameleon algorithm for document clustering. In: El-Alfy, E.-S., Thampi, S.M., Takagi, H., Piramuthu, S., Hanne, T. (eds.) Advances in Intelligent Informatics. AISC, vol. 320, pp. 335–348. Springer, Heidelberg (2015)
7. http://nlp.lsi.upc.edu/freeling/
8. Padró, L., Stanilovsky, E.: FreeLing 3.0: towards wider multilinguality. In: Proceedings of the Language Resources and Evaluation Conference (LREC 2012). ELRA, Istanbul, May 2012
9. Padró, L., Collado, M., Reese, S., Lloberes, M., Castellón, I.: FreeLing 2.1: five years of open-source language. In: Processing Tools Proceedings of 7th Language Resources and Evaluation Conference (LREC 2010), ELRA, La Valletta, May 2010

Part II
Intelligent Image Processing
and Artificial Vision

Recognizing Individuals from Unconstrained Facial Images

Radhey Shyam and Yogendra Narain Singh

Abstract This work makes an effort to address the problem of face recognition in unconstrained environments and presents a novel method of facial image representation based on local binary pattern (LBP). The method devises the appropriate descriptor that discriminates the facial features by filtering the LBP surface texture. The method, we name as augmented local binary pattern (A-LBP) works on the uniform and non-uniform patterns both. The non-uniform pattern is replaced with the majority voting of the uniform patterns which combines with the neighboring uniform patterns to extract pertinent information regarding the local descriptors. The recognition accuracy obtained by the proposed method is computed on Chi square and Bray Curtis dissimilarity metrics. The experimental results show that the proposed method performs better than the original LBP on publicly available face databases, AT & T-ORL, extended Yale B, Yale A and Labeled Faces in the Wild (LFW) containing unconstrained facial images.

Keywords Face recognition · LBP · Bray Curtis · Chi square

1 Introduction

Automatic facial image analysis is an active area of research in computer vision. Numerous face recognition methods such as PCA [1], LDA [2], Fisherface [3] have been designed that are performing satisfactorily in constrained environments. In many applications, e.g., image retrieval, biomedical image analysis, and outdoor scene analysis, the environments are not cooperative. Therefore, there is a need to devise an efficient method that accurately recognizes the individual's from their unconstrained facial images.

R. Shyam(✉) · Y.N. Singh
Department of Computer Science and Engineering,
Institute of Engineering and Technology (IET), Lucknow 226 021, India
e-mail: shyam0058@gmail.com, singhyn@gmail.com

© Springer International Publishing Switzerland 2016   383
S. Berretti et al. (eds.), *Intelligent Systems Technologies and Applications*,
Advances in Intelligent Systems and Computing 384,
DOI: 10.1007/978-3-319-23036-8_33

In literature, the methods that work in unconstrained environments are primarily based on texture representations that build several local representations of the face image and combining them into a global representation. The local features-based approaches to face recognition have drawn attention of the biometric researchers in the field of computer vision [4, 5, 6]. These methods are less sensitive to variations in pose and illumination than the traditional methods. Furthermore, an important reason for heeding the local features-based approaches are that the traditional methods which lose the local discriminatory features by averaging it, over the entire image. The major problems with the texture analysis is that, they are not uniform due to variations in orientation, scale, or, other visual appearances. A useful direction is, therefore, the development of powerful texture metrics that can be extracted and classified with a low-computational intricacy.

In unconstrained environments, the local binary pattern (LBP) is one of the most prevalent methods of face recognition which is computationally efficient [7]. This paper addresses the issues of face recognition in unconstrained environments and devises an efficient method that accurately recognizes human faces from variations in pose, illumination, and expression. A novel approach for facial image representation using LBP, called augmented local binary pattern (A-LBP) is presented. It combines the uniform and non-uniform patterns on the principle of locality which replaces the non-uniform patterns with the majority voting of uniform patterns and combined with the neighboring uniform patterns as a result. Finally, the local descriptors are generated that have shown discriminatory information to classify the facial images. The rest of the paper is organized as follows: The basics of LBP are given in Section 2. In Section 3, the theoretical and experimental demonstration of the proposed A-LBP method is presented. The performance of the A-LBP method is evaluated and the results are reported in Section 4. Finally, the conclusion is drawn in Section 5.

2 Local Binary Pattern

The Local Binary Pattern operator was first introduced in 1996 by Ojala *et al.*, for the study of texture of gray-scale images. It is a powerful means of texture representation. The intention behind using LBP operator for face representation is that, the face can be seen as a composition of various micro-patterns and it is insensitive to variations, such as pose and illumination. The LBP operator labels the pixels of an image with decimal numbers, which encodes the local structure around each pixel. Each pixel is compared with its neighborhood by subtracting the central pixel's value as a threshold. The resulting non-negative values are encoded with 1 and the others with 0. The derived binary numbers are referred to as a LBP [8, 9].

Plenty of LBP variants have been proposed in literature to improve the robustness of LBP operator in unconstrained face recognition. For instance, Liao *et al.* [10] proposed a dominant LBP which makes use of the most frequently occurred patterns of LBP. Center-symmetric LBP is used to replace the gradient operator used by the SIFT operator which is based on the strengths fusion of SIFT and LBP operators [11]. A multi-block LBP replaces the intensity values in computation of LBP with mean intensity value of image blocks [12]. Local ternary pattern was initiated by Tan and Triggs [13], to add resistance to the noise. Three-Patch LBP code is produced by comparing the values of three patches to produce a single bit value in the code assigned to each pixel. Four-Patch LBP codes compare two center symmetric patches in the inner ring with two center symmetric patches in the outer ring [14].

The pitfall of the LBP is that their insensitiveness to the monotonic transformation of the gray-scale. Moreover, LBP may not work properly for the noisy images. Furthermore, the original LBP uses only the uniform patterns, otherwise LBP would suffer from the curse of dimensionality. This reduction in size may result in the loss of important information.

3 Augmented Local Binary Pattern (A-LBP)

This section proposes a novel method that relies on the LBP, called augmented local binary pattern. Prior work on the LBP have not drawn much attention on the use of non-uniform patterns. They are either treated as noise and discarded during the texture representations, or used in combination with the uniform patterns. The proposed method considers the non-uniform patterns and extracts the discriminatory information available to them, so as to prove their usefulness. They are used in combination to the neighboring uniform patterns and extract useful information regarding the local descriptors.

The proposed method uses a grid-based region. However, instead of directly assigning all non-uniform patterns into 59th bin, it replaces all non-uniform patterns with the majority of neighboring uniform patterns. For this, we have taken a filter of size 3×3 that is moved on the entire LBP generated surface texture. In this filtering process, the central pixel's value (c_p) is replaced with the majority of a set in case of the non-uniformity of the central pixel. This set contains 8-closet neighbors of central pixel, in which non-uniform neighbors are substituted with 255. Here 255 is the highest uniform value. The proposed A-LBP method is explained in Algorithm 1. The lookup table contains decimal values of 8-bit uniform patterns are used as given in [15]. The basic steps of filtering is shown in Fig. 1.

The classification performance of the proposed method is evaluated with Chi square (χ^2) and Bray Curtis dissimilarity (BCD) metrics. which are defined as follows:

$$\chi^2(p, q) = \sum_{i=1}^{N} \frac{(p_i - q_i)^2}{(p_i + q_i)} \tag{1}$$

$$BCD(p, q) = \frac{\sum_{i=1}^{N} |p_i - q_i|}{\sum_{i=1}^{N} (p_i + q_i)} \tag{2}$$

where N is the dimensionality of the spatially enhanced histograms, p is the histogram of the test image, q is the histogram of the training image, i represent the bin

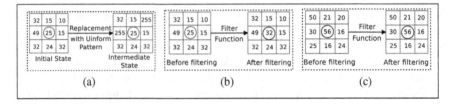

Fig. 1 Example of augmented local binary patterns (A-LBP) operator: (a) Neighboring non-uniform patterns of the central pattern are to be replaces with highest uniform pattern 255, (b) The value of central pattern 25 is a non-uniform pattern, which is replaced with majority value of 32, and (c) The value of central pattern 56 is already a uniform pattern, so it remains unchanged.

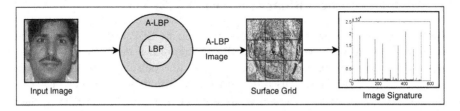

Fig. 2 Block diagram of augmented local binary patterns (A-LBP) for face recognition system.

Algorithm 1. Steps of augmented local binary pattern (A-LBP)

Input: LBP surface texture
Filtering procedure:

1. Check the uniformity of central pixel's value c_p.
2. If c_p is uniform, Then Go to step 1 with next c_p.
3. Otherwise, form a set N_8 containing 8 closet neighbors of c_p.
4. Replace all non uniform patterns in N_8 with 255.
5. Assign majority of N_8 to c_p.

Output: Augmented local binary pattern (A-LBP) surface texture

---

number and p_i, q_i are the values of the ith bin in the histograms of p and q to be compared. The concatenation of all sub histograms constitutes the image's signature.

The Block diagram of the proposed A-LBP face recognition method is shown in Fig. 2.

4 Results

The proposed method is tested on the publicly available face databases, such as AT & T-ORL [16], extended Yale B [17], Yale A [18] and Labeled Faces in the Wild [19]. The images differ in variation in pose, illumination, expression, eye glasses and

Table 1 Face recognition accuracies (%) of methods on different databases using Chi square (χ^2) distance and Bray Curtis Dissimilarity (BCD) metrics.

	Database							
	AT & T-ORL		Extended Yale B		Yale A		LFW	
Distance Metrics → Methods ↓	χ^2	BCD	χ^2	BCD	χ^2	BCD	χ^2	BCD
LPB	92.50	94.52	74.11	81.83	61.19	60.00	65.00	65.29
A-LBP	**95.00**	**95.00**	**81.22**	**86.45**	**73.33**	**71.90**	**65.00**	**67.37**

occlusion. A total of 3455 images are used to recognize 113 distinct individuals from these databases. The system is trained for each database separately, whereas the test image is selected randomly from given images for each individual and the performance is computed.

The performance of the proposed A-LBP method is analyzed using equal error rate, which is an error, where the likelihood of acceptance is assumed to be same as to the likelihood of rejection of the people who should be correctly verified. The performance of the proposed method is also confirmed by the receiver operating characteristic (ROC) curves. The ROC curve is a measure of classification performance that plots the genuine acceptance rate (GAR) against the false acceptance rate (FAR).

4.1 Recognition Performance Using Chi Square Distance Metric

The face recognition accuracy of the proposed A-LBP method is compared to the LBP method on different databases. The experimental results show that the A-LBP performs better to LBP. For example, the AT & T-ORL database the A-LBP achieves 95% recognition accuracy, whereas LBP reports an accuracy of 92.5%. Similarly for extended Yale B and Yale A databases, the A-LBP performs better than LBP. For extended Yale B database the accuracy values are reported to 81.22% for A-LBP and 74.11% for LBP. For Yale A database, the proposed method reports a better accuracy of 73.33% in comparison to the LBP accuracy value of 61.19% (See Table 1).

The ROC curve for AT & T-ORL database is plotted and shown in Fig. 3(a). It shows that the GAR is found higher for the proposed A-LBP method and reported the value 78% when the FAR is strictly nil. As FAR increases, the GAR value is also increased. Such as, the GAR is found 93% for LBP and 96% for A-LBP at 5% of the FAR. The GAR is found maximum to 100% at 32% of FAR. The ROC curve for extended Yale B database is plotted and shown in Fig. 3(b). It shows that the GAR is found higher for A-LBP method and reported the value 32% when the FAR is strictly nil. As FAR increases, the GAR value is also increased for all methods, respectively. For example, the GAR is found 62% for LBP and 82% for A-LBP at 20% of FAR. The GAR is found maximum to 100% at 83% of FAR for LBP and 78% of FAR for A-LBP. The reported results show that A-LBP method achieves better recognition accuracy due to its insensitiveness to the change of illumination.

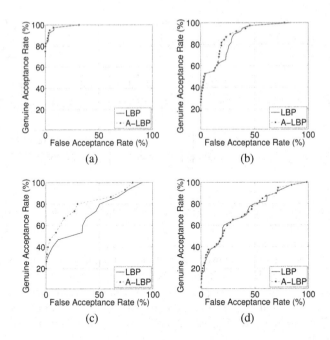

Fig. 3 ROC curves for LBP and A-LBP methods using Chi square (χ^2) distance metric on different databases: (a) AT & T-ORL, (b) Extended Yale B, (c) Yale A, and (d) LFW.

The ROC curve for Yale A database is plotted and shown in Fig. 3(c). It shows that the GAR is found higher for A-LBP method and reported value of 20% when the FAR is strictly nil. As FAR increases, the GAR value is also increased for all methods, respectively. Such as, the GAR is found 50% for LBP, 69% for A-LBP at 20% of the FAR. The GAR is found maximum 100% at 90% of FAR for LBP and 82% of FAR for A-LBP. The A-LBP method achieves better recognition accuracy due to its insensitiveness to the change of illumination. The ROC curve for LFW face database is plotted and shown in Fig. 3(d), which indicates that the GAR is found higher in A-LBP method and reported value of 15% at the zero FAR. As FAR increases, the GAR value is also increased for both methods, respectively. For example, the GAR finds 35% for LBP, 39% for A-LBP at 6% of the FAR. The GAR is found maximum 100% for both methods at 98% of FAR. A-LBP method shows the marginal recognition accuracy over the LBP on LFW database.

4.2 Recognition Performance Using Bray Curtis Dissimilarity Metric

The face recognition accuracy of the proposed A-LBP method is compared to the LBP method on different face databases. The experimental results show that the

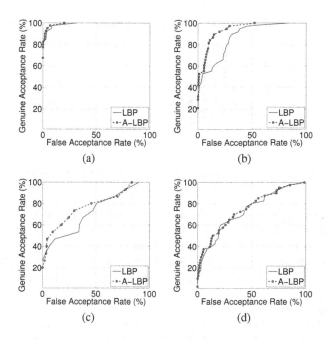

Fig. 4 ROC curves for LBP and A-LBP methods using Bray Curtis dissimilarity metric on different databases: (a) AT & T-ORL, (b) Extended Yale B, (c) Yale A, and (d) LFW.

A-LBP performs better to LBP. For example, the AT & T-ORL database the A-LBP achieved 95% recognition accuracy, whereas LBP reports an accuracy of 94.52%. Similarly for extended Yale B and Yale A databases, A-LBP performs better than the LBP. For extended Yale B the accuracy values are reported to 86.45% for A-LBP and 81.83% for LBP. For Yale A database, the proposed method reports a better accuracy of 71.9% in comparison to the LBP accuracy value of 60% (See Table 1).

The ROC curve for AT & T-ORL database is plotted and shown in Fig. 4(a). It shows that the GAR is found higher in the proposed A-LBP method and reported value of 78% when the FAR is strictly nil. As FAR increases, the GAR value is also increased. For example, the GAR finds 95.5% for LBP, 97% for the A-LBP at 7% of the FAR. The GAR is found maximum 100% at 17% for A-LBP and 21% for LBP. The A-LBP method achieves better recognition accuracy due to its insensitiveness to the mild changes in illumination, facial expression and occlusion. The ROC curve for the extended Yale B database is plotted and shown in Fig. 4(b). It shows that the GAR is found higher in A-LBP method and reported value of 32% when the FAR is strictly nil. As FAR increases, the GAR value is also increased for all methods, respectively. Such as, the GAR finds 81% for LBP, 90% for A-LBP at 14% of the FAR. The GAR is found maximum 100% at 62% of FAR for LBP and at 51% of FAR for A-LBP. The A-LBP method achieves better recognition accuracy due to its insensitiveness to the change of illumination.

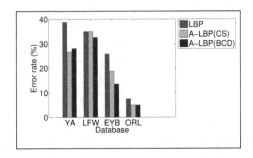

Fig. 5 Histogram of equal error rate for LBP, A-LBP(CS), and A-LBP(BCD) methods.

The ROC curve for Yale A database is plotted and shown in Fig. 4(c). It shows that the GAR is found higher in A-LBP method and reported value of 20% when the FAR is strictly nil. As FAR increases, the GAR value is also increased for all methods, respectively. Such as, the GAR finds 50% for LBP, 61% for A-LBP at 20% of the FAR. The GAR is found maximum 100% at 89% of FAR for LBP and 82% of FAR for A-LBP. The A-LBP method achieves better recognition accuracy due to its insensitiveness to the change of illumination. The ROC curve for LFW face database is plotted and shown in Fig. 4(d), which indicates that the GAR is found higher in A-LBP method and reported value of 15% at the zero FAR. As FAR increases, the GAR value is also increased for both methods, respectively. For example, the GAR finds 41% for LBP, 50% for A-LBP at 16% of the FAR. The GAR is found maximum 100% for both methods at 98% of FAR. In LFW database, the proportion of non uniform patterns is comparatively larger than the uniform patterns. As an effect, the feature descriptor of A-LBP becomes brighter that may loss the some discriminatory information. Therefore, a slight improvement in the result is only reported by our proposed method.

The histogram representation of recognition performance representing the error rate of the LBP, A-LBP (CS) using Chi square, A-LBP (BCD) using Bray Curtis dissimilarity metric on different databases, such as Yale A (YA), Labeled Faces in the Wild (LFW), extended Yale B (EYB) and AT & T-ORL is as shown in Fig. 5.

5   Conclusion

This paper has presented a novel method of face recognition under unconstrained environments. The proposed method has efficiently recognized the faces from variations, such as pose, illumination and expression. The experimental results has shown that the performance of the A-LBP method is improved to the original LBP for most of the databases. When there is a variation in illumination of images in the database, the BCD may be the better choice than the χ^2 metric. In such cases, the A-LBP shows better recognition accuracy than the original LBP. It is also found that the databases having variations in pose and facial expression, A-LBP performs superior than LBP.

Acknowledgments The authors acknowledge the Institute of Engineering and Technology (IET), Lucknow, Uttar Pradesh Technical University (UPTU), Lucknow for their partial financial support to carry out this research under the Technical Education Quality Improvement Programme (TEQIP-II) grant.

References

1. Turk, M.A., Pentland, A.P.: Eigenfaces for recognition. J. Cogn. Neurosci. **3**(1), 71–86 (1991)
2. Lu, J., Kostantinos, N.P., Anastasios, N.V.: Face recognition using LDA-based algorithms. IEEE Trans. Neural Networks **14**(1), 195–200 (2003)
3. Belhumeur, P.N., Hespanha, J.P., Kiregman, D.J.: Eigenfaces vs. Fisherfaces: Recognition Using Class Specific Linear Projection. IEEE Trans. Pattern Anal. Mach. Intell. **19**(7), 711–720 (1997)
4. Shyam, R., Singh, Y.N.: A Taxonomy of 2D and 3D Face Recognition Methods. In: Proc. of 1st Int'l Conf. on Signal Processing and Integrated Networks (SPIN 2014), pp. 749–754. IEEE, February 2014
5. Shyam, R., Singh, Y.N.: Identifying Individuals Using Multimodal Face Recognition Techniques. Procedia Computer Science, Elsevier **48**, 666–672 (2015)
6. Shyam, R., Singh, Y.N.: Face recognition using multi-algorithmic biometric systems. In: Fourth International Conference on Advances in Computing, Communications and Informatics (ICACCI-2015), Kochi, India. IEEE, TBA, August 2015
7. Ahonen, T., Hadid, A., Pietikainen, M.: Face Description with Local Binary Patterns: Application to Face Recognition. IEEE Trans. Pattern Anal. Mach. Intell. **28**(12), 2027–2041 (2006)
8. Ojala, T., Pietikainen, M., Harwood, D.: A Comparative Study of Texture Measures with Classification Based on Feature Distributions. Pattern Recogn. **29**(1), 51–59 (1996)
9. Shyam, R., Singh, Y.N.: Face recognition using augmented local binary patterns and Bray Curtis dissimilarity metric. In: Proc. of 2nd Int'l Conf. on Signal Processing and Integrated Networks (SPIN 2015), Noida, India, pp. 779–784. IEEE, February 2015
10. Liao, S., Law, M.W.K., Chung, A.C.S.: Dominant Local Binary Patterns for Texture Classification. IEEE Trans. Image Processing **18**(5), 1107–1118 (2009)
11. Heikkila, M., Pietikainen, M., Schmid, C.: Description of Interest Regions with Local Binary Patterns. Pattern Recogn. **42**(3), 425–436 (2009)
12. Zhang, L., Chu, R., Xiang, S., Liao, S., Li, S.: Face detection based on multiblock LBP representation. In: Proc. of Int'l Conf. on Biometrics. (2007)
13. Tan, X., Triggs, B.: Enhanced local texture feature sets for face recognition under difficult lighting conditions. In: Zhou, S.K., Zhao, W., Tang, X., Gong, S. (eds.) AMFG 2007. LNCS, vol. 4778, pp. 168–182. Springer, Heidelberg (2007)
14. Wolf, L., Hassner, T., Taigman, Y.: Descriptor based methods in the wild. In: Proc. of Workshop Faces in Real-Life Images: Detection, Alignment, and Recogn., Marseille, France, October 2008. https://hal.inria.fr/inria-00326729
15. Shyam, R., Singh, Y.N.: Analysis of local descriptors for human face recognition. In: Proc. of 3rd Int'l Conf. on Advanced Computing, Networking, and Informatics (ICACNI 2015), Orissa, India. Springer-Verlag, TBA, June 2015
16. Samaria, F., Harter, A.: Parameterisation of a Stochastic Model for Human Face Identification. In: Proc. of 2nd IEEE Workshop on Applications of Computer Vision, Sarasota, FL, December 1994

17. Lee, K.C., Ho, J., Kriegman, D.: Acquiring Linear Subspaces for Face Recognition under Variable Lighting. IEEE Trans. Pattern Anal. Mach. Intell. **27**(5), 684–698 (2005)
18. UCSD: Yale (2007). http://vision.ucsd.edu/content/yale-face-database
19. Huang, G.B., Ramesh, M., Berg, T., Learned-Miller, E.: Labeled faces in the wild: a database for studying face recognition in unconstrained environments. In: Tec. Report, University of Massachusetts, Amherst, pp. 07–49 (2007)

Empirical Wavelet Transform for Improved Hyperspectral Image Classification

T.V. Nidhin Prabhakar and P. Geetha

Abstract Capturing images in thousands of contiguous spectral bands has been made simpler with the emergence of technology in the field of hyperspectral remote sensing. Despite of these huge data available for analysis, Hyperspectral images (HSI) face many challenges due to high dimensionality, noise, spectral mixing and computational complexity. Several preprocessing methods can be used to overcome the above mentioned issues. In this paper, an enhancement technique using 2D-Empirical Wavelet Transform (EWT) is used as a preprocessing step for the HSI reconstruction prior to sparsity based classification (Subspace Pursuit and Orthogonal Matching Pursuit). The effectiveness of the proposed method is proved by comparing the classification results obtained with and without applying preprocessing. Experimental analysis shows a significant improvement in the classification accuracies i.e., for 40% of training samples, OMP shows an improvement in overall classification accuracy from 66.12% to 93.20% and SP shows an improvement from 66.36% to 92.74%.

1 Introduction

Hyperspectral remote sensing is concerned with collecting images of objects on the earth surface in several continuous spectral bands which are stacked together to form the hyperspectral data cube. Though hyperspectral images (HSI) are rich in spectral and spatial information, analysis of these data has become a tedious task due to issues like huge size of the data, noise and spectral mixing. By using efficient preprocessing techniques prior to data analysis like HSI classification can significantly improve the performance measures. A novel method that combines Empirical Mode Decompositions with wavelets is used for the dimensionality reduction of hyper-

T.V. Nidhin Prabhakar(✉) · P. Geetha
Centre for Excellence in Computational Engineering and Networking,
Amrita Vishwa Vidyapeetham, Coimbatore, TN, India
e-mail: nidhin89@gmail.com, p_geetha@cb.amrita.edu

© Springer International Publishing Switzerland 2016   393
S. Berretti et al. (eds.), *Intelligent Systems Technologies and Applications*,
Advances in Intelligent Systems and Computing 384,
DOI: 10.1007/978-3-319-23036-8_34

spectral images [6]. Norden E. Huang et al. [9] proposed an adaptive decomposition method to analyse non-stationary and nonlinear data which uses EMD to obtain a few number of finite Intrinsic Mode Functions (IMF) by decomposing the complex dataset. In [7], authors uses 2D EWT for images (2D data) by extending the adaptive 1D EWT [8]. Sparsity based classifiers shows better result when compared to conventional classifiers in terms of computational time and speed. In order to improve the classification performance, authors in [2] presents different approaches which includes the contextual information into sparsity based optimization problem.

In this paper, Empirical Wavelet Transform (EWT) is used for the reconstruction of HSI. This enhancement technique is followed by sparsity based classifications using Subspace Pursuit (SP) and Orthogonal Matching Pursuit (OMP) to prove the effectiveness of the proposed method. Rest of paper is organized as follows. Section 2 presents the reconstruction of HSI using EWT. An overview of sparsity based HSI classification is given in section 3. Section 4 describes the proposed method. Section 5 discuss about experimental results and analysis and section 6 concludes the paper.

2   Hyperspectral Reconstruction Using EWT

1D-Empirical Wavelet Transform (EWT) are used for reconstruction of signals [7, 8]. Unlike wavelet transform, 1D-EWT is adaptive which shows that it depends on the signal length. EWT involve two main steps. i.e., building the wavelet corresponding to the detected Fourier supports and filtering the input signal using the filter banks to obtain the different components. The transformation using EWT yields two sets of coefficients-detail and approximation coefficients.

Details coefficients can be denoted as $W_f^\varepsilon (n, t)$ and is given by (1),(2),(3).

$$W_f^\varepsilon (n, t) = \langle f, \psi_n \rangle \tag{1}$$

$$= \int f(\tau) \overline{\psi_n (\tau - t)} d\tau \tag{2}$$

$$= \left(\hat{f}(\omega) \overline{\hat{\psi}_n (\omega)} \right)^* \tag{3}$$

Approximation coefficients can be denoted as $W_f^\varepsilon (0, t)$ and is given by (4),(5),(6).

$$W_f^\varepsilon (0, t) = \langle f, \phi_1 \rangle \tag{4}$$

$$= \int f(\tau) \overline{\phi_1 (\tau - t)} d\tau \tag{5}$$

$$= \left(\hat{f}(\omega) \overline{\hat{\phi}_1 (\omega)} \right)^* \tag{6}$$

The above mentioned coefficients (approximation and detail) are used for reconstructing the original signal. Reconstruction is done using (7),(8).

$$f(t) = W_f^\varepsilon (0, t) * \phi_1 (t) + \sum_{n=1}^{N} W_f^\varepsilon (n, t) * \psi_n (t) \tag{7}$$

$$= \left(\widehat{W_f^\varepsilon} (0, \omega) \, \widehat{\phi_1} (\omega) + \widehat{W_f^\varepsilon} (n, \omega) \, \widehat{\psi_n} (\omega) \right)^* \tag{8}$$

In this paper, 2D-EWT is used for the band by band reconstruction of hyperspectral imagery. 2D-EWT is an extension of 1D-EWT which is obtained by applying EWT to both rows and columns of the 2D data (image).

3 Sparsity Based HSI Classification

In recent years, sparsity based algorithms are widely used for hyperspectral image classification. In this approach, the pixels belonging to the same class are represented in a lower dimensional subspace, which makes it a fair and fast classification method. During supervised classification, the HSI is divided into training and testing samples. The training samples (with known class labels) are randomly selected and the whole image data is given as testing samples to find the class labels.

The dictionary matrix for a D dimensional HSI is given as $A = [A_1, ..., A_N]$ with subdirectory $A_i = [a_1, a_2 ..., a_{n_i}]$. A_i is the training vector in i^{th} class, $i \in N$ and N is the total number of class. The sparse constrained optimization problem is formulated as,

$$\hat{x} = \arg \min \|x\|_0$$
$$subjected \ to \ Ax = y \tag{9}$$

where x is the sparse vector used to find the class labels of the given test pixel vectors. And minimization of $\|x\|_0$ gives the sparse solution for Ax = y.

The class of the test pixel vector is determined using,

$$Class(y) = \arg \min_{i=1,...,N} (r_i) \tag{10}$$

where r_i is residue w.r.t y. In this paper, authors use sparsity based classifiers such as Subspace Pursuit (SP) and Orthogonal Matching Pursuit (OMP) for HSI classification. The detailed description of these algorithms is given in [1, 11, 12].

4 Proposed Method

The proposed method involves reconstruction of hyperspectral data cube using 2D-EWT followed by sparsity based classification (Subspace Pursuit and Orthogonal Matching Pursuit). Standard Indian Pines dataset is chosen for doing the proposed experiment. The flow graph of the proposed method is shown in Fig. 1.

Each band of the hyperspectral image is subjected to EWT based reconstruction. This preprocessing helps to enhance each image in hyperspectral data cube and thereby improves sparsity based classification. Pixels present in the HSI are separated into training and testing samples during the classification phase. Different number of samples (10%, 20%, 30% and 40% of the whole dataset) are randomly selected for the training. The proposed sparsity based classification represents the test pixel as a linear combination of a few numbers of training samples from the entire training set. Classification accuracy is improved by increase in the training samples.

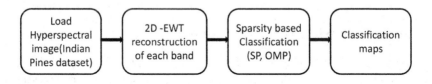

Fig. 1 Block Diagram of Proposed Method

5   Experimental Result Analysis

This section discusses about the dataset used for the proposed method, Accuracy assessment measures considered and results analysis.

5.1   Dataset Description

Indian Pines dataset is captured by the Airborne Visible Infrared Imaging Spectrometer (AVIRIS) sensor over Indiana in 1992. It has 220 bands, each with a size of 145x145. 20 bands which correspond to the water absorption regions of the spectrum were removed before processing. It consists of 16 classes which includes crops, vegetation, building, road and railway. Fig. 2(a) and Fig. 2(b) show the colour composite image and ground truth image of Indian Pines dataset.

5.2   Accuracy Assessment Measures

Quantitative assessment of HSI classification uses confusion matrix for the calculation of various classification indices such as Overall Accuracy (OA), Average Accuracy (AA), Classwise Accuracy (CA) and Kappa Coefficient. Comparison of classification results obtained with the ground truth data (reference image) helps to assess the classification accuracy. The agreement of classification is quantified using Kappa coefficient.

$$OA = \frac{Total\ number\ of\ correctly\ classified\ pixels}{Total\ number\ of\ pixels} \tag{11}$$

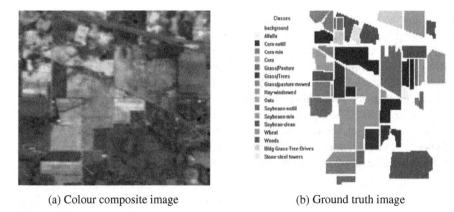

(a) Colour composite image   (b) Ground truth image

Fig. 2 Indian Pines data scene with class description

$$AA = \frac{Sum\ of\ accuracies\ of\ each\ class}{Total\ number\ of\ class} \tag{12}$$

$$CA = \frac{Correctly\ classified\ pixels\ in\ each\ class}{Total\ number\ of\ pixels\ in\ each\ class} \tag{13}$$

Kappa coefficient (Kappa statistic or Cohen's Kappa) is a measure of the classifier which is used to evaluate the performance of classifiers. It takes values in the range of [0,1]. For a 4-class classification problem with labels L=[1,2,3,4], the Kappa coefficient is given as,

$$\text{Kappa}(\kappa) = \frac{N\sum_{i=1}^{4} a_{ii} - \sum_{i=1}^{4}\left(\sum_{j=1}^{4} a_{ij} \times \sum_{j=1}^{4} a_{ji}\right)}{N^2 - \sum_{i=1}^{4}\left(\sum_{j=1}^{4} a_{ij} \times \sum_{j=1}^{4} a_{ji}\right)} \tag{14}$$

where N be the total number of pixels in the HSI image and a_{ij} be the each element.

5.3 Results and Discussion

In this paper, the experiment is conducted on standard Indian Pines dataset. The effectiveness of preprocessing using 2D-Empirical Wavelet Transform (2D-EWT) is illustrated by considering its effect on sparsity based classification. The classification accuracies obtained before and after applying preprocessing is compared to prove the efficiency of the proposed method.

Band 32 of Indian Pines dataset is chosen as a sample band to show the effect of reconstruction using EWT. Fig. 3(a) shows the original band 32 of Indian Pines datascene and the reconstructed band is shown in Fig. 3(b). Classification involves generating training and testing samples from the ground truth information. The training samples are randomly selected from each class (generally, 10%, 20%, 30% or 40%) and rest of the samples are given for testing. Table 1 shows the effect of reconstruction using 2D-EWT on OMP based classification. By analysing the table, it is inferred that, for 10% training set, the overall accuracy has been improved from 41.50% to 80.46%. Accuracy increases with increase in the training samples. Thus for 40% training set, the improvement is from 66.12% to 93.20%. Fig. 4 (a) and (b) shows the classification results of OMP based classification (for 40% training set) before and after applying preprocessing respectively. The effect of band by band

(a) Original band   (b) After EWT reconstruction

Fig. 3 Band 32 of Indian Pines dataset

Table 1 Classification results of OMP without and with EWT preprocessing

	Percentage of training	10		20		30		40	
Class	Class name	OMP	EWT+OMP	OMP	EWT+OMP	OMP	EWT+OMP	OMP	EWT+OMP
Class 1	Alfalfa	56.52	97.83	67.39	97.83	69.57	93.48	63.04	95.65
Class 2	Corn-notill	38.73	68.42	48.74	79.41	60.43	87.54	67.37	90.48
Class 3	Corn-mintill	32.41	60.96	39.04	73.13	52.29	81.69	57.35	85.18
Class 4	Corn	65.82	61.60	68.35	72.15	73.00	74.26	81.01	81.43
Class 5	Grass-pasture	28.99	90.06	44.51	92.75	46.17	95.24	60.46	94.82
Class 6	Grass-trees	27.67	95.07	41.37	98.22	50.27	99.18	53.97	99.18
Class 7	Grass-pasture-mowed	89.29	100.00	78.57	96.43	89.29	96.43	92.86	96.43
Class 8	Hay-windrowed	40.79	96.03	48.12	99.16	59.41	99.58	62.13	98.95
Class 9	Oats	100.00	100.00	100.00	100.00	100.00	100.00	100.00	100.00
Class 10	Soybean-notill	40.33	77.67	52.98	84.36	57.00	88.58	68.72	91.15
Class 11	Soybean-mintill	50.10	81.87	61.38	88.51	64.64	90.02	72.67	94.34
Class 12	Soybean-clean	37.94	65.09	44.35	74.54	54.13	78.75	60.54	84.65
Class 13	Wheat	31.71	95.61	48.78	97.56	47.32	97.56	57.56	97.56
Class 14	Woods	38.89	95.65	49.57	97.71	53.12	98.81	65.77	99.13
Class 15	Buildings-Grass-Trees-Drives	53.37	74.35	59.84	87.56	75.39	91.45	66.84	94.04
Class 16	Stone-Steel-Towers	61.29	98.92	51.61	97.85	68.82	98.92	75.27	100.00
	Average Accuracy	49.62	84.95	56.54	89.82	63.80	91.97	69.10	93.94
	Overall Accuracy	**41.50**	**80.46**	**51.64**	**87.26**	**58.61**	**90.61**	**66.12**	**93.20**
	Kappa	0.3413	0.7770	0.4516	0.8546	0.5302	0.8930	0.6142	0.9224

Table 2 Classification results of SP without and with EWT preprocessing

Percentage of training		10		20		30		40	
Class	Class name	SP	EWT+SP	SP	EWT+SP	SP	EWT+SP	SP	EWT+SP
Class 1	Alfalfa	50	97.83	63.04	97.83	71.74	97.83	73.91	100.00
Class 2	Corn-notill	39.78	79.83	50.84	80.25	60.71	88.45	69.68	90.34
Class 3	Corn-mintill	31.33	57.71	39.76	73.49	52.89	78.67	62.29	85.54
Class 4	Corn	65.40	66.67	67.09	72.57	72.15	81.01	77.64	78.90
Class 5	Grass-pasture	27.12	85.09	44.31	92.34	55.49	95.24	57.76	95.24
Class 6	Grass-trees	27.40	96.03	44.52	98.08	52.33	99.32	60.55	98.63
Class 7	Grass-pasture-mowed	89.29	96.43	78.57	96.43	92.86	96.43	85.71	92.86
Class 8	Hay-windrowed	41.00	97.07	51.05	98.74	51.88	97.91	66.32	99.37
Class 9	Oats	100.00	100.00	100.00	100.00	100.00	100.00	100.00	100.00
Class 10	Soybean-notill	46.81	77.88	50.62	84.47	59.77	90.12	62.55	90.64
Class 11	Soybean-mintill	51.20	81.91	57.47	88.59	65.30	91.04	71.24	92.22
Class 12	Soybean-clean	31.37	67.28	47.05	75.55	54.30	84.99	64.59	87.35
Class 13	Wheat	37.56	97.56	40.98	96.59	44.88	97.56	55.61	99.51
Class 14	Woods	27.91	96.52	50.67	97.79	55.02	98.58	63.08	98.66
Class 15	Buildings-Grass-Trees-Drives	61.66	80.31	60.62	88.08	65.28	92.49	70.73	94.04
Class 16	Stone-Steel-Towers	61.29	98.92	66.67	98.92	61.29	98.92	68.82	100.00
	Average Accuracy	49.32	86.07	57.08	89.98	63.49	93.03	69.40	93.96
	Overall Accuracy	**40.99**	**82.30**	**51.44**	**87.47**	**59.10**	**91.37**	**66.36**	**92.74**
	Kappa	0.3344	0.7980	0.4505	0.8570	0.5352	0.9016	0.6170	0.9173

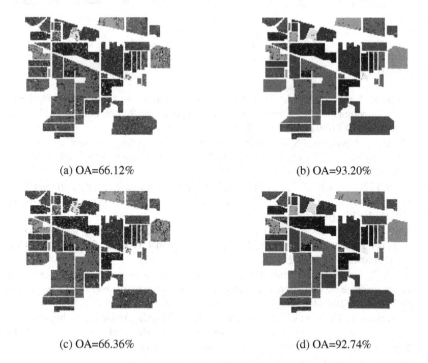

(a) OA=66.12%   (b) OA=93.20%

(c) OA=66.36%   (d) OA=92.74%

Fig. 4 Classification maps (for 40% training set) : (a) OMP based classification without applying EWT (b) OMP based classification after applying EWT (c) SP based classification without applying EWT (d) SP based classification after applying EWT

reconstruction using 2D-EWT on SP based classification is given in Table 2. With 10% training set, this classifier gives an overall accuracy of 82.30% whereas with 40% samples, it provides 92.74% overall accuracy. Classification map of SP based classification (for 40% training set) with and without preprocessing is given in Fig. 4 (c) and (d) respectively.

6 Conclusion

This paper discusses an efficient preprocessing technique using 2D-Empirical Wavelet Transform (2D-EWT) which helps to improve the classification accuracies of sparsity based classifiers (Orthogonal Matching Pursuit and Subspace Pursuit). Reconstruction of each band of HSI using EWT helps to enhance the image quality. And so significant improvements in accuracies are obtained for HSI classification by using sparsity constrained optimization problem, which are solved by greedy algorithms such as OMP and SP.

Acknowledgments The authors would like to thank Dr. Soman K P, Head of the department, Centre for Excellence in Computational Engineering and Networking, Amrita Vishwa Vidyapeetham and Ms. V. Sowmya, Assistant Prof., Centre for Excellence in Computational Engineering and Networking, Amrita Vishwa Vidyapeetham for their valuable comments and suggestions.

References

1. Chen, Y., Nasrabadi, N.M., Tran, T.D.: Hyperspectral image classification via kernel sparse representation. IEEE Transactions on Geoscience and Remote Sensing **51**, 217–231 (2013). doi:10.1109/TGRS.2012.2201730
2. Chen, Y., Nasrabadi, N.M., Tran, T.D.: Hyperspectral image classification using dictionary-based sparse representation. IEEE Trans. Geosci. Remote Sens. **49**, 3973–3985 (2011). doi:10.1109/TGRS.2011.2129595
3. Chen, Y., Nasrabadi, N.M., Tran, T.D.: Simultaneous joint sparsity model for target detection in hyperspectral imagery. IEEE Geoscience and Remote Sensing Letters **8**, 676–680 (2011). doi:10.1109/LGRS.2010.2099640
4. Chen, Y., Nasrabadi, N.M., Tran, T.D.: Sparse representation for target detection in hyperspectral imagery. IEEE J. Sel. Topics Signal Process. **5**, 629–640 (2011). doi:10.1109/JSTSP.2011.2113170
5. Demir, B., Erturk, S.: Empirical mode decomposition of hyperspectral images for support vector machine classification. IEEE Transactions on Geoscience and Remote Sensing **48**, 4071–4084 (2010). doi:10.1109/TGRS.2010.2070510
6. Gormus, E.T., Canagarajah, N., Achim, A.: Dimensionality reduction of hyperspectral images using empirical mode decompositions and wavelets. IEEE Journal of Selected Topics in Applied Earth Observations and Remote Sensing **5**, 1821–1830 (2012). doi:10.1109/JSTARS.2012.2203587
7. Gilles, J., Tran, G., Osher, S.: 2D Empirical transforms wavelets, ridgelets, and curvelets revisited. SIAM Journal on Imaging Sciences **7**, 157–186 (2014)

8. Gilles, J.: Empirical wavelet transform. IEEE Transactions on Signal Processing **61**, 3999–4010 (2013). doi:10.1137/130923774
9. Huang, N.E., Shen, Z., Long, S.R., et al.: The empirical mode decomposition and the hilbert spectrum for nonlinear and non-stationary time series analysis. In: Proceedings of the Royal Society of London A: Mathematical Physical and Engineering Sciences, vol. 454, pp. 903–995. The Royal Society, March 1998. doi:10.1098/rspa.1998.0193
10. Jensen, J.R.: Introductory Digital Image Processing A Remote Sensing Perspective. Prentice Hall Inc., Upper Saddle River (1996)
11. Tropp, J.A., Gilbert, A.C.: Signal recovery from random measurements via orthogonal matching pursuit. IEEE Transactions on Information Theory **53**, 4655–4666 (2007). doi:10.1109/TIT.2007.909108
12. Dai, W., Milenkovic, O.: Subspace pursuit for compressive sensing signal reconstruction. IEEE Transactions on Information Theory **55**, 2230–2249 (2009). doi:10.1109/TIT.2009.2016006

Prediction of Urban Sprawl Using Remote Sensing, GIS and Multilayer Perceptron for the City Jaipur

Pushpendra Singh Sisodia, Vivekanand Tiwari and Anil Kumar Dahiya

Abstract The population of India has rapidly increased from 68.33 million to 121.01 million from 1981 to 2011, respectively. It is estimated that by the year 2028 India will hold the largest population of the world. The prompt upsurge of the Indian population will force people to migrate from the rural area to the mega cities, to avail basic amenities. The enormous migration will increase the demand for more space to live in mega cities and will lead to a situation of unauthorized, unplanned, uncoordinated, and uncontrolled growth, and this condition called as urban sprawl. The key challenge for a planner is to achieve sustainable development and to predict the future urban sprawl in the city. Unfortunately, conventional techniques that predict urban sprawl are expensive and time consuming. In this paper, we have proposed a novel technique to predict the future urban sprawl. We have used an integrated approach of Remote Sensing, GIS, and Multilayer perceptron to predict the future urban sprawl for the city Jaipur up to 2021. We have compared our results with existing techniques like Linear Regression and Gaussian Process and found that the Multilayer perceptron gives better results than other existing techniques.

Keywords Remote sensing · GIS · MLP · Urban sprawl

1 Introduction

The population of India has rapidly increased from 68.33 million to 121.01 million during the year 1981 to 2011. It is estimated that by the year 2028 India will hold the largest population of the world [1-2]. This expeditious upsurge of the Indian

P.S. Sisodia(✉) · V. Tiwari · A.K. Dahiya
Manipal University Jaipur, Jaipur-Ajmer Express Highway, Dehmi Kalan,
Near GVK Toll Plaza, Jaipur 303007, Rajasthan, India
e-mail: muj@gmail.com, tiwari@jaipur.manipal.edu, dahiyaanil@yahoo.com

© Springer International Publishing Switzerland 2016   403
S. Berretti et al. (eds.), *Intelligent Systems Technologies and Applications*,
Advances in Intelligent Systems and Computing 384,
DOI: 10.1007/978-3-319-23036-8_35

population will force the Indian mega cities to populate. The process of inhabitation of mega cities will lead more urbanisation. Urbanisation is the process of converting unutilised land for habitation [3].

This conversion forced people to migrate from rural area to mega cities. This enormous migration demanded more space to live in mega cities. The demand for more space in mega cities will lead the situation of unauthorised, unplanned, uncoordinated, and uncontrolled growth, and this condition called as urban sprawl [4]. The impact of urban sprawl is observed in the mega cities in the form of lack of housing, water, electricity, sanitation, hospitals, and other basic amenities [5]. Urban sprawl is an obstruction to achieve sustainable development, and city planners are facing the problem to achieve sustainable development and predict the future urban sprawl in the city. Unfortunately, it is not possible to predict urban sprawl by conventional techniques.

Recently, researcher's interests are being directed to the prediction of urban sprawl by using new techniques such as Remote Sensing (RS) and Geographical Information System (GIS) [6]. Remote Sensing and GIS is widely used techniques for land use, land cover change detection. Many models like Cellular Automata (CA), Agent Based Modelling, and Artificial Neural Network have been successfully applied to the urban growth prediction [7-9]. Despite of all this method, more efficient and less complex method is required to predict future urban sprawl in the city. In this paper, we have proposed a novel technique to predict future urban sprawl. We have used an integrated approach of Remote Sensing, GIS, Shannon's Entropy and Multilayer perceptron (MLP).

2   Study Area

Jaipur is the capital city of the state Rajasthan and known as pink city. Jaipur is situated in 26 92'N to 75 82' E. Jaipur was one of the planned city of India and surrounded by markets and walls, but due to the increase in population it is

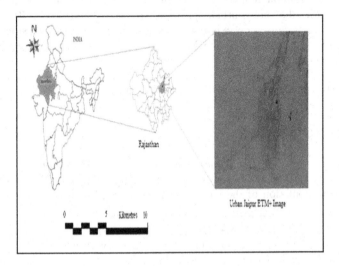

Fig. 1 Study Area [10]

continuously expanding. The urban population of the city Jaipur has increased from 1180038 to 3499204 during the year 1981 to 2011, respectively, and the urban area has expanded more than 400 km^2. The study area has been shown in the Figure 1.

2.1   Data Used in the Study

Data used in the study are listed in Table 1. We have used Remote Sensing images for decade years from 1981 to 2011 that is obtained from public domain [11] and other necessary data have been collected from various Government agencies.

Table 1 Details of data used in the study

S. No	Type of Data	Sensor/ Toposheet	Year	Scale and Resolution in meter	Path	Row
1	Landsat	MSS, TM ,ETM+	1981,1991, 2001,2011	30,80	147	041
2	Toposheets	G43D12, G43D16 G43J9, G43J13 G43J10, G43J14	2013	1:50000, 1:2,50,000	-	-
3	Zone Map	-	2011	-	-	-

3   Methodology

The methodology of entire research work has been shown by the methodology flowchart in the Figure 2.

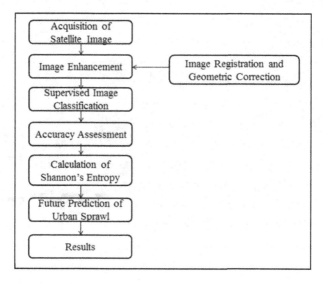

Fig. 2 Flow chart of methodology

3.1 Acquisition of Satellite Images

All satellite images have been obtained from the public domain of the United States Geological Survey. We have obtained the decadal Landsat images (MSS, TM, and ETM+) of the year 1981, 1991, 2001 and 2011.

3.2 Image Registration and Geometric Correction

Image registration and geometric correction process has been carried out with the help of toposheets, ground control points, Everest datum projection, and nearest neighbour techniques.

3.3 Image Enhancement

We have used contrast stretching technique for image enhancement for distinguishing all the features of the satellite images.

3.4 Supervised Image Classification

Maximum Likelihood Classification is the most suitable classification technique for the land use [12]. ERDAS IMAGIN 2014 software has been used for supervised classification. Classified images are shown in Figure 3.

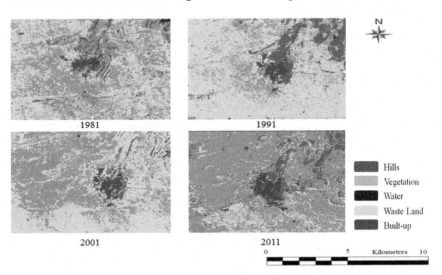

Fig. 3 Classification results of satellite images

3.5 Accuracy Assessment

We have calculated over all Kappa Statistics, User Accuracy, Overall Accuracy and Producer Accuracy for the classified satellite images of MSS, TM, and ETM+.

3.6 Shannon's Entropy

Shannon's Entropy is widely used to measure the uncertainty of random variations in Information Communication. We have used Shannon's Entropy to measure the degree of urban sprawl in our study area. Shannon's Entropy can be written as:

$$H_n = -\sum_{i=1}^{n} P_i * ln(P_i)$$

Here, H_n is the Shannon's Entropy and P_i is the ratio of built-up area of ith zone over the total sum of all zone's built-up area and n is the total number of zones.

3.7 Future Prediction of Urban Sprawl

In this paper, we have used a novel approach to predict future urban sprawl in the city. Initially, we have taken classified images of the year 1981, 1991, 2001, & 2011, and obtained the outer boundary of the urban area for each year. The boundary of each urban area is arranged according to their respective years simultaneously, and one centre point has been chosen to calculate the distance from the centre point to the urban area boundary. We have divided the urban areas with 32 lines in 360^0 clockwise as shown in Figure 4.

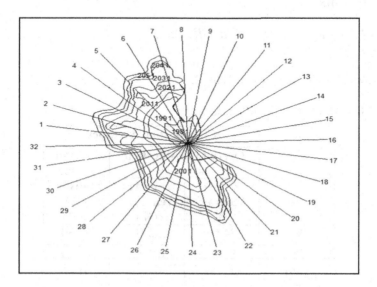

Fig. 4 Division of study area in by 32 reference lines

We have used these 32 lines to obtain the 32 reference points on each urban area boundary for each year. These points are the distance from the centre point to the urban area boundary. These points are used as the inputs in Multilayer Perceptron (MLP) as shown in Figure 5.

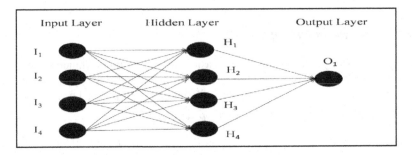

Fig. 5 Multilayer perceptron (MLP)

In Figure 5, arrows are representing the feed forward process. MLP calculates the weights for the input layer, the hidden layer and the output layer in a feed forward manner. In MPL, weights are determined by the training algorithms, and Back Propagation (BP) algorithm is the widely accepted training algorithm. The difference of the calculated output to the desired output of all observations can be summarised as the mean squared error. Now, the weights are modified according to the generalised delta rules [13] that can help to distribute the total error in all the nodes.

4   Results

4.1   Accuracy of Image Classification

Accuracy of the classified images of the year 1981, 1991, 2001, and 2011 has been calculated and as shown in Table 2.

Table 2 Details of Classification Accuracy

Year	Classification Accuracy		
	Overall Classification Accuracy		Overall Kappa Statistics (%)
	Producer Accuracy (%)	User Accuracy (%)	
1981	84.60	82.50	75.30
1991	94.30	93.70	89.90
2001	95.00	95.26	91.00
2011	93.85	94.80	89.60

4.2   Shannon's Entropy (H_n)

We have calculated the Shannon's Entropy (H_n) 1.81, 1.83, 1.86 and 1.88 of the year 1981, 1991, 2001, and 2011 respectively. If the Shannon's Entropy is near to $ln(n)$ that is 2.08 for our study, we considered it as more urban dispersal. In our study, the value of Shannon's Entropy is 1.88 that is nearer to 2.08 this continuous increment in Shannon's Entropy shows significant urban sprawl in the city.

4.3   Accuracy of Multilayer Perceptron (MLP)

We got Accuracy of Linear Regression, Gaussian Process, and Multilayer perceptron as 74.31%, 83.56 and 90.41% respectively. Multilayer Perceptron (MLP) has given better results than others. Finally, we predicted the total urban area with urban boundary for year 2021. The total urban area from the year 1981, 1991, 2001, and 2011 has been increased 40,132, 200, and 317 km^2 respectively. The predicted urban area will be 500 km^2 for year 2021. The maximum of the urban sprawl have been noticed along with the major highways like Delhi, Sikar, Ajmer and Tonk road. Whereas, only limited urban sprawl has been noticed towards Agra road as shown in Figure 6.

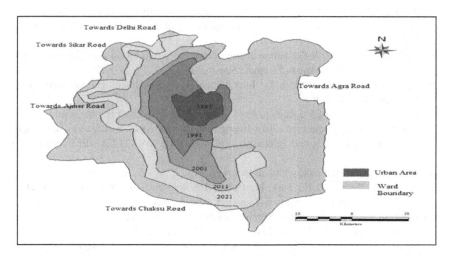

Fig. 6 Predicted Urban Sprawl

5   Conclusion

We have predicted the future urban sprawl for the year 2021. We have noticed that maximum urban sprawl will be directed towards major highways of the city. Many industries and higher education institutes are also situated in proximity of the highways that will lead more urban sprawl. In the future, this study will help the municipality and local development authorities for future evolution.

Acknowledgement This work is partially supported by Manipal University Jaipur vide file number of MUJ/REG/OO/2013-116.

References

1. Census of India (2011). http://www.census2011.co.in/city.php
2. World population (2002). http://esa.un.org/wpp/Documentation/pdf/WPP2012_Press_Release.pdf
3. Ramachandra, T.V., Aithal, B.H.: Urbanisation and Sprawl in the Tier II City: Metrics, Dynamics and Modelling Using Spatio-Temporal Data
4. Sudhira, H.S., Ramachandra, T.V., Jagadish, K.S.: Urban sprawl: metrics, dynamics and modelling using GIS. International Journal of Applied Earth Observation and Geoinformation **5**(1), 29–39 (2004)
5. Rahman, A., Aggarwal, S.P., Netzband, M., Fazal, S.: Monitoring urban sprawl using remote sensing and GIS techniques of a fast growing urban centre, India. IEEE Journal of Selected Topics in Applied Earth Observations and Remote Sensing **4**(1), 56–64 (2011)
6. Epsteln, J., Payne, K., Kramer, E.: Techniques for mapping suburban sprawl. Photogrammetric Engineering & Remote Sensing **63**(9), 913–918 (2002)
7. Torrens, P.M., O'Sullivan, D.: Cellular automata and urban simulation: where do we go from here? Environment and Planning B: Planning and Design **28**(2), 163–168 (2001)
8. Liu, X., Lathrop Jr, R.G.: Urban change detection based on an artificial neural network. International Journal of Remote Sensing **23**(12), 2513–2518 (2002)
9. Batty, M.: Urban modeling. In: International Encyclopedia of Human Geography. Elsevier, Oxford (2009)
10. Sisodia, P.S., Tiwari, V., Kumar, A.: A comparative analysis of remote sensing image classification techniques. In: 2014 International Conference on Advances in Computing, Communications and Informatics (ICACCI), pp. 1418–1421. IEEE, September 2014
11. USGS (2013). http://landsatlook.usgs.gov
12. Akgün, A., Eronat, A.H., Turk, N.: Comparing different satellite image classification methods: an application in ayvalýk district, western turkey. In: XX International Congress for Photogrammetry and Remote Sensing, Istanbul, Turkey (2004)
13. Rumelhart, D.E., McClelland, J.L., and the PDP Research Group: Parallel Distributed Processing: Explorations in the Microstructure of Cognition, vol. 1 and 2. MIT Press, Cambridge (1986)

Contourlet and Fourier Transform Features Based 3D Face Recognition System

S. Naveen and R.S. Moni

Abstract Human face recognition based on geometrical structure has been an area of interest among researchers for the past few decades especially in pattern recognition. 3D Face recognition systems are of interest in this context. The main advantage of 3D Face recognition is the availability of geometrical information of the face structure which is more or less unique for a subject. This paper focuses on the problems of person identification using 3D Face data. Use of unregistered 3D Face data for feature extraction significantly increases the operational speed of the system with huge database enrollment. In this work, unregistered Face data, i.e. both texture and depth is fed to a classifier in spectral representations of the same data. 2-D Discrete Contourlet Transform and 2-D Discrete Fourier Transform is used here for the spectral representation which forms the feature matrix. Fusion of texture and depth statistical information of face is proposed in this paper since the individual schemes are of lower performance. Application of statistical method seems to degrade the performance of the system when applied to texture data and was effective in the case of depth data. Fusion of the matching scores proves that the recognition accuracy can be improved significantly by fusion of scores of multiple representations. FRAV3D database is used for testing the algorithm.

Keywords Point cloud · Rotation invariance · Pose correction · Depth map · Contourlet transform · Fourier transform · Spectral transformations · CDF · Texture map and principal component analysis

S. Naveen(✉)
Department of ECE, LBS Institute of Technology for Women,
Trivandrum 695012, Kerala, India
e-mail: nsnair11176@gmail.com

R.S. Moni
Department of ECE, Marian Engineering College, Trivandrum 695582, Kerala, India
e-mail: monirs2006@gmail.com

© Springer International Publishing Switzerland 2016   411
S. Berretti et al. (eds.), *Intelligent Systems Technologies and Applications*,
Advances in Intelligent Systems and Computing 384,
DOI: 10.1007/978-3-319-23036-8_36

1   Introduction

3D Face recognition has been an active area of research in the past decades. The complications encountered in the enrolment phase and the huge computational requirements in the implementation phase have been the major hindrance in this area of research. The scenario has improved tremendously due to the latest innovations in 3D imaging devices and has made 3D Face recognition system a reliable option in security systems based on Biometrics. Though poor resolution is a major drawback encountered in 3D Face images the geometrical information present in 3D facial database can be exploited to overcome the challenges in 2D face recognition systems like pose variations, bad illumination, ageing etc.

In this work, focus is made on an identification problem based on 3D Face data using fusion schemes. Identification corresponds to the person recognition without the user providing any information other than the 3D facial scan. The system arrives at an identity from among the enrolled faces in the database. Use of texture information along with the geometrical information of the face seems to improve the recognition accuracy of face recognition system when pose correction is not done as a pre-processing step. Spectral transformation of depth and texture data seems to improve the recognition and authentication efficiency. Here statistical processing of depth data in spectral domain is done so as to improve the recognition accuracy.

Alexander M. Bronstein et al. [2] proposed an idea of face recognition using geometric invariants using Geodesic distances. C. Beumier [3] utilized parallel planar cuts of the facial surfaces for comparison. Gang Pan et al [4] extracted ROI of facial surface by considering bilateral symmetry of facial plane. Xue Yuan et al [5] proposed a face recognition system using PCA, Fuzzy clustering and Parallel Neural networks. Trina Russ et al [6] proposed a method in which correspondence of facial points is obtained by registering 3D Face to a scaled generic 3D reference face. AjmalMian et al [7] used Spherical Face Representation for identification. OndrejSmirg et al [8] used DWT for gender classification since the DWT best describes the features after de-correlation. HuaGao et al [9] used Active Appearance model for fitting faces with pose variations. et al [10], used 2D-PCA for getting the feature matrix vectors and used Euclidean distance for classification. , OmidGervei et al [11] proposed an approach for 3D Face recognition based on extracting principal components of range images by utilizing modified PCA methods namely 2D-PCA and bidirectional 2D-PCA. SoodamaniRamalingam [13] created a database using stereo disparity maps and the directional information from the range images were used for feature extraction. Mohammad Dawi et al [14] used a new algorithm called stereo cluster search for generating enhanced disparity maps and used it for matching purpose using PCA. GauravGoswami et al [15] proposed feature extraction using entropy of RGB-D faces along with the saliency feature obtained from a 2D face, RGB-D obtained using Kinect Sensor. Tomas Mantecon et al [16] used modified version of local binary pattern algorithm which uses high resolution Kinect sensor acquired depth data. Madal et al [17] used PCA to extract feature set from the subbands created using Contourlet Transform.

Since only a sparse set of points are available in the dataset, it is necessary to increase the data density by using multiple data representations generated from same raw data. For this the data is transformed into spectral domain using Discrete ContourletTransform(DCLT) and Discrete Fourier Transform (DFT). Use of Depth information alone is not sufficient for an efficient recognition system since pose correction is not done. So texture information is also incorporated with the fusion scheme.

2   Proposed Scheme for Face Recognition

The system aims at extracting the feature from the input data through feature extraction tools and fuses the scores to get a system with better recognition accuracy. The main feature extraction principle used in this system is the spectral transformation. The spectral transformation tool used here is 2-DDCLTand 2-D DFT. These spectral transformations transform the data to a better representation which increases the accuracy of recognition system [18]. Block diagram of the proposed scheme is shown in Fig.1. The most important part of this work lies in the pattern

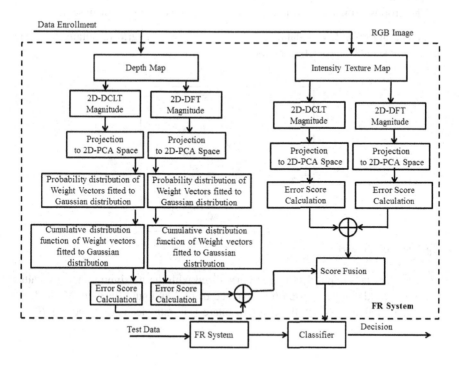

Fig. 1 Proposed Method

classification problem. A pattern of data points is available. This pattern is not sufficient for the recognition system to work since the data will be highly occluded due to pose variations in the X, Y and Z axis or in any complex plane.

The 3D Face recognition scheme is affected by pose variations of the subject (person under consideration). There are methods available in which the correction to this effect of pose variations also is included. One such method is the Iterative Closest Point (ICP) algorithm [23].

But the main disadvantage of these methods is that a reference face is to be used as a model for other rotated faces to be corrected. Also the processing time taken is very high. Therefore, in this work done, this correction to the effect of pose variation is not considered.

The idea behind spectral representation of data is that, when data is in spatial domain, comparison will be done as one to one pixel level or voxel level. So the rotation and translation of data will highly affect the result. When spectral transformation is done the distributed data will be concentrated or it may be represented in a more uniform way. The translation and rotation invariance properties of the transformations used will aid to improve the accuracy of system significantly. Here FRAV3D database is used. It contains the facial data with different face orientations and expressions. When depth information alone was considered the Face recognition accuracy (FRA) was not high. So texture data of face is also considered which significantly improves the FRA.

The proposed method involves the following steps given below in sequence.

1) The 3D Point Cloud dataFig.2 is projected on a 2D grid to obtain the depth data using standard projection method.

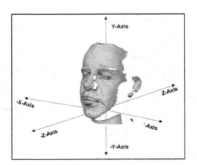

Fig. 2 D Face on space

2) The 2D face depth data is first normalized with the maximum intensity value. From this 2D depth map nose tip is detected using Maximum Intensity Method and the area around the nose (ROI-Region of Interest) is extracted (Fig. 5 and Fig. 6).

3) On this ROI data, 2-DDCLT and 2-D DFT is applied. The detailed explanations are given on following sections. Simultaneously 2-D DCLT and 2-D DFT is applied over the complete face texture data.

4) Once spectral representations are obtained, Principal Component Analysis (PCA) is applied on that data to get the corresponding weight vectors.

5) Now probability distribution of the weight vectors of depth data is computed by fitting it onto Gaussian distribution. After this fitting process corresponding CDF is calculated as a new feature vector called Cumulative Depth Feature Vector (CDFV).

6) CDFV along with weight vector of texture data is fed to classifier which uses Euclidean distance for classification. Individual error scores are calculated and then these scores are fused to get the minimum score.

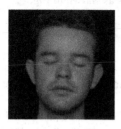

Fig. 3 Texture Map
(Frontal)

Fig. 4 Texture Map
(Pose Variation)

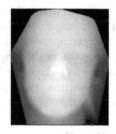

Fig. 5 Depth Map

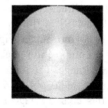

Fig. 6 ROI from Depth Map

2.1   Depth Map Normalization

Depth map obtained is normalized with the maximum intensity value to make the depth data more visible. Here the depth values are normalized between the range 0 and 255.Normalised depth map (ND) is given by

$$ND = \frac{\text{Original Depthmap} * 255}{\text{Max } Intensity (\text{Original Depthmap})} \tag{1}$$

2.2 Nose Tip Localization and Face Area Extraction

For localizing the nose tip, maximum intensity method is used. In this method assumption is made that the nose tip will be the point with maximum pixel intensity. Once the nose tip is found the circular area (ROI) around the nose tip is extracted using an optimum radius. Now the depth map will contain the face area only, all other unwanted portions are cropped away. Next face area is centralized by making the nose tip as the center pixel of the image. Otherwise the matching process will result in a lower accuracy. The face area is also normalized by the maximum intensity. The centralized face image is as shown in Fig. 6.

2.3 2D Discrete Contourlet Transform

We are familiar with the application of the conventional transformations like Fourier transform, Cosine transform, Wavelet transform etc. for feature extraction in pattern recognition.

The contourlet transform [20, 21, 22] is an extension to wavelet transform, shows a significant improvement over the former. Thetransform provides a multiresolution and multidirectional decompositionof images. This helps the transform to captureedges in various orientations within an image. The contourlettransform has a double filter bank structure. The first stage isa Laplacian pyramid (LP) and the second stage is a directionalfilter bank (DFB). The Laplacian pyramid introduced by BurtandAdelson [19] in the first stage provides a multiscaledecomposition of the input depth or texture image.Figure 7 shows the LP decompositionstage. At each level,the LP decomposition generates a downsampled version of theoriginal image and a bandpass image. The input image x isfirst filtered with the approximation filter H and downsampledto obtain a coarse image a. A prediction image is formedby upsampling this coarse image followed by filtering withthe interpolation filter G. Finally, the difference between theoriginal and this prediction image results in a bandpassimage$b$. This process can be repeated on coarse signal a.

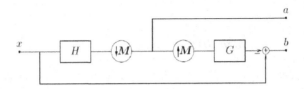

Fig. 7 Laplacian pyramid decomposition

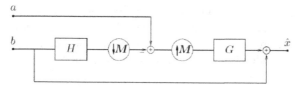

Fig. 8 Laplacian pyramid reconstruction.

The reconstruction scheme for the Laplacian pyramid isshown in Figure 8. The bandpass signal is filtered anddownsampled, and subtracted from the coarse image. Theoriginal image is reconstructed by upsampling and filtering theabove result and adding to the bandpass image. Same filtersand sampling matrices are used for LP decomposition andreconstruction.The second stage in the double filter bank structure is adirectional filter bank. The high frequency information in theimage is captured well by the DFB. The n level decompositionusing DFB results in $2n$ directional subbands. Figure 10 shows the frequency partitioning for $n=3$ and filter bankstructure of DFB. $E_k$and $D_k$are the analysis and synthesisfilters respectively, and $S_k$is the sampling matrix. The DFBis constructed with a two channel quincunx filter bank and ashearing operator. Pair of shearing operators and its inverse is added before and after the two channel filter bank. Bysuitably using the shearing operator, the frequency partitionshown in Figure 9 is obtainedand sampling matrices are used for LP decomposition andreconstruction.

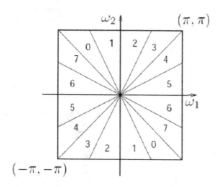

Fig. 9 Frequency Partitioning

The DFB shown in Figure 10 cannot handle the low frequency components efficiently, since the low frequency would spread into several directionalsubbands. So the low frequency components are removed by LP decomposition. The bandpass image from LPdecomposition is passed to DFB stage. The DFB decomposesit into directional subbands. The combination of Laplacianpyramid and directional filter bank results in a contourletfilter bank, which decomposes discrete images into directionalsubbands at multiple resolutions.The framework of contourlet decomposition is shown in Figure 11. The LP decomposition uses a downsamplingfactor of 2.Here the 2 level decomposition is done and face pattern obtained from optimum direction is taken as the feature information.

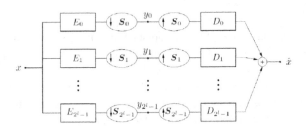

Fig. 10 Filter Bank Structure of DFB

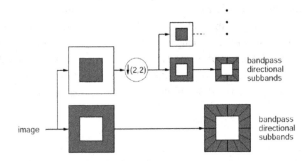

Fig. 11 Framework of contourlet decomposition

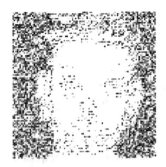

Fig. 12 D-DCLT of Texture Data

Fig. 13 D-DCLT of Depth Data

2.4   2D Discrete Fourier Transform

Face images have higher redundancy and pixel level correlation which is a major hindrance in face recognition systems. Transforming face images to spectral domain will reduce the redundancy. Here only the magnitude of spectral data is taken alone since it is not transformed back to spatial domain in any of the processing stages.

The texture image is transformed using 2D-DFT so that rotation effects are reduced. DFT is a rotation invariant transformation. So that the distributed pixel values (normalized) are properly aligned, this enables the pattern matching more efficient. DFT spectrum of texture map of face image will appear as shown in Fig.14. Transformation to spectral domain using 2-D Discrete Fourier Transformation can be done using equation (2) for a M x N depth image

$$F(U,V) = \sum_{x=0}^{M-1} \sum_{y=0}^{N-1} f(x,y) e^{-j2\pi(Ux/M + Vy/N)} , \qquad (2)$$

Now the error score is estimated using all the multiple representations separately. 1D-PCA was also checked for the 2D representations but 2D-PCA gave better result for 2D representations.

The same procedure is repeated for the depth map also to get the spectral representation using 2-DDFT as shown in Fig.15.

Fig. 14 2D-DFT of Texture Data   **Fig. 15** D-DFT of Depth Data

2.5   Principal Component Analysis

Use of spectral transformations will make the data samples almost spatially uncorrelated. Even then, some spatial dependency may exist. So Principal Component Analysis (PCA) [12] which uses the orthogonal transformations to get linear uncorrelated data sets called Principal Components is employed. Conventional covariance method is used for the above. To start with feature extraction using 1D-PCA is discussed for better understanding.

Let X_i be the spectral transformed 1D data which representations i^{th} person, it is grouped as a M' x N matrix $X=[X_1\ X_2\ ...X_N]$, where N is the number of face samples under consideration and M' is the length of each feature vector.

Standard deviation is calculated as

$$X_{SD}= \frac{1}{N}\sum_{i=1}^{N}(X_i - X_m)$$

(3)

Covariance matrix is calculated as

$$X_{COV}= X_{SD*}\ X^T_{SD}$$

(4)

Where X_m is the mean vector. Here, the covariance matrix is of size M' x M', which is of very large dimension. Also it gives M' Eigen values and M' Eigen vectors which are too large in number to process. Therefore, dimensional reduction is adopted by altering the construction of covariance matrix as follows.

$$X_{COV}= X^T_{SD*}\ X_{SD}$$

(5)

The result is a matrix of size N x N, where N is the number of subjects under consideration. It gives N Eigen values and N Eigen vectors. The Eigen values are sorted in descending order and the first N' largest Eigen values and corresponding Eigen vectors are selected as others are insignificant. The test data is projected to this lower dimension space to get the corresponding weight vectors.

Now feature extraction using 2D-PCA is considered using spectral representation of depth map. The only difference with 1D-PCA in calculating the Covariance matrix is that here a 2D matrix is used when compared to 1D Matrix in 1D-PCA. After determining the Eigen values and Eigen vectors, a 2D weight vector matrix is obtained which is then converted to a column matrix.

2.6   Cumulative Depth Feature Vector

Projection of depth data Wavelet transform spectral coefficients on 2D PCA space will give weight vectors. These weight vectors are fitted on to the probability distribution function of a Gaussian distribution. The mean μ and standard deviation σ of the Gaussian distribution is calculated as follows. If W_i is the weight vector

$$\mu = \frac{1}{M}\sum_{i=1}^{M}W_i$$

(6)

$$\sigma = \frac{1}{M}\sum_{i=1}^{M}(W_i - \mu)$$

(7)

There are different probability distributions and of which some can be fitted more closely to the observed frequency of the data than others. Difference distributions were fitted as test onto the depth weight vector and the Gaussian distribution seemed to be more effective when the data distribution is concerned.

Gaussian pdf is represented by the function

$$f(x, u, \sigma) = \frac{1}{\sqrt{2\pi\sigma^2}} e^{-\frac{1}{2\sigma^2}(x-\mu)^2} \tag{8}$$

Corresponding CDF is given by

$$F(x) = \sum_{xi \leq x} f(X = x_i) \tag{9}$$

Where F(x) is called the Cumulative Data Feature Vector (CDFV).

2.7   Score Fusion

Next the error score is estimated using all the multiple representations separately. For processing 2-D DCLTand2-D DFT representations, 2-DPCA is used. 1-DPCA was also experimented for the 2D representations but 2-D PCA gave better result for 2D representations. Error values are calculated for each data representations and all this error values are combined as a single error value using the linear expression as given in equation (10). By trial and error, the optimum value for W can be approximately obtained as 0.9.

$$Error = W * (T\_DCLT + T\_DFT) + (1 - W) * (D\_DCLT + D\_DFT) \tag{10}$$

$$T_{DCLT} = \frac{Texture_{ErrorDCLT} - min(Texture_{ErrorDCLT})}{max(Texture_{ErrorDCLT}) - min(Texture_{ErrorDCLT})} \tag{11}$$

$$D\_DCLT = \frac{Depth\_ErrorDCLT - min(Depth\_ErrorDCLT)}{max(Depth\_ErrorDCLT) - min(Depth\_ErrorDCLT)} \tag{12}$$

$$T\_DFT = \frac{Texture\_ErrorDFT - min(Texture\_ErrorDFT)}{max(Texture\_ErrorDFT) - min(Texture\_ErrorDFT)} \tag{13}$$

$$D\_DFT = \frac{Depth_{ErrorDFT} - min(Depth_{ErrorDFT})}{max(Depth_{ErrorDFT}) - min(Depth_{ErrorDFT})} \tag{14}$$

T_DCLT, D_DCLT, T_DFT and D_DFT are normalized values.

Texture_ErrorDCLT-   *Error obtained when spectrum of Texture is taken using Con-*
tourlet Transform.

Depth_ErrorDCLT-   *Error obtained when spectrum of Depth is taken using Con-*
tourletTransform.

Texture_ErrorDFT-   *Error obtained when spectrum of Texture is taken using*
Fourier Transform.

Depth_ErrorDFT-   *Error obtained when spectrum of Depth is taken using Fouri-*
er Transform.

3   Performance Analysis

3.1   Results of Fusion Scheme analysis

The Tables I shown summarize the results obtained by using Texture information alone scheme. Training images were varied from1 to 4. Accuracy was tested by using both Fourier transform and Contourlet Transform individually.

Table 1 Results Obtained Using Texture Information alone

Training Images	Using DFT	Using DCLT	Fusion of DFT and DCLT
	FRA%	FRA%	FRA%
1	61	48	66
2	65	52	70
3	70	58	73
4	75	63	78

Table 2 Results Obtained Using Fusion Scheme

Training Images	Using Texture alone	Using Depth alone	Using Both Texture and Depth
	FRA%	FRA%	FRA%
1	66	39	71
2	70	46	74
3	73	53	77
4	78	59	81

Using Texture information Fourier transform gave accuracy in the range 61% to 75%, where Contourlet gave in the range 48% to 63% for different number of training images. For both texture and depth the individual accuracies obtained using both transforms are comparable. Fusion of texture and depth feature vectors significantly increased the accuracy along different orientations. The Table II here is showing accuracy obtained with the fusion of texture and depth information. Testing is done over 1600 samples with different orientations and expressions.

3.2 Reliability Test of the Face Recognition System with Fusion Scheme

As far as a biometric system is concerned, basically four parameters are evaluated for analyzing the reliability of the system namely, TAR-True Acceptance Rate, TRR-True Rejection Rate, FAR-False Acceptance Rate, FRR-False Rejection Rate. A highly reliable biometric system should have high TAR and FRR while maintaining low TRR and FAR. The algorithm is tested for robustness against the chances of wrong user authentication (FAR) and the denial of authentication to actual enrolled users (TRR). The fusion scheme improves the rejection of unauthorized access of a user not enrolled in the database. Table 3 summarizes the averaged results of TAR, TRR, FRR and FAR analysis done along orientations along different axis.

It seems that the fusion scheme has improved the false rejection rate of individual scheme and has reduced the false acceptance rate significantly. Here only a small reduction in the TAR is observed for fusion scheme which can be ignored.

The probability of error distribution is shown in Fig. 16. There we can see that with the texture and the depth data alone the hypothesis areas are overlapping and are not separable. That is the chance of FAR are high. But with the fusion scheme the rate of FAR is reduced on an average by 32% and the FRR increased by 20 %as shown in Table 3. That is the efficiency of Texture and Depth data in FAR and FRR respectively are combined to get the better results for the fusion scheme.

Table 3. Reliability Test Results

Scheme	True Acceptance Rate	True Rejection Rate	False Acceptance Rate	False Rejection Rate
	TAR %	TRR %	FAR %	FRR %
Texture Feature	86.00	14.00	34.67	65.33
Depth Feature	84.67	15.33	38.00	62.00
Fusion of Depth and Texture	84.67	15.33	26.00	74.00

3.3   Computation Time

Testing of algorithm is done on 3GHz, Core I-5 processor; the average testing time for a single sample is approximately 770ms, 1074ms, 1701ms and 2211ms for 1, 2, 3 and 4 training images respectively. This will again increase as the number of subjects increase. By using down sampling and abstracting the data representations computational time can be further reduced. But with a dedicated system, the testing time can be further reduced to microseconds.

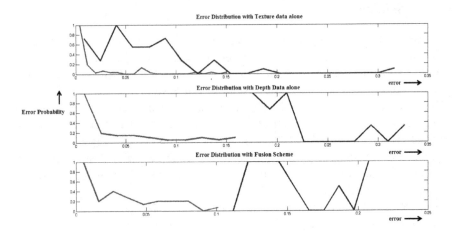

Fig. 16 Probability of Error Distribution with Texture, Depth and fusion scheme

References

1. http://www.frav.es/databases/FRAV3D
2. Bronstein, A.M., Bronstein, M.M., Kimmel, R.: Expression-invariant 3D face recognition. In: Kittler, J., Nixon, M.S. (eds.) AVBPA 2003. LNCS, vol. 2688, pp. 62–70. Springer, Heidelberg (2003)
3. Beumier, C.: 3D face recognition. In: IEEE Int. Conf. on Computational Intelligence for Homeland Security and Personal Safety (CIHSPS2004), Venice, Italy, July 21-22, 2004
4. Pan, G., Han, S., Wu, Z., Wang, Y.: 3D face recognition using mapped depth images. In: Proceedings of the 2005 IEEEComputer Society Conference on CVPR (CVPR 2005) Workshops, vol. 03, p. 175 (2005)
5. Yuan, X., Lu, J., Yahagi, T.: A Method of 3D face recognition based on principal component analysis algorithm. In: IEEE International Symposium on Circuits and Systems, May 3-26, vol. 4, pp. 3211–3214 (2005)
6. Russ, T., Boehnen, C., Peters, T.: 3D Face recognition using 3D alignment for PCA. In: Proceedings of the 2006 IEEE Computer Society Conference on CVPR (CVPR 2006), vol. 2, pp. 1391–1398 (2006)

7. Mian, A., Bennamoun, M., Owens, R.: Automatic 3D face detection, normalization and recognition. In: Proceedings of the Third International Symposium on 3DPVT (3DPVT 2006), pp. 735–742, June 2006
8. Smirg, O., Mikulka, J., Faundez-Zanuy, M., Grassi, M., Mekyska, J.: Gender recognition using PCA and DCT of face images. In: Cabestany, J., Rojas, I., Joya, G. (eds.) IWANN 2011, Part II. LNCS, vol. 6692, pp. 220–227. Springer, Heidelberg (2011)
9. Gao, H., Ekenel, H.K., Stiefelhagen, R.: Pose normalization for local appearance-based face recognition. In: Tistarelli, M., Nixon, M.S. (eds.) ICB 2009. LNCS, vol. 5558, pp. 32–41. Springer, Heidelberg (2009)
10. Taghizadegan, Y., Ghassemian, H., Naser-Moghaddasi, M.: 3D Face Recognition Method Using 2DPCA-Euclidean Distance Classification. ACEEE International Journal on Control System and Instrumentation 3(1), 5 (2012)
11. Gervei, O., Ayatollahi, A., Gervei, N.: 3D Face Recognition Using Modified PCA Methods. World Academy of Science, Engineering & Technology (39), 264, March 2010
12. Turk, M., Pentland, A.: Eigenfaces for recognition. J. Cognitive Neuroscience 3(1), 71–86 (1991)
13. Ramalingam, S.: 3D face recognition: feature extraction based on directional signatures from range data and disparity map. In: IEEE International Conference on Systems, Man, and Cybernetics (SMC), pp. 4397–4402, October 2013
14. Dawi, M., Al-Alaoui, M.A., Baydoun, M.: 3D face recognition using stereo images. In: 17th IEEE Mediterranean Electrotechnical Conference (MELECON), pp. 247–251, April 2014
15. Goswami, G., Vatsa, M., Singh, R.: RGB-D Face Recognition With Texture and Attribute Features. IEEE Transactions on Information Forensics and Security 9(10), 1629–1640 (2014)
16. Mantecon, T., del-Bianco, C.R., Jaureguizar, F., Garcia, N.: Depth-based face recognition using local quantized patterns adapted for range data. In: IEEE International Conference on Image Processing (ICIP), pp. 293–297 (2014)
17. Mandal, T., Wu, Q.M.J.: 19th International Conference on Pattern Recognition, ICPR 2008, pp.1–4, December 2008
18. Naveen, S., Moni, R.S.: Multimodal Approach for Face Recognition using 3D-2D Face Feature Fusion. International Journal of Image Processing (IJIP) 8(3), 73–86 (2014)
19. Burt, P.J., Adelson, E.H.: The Laplacian Pyramid as a Compact Image Code. IEEE Trans. on Communications, 532–540, April 1983
20. DO, M.N., Vetterli, M.: The contourlet transform: An efficient directional multiresolution image representation. IEEE Trans. Image Process. 14, 2091–2106 (2005)
21. Jiji, C.V., Chaudhuri, S., Chatterjee, P.: Single Frame Super-resolution: Should we process Locally or Globally. Multidimensional Systems and Signal Processing 18(2–3), 123–152 (2007)
22. Jiji, C.V., Krishnan, R.: Unni: fusion of multispectral and panchromatic images using nonsubsampled contourlet transform. In: IPCV 2008, pp. 608–613 (2008)
23. Besl, P.J., McKay, N.D.: A Method for Registration of 3-D Shapes. IEEE Trans. on Pattern Analysis and Machine Intelligence 14(2), 239–256 (1992)

An Efficient Ball and Player Detection in Broadcast Tennis Video

M. Archana and M. Kalaiselvi Geetha

Abstract Ball and player detection in Broadcast Tennis Video (BTV) is a critical and challenging task in tennis video semantic analysis. Informally, the challenges are due to the camera motion and the other causes such as the small size of the tennis ball and many objects resembles the ball and considering the player, the human body along with the tennis racket is not detected completely. In this paper, it is proposed an improved object detection technique in BTV. In order to detect the ball, logical AND operation is applied between the created background and image difference is performed, from that ball candidates are detected by applying threshold values and dilated. Player detection is performed from AND results by finding the biggest blob and filling the whole detected object by removing the small one. The experimental result shows that the proposed approach achieved the higher accuracy in object identification, their object the landing frames and positions. It is achieved a high hit rate and less fail rate.

Keywords Tennis ball detection · Player detection · Background subtraction · Broadcast tennis video · Hit rate · Fail rate

1 Introduction

Peoples attention in sports video analysis has been widely attracted due to the development of high-speed digital cameras and video processing. Generally in video content analysis, there is an interesting and important application for sports video analysis due to tremendous commercial value and wide viewership [2]. The development and proliferation of multimedia data, both in entertainment and professional services have increased, which leads to the generation of large amounts of

M. Archana · M. Kalaiselvi Geetha
Department of Computer Science and Engineering, Annamalai University,
Annamalai Nagar, Tamil Nadu, India
e-mail: {archana.aucse,geesiv}@gmail.com

digital videos in various fields of domain such as sports, news, movies and video surveillances. Among these videos, digital sports video has been more and more pervasive [3].

Owing to increase in the growth of videos on broadcast and internet, there is a need to access semantic events among the full-length videos arises. Instead of accessing the whole lengthy voluminous video programs, access of highlights and skipping the less interesting parts of the videos will save not only the viewers time but also the cost of the videos. To attract the users the content based views are developed based on their own preferences [4]. In Broadcast Tennis Video (BTV), the ball size is 67mm in diameter, the diagonal court is over 26m and the speed of the ball travels over 25 km/h. Since the ball is the main attention focus of viewers and players in tennis, ball-detection and tracking becomes a crucial task in tennis video analysis [5]. However, it is a very challenging task due to camera motion and other causes like as the presence of many ball-like objects, the small size and the high speed of the ball. Some of the challenges have discussed and solved as follows,

- The aimed object is occluded not fully only some part. This problem is solved by reconstructing the occluded object, then the target object reappears fully.
- The target object will be in different angles, so in order to solve this adjust various different shapes of target object to detect.
- Complete occlusion that is target object is completely occluded by some other objects.
- Dynamic background (i.e.) the background which is not static while the tennis videos which are not fully static faces false detection.
- Based on various lighting conditions, the target object looks like some other objects.
- Various shape changes of the target object due to camera position. To solve this apply various shapes which is not sensitive to camera position.

The rest of the paper is organized as follows. Section 2 reviews the related work and the methodology of the paper is discussed in section 3. The proposed model is explained in section 4 and experimental results are briefly discussed in section 5, followed by performance measures in section 6. Finally conclusion and future work are presented in section 7.

2 Related Work

Object detection in BTV is one of the key points of computer vision. Briefly speaking, object tracking is concerned, with finding and following the object of interest in a video sequence. Jui-Hsin Lai and Shao-Yi Chien [1] have proposed a semantic scalability scheme with four levels. Rather than detecting, shot categories to determine suitable scaling options for Scalable Video Coding (SVC) and provides a new dimension for scaling videos, can be extended to various video categories. Fei Yan et al. [6] have proposed a challenge of tracking a ball in BTV which is a layered data an association algorithm for tracking multiple tennis balls fully automatic and handle multiple object scenarios.

Lai and Chien [7] have proposed a video structural component, this method is sport video temporal structure decomposition, which decomposes the sport video into many video clips. Using this method the precision and recall rates are extremely high and it is also score box and additional semantic information are important clues for event judgment. Zhu G. et al. [8] have proposed a about player action recognition of multimodal features , it improves the semantic indexing and retrieval of video content and also increased highlights ranking and tactics analysis of tennis video. The feature-inconsistent problems are considered for that a novel action recognition algorithm. This method is developed this facilitates not only video indexing and retrieval based on players stroke types, but also highlights ranking and hierarchical content browsing.

Xinguo Yu et al. [9] have proposed a improving trajectory-based ball detection and tracking algorithm in tennis videos, this algorithm can obtain not only higher accuracy in ball identification, but also ball landing frames and positions with the aid of homography. Yang Wang et al. [10] have proposed the detection and tracking of player in BTV, this is obtained by support vector classification and court segmentation from that accurate player area is founded. Based on this particle filter tracking of small particles is used to improve the performance of this method. Yan et al. [11] have proposed a ball tracking using automatic annotation of tennis match for low quality video recorded with a single camera. A particle filter is used to track the tennis candidates to achieve better results. It shows the higher accuracy for tracking, while smoothing and observation of origin identification is used to refine the trajectory and it is suitable for tennis annotation.

3 Methodology

Ball detection is achieved by frame differencing between the current and consecutive images. The results are then verified against the size and shape parameters. The region is close to the expected position, which is chosen in the case of multiple detections. But this technique has the problem of double detections (i.e) more than one ball candidates appear. To solve this problem first find the regions in the current image that lies in the expected intensity range for the ball. Then perform a logical AND operation between the result generated from the image difference and the region of background images [12].

3.1 Smoothing the Image

To smoothen the image, apply median filter which is a commonly used technique for reducing small noise in an image [13]. Small noises are normally appears very distinct and its gray values is quite different from its neighbours so using this techinique eliminated the noise by changing its gray value to the median of neighbouring pixel values.

3.2 Background Model

The aim of the background model creation is to develop a standard background, because in the BTV does not possess fully static background because of the camera calibration. In this technique, first a background model is created from a collection of background images. For each queried image, edges of foreground objects are detected by background subtraction [14]. The tennis ball is identified by shape, aspect ratio and compactness. This technique is analyzed as follows:

1. Given a number of background images, a background model is created.
2. For each queried image:
   a. An edge image is created.
   b. Edges of the tennis ball are segmented from the result, our approach is based on shape recognition.
   c. Area of the tennis ball is then dilated by subjecting to the size and aspect ratio. Finally the location of the tennis ball is detected.

3.3 Frame Difference

Framee Difference (FD) is the technique to find the tennis ball candidates by considering the difference between current and next frames. In order to estimate the ball intensity levels some of the difficulties are lighting, shadows and distance variations [15]. To get better results to combine the results of image difference with background subtraction. Motion information is extracted from the video sequence by pixel-wise differencing of consecutive frames. Motion information T_k or difference image is calculated by using Eq. 1.

$$T_k(i, j) = \begin{cases} 1, & \text{if } D_k(i, j) > t; \\ 0, & \text{Otherwise}; \end{cases} \tag{1}$$

Where D_k is the difference image and t is time interval, calculated using Eq. 2.

$$D_k(i, j) = |I_k(i, j) - I_{k+1}(i, j)| \tag{2}$$
$$1 \le i \le w, 1 \le j \le h$$

3.4 Logical AND Operation

The logical AND operation is performed between the image obtained from background subtraction and image difference. While apply AND operation the commonly presented objects between these two images are obtained

3.5 Thresholding and Dilation

Finally obtained result is threshold and dilated, based on the compactness, aspect ratio and size, the remaining contour is identified and detected as ball in the upcoming frames [16].

4 Proposed Model

4.1 Modified Ball Candidate Detection Method

In order to detect the ball candidates, given the current image A and the previous image B, ball candidates from A are detected as follows:
1. Find the image difference between A and B is performed and the result is denoted as C.
2. Perform a background subtraction of image A with the created background model and the result is denoted as D.
3. Perform the logical AND operations between C and D.
4. Obtain results are the ball candidates.

4.2 Modified Player Candidate Detection Method

The player is detected using the same approach as described in our first technique. The following are,
1. Remove the smaller blobs from the AND results.
2. Detect the biggest blob and it is splitted, which is a part of the human body.
3. To construct the whole human body, find the contour near the biggest one.
4. To detect a small blobs around the biggest one are considered as one by using flood filling techniques.
5. Apply flood fill techniques, it is filled the complete region and constructed the region as human body.
6. Reconstruct the human body, it is also consider the tennis racket in together as player.

5 Experimental Results

Experiments were carried out on BTV from the different tennis tournaments. A BTV is composed of various shots such as play shots, crowd, advertisement and break. To evaluate the performance of the proposed methods, experiments were carried out on video sequence from two tennis tournament such as clay and carpet as seen in Fig 1. The top row shows the video sequences with carpet play region and the bottom row with clay play region.

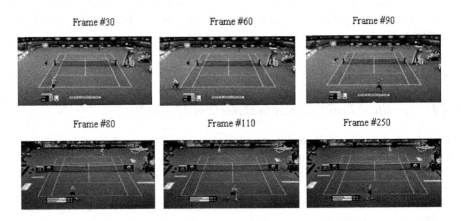

Fig. 1 Example video frames of clay and carpet play regions

The video sequence included a total of 125 shots of (3 - 5 seconds). The video format is AVI, that is image resolution of 1280 x 720 and frame rate is 25 frames/seconds. In Fig 2 (a) shows the created background model of an given frames using single

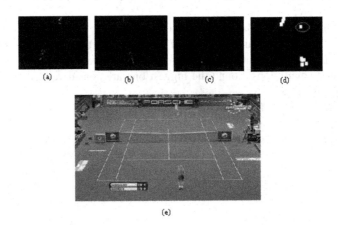

Fig. 2 Ball detected in original frame

Gaussian mixtures model. In Fig 2(b) shows the frame difference between the current and next frames. Fig 2 (c) represents the logical AND results performed between the image difference and background images. This result are removing negative values of each pixels and the threshold of image a value of 75, these same frames are considered for both the ball and player detection. In Fig 2(d). shows the dilated results, which enriches the obtained contour to some extent. Fig 2(e) shows the candidates of ball detected based on the ball size, aspect ratio and compactness. Finally, the contour is plotted in the form of rectangle in the original frame.

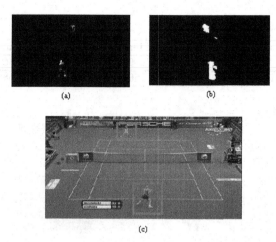

(a)   (b)

(c)

Fig. 3 Player detected in original frame

In Fig 3(a) shows the AND results and from that apply the flood filling techniques to fill the entire object is shown Fig 3(b). The detected biggest blob based on the size is shown in Fig 3(c). Finally, the biggest blob along with the tennis racket is detected and plotted in the form of rectangle in the original frame.

6   Performance Measures

The performance of the proposed model is measured by plotting the ground truth values. The ground truth of the ball and the player were plotted in the original frame. The results were obtained and compared with the detected regions of both the ball and player with 2 different BTV.

In Fig 4 shows the detected result in blue colour box while plotted ground truth box in red colour using the area of the intersection of these rectangle. To find the rate of success and location error with the reference of object centre, and the method applied in [16] is used to compute the rate of success and score is defined as given below in Eq. 3.

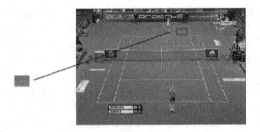

Fig. 4 Detected ball compared with ground truth

Table 1 Rate of success for ball detection

Tennis	Total Frames	Ball Frames	Hit Rate(%)	Fail Rate(%)
Clay Region	1000	600	92.64	7.36
Carpet Region	900	750	92.32	7.68

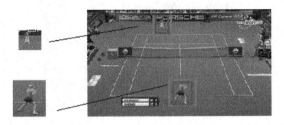

Fig. 5 Detected players compared with ground truth

$$score = \frac{area(ROI_D \bigcap ROI_G)}{area(ROI_D \bigcup ROI_G)} \tag{3}$$

Where, ROI_D is the detected bounding box and ROI_G is the ground truth bounding box . The rate of success is computed in all frames, when this score is greater than 0.5 in each frame then it is considered as a success as shown in Table 1. This work, consider 1000 frames the ball is seen in frames only 60 % of frames. The ball detection varies due to some challenges like various lighting conditions, different angles and various shapes due to camera position.

In Fig 5 shows the detected result in blue colour box while plotted ground truth box in red colour using the area of the intersection of these rectangle the rate of success is founded. To find the rate of success and location error with the reference of object centre using Eq 3. The rate of success is computed in all frames, when this score is greater than 0.5 in each frame is considered as a success as shown in Table 2. Out of 1000 frames the player occurred frames are 95 % where others are not occurred. While comparing the results of clay and carpet region the player

Table 2 Rate of success for player detection

Tennis	Total Frames	Player Frames	Hit Rate(%)	Fail Rate(%)
Clay Region	1000	950	95.32	4.68
Carpet Region	900	830	94.82	5.12

detection varies because of some challenges like the various dressing colours related to background.

7 Conclusion

In this proposed model, an improved ball and player detection for BTV, besides varying camera motion causes as the presence of many ball-like objects and the small size of the tennis ball. It is not only increases the accuracy for identifying the ball, but also improve the accuracy in determining the ball projection position. In addition, it detects the ball landing frames and landing positions based on the accurate ball candidates, whereas the player detection constructed the human body along with tennis racket. The result shows model that this model is able to precisely classify the ball detection in clay region about 92.64% and carpet region of 92.32%. Then player detection in clay region of 95.32% and carpet region also 94.82% of accuracy were founded by comparing with the ground truth values. The tracking of the detected objects such as ball and player is improved by using this proposed approach and it will be investigated in the future.

References

1. Lai, J.-H., Chien, S.-Y.: Semantic scalability using tennis videos as examples. Multimedia Tools and Applications **59**(2), 585–599 (2012)
2. Cross, R.: The footprint of a tennis ball. Sports Engineering **17**(4), 239–247 (2014)
3. Yu, X., Sim, C.-H., Wang, J.R., Cheong, L.F.: A trajectory-based ball detection and tracking algorithm in broadcast tennis video. In: Image Processing, International Conference on ICIP 2004, vol. 2, pp. 1049–1052 (2004)
4. Furht, B., Greenberg, J., Westwater, R.: Motion estimation algorithms for video compression. Springer Science and Business Media, vol. 379 (2012)
5. Cross, R.: Impact of sports balls with striking implements. Sports Engineering **17**(1), 3–22 (2014)
6. Yan, F., Christmas, W., Kittler, J.: Ball Tracking for Tennis Video Annotation. Springer International Publishing In Computer Vision in Sports (2014)
7. Aggarwal, J.K., Ryoo, M.S.: Human activity analysis: A review. ACM Computing Surveys (CSUR) **43**(3), 16 (2011)
8. Chen, H.-T., Chou, C.-L., Fu, T.-S., Lee, S.-Y., Lin, B.-S.P.: Recognizing tactic patterns in broadcast basketball video using player trajectory. Journal of Visual Communication and Image Representation **23**(6), 932–947 (2012)

9. Martn, R., Martnez, J.M.: Automatic Players Detection and Tracking in Multi-camera Tennis Videos. Springer International Publishing In Human Behavior Understanding in Networked Sensing, pp. 191–209 (2014)

10. Wang, Y., Han, Y., Zhang, D.: Research on Detection and Tracking of Player in Broadcast Sports Video. International Journal of Multimedia and Ubiquitous Engineering 9(11), 1–10 (2014)

11. Yan, F., Christmas, W., Kittler, J.: Ball Tracking for Tennis Video Annotation. Springer International Publishing In Computer Vision in Sports, pp. 25–45 (2014)

12. Sakurai, S., Reid, M., Elliott, B.: Ball spin in the tennis serve: spin rate and axis of rotation. Sports Biomechanics 12(1), 23–29 (2013)

13. Nicolaides, A., Elliott, N., Kelley, J., Pinaffo, M., Allen, T.: Effect of string bed pattern on ball spin generation from a tennis racket. Sports Engineering 16(3), 181–188 (2013)

14. Choppin, S.: An investigation into the power point in tennis. Sports Engineering 16(3), 173–180 (2013)

15. Choppin, S., Goodwill, S., Haake, S.: Impact characteristics of the ball and racket during play at the Wimbledon qualifying tournament. Sports Engineering 13(4), 163–170 (2011)

16. Spurr, J., Goodwill, S., Kelley, J., Haake, S.: Measuring the inertial properties of a tennis racket. Procedia Engineering 72, 569–574 (2014)

Steady State Mean Square Analysis of Convex Combination of ZA-APA and APA for Acoustic Echo Cancellation

S. Radhika and Sivabalan Arumugam

Abstract This paper proposes a new approach for the cancellation of acoustic echo in a loud speaker enclosed microphone system. The acoustic echo is often sparse in nature and the level of sparseness also changes with time. The input in such applications, is speech which is highly correlated. Thus there is a requirement of an adaptive filter which can work in both sparse and non sparse environment and perform well even for correlated inputs. We present convex combination of conventional affine projection algorithm (APA) with small step size and projection order and zero attraction APA (ZA-APA) for echo cancellation. Steady state excess mean square (EMSE) error analysis revealed that the proposed algorithm converges to conventional APA in non sparse environment and to ZA-APA in case of sparse environment and in semi sparse condition, the steady state EMSE value is at least same as the lesser filter's steady state error or even smaller depending on the value of the constant chosen. Simulation is performed in the context of acoustic echo cancellation and the validity of the proposed algorithm is proved from the simulation results.

Keywords Affine projection algorithm · Convex combination · Steady state mean square error · Zero attraction · Convergence

1 Introduction

Acoustic echo is a serious problem in a loudspeaker enclosed microphone system. The coupling between them results in echoes .These occurs in a hands free telephone

S. Radhika(✉)
Faculty of Electrical and Electronics Engineering,
Sathyabama University, Chennai, India
e-mail: radhikachandru79@gmail.com

S. Arumugam
NEC Mobile Networks Excellence Centre, Chennai, India

© Springer International Publishing Switzerland 2016   437
S. Berretti et al. (eds.), *Intelligent Systems Technologies and Applications*,
Advances in Intelligent Systems and Computing 384,
DOI: 10.1007/978-3-319-23036-8_38

system, hearing aids, tele-conferencing rooms, etc. The impulse response of such a system is often long and most of the coefficients are sparse is also said to be time varying due to the movement of objects ,changes in temperature etc [5].Conventional adaptive filters like least mean square (LMS), normalized least mean square (NLMS), APA, recursive least square (RLS) algorithms used for echo cancellation application does not take sparseness into consideration [9]. Hence their performance is not improved with the increase in sparseness. Out of these algorithms, APA is found to be better for echo cancellation application as it provides faster convergence with lesser steady state mean square error(MSE) especially for colored inputs such as speech signal [6],[17]. Several variants of APA have been proposed to incorporate the sparsity nature into the algorithm. The famous among them being proportionate APA and its variants [15],[20],set membership APA[19], partial update APA[8]. These algorithms work by varying either the step size or the number of update coefficients or the projection order, as per the magnitude of the impulse response so that improved performance can be achieved with the increase in sparseness .However they perform well only in sparse environment and their performance degrades when the sparsity level is decreased[4]. Another method for inclusion of sparsity is the norm based APA [10-14]. This includes the ZA-APA, reweighted zero attraction type (RZA-APA) and zero forcing type (ZF-APA) [13]. These algorithms work by inclusion of l_1 norm based penalty function in the update equation which gives faster convergence and lesser steady state error only for sparse environment .The ZA-APA has performance better that APA in sparse condition and it is worse when the condition becomes non sparse[14] .The modified version of ZA-APA were RZA-APA and ZA-APA .In RZA-APA ,a heuristic approach is adopted to distinguish between zero and non zero taps whereas in ZF-APA, a threshold is fixed to decide whether to apply zero forcing on the filter taps or not .The RZA-APA suffer from the disadvantage of deciding the shrinkage factor and ZF-APA has the difficulty in choosing the threshold value .Recently combinational approach is found to provide good results compared to single filter approach. Convex combinational of two adaptive filter is given in [2],[3]. Convex combination of LMS and ZA-LMS is discussed in [7]. Motivated by the combinational approach, in this paper we propose to combine ZA-APA and APA using a convex combiner so that the algorithm works well for both sparse and non sparse environment. It also eliminates the threshold selection problem of RZA-APA and ZF-APA and can provide better convergence and steady state error for colored inputs also. Theoretical steady state mean square error analysis is made based on excess mean square error (EMSE) and the cross EMSE is derived. Performance of the algorithm is evaluated for three different environmental conditions .It is found that in case of non sparse environment ,the algorithm is inclined towards conventional APA, and in case of sparse condition ,it converges to ZA-APA as it has least EMSE value and for semi sparse condition, our proposed system converges to the filter which has lesser steady state error or even lesser which is based on the constant chosen for ZA-APA. Through simulation, convergence and steady state error analysis is made and from the results, the suitability of the algorithm for echo cancellation application is proved. The rest of the paper is organized as follows section 2 deals with the system model, preliminary results and assumptions. Steady state analysis of the proposed algorithm is made in section 3 and simulation, results and discussion are given in section 5. Conclusion is discussed in section 6.

2   System Model

The model of the proposed adaptive filter is shown in Fig.1. It consists of convex combination of ZA-APA (filter1) and conventional APA (filter2) whose general framework is shown in [3].

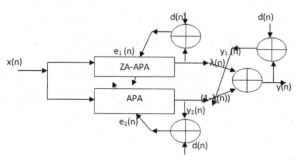

Fig. 1 Proposed system model

Each of the adaptive filter act as an independent filter with their own filter coefficients and update with their own error signal .Here x(n) is the input signal of length N with zero mean and correlation R .Throughout, n is the time index and d(n) is the desired response which is modeled as a multiple linear regression model given by d(n)=w°(n) x(n)+v(n) where w°(n) is the optimal weight vector and v(n) is the noise source which is considered as a white Gaussian noise with zero mean and variance σ_v^2 . Here $e_i(n)=d(n)-y_i(n)$ is the error signal and $y_i(n)=A(n)w_i(n)$ is the estimated output from each of the filter. A(n) is obtained by the delayed version of the input given by A(n)=[x(n) x(n-1)x(n-2)x(n-M)] and M is the number of delayed version of the input called as projection order and μ is the step size and α is the constant chosen for ZA-APA. δ is the regularization required to avoid the singularity of the inverse matrix. The update equation for ZA-APA [13] is

$$w_1(n+1) = w_1(n) + \mu A^T(n)\big(\delta I + A(n)A^T(n)\big)^{-1}e_1(n) + \alpha A^T(n)\big(\delta I + A(n)A^T(n)\big)^{-1}A(n)sgn(w_1(n)) - \alpha sgn(w_1(n)) \tag{1}$$

The weight update equation for conventional APA is given by [9]

$$w_2(n+1) = w_2(n) + \mu A^T(n)\big(\delta I + A(n)A^T(n)\big)^{-1}e_2(n) \tag{2}$$

The final output of the proposed scheme is y(n) which is obtained as the convex combination of the output obtained from filter1 and filter2 given by

$$y(n) = \lambda(n)y_1(n) + (1 - \lambda(n))y_2(n) \qquad (3)$$

$\lambda(n)$ is the convex combiner. The feature of the convex combiner is that it has a value in between 0 and 1. The overall error is given by $e(n) = \lambda(n)e_1(n) + (1 - \lambda(n))e_2(n)$.The stochastic gradient approach is used to adapt the combiner so that the overall error is minimized. As done in [3] the $\lambda(n)$ is represented as a sigmoid function a(n) where

$$\lambda(n) = sgm(a(n)) = 1/(1 + e^{-a(n)}) \qquad (4)$$

The updation of convex combiner is obtained using stochastic gradient approach given by $\lambda(n+1) = \lambda(n) - \mu_a \frac{\partial(e^2)}{\partial\lambda(n)}$ which can be simplified as

$$a(n+1) = a(n) + \mu_a e(n)(y_1(n) - y_2(n))\lambda(n)(1 - \lambda(n)) \qquad (5)$$

The range of $a(n)$ is restricted as $[-a^+, a^+]$ which restricts $\lambda(n)$ to $[(1 - \lambda^+), \lambda^+][3]$.

2.1 Assumptions

The following are the assumptions used in this algorithm which is based on [18].A1. The noise is i.i.d and is statistically independent of the regression matrix A(n) and all the Initial conditions are assumed to be zero i.e.$w_1(0)=0$, $w_2(0)=0$, $y_1(0)=0$, $y_2(0)=0$ .A2. The noise is assumed to be Gaussian with zero mean and variance σ_v^2.A3. At steady state A(n) is statistically independent of $e_{a,i}(n)$.A4. The dependency of $\tilde{w}_i(n)$ on the past noise is neglected.A5.the regularization parameter is assumed to be small and is neglected throughout the analysis.

2.2 Preliminary Results

The preliminary results are obtained from [3].For i=1, 2, the weight error vector is given by $\tilde{w}_i(n) = w^o(n) - w_i(n)$ .The a priori error vector is given by $e_{a,i}(n) = A(n)\tilde{w}_i(n)$.The a posteriori error is $e_{p,i}(n) = A(n)\tilde{w}_i(n+1)$.The error vector is given by $e_i(n) = e_{a,i}(n) + v(n)$. The convex combination of filter 1 and filter 2 results in the system whose overall a posteriori and a priori errors as $e_p(n) = \lambda(n)e_{p,1}(n) + (1 - \lambda(n))e_{p,2}(n)$, $e_a(n) = \lambda(n)e_{a,1}(n) + (1 - \lambda(n))e_{a,2}(n)$ and the final error as $e(n) = e_a(n) + v(n)$.The MSE and EMSE for the filters1 and 2 are defined as $MSE_i = \lim_{n\to\infty} E|e_i(n)|^2$, $EMSE_i = \lim_{n\to\infty} E|e_{a,i}(n)|^2$. The MSE and EMSE for the final output is given by $MSE = \lim_{n\to\infty} E|e(n)|^2$, $EMSE = \lim_{n\to\infty} E|e(n)|^2$. The cross excess mean square error ($EMSE_{1,2}$) is given by $EMSE_{1,2} = \lim_{n\to\infty} E|e_{a,1}(n)e_{a,2}(n)|$. Also from Cauchy Schwarz inequality $EMSE_{1,2} \leq \sqrt{EMSE_1}\sqrt{EMSE_2}$ which implies that the $EMSE_{1,2}$ cannot be greater than $EMSE_i$ .We also define the change in EMSE as $\Delta EMSE_1 = EMSE_1 - EMSE_{1,2}$ and $\Delta EMSE_2 = EMSE_2 - EMSE_{1,2}$.

Using the above assumptions and preliminary results we can write the equation (5) under steady state as

$$E[a(n+1)] = E[a(n)] + \mu_a E\left[\lambda(n)(1-\lambda(n))^2\right]\Delta EMSE_2 -$$

$$\mu_a E[\lambda^2(n)(1-\lambda(n))]\Delta EMSE_1 \qquad (6)$$

3   Steady State Mean Square Error Analysis

In order to analyze the convergence of the proposed algorithm, it is required to find $EMSE_{1,2}$. We know that the EMSE for filter1 with small step size is given [13] as

$$EMSE_1(\infty) = \frac{\mu\sigma_v^2}{2-\mu} + \frac{\alpha^2}{\mu}\left(\frac{sgn(w_1(n)^T(I-E(P(n))sgn(w_1(n)))}{(2-\mu)Tr(E(B(n)))}\right) \qquad (7)$$

where $P(n) = A^T(n)\left(A(n)A^T(n)\right)^{-1}A(n)$ and $B(n) = \left(A(n)A^T(n)\right)^{-1}$. Similarly EMSE for filter2 with small step size is written as defined in [18] as

$$EMSE_2(\infty) = \frac{\mu\sigma_v^2}{2-\mu} \qquad (8)$$

We use the energy conservation approach as given by [18]. The weight error update equation for filter1 is given as

$$\widetilde{w}_1(n+1) = \widetilde{w}_1(n) - \mu A^T(n)\left(\delta I + A(n)A^T(n)\right)^{-1}e_2(n) - \alpha A^T(n)\left(\delta I + A(n)A^T(n)\right)^{-1}A(n)sgn(w_1(n)) + \alpha sgn(w_1(n)) \qquad (9)$$

And the weight update equation for the filter2 is given by

$$\widetilde{w}_2(n+1) = \widetilde{w}_2(n) - \mu A^T(n)\left(\delta I + A(n)A^T(n)\right)^{-1}e_2(n) \qquad (10)$$

Taking expectation of the result obtained by multiplying transpose of $\widetilde{w}_1(n+1)$ and $\widetilde{w}_2(n+1)$ and substituting $e_i(n) = e_{a,i}(n) + v(n)$ and if the assumptions A1-A5 is used , we get

$$E\{\widetilde{w}_1^T(n+1)\widetilde{w}_2(n+1)\} =$$

$$E\{\widetilde{w}_1^T(n)\widetilde{w}_2(n)\} - 2\mu EMSE_{1,2}(n)Tr\left(S.E(B(n))\right) +$$
$$\mu^2 EMSE_{1,2}(n)Tr\left(S.E(B(n))\right) + \mu^2\sigma_v^2 Tr\left(S.E(B(n))\right) -$$
$$\alpha E\left\{\left\{(sgn(w_1(n))^T\right\}P(n)\widetilde{w}_2(n)\right\} + \alpha E\left\{\left\{(sgn(w_1(n))^T\right\}\widetilde{w}_2(n)\right\} \qquad (11)$$

At steady state $E\{\widetilde{w}_1^T(n+1)\widetilde{w}_2(n+1)\} = E\{\widetilde{w}_1^T(n)\widetilde{w}_2(n)\}$ and if the step size is small then S=I, we can write (11) as

$$EMSE_{1,2}(\infty) = \frac{\mu^2\sigma_v^2 Tr[E(B(n))] + \alpha E\left\{\left\{(sgn(w_1(\infty))^T\right\}(1-P(n))\widetilde{w}_2(\infty)\right\}}{\mu(2-\mu)Tr[E(B(n))]} \quad (12)$$

$$= \frac{\mu\sigma_v^2}{(2-\mu)} + \frac{\alpha E\left\{\left\{(sgn(w_1(\infty))^T\right\}(1-P(n))\widetilde{w}_2(\infty)\right\}}{\mu(2-\mu)Tr[E(B(n))]} = EMSE_2(\infty) + Q(\infty) \quad (13)$$

Thus the $EMSE_{1,2}$ is given by (13). Let us write $Q(\infty)$ as

$$Q(\infty) = \frac{\alpha}{\mu(2-\mu)Tr[E(B(n))]}\left\{E\left\{\sum_{k=0}^{N-1}(sgn(w_1(\infty))^T(1-P(n))\widetilde{w}_2(\infty)\right\}\right\} \quad (14)$$

In order to evaluate $Q(\infty)$ we need to know the value of $\sum_{k=1}^{N}(sgn(w_1(\infty))^T(I-P(n))\widetilde{w}_2(\infty)$. Making use of the principle followed in [7], we divide the coefficients of length N into zero and non zero coefficients such that each coefficient is either zero or non zero. Thus we can write (14) as

$$Q(\infty) = \frac{\alpha}{\mu(2-\mu)Tr[E(B(n))]}\left\{E\{\sum_{k\in Z}(sgn(w_{1,k}(\infty))^T(I-P(n))\widetilde{w}_{2,k}(\infty)\} + \right.$$
$$\left. E\{\sum_{k\in NZ}(sgn(w_{1,k}(\infty))^T(I-P(n))\widetilde{w}_{2,k}(\infty)\}\right\} \quad (15)$$

If $k \in NZ$ we know that $E(\widetilde{w}_{2,k}(\infty)) = 0$,as the input is assumed to have zero mean, thus (15) is only due to zero filter taps. Thus

$$Q(\infty) = \frac{\alpha}{\mu(2-\mu)Tr[E(B(n))]}\left\{E\left\{\sum_{k\in Z}^{N}(sgn\left(w_{1,k}(\infty)\right)^T(I-P(n))\widetilde{w}_{2,k}(\infty)\right\}\right\} \quad (16)$$

For all $k \in Z$ we know that $\widetilde{w}_{2,K}(\infty) = w^o(n) - w_2(n) = -w_2(n)$ as $w^o(n) = 0$ since the filter has only zero taps. If we further assume that $w_1(n)$ and $w_2(n)$ are jointly Gaussian ,from Price theorem[16] we can write $E[sgn(w_{1,k}(n))^T\widetilde{w}_{2,k}(n)] = -\sqrt{\frac{2}{\pi\sigma_{w_1}^2}}E[w_1(n))^Tw_2(n)]$ .In order to evaluate $E[w_1(n))^Tw_2(n)]$,we take the expectation of equation obtained by multiplying the transpose of (9) and (10). If the assumptions A1-A5 is used along with the assumption that $\widetilde{w}_i(n)$ is independent of P(n)[19]. we get

$$E\{\widetilde{w}_1^T(n+1)\widetilde{w}_2(n+1)\} = E\{\widetilde{w}_1^T(n)\widetilde{w}_2(n)\} -$$
$$2\mu E\{\widetilde{w}_1(n)\widetilde{w}_2^T(n)\}\,Tr\left(E(P(n))\right) + \mu^2 E\{\widetilde{w}_1(n)\widetilde{w}_2^T(n)\}\,Tr\left(E(P(n))\right) +$$
$$\mu^2\sigma_v^2 Tr\left(E(B(n))\right) - \alpha E\left\{\left\{sgn(w_1(n))^T\right\}P(n)\widetilde{w}_2(n)\right\} +$$
$$\alpha E\left\{\left\{sgn(w_1(n))^T\right\}\widetilde{w}_2(n)\right\} \quad (17)$$

If we let m(n)= $E\{\widetilde{w}_1^T(n)\widetilde{w}_2(n)\}$) and u(n)= $E\left\{\left\{(sgn(w_1(n))^T\right\}\widetilde{w}_2(n)\right\}$,we can write the equation (17) as

$$m(n+1) = m(n) - 2\mu E\{\widetilde{w}_1(n)\widetilde{w}_2^T(n)\}\,\mathrm{Tr}\left(E(P(n))\right) +$$
$$\mu^2 E\{\widetilde{w}_1(n)\widetilde{w}_2^T(n)\}\,\mathrm{Tr}(E(P(n))) + \mu^2\sigma_v^2\mathrm{Tr}(E(B(n))) + \alpha u(n)\,(1-(P(n))) \quad (18)$$

Let $\theta(n) = Tr[E\{\widetilde{w}_1^T(n)\widetilde{w}_2(n)\}] = Tr(m(n))$ we can write (18) as

$$m(n+1) =$$
$$m(n) - [2\mu E\{\widetilde{w}_1(n)\widetilde{w}_2^T(n)\} - \mu^2 E\{\widetilde{w}_1(n)\widetilde{w}_2^T(n)\}\,]\mathrm{Tr}(E(P(n)))\theta(n) +$$
$$\mu^2\sigma_v^2\mathrm{Tr}(E(B(n))) + \alpha u(n)(1-E(P(n))) \quad (19)$$

Taking trace on both sides of (19) and replacing m(n) into non zero and zero taps elements , we get

$$\theta(n+1) = [1 - 2\mu\mathrm{Tr}(E(P(n))) + \mu^2\mathrm{Tr}(E(P(n)))]\theta(n) + \mu^2\sigma_v^2\mathrm{Tr}(E(B(n))) +$$
$$\alpha u(n)(1 - \mathrm{Tr}(E(P(n)))). \quad (20)$$

If α is very small compared to other elements then we can write (20)as

$$= [1 - 2\mu\mathrm{Tr}(E(P(n))) + \mu^2\mathrm{Tr}(E(P(n)))]\theta(n) + \mu^2\sigma_v^2\mathrm{Tr}(E(B(n))) \quad (21)$$

As $n\to\infty$ $\theta(n)$ converges if $|1 - 2\mu\mathrm{Tr}(E(P(n))) + \mu^2\mathrm{Tr}(E(P(n)))| < 1$ [Appendix A of 1] . Therefore the step size should be such that $2\mathrm{Tr}[E(P(n)] - \mu\mathrm{Tr}[E(P(n))] > 2$. This implies $0 < \mu < \dfrac{2}{\lambda_{max}(P(n))}$ which is same as the condition for convergence of APA [18].Thus we write as

$$\theta(\infty) = \lim_{n\to\infty} \frac{\mu^2\sigma_v^2\mathrm{Tr}(E(B(n)))}{\mu\mathrm{Tr}(E(P(n)))(2-\mu)} > 0 \text{ as } 0 < \mu < \frac{2}{\lambda_{max}(P(n))} \quad (22)$$

$$m(\infty) = \lim_{n\to\infty} \frac{\mu\sigma_v^2\mathrm{Tr}(E(B(n)))}{\mathrm{Tr}(E(P(n)))(2-\mu)\theta(n)} > 0 \quad (23)$$

Thus we can write $E\left[\mathrm{sgn}\left(w_{1,k}(\infty)\right)^T \widetilde{w}_{2,k}(\infty)\right] = -E\left[\mathrm{sgn}\left(w_{1,k}(\infty)\right)^T w_{2,k}(n)\right]$

$= -\sqrt{\dfrac{2}{\pi\sigma_{w_1}^2}}E\left[\left(w_{1,k}(\infty)\right)^T w_{2,k}(\infty)\right] := -\sqrt{\dfrac{2}{\pi\sigma_{w_1}^2}}m(\infty)$. As $m(\infty) > 0$,

$E\left[\mathrm{sgn}\left(w_{1,k}(\infty)\right)^T \widetilde{w}_{2,k}(\infty)\right] < 0$. Thus we can write $Q(\infty) < 0$.For non sparse system $EMSE_{1,2} = EMSE_2$ as $Q(\infty) = 0$. So $\Delta EMSE_1 = EMSE_1 - EMSE_2$ =positive as $EMSE_1 > EMSE_2$ and $\Delta EMSE_2 = 0$ .Therefore $EMSE_{1,2} = EMSE_2 < EMSE_1$. Equation (6) reduces to $E[a(n+1)] = E[a(n)] - \mu_a E[\lambda^2(n)(1 - \lambda(n)]\Delta EMSE_1$. Thus as $n \to \infty$ the algorithm converges to filter2.For highly sparse system $EMSE_{1,2} = EMSE_2 - Q(\infty) < EMSE_2$ as $Q(\infty)$ is negative .Therefore we get either $EMSE_1 < EMSE_{1,2} < EMSE_2$ which makes $\Delta EMSE_1 = EMSE_1 - EMSE_{1,2}$ =negative and $\Delta EMSE_2 = EMSE_2 - EMSE_{1,2}$ =positive thus making the final output to converge to filter1 if the constant for ZA-APA is chosen to be small .On the other hand if α is chosen to be large ,then case 3 of [3] can occur. For semi sparse system we arrive at case 3 of [3], thus the algorithm converges to the filter which has smaller steady state error which in turn depends on the constant chosen.

4   Simulation Results and Discussion

The simulation was carried out in the context of acoustic echo cancellation application.. The impulse response generated has a length of 512 coefficients. The simulations were performed with three scenarios namely highly sparse, moderately sparse and less sparse environments. The entire experiments were carried out for a length of 10,000 samples. The system is tested for colored input which is obtained by passing white noise through a first order system $[(z)=1/(1-0.8z^{-1}]$. The values of μ is chosen to be 0.5 and the projection order is taken as 2 for both the filters .δ is chosen to be $1e^{-12}$. The noise source is a Gaussian white source with zero mean and known variance(σ_v^2).The value of $\lambda(n)$ and range of μ_a is given as in [15].For the ZA-APA the value of α is chosen to be 0.0002 as done in [10].Throughout the experiment the signal to noise ratio is maintained as 30 db .The results are averaged over 10 independent trials with single talk scenario .For a non highly sparse system , the number of zero filter taps is 12.The semi sparse system has 300 zero filter taps and highly sparse system is chosen to have 500 zero filter taps.

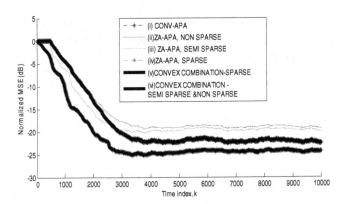

Fig. 2 Convergence analysis of the proposed algorithm with different level of sparsity

Fig 2 shows the plot of convergence of the component filters and the proposed combination filter for different impulse response. Normalized mean square error is used to obtain the convergence and steady state error .Normalized mean square error is defined as normalized mean square value of((estimated weights-desired weights)/(desired weights)).From the results it is proved that the proposed combinational approach takes the better performance of the individual filters. The combiner makes the final output to get inclined towards filter1 or filter2 depending on the impulse response so as to obtain high convergence rate and low steady state error .

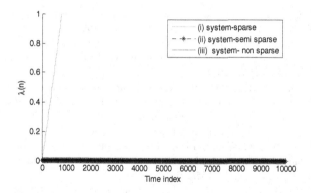

Fig. 3 Evolution of $\lambda(n)$ for difference sparsity levels

Fig 3 shows the values of $\lambda(n)$ for difference environmental conditions .As expected $\lambda(n)$ converges to one for sparse system and zero for non sparse system and to intermediate values for semi sparse system.

5   Conclusion

A new approach for the cancellation of acoustic echo is presented in this paper .The proposed algorithm makes use of convex combination of two adaptive filters namely, ZA-APA and APA with smaller step size so as to work with changing sparsity environments. Stochastic gradient update is used to update the convex combiner. Using energy conservation arguments we have evaluated the performance of the proposed algorithm without the assumption of Gaussian input .It is found from the simulation that in the steady state, the proposed system always selects the better of the two filters depending on the level of sparsity. Our future work is to analyse the tracking and convergence of the system without Gaussian assumption.

References

1. Al-Naffouri, T.Y., Sayed, A.H.: Transient analysis of data-normalized adaptive filters. IEEE Transactions on Signal Processing **51**, 639–652 (2003)
2. Arenas-García, J., Figueiras-Vidal, A.R.: Adaptive combination of proportionate filters for sparse echo cancellation. IEEE Transactions on Audio, Speech, and Language Processing **17**, 1087–1098 (2009)
3. Arenas-García, J., Figueiras-Vidal, A.R., Sayed, A.H.: Mean-square performance of a convex combination of two adaptive filters. IEEE Transactions on Signal Processing **54**, 1078–1090 (2006)
4. Benesty, J., Gay, S.L.: An improved PNLMS algorithm. In: 2002 IEEE International conference on Acoustics, Speech, and Signal Processing, vol. 2, pp. 1881–1884 (2002)
5. Benesty, J., Gänsler, T., Morgan, D.R., Sondhi, M.M., Gay, S.L.: Advances in network and acoustic echo cancellation. Springer, Heildelberg (2001)

6. Breining, C., et al.: Acoustic echo control. An application of very-high-order adaptive filters. IEEE Signal Processing Magazine **16**, 42–69 (1999)
7. Das, B.K., Chakraborty, M.: Sparse Adaptive Filtering by an Adaptive Convex Combination of the LMS and the ZA-LMS Algorithms. IEEE Transactions on Circuits and Systems **61**, 1499–1507 (2014)
8. Dogancay, K., Tanrikulu, O.: Selective-partial-update NLMS and affine projection algorithms for acoustic echo cancellation. In: 2000 IEEE International Conference on Acoustics, Speech, and Signal Processing, vol. 1, pp. 448–451 (2000)
9. Haykin, S.S.: Adaptive filter theory. Pearson Education, India (2007)
10. Li, Y., Li, W., Yu, W., Wan, J., Li, Z.: Sparse Adaptive Channel Estimation Based on-Norm-Penalized Affine Projection Algorithm. International Journal of Antennas and Propagation (2014). doi:10.1155/2014/434659
11. Lima, M.V., Martins, W.A., Diniz, P.S.: Affine projection algorithms for sparse system identification. In: 2013 IEEE International Conference on Acoustics, Speech and Signal Processing, vol. 1, pp. 5666–5670 (2013)
12. Lima, M.V., Sobron, I., Martins, W.A., Diniz, P.S.: Stability and MSE analyses of affine projection algorithms for sparse system identification. In: 2014 IEEE International Conference on Acoustics, Speech and Signal Processing, vol. 1, pp. 6399–6403 (2014)
13. Meng, R.: Sparsity-aware Adaptive Filtering Algorithms and Application to System Identification. Dissertation. University of York (2011)
14. Meng, R., de Lamare, R.C., Vitor, H.N.: Sparsity-aware affine projection adaptive algorithms for system identification. In: Sensor Signal Processing for Defence IET, pp. 1–5 (2011)
15. Paleologu, C., Ciochina, S., Benesty, J.: An efficient proportionate affine projection algorithm for echo cancellation. IEEE Signal Processing Letters **17**, 165–168 (2010)
16. Price, R.: A useful theorem for nonlinear devices having Gaussian inputs. IRE Transactions on Information Theory **4**, 69–72 (1958)
17. Radhika, S., Sivabalan, A.: A Survey on the Different Adaptive Algorithms used in Adaptive Filters. International Journal of Engineering Research & Technology **1**, 1–5 (2012)
18. Shin, H.C., Sayed, A.H.: Mean-square performance of a family of affine projection algorithms. IEEE Transactions on Signal Processing **52**, 90–102 (2004)
19. Werner, S., Apolinário Jr, J.A., Diniz, P.S.: Set-membership proportionate affine projection algorithms. EURASIP Journal on Audio, Speech, and Music Processing **1**, 10 (2007)
20. Yang, J., Sobelman, G.E.: Efficient μ-law improved proportionate affine projection algorithm for echo cancellation. Electronics letters **47**, 73–74 (2011)

An Improved DCT Based Image Fusion Using Saturation Weighting and Joint Trilateral Filter

Arun Begill and Sankalap Arora

Abstract Image fusion is used to merge important information from several images into single image. Color artifacts and noise are the two major challenges for existing transform domain methods which reduce quality of the resulting image. This paper presents a novel method for multi-focus image fusion scheme based on ACMax DCT specially designed for visual sensor networks. The proposed technique consists of multi-focus image fusion technique based on higher valued alternating current coefficients computed in DCT in combination with saturation weighting based color constancy to reduce the color artifacts. The method of fusion may affect the edges and produce noise in the digital images so to overcome this problem joint trilateral filter has been integrated with proposed algorithm to improve the results. The experimental results verify that the proposed technique gives more correct explanation and has better quality when compared with other image fusion methods.

Keywords Acmax DCT · JTF · SIDWT · DWT · Image fusion

1 Introduction

Image fusion is defined as a process used for merging two or more source images to obtain a new image which gives more detailed information [1]. Major objective of image fusion is to obtain an image which gives more detailed and accurate information, and must be more appropriate for human perception [2]. Applications of image fusion contain remote sensing, digital camera, surveillance, dangerous environment like battlefields etc [3]. In multi-focus image fusion method, multiple images of a scene captured with focused on distinct objects to produce a 'focused everywhere' image [4]. Various multi-focus image fusion algorithms have been

---

A. Begill(✉) · S. Arora
Department of Computer Science, DAV University, Jalandhar, Punjab, India
e-mail: {arunbegill490,sankalap.arora}@gmail.com

© Springer International Publishing Switzerland 2016
S. Berretti et al. (eds.), *Intelligent Systems Technologies and Applications*,
Advances in Intelligent Systems and Computing 384,
DOI: 10.1007/978-3-319-23036-8_39

developed in recent years [5-12]. The discrete wavelet transform (DWT) break down the image into four frequency bands (low-low, low-high, high-low, high-high) which contain wavelet coefficients but it suffers from translation invariance [5]. Shift-invariance is attained by eliminating the downsamplers and upsamplers in the DWT [6]. Instead of using Discrete Wavelet Transform, Stationary wavelet transform (SWT) is designed to overcome the problem of shift invariance [7]. The main merits of SWT are temporal stability and consistency verification. However, these two methods may degrade the sharpness of edges and leads to poor quality of resultant images. To improve quality, image fusion is performed by merging both wavelet and curvelet transform [8]. But the drawback of this method is that it consumes more time and has slow speed because of two multi-scale decomposition processes. Furthermore, wavelet transform is not appropriate for resource-constraint such as it consumes more memory space, high bandwidth, more battery power, more energy consumption and greater number of computations [9]. Image fusion techniques based on wavelet domain are complicated and take more time to execute on real-time images. Discrete Cosine Transform (DCT) based techniques are more suitable for resource constrained and it overcomes the problems of wavelet domain methods [10]. Moreover, when input images are coded in Joint Photographic Experts Group (JPEG) format or when the fused image is likely to be stored or given in JPEG format, then the fusion techniques which are used in DCT domain will be very effective [11]. The energy consumption for DCT based image fusion is very less when compared with wavelet domain methods. ACMax DCT based fusion method takes up the block which has higher valued AC coefficients and avoid the calculating variance using all the transformed coefficients which include multiplications, additions and floating point. It only checks the higher valued AC coefficients [12]. Thus it gives high quality output image as well as energy savings. These AC coefficients are based on two conditions: (1) The higher value of AC coefficients image details will be fine, and (2) it also results in higher value of variance.

This paper is organized as follows. In section 2, related work is discussed. Proposed methodology is illustrated in section 3. In section 4 performance measures for image fusion are explained. In section 5 experimental results are compared with state-of-arts methods including DCT + Variance + CV method, DWT with DBSS method, SWT with Haar method and ACMax DCT based method. Conclusion and Future work are discussed in section 6.

2   Related Work

Image fusion strategy may be categorized into two forms: Transform domain and spatial domain methods. The closest related work of transform domain methods are: The author has developed a new statistical sharpness evaluation discussed in the paper by exploiting the distributing of the wavelet coefficients distribution to evaluate the amount of the image's blur. Moreover, the wavelet coefficients distribution is examined utilizing a locally elastic Laplacian combination model [13]. The author had explained Shift-invariant shearlet transform (SIST) describing why the traditional Average–Maximum fusion system is not the most effective concept for medical image fusion, and thus a new system is produced, where the probability density function and standard deviation of the SIST coefficients are used to determine the fused coefficients [14].

The author has discussed multi-focus digital image fusion method in this paper, where numerous images of a given scene caught with concentration on dissimilar objects are merged in a way that most of the objects are likely to be in concentration in the resultant image [15]. In this method the author has illustrated Medical image fusion is the procedure of merging and registering various images from single or various imaging modalities to enhance the image quality and minimize randomness and redundancy in order to increase the medical applicability of medical images for analysis and evaluation of medical problems [16].

The closest related work of spatial domain methods are: The author has described Iterative fuzzy fusion can effectively protect the spectral information while enhancing the spatial resolution of medical imaging and remote sensing images [17]. The author has developed Pixel level digital image fusion describes the running and synergistic mixture of data collected by numerous imaging sources to supply a greater knowledge of a scene [18].

3   Proposed Algorithm

The proposed algorithm gives the detailed information of ACMax DCT based image fusion which has been combined with saturation weighting and Joint Trilateral Filter which give better results than existing methods. One of the biggest advantage of proposed algorithm is that it does not involve any complex computations like floating point, arithmetic operations. The proposed fusion method is performed in ACMax DCT domain and it has proven very effective. The whole process is demonstrated in Fig. 1 and step-wise details are explained below.

Step 1: First of all two images which are partially blurred are given as input.
Step 2: Then image will be divided into the Red (R), Green (G) and Blue (B) channels.
Step 3: Now fusion will be done for each channel separately using ACMax DCT based fusion.

(a) Apply the two dimensional DCT coefficient for computing each block using eq. (1).

$$\eth(\alpha,\beta) = \frac{2}{S}\xi(\alpha)(\beta)\sum_{j=0}^{S-1}\sum_{i=0}^{S-1}f(i,j)\cos\left(\frac{(2i+1)\alpha\pi}{2S}\right)*\cos\left(\frac{(2j+1)\beta\pi}{2S}\right)$$

(1)

where α, β=0, 1 .. S-1 and

$$\acute{\zeta}(\alpha) = \begin{cases} 1/\sqrt{2}, & if\ \alpha = 0 \\ 1, & if\ \alpha \neq 0 \end{cases} \tag{2}$$

the inverse transform is defined as eq. (3)

$$f(i,j) = \frac{2}{S} \sum_{\beta=0}^{S-1} \sum_{\alpha=0}^{S-1} \acute{\zeta}(\alpha)\acute{\zeta}(\beta)\eth(\alpha,\beta)\cos\left(\frac{(2i+1)\alpha\pi}{2S}\right) * \cos\left(\frac{(2j+1)\beta\pi}{2S}\right) \tag{3}$$

where i, j=0, 1, . . S-1. Here ð (0, 0) is the DC coefficient and it represents the mean value of that image block. Remaining coefficient are AC coefficients.

(b) The normalized transform coefficients are defined in eq. (4)

$$\eth\wedge(\alpha,\beta) = \frac{\eth(\alpha,\beta)}{S} \tag{4}$$

Variance (σ^2) of the image block can inferred from the transformed coefficients as follows in eq. (5):

$$\sigma^2 = \sum_{\alpha=0}^{S-1} \sum_{\beta=0}^{S-1} \frac{\eth^2(\alpha,\beta)}{S^2} - \eth\wedge(0,0) \tag{5}$$

where ð ∧ (0, 0) is the normalized DC coefficients and other ð ∧ (α,β) are normalized AC coefficients.

(c) Implies that a variance of a block 8*8 is given by the sum of the squares of the normalized coefficients.

$$\sigma^2 = \sum_{t=1}^{63} Å\wedge t^2 \tag{6}$$

(d) Max count is computed on AC coefficients in the nth block and compared with the respective blocks in other source images and image with higher Maxcount value. This process is selected for all blocks to fuse the DCT representations of multi-images into a single DCT representation image.

Step 4: Now separated Red (R), Green (G) and Blue (B) channels will be concatenated to form the fused image.

Step 5: Saturation weighting based color constancy will be applied to decrease the fusion artifacts and produce result in a new high quality vision [19].

Step 6: The process of image fusion may affect the edges in the digital images so to overcome this problem by joint trilateral filter is applied to improve the results [20].

Step 7: In the end, final image is obtained with high quality.

Table 1 Nomenclature of notations used in proposed algorithm

Symbols	Meaning
Đ	Respective DCT image
α	Horizontal DCT co-ordinates
β	Vertical DCT co-ordinates
R	Number of rows
S	Number of columns
i, j	Respective image co-ordinates
σ^2	Variance
(α)	Normalized scale factor
Ä	Alternating current coefficient

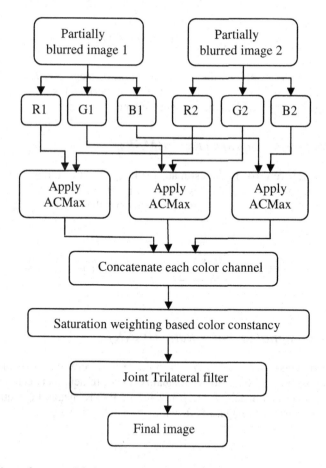

Fig. 1 Flowchart of proposed fusion scheme

4   Performance Metrics

The analysis of image quality is done by subjective evaluation along with objective evaluation. For subjective evaluation, it is hard to find difference between proposed and existing method. The human visual system (HVS) is very complex. So, performance metrics have been used to analyze the performance of color images. They are Mean square error (MSE), Root mean squared error (RMSE), Peak signal to noise ratio (PSNR), Mean Structural similarity index (MSSIM) and Spatial Frequency (SF).

4.1   Mean Square Error (MSE)

Mean square error is used to evaluate image quality index. High value of mean square indicates that quality of image is poor. Mean square error between the reference image and the fused image is computed by eq. (7).

$$MSE = \frac{1}{RS} \sum_{r=1}^{R} \sum_{s=1}^{S} [G(r,s) - L(r,s)]^2 \tag{7}$$

where $G(r, s)$ is the pixel value of the referenced image and $L(r, s)$ is the pixel value of the fused image. If $MSE = 0$ then it is perfect fusion image where the fused image exactly matches the reference image.

4.2   Root Mean Square error (RMSE)

Root Mean square error is the estimate for image quality index. Large value of root mean squared error indicates that quality of image is adverse. Root Mean squared error between the reference image and the fused image is computed as eq. (8).

$$RMSE = \sqrt{\frac{1}{RS} \sum_{r=1}^{R} \sum_{s=1}^{S} [G(r,s) - L(r,s)]^2} \tag{8}$$

4.3   Peak Signal to Noise Ratio (PSNR)

PSNR computes the peak signal-to-noise ratio between two images. This proportion is frequently used as a quality measurement between ground truth image and a fused image. The larger value of PSNR means high quality of the fused or reconstructed image. PSNR value is calculated by eq. (9):

$$PSNR = 10 \log_{10}(\frac{255^2}{MSE}) \tag{9}$$

4.4   Mean Structural Similarity Index (SSIM)

The structural similarity measure is a quality criterion used for objective evaluation of fused image. The general form of the metric that is used to evaluate the structural similarity between two signal vectors i and j is given by eq. (10).

$$SSIM(i,j) = \left(\frac{2\mu_i\mu_j + \acute{\varsigma}_1}{\mu_i^2 + \mu_j^2 + \acute{\varsigma}_1}\right)\left(\frac{2\sigma_{ij} + \acute{\varsigma}_2}{\sigma_i^2 + \sigma_j^2 + \acute{\varsigma}_2}\right) \tag{10}$$

where μ_i and μ_j are the sample means of i and j respectively, σ_i^2 and σ_j^2 are the sample variances of i and j respectively, and σ_{ij} is the sample cross-covariance between i and j. The default values for $_1$ and $_2$ are 0.01 and 0.03. The average of the SSIM values across the image (mean SSIM or MSSIM) gives the final quality measure.

4.5   Spatial Frequency (SF)

The row and column frequencies of the fused image ($\eth$) of size R*S is computed as given in eqs. (11) and (12).

$$RF = \sqrt{\frac{1}{RS}\sum_{r=0}^{R-1}\sum_{s=1}^{S-1}[\eth(r,s) - \eth(r,s-1)]^2} \tag{11}$$

$$CF = \sqrt{\frac{1}{RS}\sum_{s=0}^{S-1}\sum_{r=1}^{R-1}[\eth(r,s) - \eth(r-1,s)]^2} \tag{12}$$

Total spatial frequency of the fused image which is based on edge information is computed as eq. (13).

$$SF = \sqrt{(RF)^2 + (CF)^2} \tag{13}$$

5   Experimental Results

The proposed digital image fusion algorithm is tested on various color images and the results are calculated. Experimental images like dravid, bus, shake and calendar are taken as source images as shown in Fig. 2. It is hard to find subjectively difference between the proposed method and ACMax DCT based method. So, MSE, RMSE, PSNR, MSSIM and SF are used to calculate the performance of different techniques.

From the results given in Table 2 and 3 it can be concluded that of proposed algorithm provides best performance than others on performance metric of MSE and RMSE. From Table 4, it has been observed that the values of PSNR of proposed scheme are infinite on images such as dravid, bus, calendar images which mean ideal fusion. From Table 5 clearly shows that proposed method for Mean Structural Similarity Index is mostly near to 1. The quantitative comparisons are shown in Table 6. From Table 6 it is inferred that the proposed approach preserve more useful information than DCT + Variance + CV, DWT and SIDWT based methods.

For more comparisons, the proposed method is tested on other images and results of all images are shown in Fig. 2. Here again the proposed method yields better image quality than the other methods.

Table 2 Mean Square Error (MSE)

Image name	DCT + Variance + CV	DWT with DBSS	SIDWT with Haar	ACMax DCT	Proposed Method
Dravid	50.879450	37.856015	326.254660	2.691571	**0.357607**
Bus	77.627350	43.863675	401.151710	0.433850	**0.203186**
Shake	60.498493	34.787855	172.727725	5.617793	**0.799400**
Calendar	46.126336	55.489895	194.600220	0.404893	**0.000264**

Table 3 Root Mean Squared Error (RMSE)

Image name	DCT + Variance + CV	DWT with DBSS	SIDWT with Haar	ACMax DCT	Proposed Method
Dravid	7.132983	6.152724	18.062521	1.640601	**0.598003**
Bus	8.810638	6.622966	20.028772	0.658673	**0.450761**
Shake	7.778078	5.898123	13.142592	2.370188	**0.894092**
Calendar	6.791637	7.449154	13.949918	0.636312	**0.016257**

Table 4 Peak Signal to Noise Ratio (PSNR)

Image name	DCT + Variance + CV	DWT with DBSS	SIDWT with Haar	ACMax DCT	Proposed Method
Dravid	62.1308	64.6989	45.9905	87.6615	∞
Bus	58.4163	63.4195	44.1954	∞	∞
Shake	60.6267	65.4331	51.5144	81.2703	**98.2063**
Calendar	62.9826	61.3773	50.4787	∞	∞

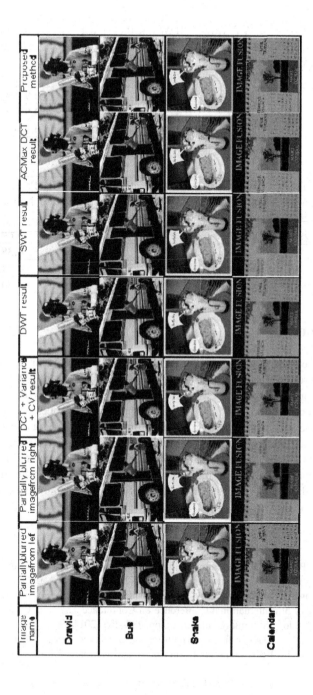

Fig. 2 Results obtained after applying various fusion methods

Table 5 Mean Structural similarity index (MSSIM)

Image name	DCT + Variance + CV	DWT with DBSS	SIDWT with Haar	ACMax DCT	Proposed Method
Dravid	0.9632	0.9788	0.8445	0.9976	**0.9985**
Bus	0.9191	0.9717	0.7831	0.9992	**0.9994**
Shake	0.9218	0.9586	0.8265	0.9971	**0.9964**
Calendar	0.9210	0.9453	0.7664	0.9987	**0.9994**

Table 6 Spatial frequency (SF)

Image name	DCT + Variance + CV	DWT with DBSS	SIDWT with Haar	ACMax DCT	Proposed Method
Dravid	5.9997	6.2465	6.1007	6.8473	**6.8473**
Bus	6.7505	7.8142	7.4662	8.3839	**8.3839**
Shake	5.4455	6.2590	5.6622	6.9126	**6.9126**
Calendar	5.2881	6.1027	5.1834	6.5929	**6.5929**

6 Conclusion

The DCT based methods of image fusion are less time consuming in real time applications. The proposed method has integrated Alternating Current (AC) DCT domain based fusion with saturation weighting based color constancy and joint trilateral filter to decrease the color artifacts and improve the sharpness of edges. The experimental results have shown that the proposed technique outperforms over the state-of-art methods on performance metrics. This work has not considered any evolutionary technique to enhance the results further. So, in near future any evolutionary algorithm to find the best alternative during image fusion process may be used. It may also prove useful in improving the spatial frequency.

References

1. Garg, R., Gupta, P., Kaur, H.: Survey on multi-focus image fusion algorithms. In: 2014 Recent Advances in Engineering and Computational Sciences (RAECS). IEEE (2014)
2. Gupta, R., Awasthi, D.: Wave-packet image fusion technique based on genetic algorithm. In: 2014 5th International Conference on Confluence The Next Generation Information Technology Summit (Confluence), pp. 280–285 (2014)
3. Drajic, D., Cvejic, N.: Adaptive fusion of multimodal surveillance image sequences in visual sensor networks. IEEE Transactions on Consumer Electronics **53**(4), 1456–1462 (2007)
4. TWan, T., Zhu, C., Qin, Z.: Multifocus image fusion based on robust principal component analysis. Pattern Recognition Letters **34**(9), 1001–1008 (2013)

5. Prakash, O., Srivastava, R., Khare, A.: Biorthogonal Wavelet Transform Based Image Fusion Using Absolute Maximum Fusion Rule. In: Image processing, 2013 International Conference on Information and Communication Technologies, pp. 577–582. IEEE (2013)

6. Harrity, K., et al.: Double-density dual-tree wavelet-based polarimetry analysis. In: Aerospace and Electronics Conference, NAECON 2014-IEEE National, pp. 121–126. IEEE (2014)

7. Rockinger, O.: Image sequence fusion using a shift-invariant wavelet transform. In: International Conference on Proceedings of the Image Processing, vol. 3. IEEE (1997)

8. Li, S., Yang, B.: Multifocus image fusion by combining curvelet and wavelet transform. Pattern Recognition Letters **29**(9), 1295–1301 (2008)

9. Kociołek, M., Materka, A., Strzelecki, M., Szczypiński, P.: Discrete wavelet transform –derived features for digital image texture analysis. In: Proc. of International Conference on Signals and Electronic Systems, Lodz, Poland, pp. 163–168 (2001)

10. Haghighat, M.B.A., Aghagolzadeh, A., Seyedarabi, H.: Multi-focus image fusion for visual sensor networks in DCT domain. Computers & Electrical Engineering **37**(5), 789–797 (2011)

11. Haghighat, M.B.A., Aghagolzadeh, A., Seyedarabi, H.: Real-time fusion of multi-focus images for visual sensor networks. In: 2010 6th Iranian of the Machine Vision and Image Processing (MVIP). IEEE (2010)

12. Phamila, Y.A.V., Amutha, R.: Discrete Cosine Transform based fusion of multi-focus images for visual sensor networks. Signal Processing 161–170 (2014)

13. Tian, J., Chen, L.: Adaptive multi-focus image fusion using a wavelet-based statistical sharpness measure. Signal Processing **92**, 2137–2146 (2012)

14. Liu, Y., Liu, S., Wang, Z.: Multi-focus image fusion with dense SIFT. In: Information Fusion, vol. 23, pp. 139–155. IEEE (2015)

15. Cao, L., et al.: Multi-focus image fusion based on spatial frequency in discrete cosine transform domain, pp. 220–224 (2015)

16. James, A.P., Dasarathy, B.V.: Medical image fusion: A survey of the state of the art. Information Fusion **19**, 4–19 (2014)

17. Dammavalam, S.R., Maddala, S., Krishna Prasad, M.H.M.: Iterative image fusion using fuzzy logic with applications. In: Advances in Computing and Information Technology. Springer, Heidelberg, pp. 145–152 (2013)

18. Anita, S.J.N., Moses, C.J.: Survey on pixel level image fusion techniques. In: 2013 International Conference on Emerging Trends in Computing, Communication and Nanotechnology (ICE-CCN). IEEE (2013)

19. Ahn, H., Lee, S., Lee, H.S.: Improving color constancy by saturation weighting. IEEE International Conference on Acoustics, Speech and Signal Processing (ICASSP) (2013)

20. Serikawa, S., Huimin, L.: Underwater image dehazing using joint trilateral filter. Computers & Electrical Engineering **40**(1), 41–50 (2014)

Improved Fuzzy Image Enhancement Using L*a*b* Color Space and Edge Preservation

Shruti Puniani and Sankalap Arora

Abstract Image enhancement is a process of improving the perceptibility of an image so that the output image is better than input image. The traditional image enhancement techniques may affect the edges of an image which leads to loss of perceptual information. The existing techniques use primary/secondary color spaces which are device-dependent. This research paper works on these two issues. It uses L*a*b* color space which is device independent. To evaluate fuzzy membership values, L component is stretched while preserving the chromatic information a and b. Moreover, an edge preserving smoothing has been integrated with fuzzy image enhancement so that edges are not affected and remain preserved. The proposed technique is compared with existing techniques such as Histogram equalization, Adaptive histogram equalization and fuzzy based enhancement. The experimental results indicate that the proposed technique outperforms the existing techniques.

Keywords Image enhancement · Fuzzy-logic · L*a*b* color space (CIELAB) · Color images · Edge preservation, histogram equalization

1 Introduction

Image enhancement is defined as a process in which the pixel's intensity is transformed in such a way that the subjective quality of resultant image is better than the original image which provides better processing and machine analysis [1]. The major objective of image enhancement is to provide an image with fine details and information, so that further processing such as segmentation and edge detection becomes fairly easy [2]. In the past, image enhancement has been applied to various fields like medical imaging, astronomical imaging, camera and

S. Puniani(✉) · S. Arora
Department of Computer Science, DAV University, Jalandhar, Punjab, India
e-mail: {puniani.shruti,sankalap.arora}@gmail.com

© Springer International Publishing Switzerland 2016   459
S. Berretti et al. (eds.), *Intelligent Systems Technologies and Applications*,
Advances in Intelligent Systems and Computing 384,
DOI: 10.1007/978-3-319-23036-8_40

video processing, geographical prospecting, ocean imaging etc. Various techniques for image enhancement have been developed so far. These techniques are categorized into two domains: (1) Transform domain methods (2) Spatial domain methods [3]. In the transform domain method, the frequency transform of an image is modified which is achieved by calculating 2-D transform but it consumes more time thus making it unsuitable for real time image processing [4]. In spatial domain methods, the operation is performed on the pixels itself. These include intensity transformation and spatial filtering [5]. Histogram equalization (HE) is a spatial domain technique which is very popular as it is very simple and easy to implement. Though the technique is simple, yet its conventional nature leads to unnatural look of over enhancement. The major drawback of HE is that it gives washed out appearance to an image and focuses on the global contrast enhancement of the image leading to loss of local details which can also lead to over enhancement of an image [6][7]. Adaptive Histogram equalization (AHE) works by transforming every pixel in an image using a transformation function which is produced from a neighborhood area. It improves the local contrast of an image thus preserving all details. Dynamic Histogram Equalization (DHE) technique is a modification of traditional HE in which enhancement is performed without any loss of information in the image [8]. In this, firstly the original histogram is divided into multiple sub-histograms. Then, a dynamic gray level (GL) is allotted to each sub-histogram. This hampers contrast from being dominated and produces a washed out effect. Though these techniques undergo local image enhancement, yet they suffer from drawbacks.

The traditional image enhancement techniques discussed above produce poor illumination in images thus resulting in vagueness [9]. This vagueness is produced due to uncertain boundaries and color values [10]. Thus, to solve this vagueness in images, fuzzy logic was implemented. Fuzzy-logic has been successfully employed in various areas of image processing. According to recent studies [11], fuzzy-logic based techniques have proven to be better than traditional methods of image enhancement. Fuzzy image processing consists of three main steps: (1) Image fuzzification in which spatial domain is converted into fuzzy membership domain, (2) Modification of membership values using Fuzzy rules and (3) Image defuzzification in which fuzzy membership domain is re-converted into spatial domain. After the data of image is transformed to fuzzy membership domain, membership values are altered using fuzzy methods. Examples of such fuzzy methods are fuzzy rule based method, fuzzy clustering, etc. In the fuzzy rule based method, there is an antecedent and consequent and they are modified based on neighborhood pixels [12]. Image enhancement using fuzzy logic helps to overcome shortcomings of traditional enhancement methods discussed earlier. Fuzzy logic handles the uncertainties of an image which a machine cannot understand and helps in automatic contrast enhancement of images.

This paper is organized as: Section 2 gives a detailed explanation about motivations of this research. Section 3 presents the proposed methodology of this paper while a performance evaluation criterion is discussed in section 4. The proposed method is compared with existing techniques such as Histogram equalization, Adaptive histogram equalization and Fuzzy based technique in section 5. Finally, the paper is concluded in section 6.

2   Related Work

Image enhancement is the most important step in digital image processing. It improves the visual quality of an image so that the final image subjectively looks better than the input image thus producing fine details. Image enhancement techniques are categorized as: (1) Transform domain (2) Spatial domain. Since transform domain methods are time consuming as discussed earlier, they are rarely used nowadays. Spatial domain techniques include traditional and advanced techniques. Histogram equalization and adaptive histogram equalization are traditional image enhancement techniques while techniques based on fuzzy logic are advanced techniques. Hasikin, Khairunnisa and Nor Ashidi Mat Isa have discussed an enhancement technique based on fuzzy-logic in which pixels in spatial domain are transformed into fuzzy membership domain using a Gaussian membership function. This function calculates the membership values of pixels so that intensity of dark pixels is increased above a threshold value and that of brighter pixels is decreased [14]. Raju, G., and Madhu S. Nair have discussed a fast and reliable method for enhancing low contrast and low bright color images using fuzzy-logic and histogram. They have mainly focused on converting skewed histogram into a uniform histogram. The input RGB image is converted into HSV so as to stretch the V component preserving the chromatic information. The enhancement of V is done using two parameters i.e. contrast intensification parameter and average intensity value of the image. The technique is comparatively fast when compared to existing techniques [2]. Liejun, Wang, and Yan Ting have introduced fuzzy shrink image enhancement algorithm which enhances remote sensing images suffering from Gaussian noise and edge degradation. The technique is made up of non-sub sampled contour-let transform domain and fuzzy domain. It shows optimal de-noising effect and the finally produced image is almost similar to input image [15]. Alajarmeh, Ahmad et al. have presented a fuzzy based method for video enhancement using dark channel. The technique helps to improve the quality of images and videos taken during natural phenomenon such as rain, fog and haze. This technique is fast and efficient and also suitable for real time applications [16].

Hanmandlu, M. and Jha, D. have produced a Gaussian membership function to convert the spatial domain into fuzzy membership domain. A global contrast intensification parameter contains three parameters t, intensification parameter, f_h, the fuzzifier and u_c, the crossover point to enhance color images. But the major limitation of this method is that it is more time consuming as compared to other fuzzy based methods [1].

Image enhancement has a wide application in medical imaging. Different methods have been proposed for this. Aggarwal, Anshita, and Amit Garg have discussed an enhancement technique for medical images using Adaptive multi-scale thresholding by reducing noise while preserving edges. It removes salt and pepper noise, speckle and random noise while preserving the edges which are damaged due to its grainy appearance [17].

A lot of techniques have been developed based on histogram. Kotkar V.A et al. have proposed Weighted of local and bidirectional smooth histogram stretching (WLBSHS) and Local Bi-histogram smooth Histogram Stretching (LBSHS) which focuses on local and global image enhancement while preserving brightness of an image [18]. Humied I.A et al. have proposed a combined technique for automatic contrast enhancement of digital images to enhance low contrast images by balancing the amplitude of histogram at both the ends. It is achieved by combining Histogram Equalization and gray level grouping. It is fully automatic technique unlike other traditional techniques [19].

3   Proposed Methodology

3.1   *Modification of Membership Functions*

The existing color image enhancement method based on fuzzy logic considers RGB and HSV color spaces which are both   machine dependent which means that a set of parameters produce different colors on different machines [20]. Also, existing image enhancement methods may degrade the edges too. The proposed method deals with these two problems. The RGB color model is converted into L*a*b* color space, (where L is Lightness, a and b are chromatic components) because RGB image contains only color channels and the light channel is mixed in it. The L component is stretched on the basis of X and Y where X is calculated from histogram G(x) using eq.(1), G(x) is number of pixels with intensity value x. Y=128 according to experiments

$$X=\frac{\sum_x xG(x)}{\sum_x G(x)} \tag{1}$$

Histogram is divided into two classes X_1 and X_2 and fuzzy membership values α_1 and α_2 are calculated based on two fuzzy rules.

Rule 1: If the difference between x and X is large, then the intensity of stretching should be small. To implement this rule and to calculate value of α_1 , eq. (2) is used.

$$\alpha_1 = \frac{1 - ((X - x)}{X} \tag{2}$$

Rule 2: If the difference between x and D is large, then the intensity of stretching should be large. (D is extreme value e.g D =255 for 8-bit images). The value of α_2 is calculated using eqn. (3)

$$\alpha_2 = \frac{D-x}{D-X} \tag{3}$$

Then fuzzy enhanced values are evaluated using eq. (4) and (5) where α_e is the enhanced intensity value and original intensity is replaced with enhanced intensity.

$$\alpha_e = x + \alpha_1(Y) \tag{4}$$
$$\alpha_e = x\alpha_2 + D - x\alpha_2 \tag{5}$$

3.2 Edge Preserving Smoothing

Take a small square sliding window of length Z. It has been proved that a small square window yields best possible results in terms of preserving edges and fine details [21]. The value of Z can be evaluated using eq. (6)

$$Z = \begin{cases} 3, & n < 0.5 \\ 5, & 0.5 \leq n \leq 0.6 \end{cases} \tag{6}$$

where n is noise density which can be calculated using eq. (7)

$$n = \frac{L}{RC} \tag{7}$$

where L is total number of zeros and 255 in the image and product RC is number of pixels in image. Then, the most optimum value of a threshold β is evaluated by using eq. (8) where β is a threshold which depends on noise density and characteristics of an image and $\beta \leq Z^2$.

$$\beta = \begin{cases} \lfloor Z^2(n + 0.50) \rfloor & Z = 3 \\ \lfloor Z^2(n + 0.15) \rfloor & Z = 5 \end{cases} \tag{8}$$

where $\lfloor . \rfloor$ is the floor operation.

Table 1 Nomenclature of symbols used in algorithm

Symbols	Meaning
X	Average intensity value of input image
Y	Contrast intensification parameter
α_1, α_2	Fuzzy membership value
α_e	Enhanced Fuzzy membership value
D	Extreme value of intensity
G(x)	No. of pixels with intensity value x
Z	Size of square sliding window
n	Noise density
L	Number of 0s and 255s in image
RC	Number of pixels in image
β	Optimum threshold value
A	Average value of local contrast
T(x, y)	Gradient
j_x, j_y	Convolution kernels

The whole process is demonstrated in Figure 1, Table 1 explains all the symbols used in proposed method and the algorithm is explained below:

Algorithm for Image enhancement:
1) Input the given RGB image and convert it to L*a*b*.
2) Calculate the histogram G(x) where x ∈ L.
3) Calculate X using (1).
4) Divide G(x) into two classes X_1 [0, X-1] and X_2[X, 255].
5) Fuzzify L and modify membership functions using (2) and (3).
6) Calculate fuzzy enhanced values using (4) and (5).
7) Convert the enhanced L*a*b* image to RGB.
8) Apply edge preserving smoothing using (6), (7) and (8).
9) Obtain the enhanced output image.

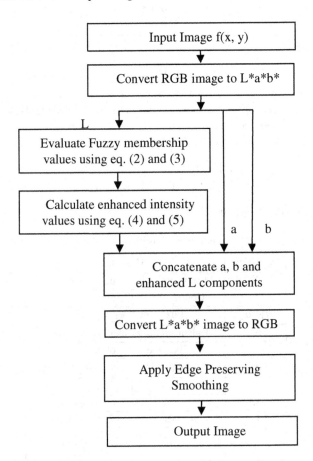

Fig. 1 Flowchart of the proposed method

4   Performance Metrics

To compare different image enhancement techniques, two quantitative performance measures have been taken which are discussed below. Along with these, visual results are also shown.

4.1   Contrast Improvement Index (CII)

It is the most important performance measure for contrast of an image [2]. The higher value of CII indicates higher contrast of an image. The value of CII can be calculated using Eq. (9)

$$\text{CII} = \frac{A_{proposed}}{A_{original}} \tag{9}$$

where A is average value of local contrast which is calculated using 3×3 window using eq. (10)

$$\frac{max - min}{max + min} \tag{10}$$

Where $A_{proposed}$ is average value of local contrast in output image and $A_{original}$ is average value of local contrast in original image.

4.2   Tenengrad Measure

The Tenengrad measure is one of the most robust and precise image performance measures based on gradient magnitude maximization [2]. Its value can be calculated from gradient where partial derivative is calculated using a sobel filter with convolution kernels j_x and j_y. The equation of gradient is given using eq. (11)

$$T(x,y) = \sqrt{(j_x \otimes J(x,y))^2 + (j_y \otimes J(x,y))^2} \tag{11}$$

The value of Tenengrad is given using eq. (12)

$$TGD = \sum_x \sum_y T(x,y)^2 \tag{12}$$

The higher value of Tenengrad shows that the structural information of an image has been improved.

5   Results and Discussions

The performance of the proposed method has been tested on various color images. The values of performance measures prove that proposed method is superior to traditional methods. In order to prove this fact, two quantitative performance measures have been used. Subjectively, it is hard to find difference between the proposed method and traditional image enhancement methods. Thus, CII and Tenengrad are used to evaluate the performance of different methods.

Figure 2 shows the enhanced color images after applying Histogram Equalization, Adaptive Histogram Equalization, Fuzzy based image enhancement and proposed method on img1.jpg, img2.jpg, img3.jpg, img4.jpg and img5.jpg.

Tables 2-6 show values of performance measures obtained after applying various enhancement techniques on img1.jpg, img2.jpg, img3.jpg, img4.jpg and img5.jpg.

From the tables, it is evident that the proposed method produces higher values of CII and Tenengrad as compared to the traditional techniques. Histogram equalization have yielded very less value of CII and Tenengrad because it focuses on enhancing the global contrast of an image whereas Adaptive histogram equalization has produced more values than Histogram equalization because it focuses on the local contrast of an image. But its value is less than fuzzy method. This is because of the fact that it can't handle vagueness which is introduced in images in the form of imprecise boundaries and color values. The proposed method has proved to be the best out of all these methods because it uses a device independent color space i.e L*a*b* and it also works on preservation of edges.

Table 2 Performance measure values obtained after applying different techniques on img1.jpg

Performance measures	Histogram equalization	Adaptive histogram equalization	Fuzzy method	Proposed Method
CII	0.2398	0.0095	6.3811	**7.6752**
Tenengrad($\times 10^4$)	1.1479	1.7779	3.3985	**4.4017**

Table 3 Performance measure values obtained after applying different techniques on img2.jpg

Performance measures	Histogram equalization	Adaptive histogram equalization	Fuzzy method	Proposed method
CII	0.3187	0.3534	4.0771	**5.7261**
Tenengrad($\times 10^4$)	1.4508	2.6969	3.0366	**3.8710**

Table 4 Performance measure values obtained after applying different techniques on img3.jpg

Performance measures	Histogram equalization	Adaptive histogram equalization	Fuzzy method	Proposed method
CII	0.7187	0.1450	1.1740	**4.9519**
Tenengrad($\times 10^4$)	1.4252	2.1144	4.9108	**5.9423**

Table 5 Performance measure values obtained after applying different techniques on img4.jpg

Performance measures	Histogram equalization	Adaptive histogram equalization	Fuzzy method	Proposed method
CII	0.7881	2.3883	3.3753	**4.8596**
Tenengrad $(\times 10^4)$	1.3341	3.1187	4.7722	**5.8585**

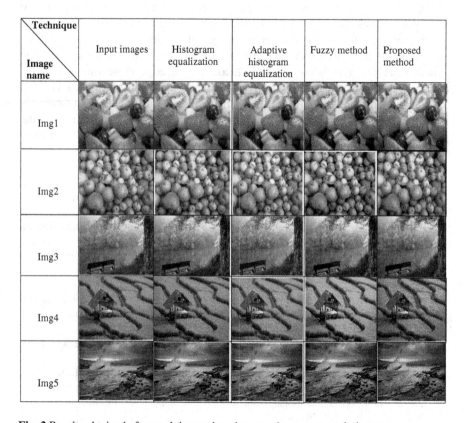

Technique / Image name	Input images	Histogram equalization	Adaptive histogram equalization	Fuzzy method	Proposed method
Img1					
Img2					
Img3					
Img4					
Img5					

Fig. 2 Results obtained after applying various image enhancement techniques

Table 6 Performance measure values obtained after applying different techniques on img5.jpg

Performance measures	Histogram equalization	Adaptive histogram equalization	Fuzzy method	Proposed method
CII	0.9690	1.4612	2.2261	**4.2930**
Tenengrad$(\times 10^4)$	1.6093	3.5179	5.4371	**6.4799**

6   Conclusion

An efficient L*a*b* based Fuzzy image enhancement has been proposed in this paper. The method uses L*a*b* color space which is device independent unlike other color models like HSV, HSI, etc. Since image enhancement may introduce noise in the image leading to degradation of edges, the technique is integrated with an edge preserving smoothing algorithm. The comparison of the proposed method has also been drawn with existing image enhancement techniques to prove effectiveness of contrast enhancement. The results have shown that proposed method improves the values of CII and Tenengrad measure. The L*a*b* based membership function adds to it an advantage of its device independent nature. According to this, the color and contrast of an image will not depend upon whatever device it is being used. Furthermore, an edge preserving smoothing algorithm has been integrated with it so that while applying image enhancement, the edges remain preserved. In this paper the value of Y is fixed. So, in future the value of Y will be calculated adaptively using an optimization algorithm. Also, the proposed method is a little time consuming. Thus, we will try to make it fast in future.

References

1. Hanmandlu, M., Jha, D.: An Optimal Fuzzy System for Color Image Enhancement. IEEE Transactions on Image Processing 2956–2966 (2006)
2. Raju, G., Nair, M.S.: A fast and efficient color image enhancement method based on fuzzy-logic and histogram. AEU- International Journal of Electronics and Communications 237–243 (2014)
3. Gao, M.-Z., Wu, Z.-G., Wang, L.: Comprehensive evaluation for HE based contrast enhancement techniques. In: Advances in Intelligent Systems and Applications, vol. 2, pp. 331–338. Springer, Heidelberg (2013)
4. Celik, T.: Spatial Entropy-Based Global and Local Image contrast Enhancement. IEEE Transactions on Image Processing 5298–5308 (2014)
5. bt. Shamsuddin, N., bt. Wan Ahmad, W.F., Baharudin, B.B., Kushairi, M., Rajuddin, M., bt. Mohd, F.: Significance level of image enhancement techniques for underwater images. In: 2012 International Conference on Computer & Information Science (ICCIS), pp. 490–494 (2012)
6. Senthilkumaran, N., Thimmiaraja, J.: Histogram equalization for image enhancement using MRI brain images. In: 2014 World Congress on Computing and Communication Technologies (WCCCT), pp. 80–83. IEEE (2014)
7. Kaur, M., Kaur, J., Kaur, J.: Survey of contrast enhancement techniques based on histogram equalization. International Journal of Advanced Computer Science and Applications (2011)
8. Kim, B., Gim, G.Y., Park, H.J.: Dynamic histogram equalization based on gray level labeling. IS&T/SPIE Electronic Imaging. International Society for Optics and Photonics (2014)

9. Cheng, D., Shi, D., Tang, X., Liu, J.: A local-context-based fuzzy algorithm for image enhancement. In: Yang, J., Fang, F., Sun, C. (eds.) IScIDE 2012. LNCS, vol. 7751, pp. 165–171. Springer, Heidelberg (2013)

10. Hassanien, A.E., Soliman, O.S., El-Bendary, N.: Contrast enhancement of breast MRI images based on fuzzy type-II. In: Corchado, E., Snášel, V., Sedano, J., Hassanien, A.E., Calvo, J.L., Ślęzak, D. (eds.) SOCO 2011. AISC, vol. 87, pp. 77–83. Springer, Heidelberg (2011)

11. Sudhavani, G., et al.: K enhancement of low contrast images using fuzzy techniques. In: 2015 International Conference on Signal Processing and Communication Engineering Systems (SPACES). IEEE, pp. 286–290 (2015)

12. Tehranipour, F., et al.: Attention control using fuzzy inference system in monitoring CCTV based on crowd density estimation. In: 2013 8th Iranian Conference on Machine Vision and Image Processing (MVIP), pp. 204–209. IEEE (2013)

13. Nair, M.S., et al.: Fuzzy logic-based automatic contrast enhancement of satellite images of ocean. Signal, Image and Video Processing 5(1), 69–80 (2011)

14. Hasikin, K., Isa, N.A.M.: Fuzzy image enhancement for low contrast and non-uniform illumination images. In: IEEE International Conference on Signal and Image Processing Applications (ICSIPA), pp. 275–280. IEEE (2013)

15. Liejun, W., Ting, Y.: A new approach of image enhancement based on improved fuzzy domain algorithm. In: International Conference on Multisensor Fusion and Information Integration for Intelligent Systems (MFI), pp. 1–5. IEEE (2014)

16. Alajarmeh, A., et al.: Real-time video enhancement for various weather conditions using dark channel and fuzzy logic. In: 2014 International Conference on Computer and Information Sciences (ICCOINS), pp. 1–6. IEEE (2014)

17. Aggarwal, A., Garg, A.: Medical image enhancement using Adaptive Multiscale Product Thresholding. In: 2014 International Conference on Issues and Challenges in Intelligent Computing Techniques (ICICT), pp. 683–687. IEEE (2014)

18. Kotkar, V.A., Gharde, S.S.: Image contrast enhancement by preserving brightness using global and local features, pp. 262–271 (2013)

19. Humied, I.A., Abou-Chadi, F.E.Z., Rashad, M.Z.: A new combined technique for automatic contrast enhancement of digital images. Egyptian Informatics Journal 27–37 (2012)

20. Ganesan, P., Rajini, V., Rajkumar, R.I.: Segmentation and edge detection of color images using CIELAB color space and edge detectors. In: 2010 International Conference on Emerging Trends in Robotics and Communication Technologies (INTERACT), pp. 393–397. IEEE (2010)

21. Ramadan, Z.M.: A New Method for Impulse Noise Elimination and Edge Preservation. Canadian Journal of Electrical and Computer Engineering 2–10 (2014)

Partial Image Scrambling Using Walsh Sequency in Sinusoidal Wavelet Transform Domain

H.B. Kekre, Tanuja Sarode and Pallavi N. Halarnkar

Abstract Information security is a major concern. Several methods are proposed for securing digital images by either encrypting or scrambling them. As digital images are huge in size, scrambling the entire image would be a time consuming process. To overcome this constraint, partial image scrambling method has been proposed in this paper. The proposed method has been rigorously tested so as to find out which components in the wavelet domain can be removed from scrambling process at the cost of quality in descrambled images. A number of experimental parameters have been used for analysis and it has been observed that DCT and Hartley wavelet transforms are the best performers.

Keywords Partial image encryption · Wavelet · Kekre's walsh sequency

1 Introduction

Information security is an important aspect. The traditional information security algorithms are not suitable for digital images as they are bulky in size. Digital images may be secured with scrambling or encrypting the image pixels. Scrambling involves shifting the positions of pixels, whereas encryption involves changing the pixel values. Yang Fengxia has proposed an image encryption algorithm in DCT domain [1]. The said algorithm makes use of three dimensional Arnold Mapping. The encryption process includes shifting the original values, applying Arnold scrambling and symbol encryption of DCT coefficients.

H.B. Kekre · P.N. Halarnkar(✉)
MPSTME, NMIMS University, Mumbai, India
e-mail: pallavi.halarnkar@gmail.com

T. Sarode
TSEC, Mumbai University, Mumbai, India

© Springer International Publishing Switzerland 2016
S. Berretti et al. (eds.), *Intelligent Systems Technologies and Applications*,
Advances in Intelligent Systems and Computing 384,
DOI: 10.1007/978-3-319-23036-8_41

471

[4] PENG Jing-yu scrambles the pixel values using Arnold transformation and maps it to a different color space.

Narendra K. Pareek et al. [2] used 128 bits key for image encryption process. Firstly the image is scrambled using a mixing process, then the image is divided into blocks, these blocks are passed through diffusion and substitution process. Gaurav Bhatnagar et al. proposed an image encryption scheme suitable for securing digital images over the transmission channel[3]. The proposed technique is a combination of fractional wavelet transform and chaotic maps.

An extended variation to TJ-ACA image encryption algorithm is been proposed by Taranjit Kaur et al. [5]. The method involves no preview of the encrypted image thus making it impossible for the attacks and the intruder to decrypt the image. An image encryption scheme that involves scrambling the amplitude which enhances the security of the double random phase encoding is proposed by Zhengjun Liu et al. [6]

A novel image encryption scheme based on finite field cosine transform is proposed [7] J.B. Lima et al. The image is firstly divided into blocks, finite field cosine transform is applied over the blocks, a secret key is then used to decide the positions of these blocks in the scrambled image. S. Behnia et al. presented an image encryption scheme based on Jacobian elliptic maps[8]. The proposed approach is not only suitable for images but can be extended to video encryption also.

Image encryption using the concept of Data hiding is proposed by Long Bao et al. [9]. Traditional image encryption techniques produces a noise like image by applying image encryption technique, the aim in the proposed method was to transform the encrypted image into another useful image which would visually represent the cover image. A hybrid genetic based approach for image encryption is presented by Shubhangini P.Nichat et al. [10]. The method makes use of genetic algorithm and chaotic map. A number of encrypted images are given as an initial population to the genetic algorithm, using genetic algorithm the optimum result is obtained. The best cipher image is selected based on correlation and entropy.

Mohammed A. Shreef et al. proposed an image encryption algorithm based on XLLs, which involves XOR operation, Lagrange Process and Least Square Process[11]. The proposed approach also discusses about the key used in the encryption process. Two alternatives are discussed, in the first a 128 bit long key can be used and as the second one an image is itself used as a key. A modification to the AES algorithm to make it suitable for image encryption is proposed by Salim M. Wadi et al. [12]. The modifications include the number of rounds to one and replacing the S-box technique with a new S-box technique. The proposed AES also worked on enhancing the time ciphering and pattern appearance. Discrete fractional Fourier transform using exponential random phase mask is been used for image encryption by Ashutosh et al. [13]. Two keys used in encryption process makes the decryption impossible without the right keys. Lini Abraham et al. [14] compared two image encryption algorithms one of them based on Rubik's cube principle and the other on new chaos based fast image encryption using parameters like NPCR, UACI, Entropy and correlation. An RGB based image encryption is proposed by Manish Kumar et al. [15]. The proposed method discusses about the keys and the arrangement of RMAC parameters. A formula is formulated for choosing the keys for encryption and decryption. An image encryption method using wave perturbations is been proposed by Yue Wu et al. [16]. The method includes pseudorandom wavefronts and additional salt and pepper bits. A spatio-temporal chaos of mixed linear- non linear coupled map lattices are used for image encryption Zhang Ying-Qian et al. [17]. A bit level pixel permutation that enables the lower and higher bit planes permute mutually.

Muhammad Rafiq Abururab [18] proposed and Asymmetric image encryption algorithm using Schur decomposition in gyrator transform domain. Experimental results prove the validity of the proposed approach. A novel image encryption scheme utilizing Julia sets and Hilbert curves is been proposed Yuanyuan Sun et al. [19]. A random sequence is generated by Julia sets this is then encrypted by Hilbert curves to get the final encrypted keys, the image cipher is produced using the encrypted keys and modulo arithmetic and diffuse operation. Image encryption scheme based on rotation matrix bit level permutation and block diffusion is presented Yushu Zhang et al. [20]. The proposed method provides security along with robustness against noise.

A variation of image encryption schemes are proposed Yue Wu et al. [21]. The encryption scheme uses Latin square whitening, Latin square S-box and Latin Square P-box. A probabilistic image encryption scheme is developed by embedding noise into LSB planes of the image. A novel approach to encrypt 3D images and real time videos is proposed by Massimiliano Zanin et al. [22]. The proposed method P-Box is fast and is optimised for integer q-bit operations which makes it suitable for any hardware. The P-Box method is combined with S-Box chaotic method for fast and a highly secure system. An image cryptographic technique based on Haar transform is proposed Sara Tedmori et al. [23]. The image is first converted to frequency domain using Haar transform and its sub bands are encrypted in such a way that the data is unbreakable.

2   Wavelet Generation Using Kronecker Product

For Wavelet generation of Sinusoidal Wavelet, Kronecker Product method is used. The image size used for experimental purpose is 256x256. Hence to generate a wavelet transform, the transform matrix of size 16x16 is used. The Kronecker Product of the matrix is taken with itself to generate a wavelet having four components LL, LH , HL and HH.

The Kronecker Product can be applied as follows

$$A \otimes A = a_{ij} [A] \tag{1}$$

Where size of A is 16x16 and is used to generate a wavelet transform of size 256x256.

A (16x16)	⊗	A (16x16)	=	LL	LH
				HL	HH

Wavelet(256X256)

Fig. 1 An Example of Wavelet Generation

In the above Figure 1. LL represents the component with maximum energy of the original image, LH, HL and HH represents the components with some amount of image energy. The above concept is used to generate the sinusoidal wavelets, the transforms used are DCT, DST, Hartley and Real Fourier.

3 Proposed Approach

The Figure below explains the step by step procedure used for scrambling image using walsh sequency in wavelet domain. Figure 2(a) shows the scrambling process and (b) shows the all the different combinations that are used for scrambling the image in wavelet domain using Walsh Sequency [25].

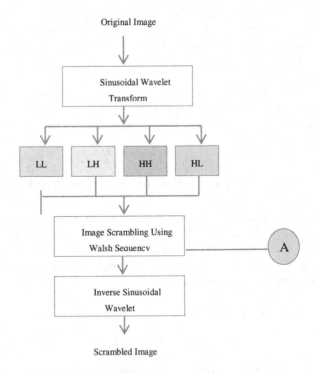

(a) Scrambling Process

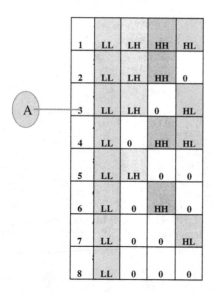

1	LL	LH	HH	HL
2	LL	LH	HH	0
3	LL	LH	0	HL
4	LL	0	HH	HL
5	LL	LH	0	0
6	LL	0	HH	0
7	LL	0	0	HL
8	LL	0	0	0

(b)   Different combinations of Components scrambled

Fig. 2 Partial Image Scrambling Process

The proposed approach focuses towards partial image scrambling. The Walsh sequence is applied on all the combinations of wavelet components , so as to do an in depth study of which components of wavelet (i.e. LL, LH, HH and HL) results in a higher error in the resultant image. The different combinations explored for the proposed approach includes (1) LL,LH,HH & HL (2) LL, LH, HH (3) LL,LH,HL (4) LL,HH,HL (5) LL,LH (6) LL, HH (7) LL, HL(8) LL. For e.g. (2) LL, LH and HH (HL component is removed from the wavelet domain and only LL,LH and HH components are scrambled by applying Walsh sequence), which reduces the computational complexity in the scrambling and descrambling process , as compared to the traditional approach in which the scrambling and descrambling process needs to be applied to all the pixels of the image in the spatial domain.

4   Experimental Results

For experimental purpose a custom database of 15 (24 bit color) images of size 256x256 were used. The results displayed below are averaged over fifteen images. The parameters used for experimental analysis used include Adjacent row pixel correlation (ARPC), Adjacent column pixel correlation (ACPC), Adjacent diagonal pixel correlation (ADPC), Adjacent anti diagonal pixel correlation(AADPC), Structural similarity index measure (SSIM), Peak average fractional change in pixel value (PAFCPV)[24] and Mean Squared Error(MSE).

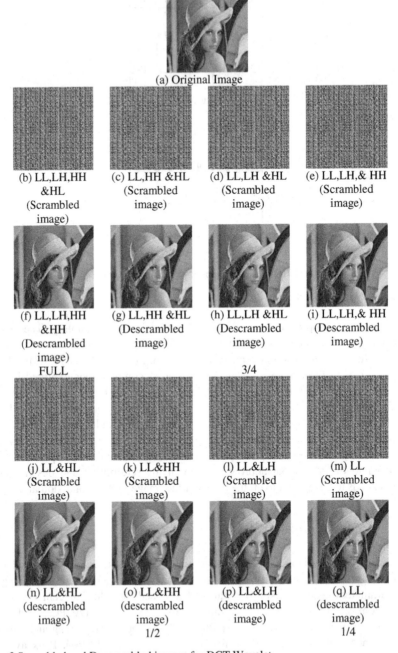

Fig. 3 Scrambled and Descrambled images for DCT Wavelet

Hartley Wavelet

(a) Original Image

(b) LL,LH,HH &HL (Scrambled image)

(c) LL,HH &HL (Scrambled image)

(d) LL,LH &HL (Scrambled image)

(e) LL,LH,& HH (Scrambled image)

(f) LL,LH,HH &HL (Descrambled image)

(g) LL,HH &HL (Descrambled image)

(h) LL,LH &HL (Descrambled image)

(i) LL,LH,& HH (Descrambled image)

FULL

3/4

(j) LL&HL (Scrambled image)

(k) LL&HH (Scrambled image)

(l) LL&LH (Scrambled image)

(m) LL (Scrambled image)

(n) LL&HL (descrambled image)

(o) LL&HH (descrambled image)

1/2

(p) LL&LH (descrambled image)

(q) LL (descrambled image)

1/4

Fig. 4 Scrambled and Descrambled images for Hartley Wavelet

Figure 3 & 4, shows the different scrambled images obtained for scrambling different combinations of wavelet components for DCT and Hartley Wavelet. Figure 3(a) shows the original image, (b)–(e) & (j)-(m) scrambled images, (f)-(i) &(n)-(q) descrambled images. Figure 5-8 (a) Shows the Reduction in correlation obtained in row, column, diagonal and anti-diagonal pixels, (b) shows the structural similarity index measure, (c) shows the Peak average fractional change in pixel value, and (d) shows the Mean squared error.

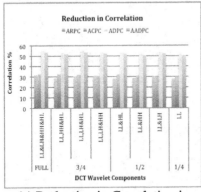

(a) Reduction in Correlation in scrambled images for DCT Wavelet

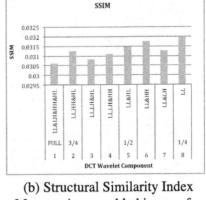

(b) Structural Similarity Index Measure in scrambled images for DCT Wavelet

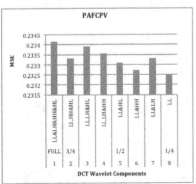

(c) Peak average fractional change in pixel value in scrambled images for DCT wavelet

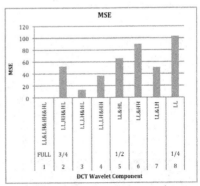

(d) Mean squared error in Descrambled images for DCT wavelet

Fig. 5 Analysis for DCT Wavelet

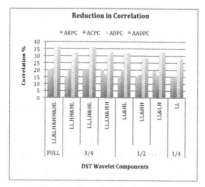

(a) Reduction in Correlation in
scrambled images for DST Wavelet

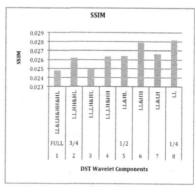

(b) Structural Similarity Index
Measure in scrambled images for
DST Wavelet

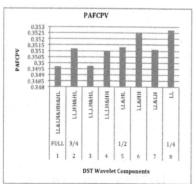

(c) Peak average fractional change in
pixel value in scrambled images for
DST wavelet

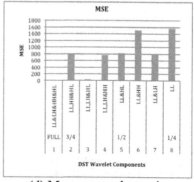

(d) Mean squared error in
Descrambled images for DST
wavelet

Fig. 6 Analysis for DST Wavelet

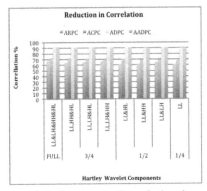

(a) Reduction in Correlation in
scrambled images for Hartley
Wavelet

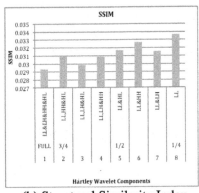

(b) Structural Similarity Index
Measure in scrambled images for
Hartley Wavelet

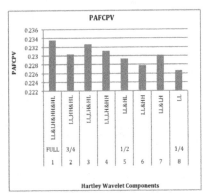

(c) Peak average fractional change
in pixel value in scrambled images
for Hartley wavelet

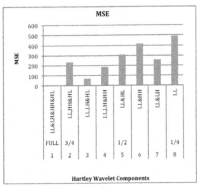

(d) Mean squared error in
Descrambled images for Hartley
wavelet

Fig. 7 Analysis for Hartley Wavelet

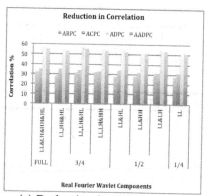

(a) Reduction in Correlation in scrambled images for Real Fourier Wavelet

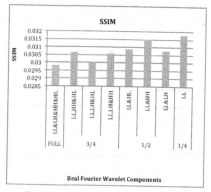

(b) Structural Similarity Index Measure in scrambled images for Real Fourier Wavelet

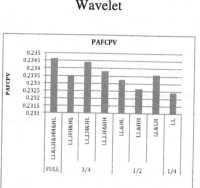

(c) Peak average fractional change in pixel value in scrambled images for Real Fourier wavelet

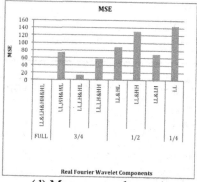

(d) Mean squared error in Descrambled images for Real Fourier wavelet

Fig. 8 Analysis for Real Fourier Wavelet

Table 1 Overall Analysis of Sinusoidal Wavelet Transforms

Wavelet	Full		¾			½		¼
	LL&L H&H H&HL	LL, HH &H L	LL,LH &HL	LL,L H&H H	LL &H L	LL& HH	LL&L H	LL
DCT WAVELET								
Reduction in Correlation	ADPC ✕		ADPC ✔				ADPC ✔	
SSIM	✕		✔				✔	
PAFCPV	✕		✔				✔	
MSE	✕		✔				✔	
DST WAVELET								
Reduction in Correlation	ADPC		ADPC				ADPC	
SSIM	✕		✔		✔			
PAFCPV		✔				✔		✕
MSE	✕		✔				✔	
HARTLEY WAVELET								
Reduction in Correlation	ADPC ✕		ADPC ✕				ADPC ✕	ADPC ✕
SSIM	✕		✔				✔	
PAFCPV	✕		✔				✔	
MSE	✕		✔				✔	
REAL FOURIER WAVELET								
Reduction in Correlation	ADPC ✕		ADPC ✕				ADPC	
SSIM	✕		✔				✔	
PAFCPV	✕		✔				✔	
MSE	✕		✔				✔	

5   Conclusion

A Partial image scrambling technique in wavelet domain has been successfully implemented in this paper. From the experimental results, a summary of the same is presented in Table No 1. Based on subjective fidelity criteria the descrambled images obtained in DCT and Real Fourier wavelet are good. The highest reduction in correlation is obtained in Hartley wavelet above 90%. The least MSE is obtained in DCT wavelet. When ¾ of the components are used, removal of HH components results in best performance across all the sinusoidal wavelet transforms. When ½ of the components are used, removal of HH & HL results in good performance across all the transforms except DST. In case of the last category i.e. when only LL component is maintained in case of DST highest PAFCPV is obtained compared to the rest of the combinations. Good results are obtained in those categories where low frequency components are maximum.

Hence it can be concluded that the proposed method is suitable for partial image encryption at the cost of degradation of descrambled images, which is minimum. The best performers are DCT and Hartley with respect to MSE and reduction in correlation and based on subjective observation of descrambled images DCT and Real Fourier are the best performers.

References

1. Fengxia, Y.: DCT Domain Color Image Block Encryption Algorithm Based on Three-Dimension Arnold Mappin. In: Fifth International Conference on Computational and Information Sciences (ICCIS), pp. 682–685. IEEE (2013)
2. Pareek, N.K., Patidar, V., Sud, K.K.: Diffusion- substitution based gray image encryption scheme. Digital Signal Processing 23, 894–901 (2013)
3. Bhatnagar, G., Jonathan Wu, Q.M., Raman, B.: Discrete fractional wavelet transform and its application to multiple encryption. Information Sciences 223, 297–316 (2013)
4. Jing-yu, P.E.N.G.: Efficient Color Image Encryption and Decryption Algorithm. International Journal of Digital Content Technology & its Applications 7(6) (2013)
5. Kaur, T., Sharma, R.: Image Cryptography by TJ-SCA: Supplementary Cryptographic Algorithm for Color Images. International Journal of Scientific & Engineering Research (IJSER) 4 (2013)
6. Liu, Z., Li, S., Liu, W., Wang, Y., Liu, S.: Image encryption algorithm by using fractional Fourier transform and pixel scrambling operation based on double random phase encoding. Optics and Lasers in Engineering 51(1), 8–14 (2013)
7. Lima, J.B., Lima, E.A.O., Madeiro, F.: Image encryption based on the finite field cosine transform. Signal Processing: Image Communication 28(10), 1537–1547 (2013)
8. Behnia, S., Akhavan, A., Akhshani, A., Samsudin, A.: Image encryption based on the Jacobian elliptic maps. Journal of Systems and Software 86(9), 2429–2438 (2013)
9. Bao, L., Zhou, Y., Chen, C.L.P.: Image encryption in the wavelet domain. In: SPIE Defense, Security, and Sensing, pp. 875502–875502. International Society for Optics and Photonics (2013)
10. Nichat, S.P., Sikchi, S.S.: Image Encryption using Hybrid Genetic Algorithm. International Journal of Advanced Research in Computer Science and Software Engineering, 3(1) (2013)
11. Shreef, M.A., Hoomod, H.K.: Image Encryption Using Lagrange-Least Squares Interpolation. International Journal of Advanced Computer Science and Information Technology (IJACSIT) 2, 35–55 (2013)

12. Wadi, S.M., Zainal, N.: Rapid Encryption Method based on AES Algorithm for Grey Scale HD Image Encryption. Procedia Technology **11**, 51–56 (2013)

13. Sharma, D.: Robust Technique for Image Encryption and Decryption Using Discrete Fractional Fourier Transform with Random Phase Masking. Procedia Technology **10**, 707–714 (2013)

14. Abraham, L., Daniel, N.: Secure image encryption algorithms: A review. Entropy **100** (2013)

15. Kumar, M., Mishra, D.C., Sharma, R.K.: A first approach on an RGB image encryption. Optics and Lasers in Engineering **52**, 27–34 (2014)

16. Wu, Y., Zhou, Y., Agaian, S., Noonan, J.P.: A symmetric image cipher using wave perturbations. Signal Processing **102**, 122–131 (2014)

17. Zhang, Y.-Q., Wang, X.-Y.: A symmetric image encryption algorithm based on mixed linear–nonlinear coupled map lattice. Information Sciences **273**, 329–351 (2014)

18. Abuturab, M.R.: An asymmetric color image cryptosystem based on Schur decomposition in gyrator transform domain. Optics and Lasers in Engineering **58**, 39–47 (2014)

19. Sun, Y., Chen, L., Xu, R., Kong, R.: An Image Encryption Algorithm Utilizing Julia Sets and Hilbert Curves. PloS one **9**(1) (2014)

20. Zhang, Y., Xiao, D.: An image encryption scheme based on rotation matrix bit-level permutation and block diffusion. Communications in Nonlinear Science and Numerical Simulation **19**, 74–82 (2014)

21. Wu, Y., Zhou, Y., Noonan, J.P., Agaian, S.: Design of image cipher using latin squares. Information Sciences **264**, 317–339 (2014)

22. Zanin, M., Pisarchik, A.N.: Gray code permutation algorithm for high-dimensional data encryption. Information Sciences **270**, 288–297 (2014)

23. Tedmori, S., Al-Najdawi, N.: Image cryptographic algorithm based on the Haar wavelet transform. Information Sciences **269**, 21–34 (2014)

24. Kekre, H.B., Sarode, T., Halarnkar, P.N.: Symmetric Key image Encryption using continuous distributions with MOD operator. International Journal of Engineering Science and Technology **6**(6), 316–330 (2014)

25. Kekre, H.B., Mishra, D.: Performance Comparison of Density Distribution and Sector mean of sal and cal functions in Walsh Transform Sectors as Feature Vectors for Image Retrieval. International Journal Of Image Processing (IJIP) **4**(3), 205–217 (2010)

Level Based Anomaly Detection of Brain MR Images Using Modified Local Binary Pattern

Abraham Varghese, T. Manesh, Kannan Balakrishnan and Jincy S. George

Abstract The medical imaging technology plays a crucial role in visualization and analysis of the human body with unprecedented accuracy and resolution. Analyzing the multimodal for disease-specific information across patients can reveal important similarities between patients, hence their underlying diseases and potential treatments. Classification of MR brain images as normal or abnormal with information about the level at which it lies is a very important task for further processing, which is helpful for the diagnosis of diseases. This paper focuses on the abnormality detection of brain MR images using search and retrieval technique performed on similar anatomical structure images. Similar anatomical structure images are retrieved using Modified Local Binary Pattern (MOD-LBP) features of the query and target images and the level of the image is identified. The query image is compare with images in the same level and classification is done using the SVM classifier. The result reveals that the classification accuracy is improved significantly when the query image is compared with similar anatomical structure images.

Keywords MOD-LBP · Level identification · Classification

1 Introduction

Magnetic resonance imaging of the brain helps the radiologists to identify abnormal tissues due to bleed, clot, Acute-infarct, tumor, trauma etc. Since brain controls and coordinates most movement, behavior and homeostatic body functions such as heartbeat, blood pressure, fluid balance and body temperature, any injuries to brain affect

A. Varghese(✉) · J.S. George
Adi Shankara Institute of Engineering and Technology, Kalady, India
abraham.cs@adishankara.ac.in

T. Manesh
Salman University, Al-Kharj, Saudi Arabia

K. Balakrishnan
Cochin University of Science and Technology, Cochin, India

© Springer International Publishing Switzerland 2016
S. Berretti et al. (eds.), *Intelligent Systems Technologies and Applications*,
Advances in Intelligent Systems and Computing 384,
DOI: 10.1007/978-3-319-23036-8_42

the entire organ. The properties like soft tissue contrast and non-invasiveness of MRI are quite useful to identify any such abnormalities. The purpose of classification of MRI image into normal and abnormal is to find out the subjects with the possibility of having abnormalities or tumors. Many techniques for the classification of MR brain images are depicted in the literature. The feature extraction method for the classification of textures using a GMRF model on linear wavelets is presented (Ramana et al. 2010). These approaches to the texture analysis are restricted to spatial interactions over relatively small neighborhoods. An approach for texture image retrieval is performed in transform domain by computing standard deviation, energy and their combination on each sub band of the decomposed images (Prakash et al. 2012). Ruchika et al (2012) discussed DCT, DWT and Hybrid DCT-DWT based image compression and their performance in terms of Peak Signal to Noise Ratio (PSNR), Compression Ratio (CR) and Mean Square Error (MSE). In all these methods, classification of normal versus abnormal slices is performed without considering the features relevant to similar anatomical structures. In this paper, images are classified as normal or abnormal by giving importance to its anatomical structures so that the accuracy of the classification is improved. MOD-LBP (Abraham et al. 2014) descriptor has been used to retrieve similar MR images from a large database based on a query image. Depending on the 10% of the images retrieved, the level at which the given image lies is determined. Once desired level is identified, the images in that particular level are compared in order to classify it as a normal or abnormal. As the classification is performed by considering the features in the similar anatomical structures too, it is possible to predict the abnormality with higher accuracy.

2   Methodology

The overview of the methodology is given in Fig 1.

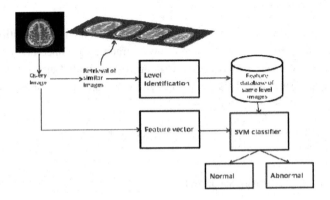

Fig. 1 Overview of Methodology

The similar images are retrieved correspond to a query image using histogram of MOD-LBP. Based on the images retrieved, level of the images is identified. The features of the same level images are given as input to the Support Vector Machine (SVM) and query image is classified as normal or abnormal.

2.1 Retrieval of Similar Images Using MOD-LBP

The procedure for retrieving similar images from the large image database is shown in Fig 2. In the pre-processing stage, the region of interest is extracted using morphological operations. The texture features are extracted using the descriptor MOD-LBP (Abraham et al. 2014) which is computed using the equation

$$MOD\text{--}LBP(P,R) = \frac{1}{P}\sum_{i=0}^{P-1} s(g_i - g_c)(g_i - \mu)^2 ,\qquad (1)$$

on a circular neighborhood with P neighboring pixels and Radius R , where μ is the mean and $\begin{cases} s(x)=1, x\geq 0 \\ \quad\ \ 0, x<0 \end{cases}$. The MOD-LBP image is converted to polar form (r,θ) in order to compensate for rotation and translation in the spatial description of the features by taking centroid as the origin of the image. The output image obtained is of size $N \times N$ with N points along the r-axis and N points along the θ-axis. The pixel value of the non-integer coordinate of the image is estimated using bilinear interpolation. The histogram of MOD-LBP is computed spatially, where the entries of each bin are indexed over angularly partitioned regions. The pixel intensities are brought into the range [0, L], where L is a positive integer and normalized histogram of the image is taken as feature vectors for similarity computation. The similarity computation of 2 images in the database is computed using Bhattacharya coefficient, $d = 1 - \sum_{i=0}^{L-1}\sqrt{p(i)q(i)}$ where p and q are normalized histograms with L1-bins. The Bhattacharya coefficient of two exactly similar images is 1, and the dissimilarity increases as it is different from 1. The images in the database are ranked based on this distance measure and the accuracy of the retrieval is computed.

Furthermore, the moment features of MOD-LBP are computed spatially over an angularly partitioned area and performance of the retrieval system is evaluated using the distance function of moment features. In order to achieve an optimum performance, relevance feedback mechanism has been applied by incorporating the user's input in the retrieval process. This is achieved by reweighting the moment features of MOD-LBP based on the relevance of individual features in the retrieval process. An average rank, which shows the closeness of the system performance, is calculated using the formula,

$$Average\ \ Rank = \frac{1}{N_R}(\sum_{i=1}^{N_R} R_i - \frac{N_R(N_R-1)}{2}) \tag{2}$$

N_R represents number of relevant images and R_i represents the rank at which the ith relevant image is retrieved [Henning & Wolfgang (2001)]. The Accuracy of the retrieval system for a set of queries is also calculated using the formula,

$$Accuracy = (1 - \frac{No\ of\ irrelevant\ images\ retrieved}{Total\ no\ of\ irrelevant\ images}) \times 100 \tag{3}$$

In the results section, we will illustrate how the average ranking and accuracy make use of, in classifying the retrieval performance at different levels. A comparison has been made between different local measure like LBP (Unay et al 2010), Rotational invariant LBP ($LBP^{ri}_{P,R}$), Rotational invariant uniform LBP ($LBP^{riu2}_{P,R}$) in retrieving 10 relevant images from 4 different levels. LBP has been used with a window of size 3 (P = 8, R = 1) and window of size 5 (P = 16, R = 2). Bashier et.al. (2012) proposed Local Graph Structure (LGS (8,1)) which is formed with 8 neighbors of a pixel, obtained by moving anticlockwise at the left region of the centre pixel and then right region of the central pixel. If the neighborhood pixel has a higher gray level value, assign the value 1 to the edge connecting the two vertices, else assign a value 0. The MOD-LBP defined in equation (1) has been used in 3 different ways (Histogram features, Moment features, reweighted moment features) in the retrieval process.

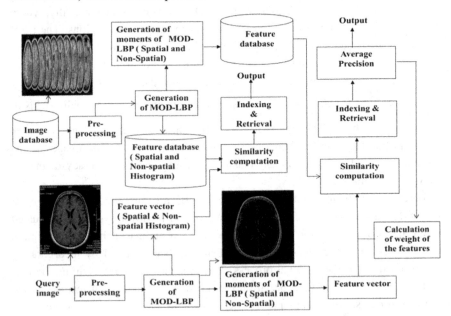

Fig. 2 Image retrieval scheme Level Identification

Based on the most similar and dissimilar images retrieved, the level of the query image is identified using the weights w and 1-w respectively (w $\in$ [0,1]). The level of the query image is computed as the weighted combination of the R_h most similar and R_l most dissimilar retrieval images.

If the database consists of N items ranked in decreasing order of similarity as i = 1, 2 ... N given the query, then it calculates four values for each query.

$$c_{l1} = w \times \frac{1}{R_h} \sum_{i=1}^{R_h} c1_i + (-1)(1-w)\frac{1}{R_l}\sum_{i=1}^{Rl} c1_i{}'$$

$$c_{l2} = w \times \frac{1}{R_h} \sum_{i=1}^{R_h} c2_i + (-1)(1-w)\frac{1}{R_l}\sum_{i=1}^{Rl} c2_i{}'$$

$$c_{l3} = w \times \frac{1}{R_h} \sum_{i=1}^{R_h} c3_i + (-1)(1-w)\frac{1}{R_l}\sum_{i=1}^{Rl} c3_i{}' \tag{4}$$

$$c_{l4} = w \times \frac{1}{R_h} \sum_{i=1}^{R_h} c4_i + (-1)(1-w)\frac{1}{R_l}\sum_{i=1}^{Rl} c4_i{}'$$

where c1, c2, c3,& c4 assumes the value 1, if the R_h most similar images belong to the respective levels, and it assumes the value -1, if the R_l most dissimilar images belong to the respective levels. The level of the query image is identified by the maximum value of c_{l1}, c_{l3}, & c_{l4}. i.e. if c_{l1} has got the maximum value, the query image belongs to the level 1($l1$).

2.2 Classification

Support Vector Machines (SVMs) are feed forward networks with a single layer of nonlinear units. It is capable of solving complex nonlinear classification problems. It solves these problems by means of convex quadratic programming (QP) and also the sparseness resulting from this QP problem. The learning is based on the principle of structural risk minimization. Instead of minimizing an objective function based on the training samples (such as mean square error), the SVM attempts to minimize the bound on the generalization error (i.e., the error made by the learning machine in the test data not used during training). As a result, an SVM tends to perform well when applied to data outside the training set. SVM achieves this advantage by focusing on the training examples that are most difficult to classify. These "borderline" training examples are called support vectors. SVM provides an accurate classification although the training time is very high (Othman et al. 2011).

Once the level of the image is identified using similar and dissimilar images retrieved, the query image is compared with images in the same level. The 50% of the images from both normal and abnormal images has been used for training using SVM classifier and remaining for testing. It improves the accuracy of classification as it gives importance to the similar anatomical structure of the images

The performance of the proposed method is evaluated in terms Sensitivity, specificity and Accuracy.

Sensitivity=TP/TP+FN
Specificity=TN/TN+FP
Accuracy=TP+TN/TP+TN+FP+FN   where,
TP (True Positives) – correctly classified positive cases,
TN (True Negative) – correctly classified negative cases,
FP (False Positives) – incorrectly classified negative cases,
FN (False Negative) – incorrectly classified positive cases

3   Results

The proposed system has been implemented on a Brain Web dataset [BrainWeb]. Input dataset consists of axial T1 weighted bias 0% and bias 40% images (290 normal and 290 abnormal images). The 52 images from level 1, 64 images from level 2, 44 images from level 3, and 72 images from level 4 are used for level identification.

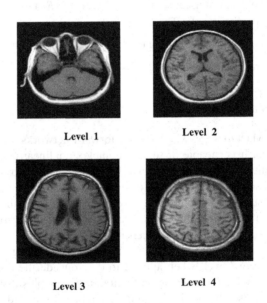

Level 1   Level 2

Level 3   Level 4

Fig. 3 Different Levels of T1 weighted axial MR slices

3.1 Similarity Retrieval

Ten images from each levels as shown in Fig 3 are randomly chosen as query image and accuracy of the retrieval is calculated based on the first 10 mostly relevant images retrieved. The performance of the retrieval is compared with different local measures and the result is shown in Fig 4. It is observed that LGS performs better compared to LBP (8,1) and LBP (16,2) and moment features of MOD-LBP. But spatial histogram of MOD-LBP outperforms LGS. An optimum performance can be achieved by incorporating users feedback into the retrieval system. An Average rank of 3.86 is obtained by reweighting the moment features based on the performance of the individual features in the retrieval process.

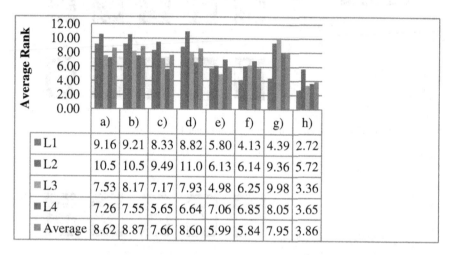

	a)	b)	c)	d)	e)	f)	g)	h)
■ L1	9.16	9.21	8.33	8.82	5.80	4.13	4.39	2.72
■ L2	10.5	10.5	9.49	11.0	6.13	6.14	9.36	5.72
■ L3	7.53	8.17	7.17	7.93	4.98	6.25	9.98	3.36
■ L4	7.26	7.55	5.65	6.64	7.06	6.85	8.05	3.65
■ Average	8.62	8.87	7.66	8.60	5.99	5.84	7.95	3.86

Fig. 4 Comparison of different local measures for retrieving 10 relevant images from 4 different levels. a) $LBP^{ri}_{8,1}$ b) $LBP^{riu2}_{8,1}$ c) $LBP^{ri}_{16,2}$ d) $LBP^{riu2}_{16,2}$ e) $LGS(8,1)$ f) Spatial Histogram Of MOD-$LBP(8,1)$ using 18 angular regions g) Spatial Moments of MOD-$LBP(8,1)$ using 18 angular regions h) Reweighting the moment features 5 times.

The Fig. 5 (a),(b), (c) ,(d), show the most similar 10 images and most dissimilar 10 images correspond to a randomly chosen image from each level based on histogram of MOD-LBP computed spatially on 4 angular regions.

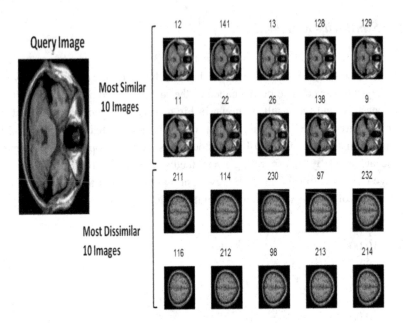

Fig. 5(a) Most Similar and Dissimilar images in Level 1

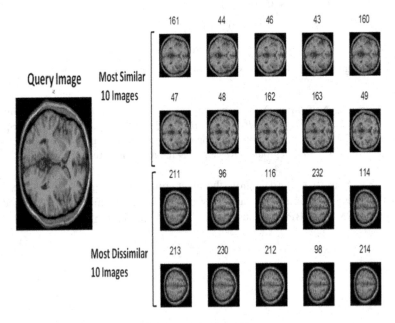

Fig. 5(b) Most Similar and Dissimilar images in Level 2

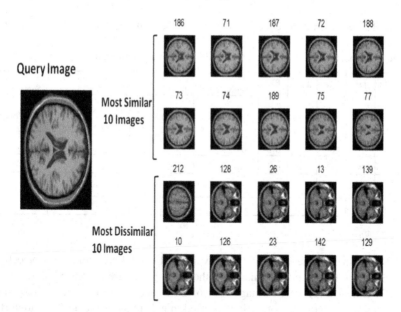

Fig. 5(c) Most Similar and Dissimilar images in Level 3

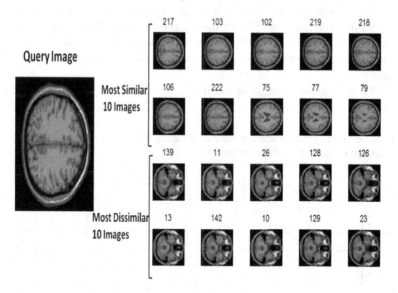

Fig. 5(d) Most Similar and Dissimilar images in Level 4

Accordingly, the level of the query image is identified using 10 retrieved similar and dissimilar images with weight w = 0.5.

The Table 1 shows the accuracy of level identification on BrainWeb data set.

Table 1 The accuracy,sensitivity,and specificity of Level Identification

Level	TP	TN	FP	FN	\Sensitivity	Specificity	Accuracy	Sensitivity
					TP/(TP+FN)	TN/(TN+FP)	(TP+TN)/ (TP+FN+TN+FP)	TP/(TP+FN)
1	26	90	0	0	100%	100%	100%	1
2	32	84	0	0	100%	100%	100%	2
3	20	94	0	2	91%	100%	98%	3
4	36	78	2	0	100%	98%	98%	4

It shows that all the images in the level 1 and level 2 are identified correctly in the respective levels and no images in the other levels are identified as level 1 or level 2. But in level 3, two images out of 22 images are not correctly identified. At level 4 all images are correctly classified, but 2 images in level 3 are identified as level 4 images. Once the level is identified, the query image is compared with images in the same level for classification as normal or abnormal using SVM classifier with linear kernel. The 50% of the images randomly chosen for training and remaining is used for testing. This has been repeated 100 times and mean of the accuracy is calculated using histogram of MOD-LBP and Moment of MOD-LBP as features.

The Table 2 shows the table of sensitivity, specificity and accuracy of the classification.

Table 2 The Mean of the accuracy of the classification on level based and non-level based image dataset using SVM with linear kernel.

Features	Measures	Non-level based images	Levels				
			1	2	3	4	Total
Histo- gram of MOD- LBP	Sensitivity	74.69%	85.15%	97.12%	99.00%	98.78%	95.01%
	Specificity	76.00%	87.31%	99.06%	99.73%	98.33%	96.11%
	Accuracy	75.34%	86.23%	98.09%	99.36%	99.22%	95.73%
Moments of MOD- LBP	Sensitivity	84.50%	85.46%	99.62%	99.36%	100%	96.11%
	Specificity	84.52%	90.62%	98.12%	99.00%	99.97%	96.93%
	Accuracy	84.51%	88.04%	98.80%	99.18%	99.94%	96.49%

The results show that the accuracy of the anomaly detection on a level based image dataset is much higher compared to non-level based image data set. The mean of the accuracy of the classification using histogram of MOD-LBP features improve the accuracy of the classification from 75.34% to 95.73% when the image database is switched to level based from non-level based image data set. A further improvement is observed using moments of MOD-LBP features.

4   Conclusion

This paper focuses on the application of similar slice retrieval to efficient abnormality detection. The level of the image is identified from the most similar and most dissimilar images retrieved correspond to a query image. Once the level is identified, the query image is compared with images in the same level for normal-abnormal classification. The method is robust to rotation and translation misalignment of images. The results show that there is a significant improvement in the accuracy of classification if the query image is compared images with similar anatomical structures. This can be useful to identify the possibilities of certain diseases at the early stage itself provided enough data set is available for comparison.

References

Ramana Reddy, B.V., Radhika Mani, M., Subbaiah, K.V.: Texture Classification Method using Wavelet Transforms Based on Gaussian Markov Random Field. International Journal of Signal and Image Processing 1(1) (2010)

Prakash, I.S.K.N., Satya Prasad, K.: M-Band Dual Tree Complex Wavelet Transform For Texture Image Indexing and Retrieval. IJAET (July 2012), ISSN: 2231–1963

Ruchika, M.S., Singh, A.R.: Compression of Medical Images Using Wavelet Transforms. (IJSCE) 2(2), May 2012, ISSN: 2231–2307

Varghese, A., Balakrishnan, K., Varghese, R.R., Paul, J.S.: Content Based Image Retrieval of Brain MR Images across different classes. World Academy of Science, Engineering and Technology 80 (2013)

Müller, H., Müller, W.: Automated Benchmarking in Content-Based Image Retrieval, work supported by Swiss National Foundation for Scientific Research (grant no. 2000–052426.97)

Unay, D., Ekin, A., Jasinsch, R.: Local Structure-Based Region-Of- Interest Retrieval in brain MR Images. IEEE Transactions on Information technology in Biomedicine 14(4), July 2010

Bashier, H.K., Abdu Abusham, E., Khalid, F.: Face detection based on graph structure and neural networks. Trends in Applied Sciences Research 7(8), 683–691 (2012), ISSN 1819 3579/DOI:10.3923/tasr.2012.683.691

Othman, M.F.B., Abdullah, N.B., Kamal, N.F.B.: MRI brain classification using support vector machine. In: 4th International Conference on Modeling, Simulation and Applied Optimization, ICMSAO 2011 (2011)

http://www.bic.mni.mcgill.ca/brainweb/

Multiple Image Characterization Techniques for Enhanced Facial Expression Recognition

Mohammed Saaidia, Narima Zermi, and Messaoud Ramdani

Abstract This paper describes an enhanced facial expression recognition system. In the first step, the face localization is done using a simplified method, then the facial components are extracted and described by three feature vectors: the Zernike moments, the spectral components' distribution through the DCT transform and by LBP features. The different feature vectors are used separately then combined to train back-propagation neural networks which are used in the facial expression recognition step. A subset feature selection algorithm was applied to these combined feature vectors in order to make dimensionality reduction and to improve the facial expression recognition process. Experiments performed on the JAFFE database along with comparisons to other methods have affirmed the validity and the good performances of the proposed approach.

Keywords Face detection · Expression recognition · DCT transform · Zernike moments · LBP · MIFS · NMIFS

1 Introduction

Facial expression recognition is still a hard research area with many challenges to meet. This is mainly due to several semantic concepts related to, the amalgam between the terms 'expression' and 'emotion' ; the latter one being only a semantic interpretation of the former as the term 'happiness' to 'smile'[1]. So a facial expression is a simple physiological activity of one or more parts of the face (eyes, nose, mouth, eyebrows,...) [2], while an emotion is our semantic interpretation to this activity.

M. Saaidia(✉)

Département de Génie-Electrique, Université M.C.M Souk-Ahras, Souk-Ahras, Algeria
mohamed.saaidia@univ-soukahras.dz

N. Zermi · M. Ramdani

Département d' Electronique, Université Badji-Mokhtar de Annaba, Annaba, Algeria
{narima.naili,messaoud.ramdani}@univ-annaba.dz

© Springer International Publishing Switzerland 2016

S. Berretti et al. (eds.), *Intelligent Systems Technologies and Applications*,
Advances in Intelligent Systems and Computing 384,
DOI: 10.1007/978-3-319-23036-8_43

This problem is still difficult to disentangle especially as our mathematical models do not have the ability to create the semantic dimension in the representation of our environment.

Another serious constraint which is also pointed out is the cultural background related to the ethnicity of the person [3], [4]. Technical problems related to the acquisition devices and variation in illumination conditions also make the problem harder [5]. Most researches have been turned towards the study and classification of the six basic facial expressions (universally recognized) summarized on Fig 1.

Developed methods can be classified according to several criteria. Taking into account the way how to characterize the face, methods are "motion extraction" [8] or "deformation extraction" [9]. According to classification techniques, methods are "spatial methods" [10] or "spatiotemporal methods" [11]. Facial expression recognition methods can also be classified according to the way how to consider the face; as a single entity which will lead to "global process" or as a set of features (eyes, nose and mouth) which have to be extracted before performing characterization step.

neutral   anger   disgust   happiness   fear   surprise   sadness

Fig. 1 Example of Basic facial expressions from the JAFFE database

In the present work, a multi-feature representation with a subset feature selection was done on the face features (eyes, nose and mouth) instead of the whole face. This choice is justified by psychological [12].

Thus, three different representations were computed. The first one aims to extract the geometric information, while the second one exploits the well known LBP technique to resume the statistical luminance information and the last one computes the spectral source model of the face by the DCT transform. Finally, combined feature vectors were used to recognize the six basic facial expressions.

The paper is structured as follows: Section 2 outlines the facial feature extraction. Section describes the various characterization methods. Then, Section 4 presents the subset feature selection procedure based on mutual information. Experimental results are given in section 5. Finally, a general conclusion is given.

2   Facial Components Detection

Face expression recognition can be done on different information supports like images with single face, multi-face images, video, etc. Abstracting the semantic information, processed by human brain; a face in an image remains a common object with specific geometric and color characteristics so we need to isolate the target which will be subject to the expression processing ("face"). Face detection is one of the most studied problems in computer vision [13]. However, we are interested in face features of the face itself since Psychologists have demonstrated that the eyes, eyebrows, mouth and nose are the areas of the face on which appear the overall facial expressions [12]. So, important processing time will be avoided by characterizing facial components instead of the whole face.

Facial components detection was widely studied and different techniques were developed beginning by Yuille et al. [14] until recent works as Jongju Shin et al. [15]. In our work we made use of the well known and efficient method developed by viola and Jones [16] based on the evaluation of Haar-like features through the compilation of the integral image then the huge number of obtained features is processed by an AdaBoost algorithm [17] to focus on the main important features only. At the last stage, a structure of complex classifiers is used to enhance the speed of the election process by focusing attention on promising regions of the image.

To perform efficient facial components detection, the previous technique was used for both face detection and facial features localization. Fig. 2 gives some examples.

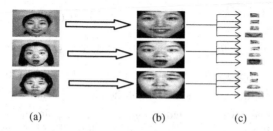

(a)   (b)   (c)

Fig. 2 Examples of processed images from JAFEE database (a) original image, (b) detected face and (c) detected facial features

Face detection and facial features detection were performed with high rated results. Indeed, 95% of True Positive rate (TPR) was obtained for face detection and up to 88.7% of TPR average rate for all facial features. This is due not only to the technique's efficiency but also to the simplicity of the database which is composed by images containing a unique person with the same gender in frontal pose.

3   Face Features' Characterization

In pattern recognition, the dimensionality problem is one of the most difficult constraints that rises when we try to directly process the acquired environmental information [18]. So we have to find an alternative representation of the processed information object; like face features in our case. Instead of the matrix of pixels, we have to find a reduced representative vector which compact the information needed for our processing task. This is what we call a characterization phase.

In the present work, we tried to exploit different ways to perform this characterization in order to enrich the representative feature vector. Three different types of information are used ; Zernike moments, to compact image's geometric characteristics; LBP method which is considered as the most significant way to characterize texture information of the image and DCT transform to obtain its spectral components distribution.

3.1 Face Features' Characterization by Zernike Moments

Zernike moments form part of the general theory of the geometrical moments. They were introduced initially by F. Zernike [19]. At the difference of the general geometrical moments, those of Zernike are built on a set of orthogonal polynomials which were used as the basic elements of the construction of an orthogonal base given by the equation (1)

$$V_{n,m}(x, y) = V_{n,m}(\rho,\theta) = R_{n,m}(\rho) \cdot e^{j.m.\theta} \tag{1}$$

$$\begin{cases} R_{n,m}(\rho) = \sum_{k=|m|}^{n} \dfrac{(-1)^{(n-k)/2}.(n+k)!}{(\dfrac{n-k}{2})!(\dfrac{k+m}{2})!(\dfrac{k-m}{2})!} . \rho^k \\[6mm] \rho = \sqrt{x^2 + y^2} \qquad \text{and} \qquad \theta = arctg\,(y/x) \end{cases} \tag{2}$$

with: $n \geq 0$, $m \neq 0$, $m < n$, $n - m < n$ and $(n-k)$ even.

$R_{n,m}(\rho)$ is the orthogonal radial polynomial, n is the order of the moment and m the factor of repetition (the smoothness of the required details) at this order. ρ and θ are respectively the radius and the angle of treated point of the function.

To implement it, we use the fast algorithm developed by G. Amayeh et all [20] and given in (3) for face characterization through Zernike moments and a trained back-propagation neural network for the classification step.

$$Z_{n,m} = \frac{n+1}{\pi} \sum_{x^2+y^2\leq 1} \sum \left(\sum_{k=|m|}^{n} \beta_{n,m,k} \cdot \rho^k \right) \cdot e^{-j.m.\theta} . f(x_j, y_i)$$

$$= \frac{n+1}{\pi} \sum_{k=|m|}^{n} \beta_{n,m,k} \cdot \left(\sum_{x^2+y^2\leq 1} \sum e^{-j.m.\theta} \cdot \rho^k . f(x_j, y_i) \right) = \frac{n+1}{\pi} \sum_{k=|m|}^{n} \beta_{n,m,k} . X_{m,k} \tag{3}$$

Zernike moments are known for their capacity to compress the geometric information of the image into a vector of reduced dimensions depending on the parameters m and n (see equations 1, 2 and 3). The obtained feature vector compacts the geometric characteristics of the image such as the surface, the vertical symmetry and distribution centers masses in the horizontal and vertical directions and other image characteristics which deal with information required for such type of classification problem.

In Fig. 3 we give some samples of Zernike moments feature vectors for two different expressions for images of JAFFE database.

Each curve is a concatenation of four zernike moments feature vectors of the four regions containing principal components of the face (the two eyes, the noze and the mouth). There are many apparent differences between the curves representing images of different persons. This justifies the different research that has been conducted on the use of characterization by such attributes for recognizing faces and facial expressions. [21], [22].

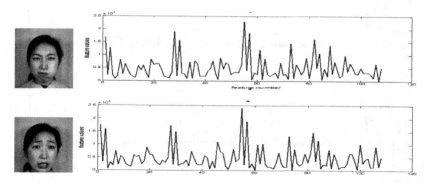

Fig. 3 Zernike moments feature vectors for 2 different expressions (Anger and fear) from images of JAFFE database

3.2 Face Features' Characterization by LBP

Initially introduced by Ojala et al. [23] for image texture analysis, the LBP was rapidly adopted as an efficient tool for image analysis in general and pattern recognition in particular [23], [24], [25]. The success of this method in image analysis and representation generates a multitude of extended versions adapted to the encountered problems.

Fig. 4, gives a representative scheme of the compiling process in the Basic LBP method. The most important extensions of basic LBP were the extension of the operator to use neighborhood of different sizes, to capture dominant features at different scales and the definition of "uniform patterns" used to reduce the representative feature vectors

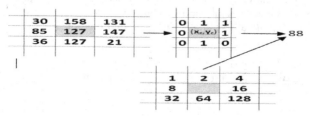

Fig. 4 Basic LBP value for the central pixel (xc,yc)

On figure 5, we give a sample of LBP feature vectors compiled for some faces from JAFFE database.

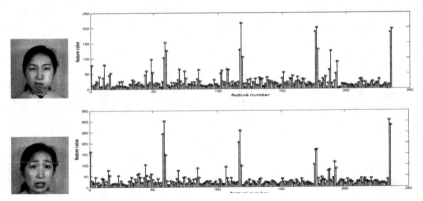

Fig. 5 LBP feature vectors for 2 different expressions (Anger and fear) from images of JAFFE database

This type of characterisation was largely used by researchers for facial expression recognition [26], [27] especially for its capacity to encapsulate the texture information of the image, its robustness to monotonic gray-scale changes caused by illumination variations and its computational simplicity.

3.3   Face Features' Characterization by DCT

The idea here is to exploit another type of information brought by the image. Indeed, spectral components distribution is an important characteristic of physical signals like speech, image, etc. It was largely used in different types of signal processing domains; like communication, medicine, human-machine interaction, and so on. Despite the existence of several methods for spectral analysis, the DCT seems to be the most preferred and most used by the researchers [28]. This is due to the fact that it concentrates the most important visual information on a set of few coefficients which explains why it was largely used for image and video compression. In addition it leads to real coefficients.

2D-DCT mathematical formulation is given on Equation 4.

$$
\begin{cases}
X_C(k_1,k_2) \triangleq \sum_{n_1=0}^{N_1-1}\sum_{n_2=0}^{N_2-1} 4x(n_1,n_2)\cos\frac{\pi k_1}{2N_1}(2n_1+1)\cos\frac{\pi k_2}{2N_2}(2n_2+1), \\
\text{where:} \\
(k_1,k_2) \in [0, N_1-1] \times [0, N_2-1]; Otherwise, X_C(k_1,k_2) \triangleq 0.
\end{cases}
\tag{4}
$$

However, different ways can be used such as using FFT algorithm or optimized DCT algorithms to enhance DCT coefficients compilation.

In many applications which use DCT transform, like in image and video compression, researchers exploit the fact that most of the energy at the spectral domain is localized at the low frequencies for dimensionality reduction. Here, it was not the case due to the fact that we are working on the face features where the energy at the high frequencies is predominant especially in the case of the eyes region (see Fig. 6).

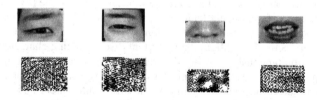

Fig. 6 DCT transforms for different face features

So, we restored to use modified feature vectors by concatenating matrices lines and sorting feature values (see Fig. 7). These vectors will be then truncated eliminating monotone parts.

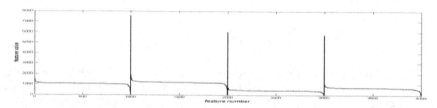

Fig. 7 Original concatenated feature vectors

Examples of applying this type of characterization to images from JAFEE database is given on Fig. 8

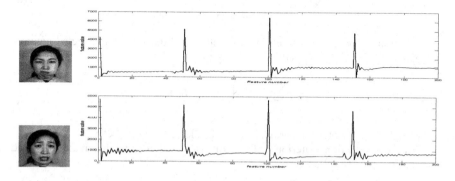

Fig. 8 DCT feature vectors for 2 different expressions (Anger and) from images of JAFFE database

4   Mutual Information Feature Selection

Feature vectors described previously will be used separately, and then combined in a common feature vector. However, due to the rising of dimensionality problem in our classification process, we propose here to introduce a selective tool to compact the combined feature vectors in order to improve the classification performance, optimize the computational cost and reduce the classifier complexity.

The extraction of a compact feature set, which can still capture most of the useful information inherent in the original signal, is thus very important. Suitable feature extraction methods highlight the important discriminating characteristics of the data, while simultaneously ignoring the irrelevant attributes. This type of process can be implemented in two different ways; feature selection or feature extraction [29]. Feature selection algorithms can be classified into filters and wrappers; the formers being used as a preprocessing operation before classification step while last ones are used as part of the classification process. In this work, the dimensionality reduction is performed by an efficient feature selection method called Normalized Mutual Information Feature Selection(NMIFS) [29]; a filter feature selection type.

Consider two discrete random variables x and y, with alphabets X and Y, respectively. The MI between x and y with a joint probability mass function p(x,y) and marginal probabilities p(x) and p(y) is defined as follows:

$$I(x,y) = \sum_{x \in X} \sum_{y \in Y} P(x,y).\log(\frac{P(x,y)}{P(x).P(y)}) \tag{5}$$

So, given an initial F set with n features, find subset S in F with k features that maximizes the mutual information I(C,S) between the class variable C , and the subset of selected features S. Different methods were developed to solve this problem. In the case of NMIFS, ones define the normalized mutual information between f_i and f_s , $NI(f_i , f_s)$ as the mutual information normalized by the minimum entropy of both features like given in (6).

$$NI(f_i, f_s) = \frac{I(f_i, f_s)}{\min(H(f_i), H(f_s))} \tag{6}$$

Redundancy measure between the ith feature and the subset of selected features S={ f_s } with s=1,2,...,|S| will be given by (7)

$$\frac{1}{|s|} \sum_{f_s \in S} NI(f_i, f_s) \tag{7}$$

Where |S|, denotes the cardinality of the selected features set S.

Thus, the selection criterion will be to select the feature that maximizes the measure G given by (8).

$$G = I(C, f_i) - \frac{1}{|s|} \sum_{f_s \in S} NI(f_i, f_s) \tag{8}$$

5 Experimental Results

Several classifiers have been proposed and used in facial expression recognition systems to handle the issues of non-linearity, dimensionality and generalization, and three main classification can be distinguished, namely, the HMM (Hidden Markov Models) [30], the SVM (Support Vector Machines) [31] and the feedforward neural networks (FFNN) [32].

In this work, the recognition step is carried out by three feed-forward neural network; each one uses a vector type, as explained previously. Also, the training process is done on combined feature vectors directly and after applying the subset feature selection based on mutual information. To check the validity of the proposed approach, experimental studies were carried out on the well known facial expression JAFFE database [33].

By taking 2 images in random for each person with each expression, a training dataset of 140 couples (Fi, T) examples is formed. The Zernike moments were computed with couple values (m=10, n=5) which seems to give the best results. A DCT feature vector of 50 coefficients was computed for each face's component, and for LBP we used LBP histogram in (8,1) neighborhood. All recorded results were obtained through a FFNN with 25 units in hidden layer, 7 units for the output layer and a number of units adapted to the feature vector used in input layer. "Purelin" was used as transfer function for input and output layers while "Tansig" function was used for hidden one. The training function was "Traingda".

The obtained results will be detailed in the following subsections.

5.1 Individual Feature Vectors

This section reports the recognition performance using the three different types of features separately. The training images were chosen randomly and reported results are the average of 10 experiments. More accurate results were obtained, (a Global TPR about 90.79 for Zernike moments, 93. 52 for LBP and 88.62 for DCT transform) when we choose manually the images to be used for training and testing the neural networks. This is due to the fact that there are several images where the expressions are not very apparent and it's even difficult for humans to process them correctly.

Table 1 Global TPR for Zernike moments, LBP and DCT with and without applying NMIFS

Feature vectors Expression	Global TPR %					
	Zernike	LBP	DCT	NMIFS+ Zernike	NMIFS+ LBP	NMIFS+ DCT
Neutral	88.5	90	83	89.23	90	85
Happiness	90	92	81	90	93	85
Surprise	85	91	87	87	93	87
Anger	78	83	67	77.5	83.5	70
Disgust	80	81	78	82	83	78
Fear	73	79	81	74.6	79	80
Sadness	81.5	78	83	83	78	83.5
Global TPR %	**82.29**	**84.85**	**80.00**	**83.35**	**85.64**	**81.21**

5.2 Combined Feature Vectors

Table 2 reports the results recorded using combined feature vectors.

Table 2 Global TPR recorded for combined feature vectors

Feature Vectors / Expression	Global TPR %			
	Zernike-LBP	LBP-DCT	Zernike-AR	Zernike-LBP-AR
Neutral	83	81	75	78
Happiness	83.5	83	70	77.5
Surprise	79	81	78	75
Anger	72	76	61	71
Disgust	78	73	76	69
Fear	67	72	72	72
Sadness	75	77	78	71
Global TPR %	**76.79**	**75.28**	**77.57**	**73.36**

It is clear when using the whole features that not only there is no improvement but the degradation is well apparent. This will be due in one hand to the great number of features which makes the neural networks converge to local minima instead of global one and in another hand to divergent directions of the feature's influences on the decision of class's membership.

5.3 Combined Feature Vectors with NMIFS

NMIFS was used to retain the strongly relevant features. In addition to the enhancement of facial expression recognition results, the classifiers converge faster due to the reduction of its complexity. For all the reduced combined feature vectors, we found that the Zernike moment features are preponderant against other types of features while DCT features are the less preponderant.

Table 3 Global TPR recorded for combine feature vectors processed with NMIFS

Feature vectors / Expression	Global TPR %			
	NMIFS+ (Zernike-LBP)	NMIFS+ (LBP-DCT)	NMIFS+ (Zernike-DCT)	NMIFS+ (Zernike-LBP-DCT)
Neutral	93.50	94.25	92.00	95.00
Happiness	94.78	95.38	93.00	96.37
Surprise	94.40	95.00	94.20	95.82
Anger	88.70	89.00	84.36	93.65
Disgust	91.47	91.00	90.52	92.85
Fear	85.35	85.35	83.00	89.00
Sadness	88.39	89.40	89.25	92.50
Global TPR %	**90.94**	**91.34**	**89.48**	**93.59**

5.4   Comparison with Other Techniques

Finally, a comparison with common techniques is also given to demonstrate the validity of the proposed approach. The comparison was done on the same database (JAFFE) and according to the same measurement strategy, namely Leave-One-Image-Out (LOIO) [34], [35] and [36]. Table 4 summarizes the results.

Table 4 Global rates' comparison between the proposed technique and former techniques

Applied technique	Global rate
R.S. El-Sayed et al. [38]	90.00 %
S. Zhang et al. [39]	90.70 %
R. Hablani et al. [40]	94.44 %
Proposed technique	93.59 %

It is clear that the proposed technique outperforms the former techniques except the one presented in [36]. However, its main advantage is still the reduced feature vectors which lead to the construction of simplified, fast and accurate classifiers. In the case of the Neural one used in this study, the number of internal parameters became at the ratio of one to five and it converges three times faster.

6   Conclusion

Facial expression recognition using three different types of characterization feature vectors to train neural network classifiers were studied in this paper. The characterization step was done on faces' components instead of the whole face. First, we found that uniform LBP feature vectors provide better results compared to those obtained using Zernike or DCT transform. Training classifiers using combined feature vectors was also studied and worse results were reported. We also demonstrate that the introduction of a feature selection technique, as a post-processing operation before the classification step, on single and combined feature vectors permits not only a significant enhancement of the classification results but also an improvement of the neural networks classifiers' speed and simplicity. The present technique needs deepest study concerning the appropriate way of combining the different parameters and the best classifier architecture.

References

1. Decety, J., Meyer, M.: From emotion resonance to empathic understanding: A social develomental neuroscience account. Development and Psychopathology **20**, 1053–1080 (2008)
2. Ekman, P., Friesen, W.V.: Facial Action Coding System: A technique for the Mesurement of Facial Movement. Consulting Psychologists Press, Palo Alto (1978)
3. Kilbride, J.E., Yarczower, M.: Ethnic bias in the recognition of facial expressions. Journal of Nonverbal Behavior FALL **8**(1), 27–41 (1983)

4. Boucher, J.D., Carlson, G.E.: Recognition of facial expression in three cultures. Journal of Cross-Cultural Psychology 11, 263–280 (1980)
5. Jung-Wei, H., Kai-Tai, S.: Facial expression recognition under illumination variation. In: IEEE Advanced Robotics and its Social Impacts, ARSO 2007, pp. 1–6 (2007)
6. Wu, S., Lin, W., Xie, S.: Skin heat transfer model of facial thermograms and its application in face recognition. Elsevier Pattern Recognition 41(8), 2718–2729 (2008)
7. Graf, H., Cosatto, E., Ezzat, T.: Face analysis for the synthesis of photo-realistic talking heads. In: Proc. Fourth IEEE Int. Conf. Automatic Face and Gesture Recognition, Grenoble, France, pp. 189–194 (2000)
8. Reisfeld, D., Yeshurun, Y.: Preprocessing of face images: Detection of features and pose normalization. Comput. Vision Image Understanding 71(3), 413–430 (1998)
9. Yokoyama, T., Yagi, Y., Yachida, M.: Facial contour extraction model. In: IEEE Proc. of 3rd Int. Conf. On Automatic Face and Gesture Recognition (1998)
10. Yuille, A.L., Hallinan, P.W., Cohen, D.S.: Feature extraction from faces using deformable templates. Int. J. Comput. Vision 8, 99–111 (1992)
11. Smith, M.L., Cottrell, G.W., Gosselin, F., Schyns, P.G.: Transmitting and decoding facial expressions. Psychol. Sci. 16, 184–189 (2005)
12. Hjelmas, E., Low, B.K.: Face detection: A survey. Computer Vision and Image Understanding 83(3), 236–274 (2001)
13. Yuille, A.L., Hallinan, P.W., Cohen, D.S.: Feature Extraction From Faces Using deformable Templates. Int. J. Comput. Vision 8, 99–111 (1992)
14. Shin, J., Kim, D.: Hybrid Approach for Facial Feature Detection and Tracking under Occlusion. IEEE Signal Processing Letters 21(12) (2014)
15. Viola, P., Jones, M.: Robust real-time face detection. International Journal of Computer Vision (IJCV) 57(2), 137–154 (2004)
16. Freund, Y., Schapire, R.E.: A decision-theoretic generalization of on-line learning and an application to boosting. In: Vitányi, P. (ed.) EurocoLT 1995, pp. 23–37. Springer, Heidelberg (1995)
17. Saaidia, M., Chaari, A., Lelandais, S., Vigneron, V., Bedda, M.: Face localization by neural networks trained with Zernike moments and Eigenfaces feature vectors. A comparaison. In: AVSS 2007, pp. 377–382 (2007)
18. Kotropoulos, C., Pitas, I.: Rule-based face detection in frontal views. In: Proc. Int. Conf. on Acoustic, Speech and Signal Processing (1997)
19. Amayeh, G., Erol, A., Bebis, G., Nicolescu, M.: Accurate and efficient computation of high order zernike moments. In: First ISVC, Lake Tahoe, NV, USA, pp. 462–469 (2005)
20. Akkoca, B.S, Gokmen, M.: Facial expression recognition using local zernike moments. In: Signal Processing and Communications Applications Conference (SIU), pp. 1–4 (April 2013)
21. Alirezaee, S., Ahmadi, M., Aghaeinia, H., Faez, K.: A weighted pseudo-zernike feature extractor for face recognition. IEEE, ICSMC 3, 2128–2132 (2005)
22. Ojala, T., Pietikäinen, M., Harwood, D.: A comparative study of texture measures with classification based on featured distribution. Pattern Recognition 29(1), 51–59 (1996)
23. Feng, X.-Y., Hadid, A., Pietikäinen, M.: A coarse-to-fine classification scheme for facial expression recognition. In: Campilho, A.C., Kamel, M.S. (eds.) ICIAR 2004. LNCS, vol. 3212, pp. 668–675. Springer, Heidelberg (2004)
24. Shan, C., Gong, S., McOwan, P.W.: Facial expression recognition based on Local Binary Patterns: A comprehensive study. Image and Vision Computing 27, 803–816 (2009)

25. Zhang, S., Zhao, X., Lei, B.: Facial Expression Recognition Based on Local Binary Patterns and Local Fisher Discriminant Analysis. WSEAS Transactions on Signal Processing **8**(1), 21–31 (2012)
26. Luoa, Y., Wub, C.-M., Zhang, Y.: Facial expression recognition based on fusion feature of PCA and LBP with SVM. Optik - International Journal for Light and Electron Optics **124**(17), 2767–2770 (2013)
27. Kharat, G.U., Dudul, S.V.: Neural Network Classifier for Human Emotion Recognition from Facial Expressions Using Discrete Cosine Transform. IEEE, 653–658 (2008)
28. Estévez, P.A., Tesmer, M., Perez, C.A., Zurada, J.M.: Normalized Mutual Information Feature Selection. IEEE Transactions on Neural Networks **20**(2), 189–201 (2009)
29. Terrillon, J.-C., Shirazi, M.N., Fukamachi, H., Akamatsu, S.: Comparative performance of different skin chrominance models and chrominance spaces for the automatic detection of human faces in colour images. In: Proc. of the International Conference on Face and Gesture Recognition, pp. 54–61 (2000)
30. Kovac, J., Peer, P., Solina, F.: Human skin color clustering for face detection. In: EUROCON 2003. Computer as a Tool. The IEEE Region 8, vol. 2, pp. 144–148 (September 2003)
31. Brown, D., Craw, I., Lewthwait, J.: A SOM based approach to skin detection with application in real time systems. In: Proc. of the British Machine Vision Conference (2001)
32. Lyons, M.J., Akamatsu, S., Kamachi, M., Gyoba, J.: Coding facial expressions with gabor wavelets. In: Proc. of the Third IEEE Int. Conf. on Automatic Face and Gesture Recognition, pp. 200–205. IEEE Computer Society, Nara, Japan (1998)
33. El-Sayed, R.S., El-Kholy, A., El-Nahas, M.Y.: Robust Facial Expression Recognition via Sparse Representation and Multiple Gabor filters. International Journal of Advanced Computer Science and Applications **4**(3), 82–87 (2013)
34. Zhang, S., Zhao, X., Lei, B.: Facial Expression Recognition Using Sparse Representation. Wseas Transactions On Systems **11**(8), 440–452 (2012)
35. Hablani, R., Chaudhari, N., Tanwani, S.: Recognition of Facial Expressions using Local Binary Patterns of Important Facial Parts. International Journal of Image Processing (IJIP) **7**(2), 163–170 (2013)

Combined MFCC-FBCC Features for Unsupervised Query-by-Example Spoken Term Detection

Drisya Vasudev, Suryakanth V. Vasudev, K.K. Anish Babu and K.S. Riyas

Abstract A new set of features for addressing the problem of unsupervised spoken term detection is proposed in this paper. If we have a large audio database, the objective of this system is to find a spoken query in the databases. In unsupervised audio search, language specific resources are not required. Thus this system is more appropriate in cases where enough training data is not available for creating an Automatic Speech Recognition(ASR). Current state-of-the-art techniques use Mel Frequency Cepstral Coefficients(MFCC), Linear Predictive Cepstral Coefficients(LPCC) etc. as the features. For improving the performance of the system, FBCC (Fourier Bessel Cepstral Coefficients) combined with MFCC is used in this paper. Here, from the spoken example of a keyword, segmental Dynamic Time Warping is used to compare the Gaussian Posteriorgrams (GP),which are created from the feature vectors. By combining the GPs of MFCCs and FBCCs, a new set of feature representation is adapted in this work. The keyword detection result obtained using MediaEval 2012 database shows that this system outperforms the one that uses MFCC alone.

Keywords Spoken term detection · Query · FBCC · Gaussian mixture · Gaussian posteriorgram · Dynamic time warping

1 Introduction

In this digital era, the volume of speech data stored in audio repositories has got tremendous increase. Thus, the need for extracting the stored information is high.

D. Vasudev(✉) · K.K.A. Babu · K.S. Riyas
Department of Electronics and Communication Engineering,
Rajiv Gandhi Institute of Technology, Kerala, India
e-mail: drisvas@gmail.com, anishkochi@yahoo.co.uk, riyas@rit.ac.in

S.V. Vasudev
Speech and Vision Lab, IIIT Hyderabad, Hyderabad, India
e-mail: svg@iiit.ac.in

© Springer International Publishing Switzerland 2016
S. Berretti et al. (eds.), *Intelligent Systems Technologies and Applications*,
Advances in Intelligent Systems and Computing 384,
DOI: 10.1007/978-3-319-23036-8_44

Automatic systems for indexing, archiving, searching and browsing of large amounts of spoken communications have become a reality in the last decade. Such systems use automatic speech recognition (ASR) modules. They convert the speech signal into text. Then the information can be retrived using lattice based searching. This technique can be used when the output of speech recognition system is mostly correct or the documents are long enough so that some occurrences of the query terms are recognized correctly.

Spoken term detection (STD) is the task of finding the occurence of a particular query term in a large audio archive. The technique for overcoming the effect of out of vocabulary problems is Query-by-example (QbE) STD. It uses a posteriorgram-based template matching framework. Here speech segments are represented using Gaussian posteriorgrams, and matches query posteriorgrams with test posteriorgrams. Dynamic time warping techniques are used for this purpose.

The paper is organised as follows. Section 2 explains the literature survey. The details of the database used are explained in section 3. The feature representation process is given in section 4. It includes the feature extraction process and the posteriorgram creation. Section 5 contains the information about the segmental DTW algorithm. The evaluation process including the experimental set up and the scoring results are given in section 6, before concluding in section 7.

2 Literature Survey

In this generation of digital world, an exponential growth in the audio data is being noticed in the internet. Different forms of audio data such as audio books, films, news articles, lectures etc are tremendously used in this era. Effective utilization of audio data requires the use of searching techniques. This is very essential for the fast retrieval of audio data. An effective approach for solving this problem is the use of Large Vocabulary Continuous Speech Recognition Systems (LVCSR) [1, 2]. Such systems convert the audio data into textual representations. Then the task of audio retrieval becomes the string matching problem. In [3], the performance of a system that uses unsupervised techniques rather than the conventional supervised techniques is analysed.

The drawback of LVCSR systems is the problem related to out-of-vocabulary (OOV) words. Large vocabulary coverage is required in the training phase to avoid OOV problems. Since, these methods are practically impossible, word error rate increases which results in the degradation of the performance of the system. As a solution for the OOV problems, searching techniques based on sub-word units like syllables or phones are introduced [4]. Creating acoustic models for resource rich languages and then adapting such systems for minimal resource languages are also introduced in [5].

Recently, many studies have given contributions to unsupervised spoken term detection. The work in [4] uses a Gaussian Mixture Model (GMM) trained to represent the speech frames. Gaussian Posteriorgrams (GP) are obtained by this process. Given the spoken keyword, Segmental Dynamic Time Warping (SDTW) algorithm is used here for comparing the GPs between the query utterances and the test utterances. The detection scores of the most reliable warping paths are considered as the output of the system.

For the effective spoken term detection task, an effective representation of the speech signal is very essential. The state-of-art systems use linear prediction cepstral coefficients (LPCC) or Mel frequency cepstral coefficients (MFCC) for the representation of the speech signal.

In this work, as an alternative to the MFCC and LPCC, Fourier-Bessel cepstral coefficients (FBCC) are used. [6-8] describe the use of FBCC for speaker identification tasks. Due to the damping nature of Bessel functions, it is obtained that FBCC can be used for the effective representation of the speech signals. Several experiments are being conducted using FBCC in the areas of speech recognition and speaker identification. But the use of FBCC for spoken keyword detection is not exploited yet.

The main aim of this is work is to exploit the advantages of FBCC for audio information retrieval tasks. The performance of the keyword detection system that uses FBCC is analysed [8]. For a better performance, the feature representation using combined MFCC-FBCC features are also included in this work.

3   Database

The proposed work is based on the experiments conducted on MediaEval 2012 database which is a subset of Lwazi database. The database consist of two sets of data for development(dev) and evaluation(eval), both contain reference spoken data and query data. Table 1 shows the details of the database.

Table 1 Details of Mediaeval 2012 Database

	Dev. Data		Eval. Data	
	Reference	Query	Reference	Query
Number of Ut-terances	1580	100	1660	100
Total Duration (min)	221.9	2.4	232.5	2.5
Average Dura-tion (sec)	8.42	1.44	8.40	1.50

This database is used for Spoken Web Search (SWS). It performs audio audio search in multiple languages and acoustic conditions where there are very few resources available to develop a solution for each individual language.

4   Feature Representation

The general framework of posteriorgram-based template matching for QbE STD is depicted in Fig.1. The first block is a tokenizer. It is obtained from the data input.

If the training data is given with training transcriptions, this tokenizer is referred to as supervised, otherwise as unsupervised. The output of the tokenizer is posteriorgram. It converts the query example and test utterance in to query posteriorgram and test posteriorgram.

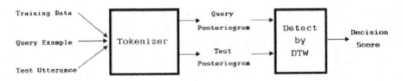

Fig. 1 Spoken Term Detection System

For obtaining the posteriorgrams, the features are extracted from the speech signals. In this paper, the Mel Frequency Cepstral Coefficients (MFCC) and Fourier Bessel Cepstral Coefficients (FBCC) features are extracted. The next step is training a GMM. In the training mode, we will use the feature vectors for obtaining the model. The model is characterized by the value of their parameters. The parameters are mean, variance and the weights.

The initial value for the mean can be obtained by using k-means algorithm. Here, a single Gaussian Mixture Model is trained using all the available data. The feature vectors are combined after removing the silent frames. From these feature vectors, the parameters of GMM are obtained. The number of mixtures is taken as 128. Thus mean, variance and weights are obtained. This is the basic model of the spoken term detection system.

Once the parameters are extracted, the next step is the generation of Gaussian posteriorgram (GP) vectors. For each of the test utterance and query utterance, we will obtain GP vectors. Here, features were pooled for each Gaussian having a maximum likelihood. GP vectors are obtained as a probability vector representing the posterior probabilities of the set of 128 number of Gaussian mixtures. Thus for each vector, we will obtain the probability for the occurence of the vector in each of the 128 mixtures. In this paper, a new kind of feature representation is considered. The GPs of FBCC and MFCC are combined to get a new set of features.

5   Segmental Dynamic Time Warping

The next step in spoken term detection is the searching process. DTW algorithms are used for this task. DTW is applied to scan through the test posteriorgrams and determine the best-matching region, which has the smallest distortion with respect to the query posteriorgrams.

In this paper, Segmental DTW algorithm is used for comparing the query and test posteriorgrams. The different possible warping paths and the scores are obtained in this step. Once we obtain all the warping paths and the corresponding detection scores for each utterances, we will choose the warping region with the minimum distortion score as the region where the query is present.

6   Evaluation Process

A. Scoring Method

For finding the 'value' of a system, we have to measure the usefulness of a system to a user. A perfect system always responds correctly to a stimulus, however an omitted response or a misleading response reduces the value of a system to v a user. Maximum Term-Weighted Value(MTWV) is used for expressing the performance of the system. Term-Weighted Value (TWV) is one minus the average value lost by the system per term. The value lost by the system is a weighted linear combination of probability of miss and the probability of false alarm. The weight β, takes into account both the prior probability of a term and the relative weights for each error type.

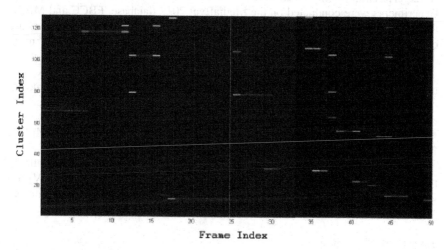

Fig. 2 Gaussian Posteriorgram

The miss and false alarm probabilities are functions of the detection threshold, θ. This (θ) is applied to the systems detection scores, which are computed separately for each search term, then averaged to generate a DET line trace. The formulae for a single term's probabilities are:

$$P_{Miss}(term, \theta) = 1 - \frac{N_{correct}(term, \theta)}{N_{true}(term)} \tag{1}$$

$$P_{FA}(term, \theta) = \frac{N_{spurious}(term, \theta)}{N_{NT}(term)} \tag{2}$$

where:

Ncorrect(term; θ) is the number of correct detections of term with a detection score greater than or equal to θ.

Nspurious(term; θ) is the number of spurious (incorrect) detections of term with a detection score greater than or equal to θ.

Ntrue(term) is the true number of occurrences of term in the corpus.

NNT (term) is the number of opportunities for incorrect detection of term in the corpus.

Thus, the term weighted value (TWV) can be found using Eqn.3.

$$TWV(\theta) = 1 - average_{term}\{P_{Miss}(term, \theta) + \beta.P_{FA}(term, \theta)\} \tag{3}$$

MTWV is the TWV at the point on the DET curve where a value of θ yields the maximum TWV.

B. Experimental Setup and Results

Experiments were conducted on the MediaEval 2012 database. FBCC and MFCC features are extracted from the speech signals. 39 dimensional feature vectors are obtained by this manner. Feature extraction process uses frames of 25 msec duration and 10 msec shift.

The next step is the training of GMM. 128 number of mixtures are used for training using the developoment reference signals. Once we obtain the parameters of GMM, GPs are created for each of the files. The GP obtained for a reference file is shown in Fig.2.

New set of GPs are obtained by combining the GPs of FBCC and MFCC. Different experiments are done by providing different weightage to the GPS.

Then using SDTW algorithm, detection scores are obtained for all the available pairs of queries and references. Maximum Term-Weighted Values (MTWV) are computed for finding the performance of the system. MTWV scores for the systems are given in Table 2.

Table 2 MTWV Scores

Feature Vector	MTWV Score
MFCC	0.2736
FBCC	0.2796
Combined FBCC-MFCC	0.2819

Table 3 shows the MTWV scores obtained for different combinations of GPs. GP1 indicates the GP vector obtained using MFCC and GP2 is that created using FBCC.

Table 3 GP Combining

GP Combination	MTWV Scores
GP1 + GP2	0.2522
GP1 + 0.5 GP2	0.269
0.7 GP1 + 0.3 GP2	0.2735
0.8 GP1 + 0.2 GP2	0.2819

Consider one reference signal and 20 number of query signals. The warping path obtained using SDTW algorithm is shown in Fig.3. The figure shows the possible region of occurence of a particular query in the reference signal region.

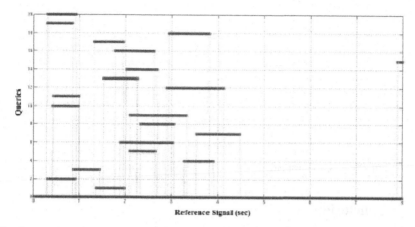

Fig. 3 Warping Path

The DET curves obtained for the combined FBCC-MFCC feature set is shown in Fig.4. Fig.5 shows the term weighted threshold plot.

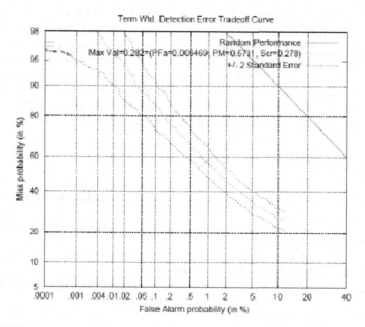

Fig. 4 DET Curve

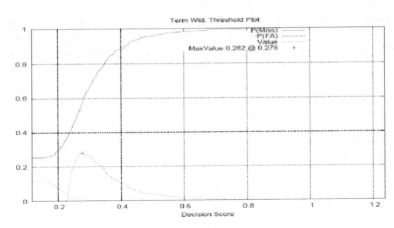

Fig. 5 Term Weighted Threshold Plot

7   Conclusion

In this paper, a new technique for effective representation of the speech signal is proposed. Segment-based DTW algorithm is used for the unsupervised spoken term detection. For representing the feature, FBCC features are used. A new set of features are obtained by using the Bessel features. This system outperfoms the one that uses MFCC. A completely unsupervised GMM learning framework is used to generate the Gaussian posteriorgrams for the query and reference speech. SDTW algorithm compares GPs between keywords and test utterances. The warping paths obtained by comparing all the possible query-reference pairs are collected for the scoring process. MTWV scoring technique is used for comparing the results. Experiments are conducted using FBCC feature set and MFCC feature alone. The experimental results indicate the improvement in the MTWV score while expressing the features with the FBCC feature vectors.

References

1. Miller, D.R.H., et. al.: Rapid and accurate spoken term detection. In: INTERSPEECH 2007, pp. 314–317 (2007)
2. Saraclar, M., Sproat, R.: Lattice-based search for spoken utterance retrieval. In: HLT-NAACL, pp. 129–136 (2004)
3. Ng, K.: Subword-based approaches for spoken document retrieval, Ph.D. dissertation, Massachusetts Institute of Technology (2000)
4. Zhang, Y., Glass, J.R.: Unsupervised spoken keyword spotting via segmental dtw on Gaussian posteriorgrams. In: ASRU, pp. 398–403 (2009)
5. Zgank, A., Kacic, Z., Vicsi, K., Szaszak, G., Diehl, F., Juhar, J., Lihan, S.: Crosslingual transfer of source acoustic models to two different target languages. Robust (2004)

6. Prakash, C., Gangashetty, S.V.: Fourier Bessel based Cepstral Coefficient features for Text-Independent Speaker Identification. In: International Conference on Articial Intelligence (December 2011)

7. Vasudev, D., Babu, K.K.A.: Speaker Identification using FBCC in Malayalam language. in: ICACCI 2014 (September 2014)

8. Vasudev, D., Babu K.K.A.: Query-by-example Spoken Term Detection using Bessel Features. In: IEEE SPICES 2015 (February 2015)

Hyperspectral Image Denoising Using Legendre-Fenchel Transform for Improved Sparsity Based Classification

Nikhila Haridas, C. Aswathy, V. Sowmya and K.P. Soman

Abstract A significant challenge in hyperspectral remote sensing image analysis is the presence of noise, which has a negative impact on various data analysis methods such as image classification, target detection, unmixing etc. In order to address this issue, hyperspectral image denoising is used as a preprocessing step prior to classification. This paper presents an effective, fast and reliable method for denoising hyperspectral images followed by classification based on sparse representation of hyperspectral data. The use of Legendre-Fenchel transform for denoising is an effective spatial preprocessing step to improve the classification accuracy. The main advantage of Legendre-Fenchel transform is that it removes the noise in the image while preserving the sharp edges. The sparsity based algorithm namely, Orthogonal Matching Pursuit (OMP) is used for classification. The experiment is done on Indian Pines data set acquired by Airborne Visible Infrared Imaging Spectrometer (AVIRIS) sensor. It is inferred that the denoising of hyperspectral images before classification improves the Overall Accuracy of classification. The effect of preprocessing using Legendre Fenchel transformation is shown by comparing the classification results with Total Variation (TV) based denoising. A statistical comparison of the accuracies obtained on standard hyperspectral data before and after denoising is also analysed to show the effectiveness of the proposed method. The experimental result analysis shows that for 10% training set the proposed method leads to the improvement in Overall Accuracy from 83.18% to 91.06%, Average Accuracy from 86.17% to 92.78% and Kappa coefficient from 0.8079 to 0.8981.

N. Haridas(✉) · C. Aswathy · V. Sowmya · K.P. Soman
Center for Excellence in Computational Engineering and Networking,
Amrita Vishwa Vidyapeetham, Coimbatore, Tamil Nadu, India
e-mail: {nikhila.haridas92,aswathy0257}@gmail.com, v_sowmya@cb.amrita.edu,
  kp_soman@amrita.edu

© Springer International Publishing Switzerland 2016   521
S. Berretti et al. (eds.), *Intelligent Systems Technologies and Applications*,
Advances in Intelligent Systems and Computing 384,
DOI: 10.1007/978-3-319-23036-8_45

1 Introduction

With the advent of new technologies in the field of remote sensing, hyperspectral image analysis has emerged as one of the prominent areas of research. These images with rich spectral, spatial and temporal resolutions can be used for wide range of industrial, laboratory and airborne applications. The major challenges in hyperspectral data analysis is the large dimensionality of data, spectral mixing, noise corruption etc [1].

Hyperspectral images are often prone to noise which, affects the quality of the obtained image and thereby, reduces the accuracy of the image classification. In order to address this issue, several denoising techniques are done prior to classification. Adam C. Zelinski et al.[2] proposed a novel algorithm to effectively denoise the hyperspectral data using wavelet decompositions and sparse approximation techniques by exploiting the high correlation between the spectral bands of hyperspectral images. By using a spectral and spatial adaptive Total Variation (TV) along with the split Bregman iteration algorithm, Yuan et al. introduced a fast method for hyperspectral image denoising[3]. Kavitha Balakrishnan et al. in [4] used a spatial preprocessing method called Perona-Malik diffusion for hyperspectral image classification, which increases the separability of the classes by smoothening the homogenous areas of hyperspectral imagery. A novel sparsity based algorithm developed by Yi Chen et al. [5] sparsely represents the hyperspectral image as the linear combination of a few training samples. Santhosh et al. in [6] proposed the use of Legendre Fenchel transform for denoising coloured remote sensing images.

Rest of the paper is organized as follows. Section 2 gives an overview of Legendre-Fenchel transformation. Orthogonal Matching Pursuit(OMP) algorithm for hyperspectral image classification is presented in section 3. Section 4 describes about the methodology used in this work. Section 5 presents the experimental results and analysis. Section 6 concludes the paper.

2 Legendre-Fenchel Transformation

Transformation maps functions from one space to another space where, better and easy ways of understanding the function emerges out. Legendre-Fenchel transform[7] is one such transform which, plays a significant role in convex analysis. It maps any function to a new space, which is the space of its slope. The transformed function is the dual(conjugate) of the original function and the transformed space is its dual space. It is a simple and efficient algorithm that can be used for denoising remote sensing images. The advantage of using Legendre-Fenchel transform lies in its ease of divergence calculation and less processing time.

2.1 Concept of Duality

The Legendre-Fenchel transform (LFT) is based on the concept of duality. The Legendre-Fenchel transform of a function $f : R \rightarrow R \cup \{\infty\}$ is a function defined in the dual space $f^*(p)$ (i.e for every $(x, f(x))$ there exists a $(p, f^*(p))$ where $f^*(p)$ is the dual of $f(x)$).

For a continuous but not necessarily differentiable function $f : R \to R \cup \{\infty\}$, the Legendre-Fenchel transform can be defined as,

$$f^*(p) = \sup_{x \in R} \{px - f(x)\} \qquad (1)$$

The supremum is used for transformation. The Legendre-Fenchel conjugate of the function $f^*(p)$ is always convex irrespective of $f(x)$ is convex or not. The mapping involves finding a point x on the function $f(x)$ where the slope of the line p, which pass through the point $(x, f(x))$ has maximum intercept on the y axis. The line with slope p on the function $f(x)$ at point x is nothing but tangent to the function as illustrated in Fig. 1. i.e.,

$$p = f'(x) \qquad (2)$$

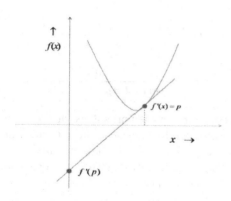

Fig. 1 Illustration of Legendre-Fenchel transform

2.2 Dual Formulation of ROF Model

Based on the concept of duality discussed above, a dual ROF model is formulated which overcomes the drawbacks of commonly used Euler lagrangian ROF model[8]. This method uses the perspectives of primal and dual concept for better understanding of the problem. A standard form of ROF model can be expressed as,

$$\min_{u \in X} \|\nabla u\|_1 + \frac{\lambda}{2} \|u - g\|_2^2 \qquad (3)$$

The convex conjugate of the norm $\|.\|$ is indicator function which is given as

$$\delta(p) = \begin{cases} 0 & if \ \|p\| \le 1 \\ \infty & otherwise \end{cases}$$ (4)

By Legendre-Fenchel transform,

$$\|\nabla u\|_1 = \max_{p \in P} \left(\langle p, \nabla u \rangle - \delta_p(P) \right)$$ (5)

where P is the domain of inidicator function.

Hence the dual formulation of ROF by Legendre-Fenchel transform can be expressed as

$$\min_{u \in X} \max_{p:=\|p\|_\infty \le 1} \left(\langle p, \nabla u \rangle - \delta_p(P) + \frac{\lambda}{2} \|u - g\|_2^2 \right)$$ (6)

On proceeding with the computation, the updated primal (u^{n+1}) and dual form (p^{n+1}) is given as

$$u^{n+1} = \frac{u^n + \tau div \, p^{n+1} + \tau \lambda g}{1 + \tau \lambda}$$ (7)

$$p^{n+1} = \frac{p^n + \sigma \nabla u^n}{\max (1, |p^n + \sigma \nabla u^n|)}$$ (8)

In the denoising procedure, the updated primal and dual is applied to each band of hyperspectral image. The control parameter λ, lipchitz constant τ and the number of iterations n are fixed experimentally. Gradient and divergence computations are made simpler with the use of nabla matrix. Hence the computational complexity is reduced which makes the process faster.

3 Orthogonal Matching Pursuit for Hyperspectral Image Classification

Orthogonal Matching Pursuit (OMP) is one among the various sparse approximation algorithms [9]. It is often referred as greedy approach as it picks up only the necessary columns in a greedy manner to reconstruct the signal.

Let the hyperspectral data be D-dimensional. The dictionary matrix is formed by concatenating randomly selected training samples from each class which can be represented as $A = [A_1, A_2, ..., A_N]$. Sub-directory $A_i = [a_1, a_2, ..., a_{n_i}]$ represents the training data vectors belonging to i^{th} class, where $i \in N$, N denotes the total number of class and n_i pixel samples from each class. Each pixel vector is of size $D \times 1$. The training samples are randomly selected data feature vectors whose class labels are known. The problem formulation for obtaining the sparse vector x using OMP algorithm is given by

$$\min \|x\|_0 \quad subject\ to\ Ax = y$$

where the dictionary matrix A, and the sparse vector x is used to find the class label of the given test pixel vector [10].

The residue is calculated as $r_i = \|y - A_i x\|_2$ where $i \in N$ and A_i consist of the data vectors belonging to class i. The test pixel vector is assigned to the class with the minimum residue.

$$Class(y) = \arg\min_{i=1,2,...,N}(r_i)$$

4 Methodology

The experiment is performed on Indian Pines dataset captured by AVIRIS system on June 12, 1992 over a 2 x 2 mile portion of Northwest Tippecanoe, Indiana. The water absorption bands 104-108, 150-163, 220 are considered as noisy bands and are removed before preprocessing. The remaining 200 bands are used for the experiment. a The proposed method consists of two main stages: spatial preprocessing and classification.

Spatial Preprocessing: Noise affected images have low signal-to-noise ratio which reduces the classification accuracy of various algorithms. Hence, bandwise denoising using Legendre-Fenchel transform which is an efficient preprocessing step is performed to smoothen the image to get a better accuracy for classification. The advantage of this method is that, it consumes less time and is easy to understand.

Classification: Classification involves separating the data into training and testing samples. The proposed method performs sparsity based classification where, a sparse representation of a test sample is found with respect to the training samples obtained from a dictionary. During training stage, dictionary matrix is formed by random selection of pixels from the hyperspectral data. All the pixels excluding the background pixels are used in the testing phase for validation. In order to show the effectiveness of the proposed method, comparison is done based on the classification accuracy assessment measures obtained for the standard hyperspectral dataset before and after applying preprocessing technique.

5 Results and Discussion

In order to show the effectiveness of proposed method a comparative analysis with state of art method such as Total Variation (TV) denoising is performed before further experimentation. The obtained results are shown in Table 1 .The result analysis shows that Legendre Fenchel denoising outperforms TV denoising for OMP based classification.

The experiment also compares the classification results obtained before and after applying the denoising algorithm with different training samples. Each band of the

(a) Original image(Band 165)   (b) Denoised image(Band 165)

Fig. 2 Preprocessing on Indian Pines dataset using Legendre-Fenchel transform for for $\lambda=15$

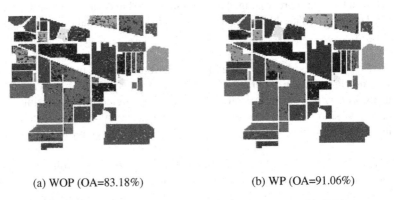

(a) WOP (OA=83.18%)   (b) WP (OA=91.06%)

Fig. 3 Classification map for Indian Pines dataset without (WOP) and with (WP) preprocessing (10% training samples)

hyperspectral image is denoised using Legendre-Fenchel transformation algorithm. The performance is evaluated by visual interpretation. Visual evaluation is conducted for different values of control parameter, $\lambda(5,10,15,20,25,50)$ subjected to 20 persons to select the best value. Number of iterations, n and lipchitz constant, τ are chosen experimentally by trial and error method to give the better result. The parameter values used for denoising step are $\lambda=15$, $n=30$ and $\tau = \sqrt{8}$.

Fig.2 shows the effect of preprocessing on Indian Pines dataset using Legendre-Fenchel transform. Fig.3 shows the classification maps obtained before and after applying preprocessing using 10% training samples. Comparison of classification accuracies obtained without(WOP) and with preprocessing(WP) on Indian Pines dataset is given in Table 2. The experimental result analysis shows that, for 10% training samples, the proposed method leads to the improvement in Overall Accuracy (OA) from 83.18% to 91.06%, Average Accuracy (AA) from 86.17 % to 92.78%

and Kappa coefficient (K) from 0.8079 to 0.8981 and for 40 % training samples, the proposed method leads to the improvement in overall accuracy from 96.86% to 99.07%, average accuracy from 97.11 % to 99.27% and kappa coefficient from 0.9645 to 0.9894.

Table 1 Comparison of classification accuracies obtained for TV denoising and proposed method on Indian Pines dataset

Accuracy Assessment Measures	Without Preprocessing	Total Variation denoising	Legendre Fenchel denoising
Average Accuracy	86.17	87.27	92.78
Overall Accuracy	83.18	84.14	91.06
Kappa Coefficient	0.8079	0.8130	0.8981

Table 2 Comparison of classification accuracies obtained without(WOP) and with preprocessing(WP) on Indian Pines dataset

Percentage of training		10		20		30		40	
Class	Class Name	WOP	WP	WOP	WP	WOP	WP	WOP	WP
Class 1	Alfalfa	93.48	93.48	100.00	97.83	97.83	97.83	93.48	**100.00**
Class 2	Corn-notill	76.61	86.90	**82.00**	92.93	88.31	96.50	94.47	98.81
Class 3	Corn-mintill	**69.16**	85.30	**83.98**	93.01	92.89	97.47	95.06	98.80
Class 4	Corn	**64.56**	83.97	82.28	93.25	83.97	**96.62**	92.83	97.05
Class 5	Grass-pasture	92.75	94.41	96.89	97.31	98.14	99.17	99.38	99.59
Class 6	Grass-trees	96.58	98.63	98.08	99.18	99.18	99.45	99.59	99.73
Class 7	Grass-pasture-mowed	100.00	100.00	100.00	100.00	100.00	96.43	100.00	100.00
Class 8	Hay-windrowed	97.70	99.58	99.37	100.00	99.58	99.79	100.00	100.00
Class 9	Oats	100.00	100.00	100.00	100.00	100.00	100.00	100.00	100.00
Class 10	Soybean-notill	**81.79**	90.53	88.89	92.80	92.39	97.02	96.60	98.87
Class 11	Soybean-mintill	**83.42**	**91.00**	90.18	96.21	93.40	98.04	97.35	98.98
Class 12	Soybean-clean	70.32	78.25	81.28	89.88	89.21	98.04	94.44	97.81
Class 13	Wheat	98.05	100.00	99.51	99.51	100.00	99.51	100.00	100.00
Class 14	Woods	94.55	98.02	96.05	98.42	97.71	98.74	98.97	99.68
Class 15	Buildings-Grass-Trees-Drives	63.99	84.46	77.46	**93.26**	84.46	**95.60**	92.75	**98.96**
Class 16	Stone-Steel-Towers	95.70	100.00	94.62	100.00	98.92	100.00	98.92	100.00
	Average Accuracy	86.17	92.78	91.91	96.47	94.75	97.87	97.11	99.27
	Overall Accuracy	83.18	91.06	89.59	95.46	93.44	97.68	96.89	99.07
	Kappa Coefficient	0.8079	0.8981	0.8812	0.9483	0.9252	0.9735	0.9645	0.9894

6 Conclusion

In this paper, the use of Legendre-Fenchel transform as an effective spatial preprocessing technique for denoising hyperspectral images prior to sparsity based classification (OMP) is discussed. The experiment is performed on standard hyperspectral dataset Indian Pines and the effectiveness of the proposed method is analyzed using the performance measures namely, Overall Accuracy, Classwise Accuracy, Average Accuracy and Kappa coefficient. The experimental results shows that denoising of hyperspectral images using proposed method prior to classification, has enhanced the quality of the hyperspectral image and led to the significant improvement in classification accuracy compared to the state-of-art method (TV denoising).

Acknowledgments The authors also wish to express sincere gratitude to all, who gave their useful suggestions and comments, which helped to improve the quality of this manuscript.

References

1. Bioucas-Dias, J., Plaza, A., Camps-Valls, G., Scheunders, P., Nasrabadi, N., Chanussot, J.: Hyperspectral remote sensing data analysis and future challenges. IEEE Geoscience and Remote Sensing Magazine 1(2), 6–36 (2013)
2. Zelinski, A.C., Goyal, V.K.: Denoising hyperspectral imagery and recovering junk bands using wavelets and sparse approximation. In: IEEE International Conference on Geoscience and Remote Sensing Symposium, IGARSS 2006, pp. 387–390. IEEE (2006)
3. Yuan, Q., Zhang, L., Shen, H.: Hyperspectral image denoising employing a spectral-spatial adaptive total variation model. IEEE Transactions on Geoscience and Remote Sensing 50(10), 3660–3677 (2012)
4. Soman, K.P., Kavitha, B., Sowmya, V.: Spatial preprocessing for improved sparsity based hyperspectral image classification. International Journal of Engineering Research and Technology 1. ESRSA, July 2012
5. Chen, Y., Nasrabadi, N.M., Tran, T.D.: Hyperspectral image classification using dictionary-based sparse representation. IEEE Transactions on Geoscience and Remote Sensing 49(10), 3973–3985 (2011)
6. Santhosh, S., Abinaya, N., Rashmi, G., Sowmya, V., Soman, K.P.: A novel approach for denoising coloured remote sensing image using legendre fenchel transformation. In: 2014 International Conference on Recent Trends in Information Technology (ICRTIT), pp. 1–6. IEEE (2014)
7. Handa, A., Newcombe, R.A., Angeli, A., Davison, A.J.: Applications of legendre-fenchel transformation to computer vision problems, Tech. Rep., Tech. Rep. DTR11-7, Department of Computing at Imperial College London (2011)
8. Rudin, L.I., Osher, S., Fatemi, E.: Nonlinear total variation based noise removal algorithms. Physica D: Nonlinear Phenomena 60(1), 259–268 (1992)
9. Suchithra, M., Sukanya, P., Prabha, P., Sikha, O.K., Sowmya, V., Soman, K.P.: An experimental study on application of orthogonal matching pursuit algorithm for image denoising. In: 2013 International Multi-Conference on Automation, Computing, Communication, Control and Compressed Sensing (iMac4s), pp. 729–736. IEEE (2013)
10. Tropp, J.A., Gilbert, A.C.: Signal recovery from random measurements via orthogonal matching pursuit. IEEE Transactions on Information Theory 53(12), 4655–4666 (2007)

Analysis of Various Color Space Models on Effective Single Image Super Resolution

Neethu John, Amitha Viswanath, V. Sowmya and K.P. Soman

Abstract Color models are used for facilitating the specification of colors in a standard way. A suitable color model is associated with every application based on color space. This paper mainly focuses on the analysis of effectiveness of different color models on single image scale-up problems. Single image scale-up aims in the recovery of original image, where the input image is a blurred and down- scaled version of the original one. In order to identify the effect of different color models on scale-up of single image applications, the experiment is performed with the single image scale-up algorithm on standard image database. The performance of different color models (YCbCr, YCoCg, HSV, YUV, CIE XYZ, Photo YCC, CMYK, YIQ, CIE Lab, YPbPr) are measured by quality metric called Peak Signal to Noise Ratio (PSNR). The experimental results based on the calculated PSNR values prove that YCbCr and CMYK color models give effective results in single image scale-up application when compared with the other available color models.

Keywords Super resolution · Bicubic interpolation · Color spaces · K-SVD · OMP · Sparse representation

1 Introduction

In modern image processing and computer vision applications, improvisation of the resolution of blurry images is a challenging task. The hunt for achieving high resolution imaging systems ends in the problem of diminishing returns. Particularly, the expense of optical components needed for capturing high resolution images are extremely high. The class of image processing algorithms, that overcomes the limitations of optical imaging systems at relatively low cost can be represented using

N. John(✉) · A. Viswanath · V. Sowmya · K.P. Soman
Centre for Excellence in Computational Engineering and Networking,
Amrita Vishwa Vidyapeetham, Coimbatore, India
e-mail: {jneethu.john,amithaviswanath,sowmiamrita}@gmail.com, kp_soman@amrita.edu

© Springer International Publishing Switzerland 2016
S. Berretti et al. (eds.), *Intelligent Systems Technologies and Applications*,
Advances in Intelligent Systems and Computing 384,
DOI: 10.1007/978-3-319-23036-8_46

the term Super-resolution [8]. Super-resolution interpolates lost data from available evidence with minimal blurring and aliasing of images. It aims to uncover fine-scale structure from coarse-scale measurements [2]. Multi frame super resolution focuses on fusing a sequence of noisy blurred low resolution images in order to obtain high quality image whereas, in single frame super resolution a high resolution image is obtained as a result of up-sampling a single blurred image [10] [9]. Single image scale-up algorithm [25]used for this work uses the Sparse-Land local model [5] [6] [1] [4]. The assumption derived from Sparse-Land local model is that, multiplication of sparse vector coefficient with a dictionary generates each patches of the image. Image scale-up algorithm is based on the above specified assumption. The scale-up algorithm used in this paper consists of two phases: training and reconstruction. During the training phase, scale-down operation is used to construct a set of images of low resolution from the training set of images of high resolution. For high and low resolution patches separate dictionaries are trained. By making use of the model trained from the previous phase, reconstruction phase perform scale-up operation on the test images [25].

Many image processing and machine vision applications make use of super-resolution methods. The advancement in computer with more processing power and better accuracy increases the importance of software based super-resolution methods. Super-resolution is used in face recognition [24], automatic target recognition [3], monitoring systems such as the identification and recognition of license plates [12], remote sensing [17], converting video to different standards like converting National Television System Committee (NTSC) to High-definition television (HDTV), image enhancing [22], medical image processing applications like Magnetic Resonance Imaging (MRI) [18] [21], processing of satellite images [26], microscopic image processing [19], image mosaicking [16] and astronomical image processing [23] [7].

Brains reaction to a specific visual stimulus is termed as color, which is extremely subjective and personal. In other words, the property which causes human vision to be sensitive to spectral content of light is termed as color. Assigning numbers to the brain reaction to visual stimuli is a tedious task. The main goal of color spaces is to assist the process of describing color either between human or between programs or machines. The mathematical way of representing a set of colors forms color space. RGB, YCbCr and CMYK are the most accepted three color models. In computer graphics RGB color model is mainly used, video system applications uses YCbCr, CMYK finds its application mainly in the area of color printing. Since none of this color spaces are not particularly assigned to any specific application , this results in temporary search of other models for the purpose of simplification of processing, programming and end-user manipulation in every applications [15] [20] [14] [11]. In this work, the effect of different color spaces like YCbCr, YCoCg, HSV, YUV, CIE XYZ, Photo YCC, CMYK, YIQ, CIE Lab and YPbPr on single image scale-up problem is analyzed by transforming the scale-down image into each of these color spaces prior to the reconstruction. The experimental result analysis based on PSNR measurement shows that YCbCr and CMYK models give effective results in Single image scale-up applications.

This paper is organized as follows, section 2 contain detailed description of materials and methods used for single image scale-up and the effective color spaces YCbCr and CMYK for enhancing the resolution of the output. Analysis of the effect of different color spaces in terms of measurements like visual perception and PSNR value is given in section 3 and finally, section 4 concludes the paper.

2   Materials and Methods

2.1   Algorithm for Scale-up Operation on Single Image

The algorithm for performing scale-up operation on single image, used in this paper is proposed by Matan Protter, Roman Zeyde and Michael Elad [25]. This algorithm consists of two phases: training and reconstruction. The steps to be followed in the Training phase are given below:

1. Initially, the construction of training set is performed. From a set of training images of high-resolution s y_h,images of low-resolution y_l are constructed using scale-down operator L_{all} and the training database P $= \{p_h^k, p_l^k\}_k$ which is formed from the pairs of matching patches are extracted. p_h^k and p_l^k corresponds to the high and low resolution image patch extracted from the image y_h and y_l in location k.
2. This database passes through a pre-processing stage where the removal of low-frequencies from p_h^k and extraction of features from p_l^k take place.
3. In order to make the dictionary training step faster, reduction of dimension is performed on the features of low-resolution patches p_l^k.
4. From the low-resolution patches, a dictionary A_l is trained which gives sparse representation of these patches.
5. From the high-resolution patches, a dictionary A_h is trained and it matches the dictionary corresponds to the low resolution patches.

By making use of the trained model generated from the previous phase, reconstruction phase perform scale-up on the test images. The steps followed for reconstruction phase are listed below:

1. The input is a low-resolution image z_l. By using bicubic interpolation Q input must be upscale by a factor of s, resulting in $y_l \in R^{n_l}$, where n_l is the dimension of low-resolution patches.
2. In order to sharpen this image, it undergoes a spatial non-linear filtering, i.e. using R high-pass filters image y_l is filtered, obtaining $f_k * y_l$, where f_k is the filter coefficient.
3. From each location $k \in \Omega$ pre-processed patches p_l^k of size $\sqrt{n} \times \sqrt{n}$ are generated. Here Ω represents the set of locations centered around locations correspond to the true pixels in y_l, which is the low-resolution image. A patch vector is formed $\tilde{p}_l^k$ by concatenating every R such patches from the same location. The set $\{\tilde{p}_l^k\}_k$ is formed from this collection.

4. For attaining the dimensionality reduction, the projection operator B is multiplied with the patches $\{\tilde{p}_l^k\}_k$, resulting with the set $\{p_l^k\}_k$.
5. By applying the Orthogonal Matching Pursuit (OMP) algorithm to $\{p_l^k\}_k$, the sparse representation vector $\{q^k\}_k$ is obtained.
6. Multiplication of $\{q^k\}_k$ with A_h recovers the High-resolution patches $\{\tilde{p}_h^k\}_k$.
7. The final result i.e. super-resolved image $\hat{y}_h^*$ is created by merging the recovered $\{\tilde{p}_h^k\}_k$ by taking the average in the overlap area, i.e. solution of the following minimization problem with respect to $\hat{y}_h$ gives the final result:

$$\hat{y}_h^* = \arg\min_{\hat{y}_h} \sum_k \left\| R_k(\hat{y}_h - y_l) - \hat{p}_h^k \right\|_2^2 \tag{1}$$

For the above optimization problem to be minimum, the difference between the low and high resolution image should be as equal as possible to the approximated patches. A Least Square solution for the above problem is given by:

$$\hat{y}_h^* = y_l + \left[\sum_k R_k^T R_k \right]^{-1} \sum_k R_k^T \hat{p}_h^k \tag{2}$$

Even though the above solution seems to be complex, the basic idea is very simple- it is equivalent to place $\hat{p}_h^k$ in their exact locations, taking average in the overlap region and adding y_l to get $\hat{y}_h^*$, the final image.

2.2 Color Spaces

The idea of color space provides procedure for visualizing and specifying color. Brightness, hue and colorfulness are the attributes with which humans define a color. A computer may describe a color by using the amount of red, green and blue phosphor emission required to watch a color. For a printing press, a color is the reflectance and the absorbents of cyan, magenta, yellow and black inks on the printing paper [11]. Color space contains set of rules that allows representing colors with numbers. Different color spaces are available for variety of applications. Some color spaces are perceptually linear out of which, few are sensitive to use, i.e. user can easily steer within them and create required colors, where as some other color spaces are very complex and confusing. Finally, some color spaces dependents on devices, while others are equally portable to all devices.

For analyzing the effect of color spaces on super resolution application, the super resolution algorithm [25] is applied on the same image in ten different color spaces such as YCbCr, YCoCg, HSV, YUV, CIE XYZ, Photo YCC, CMYK, YIQ, CIE Lab and YPbPr. The super resolution algorithm [25] used for this analysis already follows YCbCr color model, i.e. the input low-resolution RGB image in the reconstruction phase is transformed to YCbCr space prior to the reconstruction. The comparison of different color models are carried out by visual perception and PSNR measurements. From this analysis, it is observed that YCbCr and CMYK color models gives better results for single image scale-up applications.

2.2.1 YCbCr Color Space

YUV color space's offset and scaled version forms YCbCr color space. This color space splits RGB into luminance and chrominance information. It was created as a part of the recommendation of ITU-R BT.601. YCbCr is a digital standard used for compression applications and each color is represented with three numbers. In this model, information regarding luminance is stored in Y component , and the Cb and Cr components hold the chrominance information. Cb indicates the intensity of blue component with respect to green and Cr indicates the intensity of red component relative to green component [13]. This color space concentrates more on exploiting the human eye properties. The human eye is more intuitive to light intensity changes compared to hue changes. One can store the Y (intensity) component more accurately than the Cb and Cr component when the amount of information is to be minimized. The JPEG file format makes use of this color space to filter out unimportant information. The value of Y is within an 8-bit range of 16-235; 16-240 is the nominal range of Cb and Cr. There exist several sampling formats for YCbCr such as 4:4:4, 4:2:2, 4:1:1 and 4:2:0. RGB color cube in the YCbCr color space is illustrated in Fig. 1.

In MATLAB the function 'rgb2ycbcr' converts RGB images or color maps to the YCbCr color space and 'ycbcr2rgb' performs the reverse operation. In order to provide the conversion between $R^*G^*B^*$ in the range of 0-255 and $Y^*Cb^*Cr^*$ following are the basic equations used by the Intel IPP functions[13]:

$$Y^* = 0.257 * R^* + 0.504 * G^* + 0.098 * B^* + 16 \tag{3}$$

$$Cb^* = -0.148 * R^* - 0.291 * G^* + 0.439 * B^* + 128 \tag{4}$$

$$Cr^* = 0.439 * R^* - 0.368 * G^* - 0.071 * B^* + 128 \tag{5}$$

$$R^* = 1.164 * (Y^* - 16) + 1.596 * (Cr^* - 128) \tag{6}$$

$$G^* = 1.164 * (Y^* - 16) - 0.813 * (Cr^* - 128) - 0.392 * (Cb^* - 128) \tag{7}$$

$$B^* = 1.164 * (Y^* - 16) + 2.017 * (Cb^* - 128) \tag{8}$$

The color conversion functions of Intel IPP specific for the JPEG codec uses different equations:

$$Y' = 0.299 * R' + 0.587 * G' + 0.114 * B' \tag{9}$$

$$Cb' = -0.16874 * R' - 0.33126 * G' + 0.5 * B' + 128 \tag{10}$$

$$Cr' = 0.5 * R' - 0.41869 * G' - 0.08131 * B' + 128 \qquad (11)$$

$$R' = Y' + 1.402 * Cr' - 179,456 \qquad (12)$$

$$G' = Y' - 0.34414 * Cb' - 0.71414 * Cr' + 135.45984 \qquad (13)$$

$$B' = Y' + 1.772 * Cb' - 226.816 \qquad (14)$$

2.2.2 CMYK Color Space

CMYK color space is the counter part of RGB color space and is mainly used in the production of color print. The name CMYK denotes the colors cyan, magenta and yellow along with black (noted as K). CMYK is a subtractive color space, i.e. final color is created by removing some colors from white surface by applying cyan, magenta, yellow and black pigments to this white surface. For example cyan is formed by subtracting red from white, magenta is the difference between white and green and yellow is white minus blue, as illustrated in Fig. 2. Theoretically subtraction of the entire colors by the combination of full saturation CMY should render black. In order to improve both the available color gamut and the density range, the fourth color black is included [11]. Fig. 3 is the illustration of CMY cube, three corners are occupied by CMY values and the opposite three corners are occupied by red, green and blue, the origin contains white; and the corner furthest from the origin contains black.

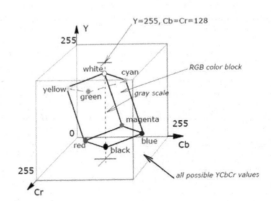

Fig. 1 RGB color cube in the YCbCr space [13]

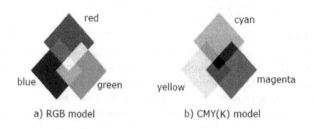

Fig. 2 RGB and CMYK model's primary and secondary colors. [13].

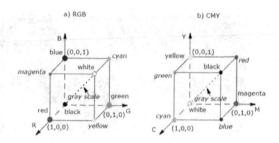

Fig. 3 RGB and CMY color models [13].

CMYK is comparatively easy to implement but proper transfer from RGB to CMYK is very tough task. Also CMYK depends on device, reasonably un-intuitive and non-linear with visual perception [11]. Following are the equations providing the conversion between RGB and CMYK:

$$\begin{bmatrix} C \\ M \\ Y \end{bmatrix} = \begin{bmatrix} 1 - R \\ 1 - G \\ 1 - B \end{bmatrix} \tag{15}$$

$$\begin{bmatrix} K \\ C \\ M \\ Y \end{bmatrix} = \begin{bmatrix} \min(C, M, Y) \\ \frac{C-K}{1-K} \\ \frac{M-K}{1-K} \\ \frac{Y-K}{1-K} \end{bmatrix} \tag{16}$$

$$\begin{bmatrix} R' \\ G' \\ B' \end{bmatrix} = \begin{bmatrix} 1 + C \\ 1 + M \\ 1 + Y \end{bmatrix} \tag{17}$$

In MATLAB the function 'rgb2cmyk' converts RGB images or color maps to the CMYK color space and 'cmyk2rgb' performs the reverse operation.

3　Experimental Results and Discussion

The effect of different color spaces like YCbCr, YCoCg, HSV, YUV, CIE XYZ, Photo YCC, CMYK, YIQ, CIE Lab and YPbPr on single image scale-up problem is analyzed by transforming the scale-down image into each of these color spaces prior to the reconstruction. Results obtained are illustrated in Fig. 4.

From PSNR measurement illustrated in Fig. 5, it is observed that YCbCr and CMYK models give almost equal and effective results in applications, that needs to perform scale-up operation of Single image. The PSNR value is calculated by the following equation[25]:

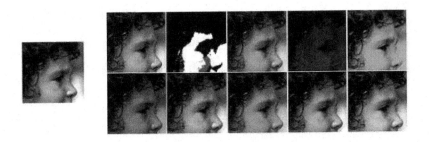

Fig. 4 Visual comparison of reconstructed Face image after applying scale-up algorithm in different color spaces. (Left image: The original image. Right image- from top left to bottom right: Results in YCbCr, HSV, YUV, YCoCg, CIE XYZ, Photo YCC, CMYK, YIQ, CIE Lab and YPbPr color spaces.

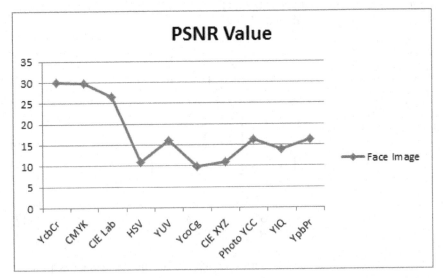

Fig. 5 Graph representation of PSNR values for the reconstructed face image in Fig. 4.

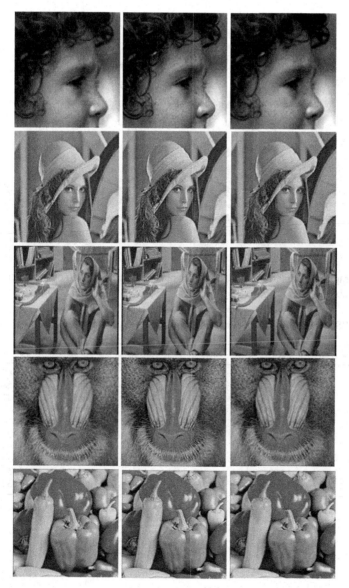

Fig. 6 Visual comparison of various reconstructed images after applying single scale-up algorithm (from top to bottom: Face, Lenna, Barbara, Baboon, Pepper): Left to right: The original image, YCbCr color space, CMYK color space.

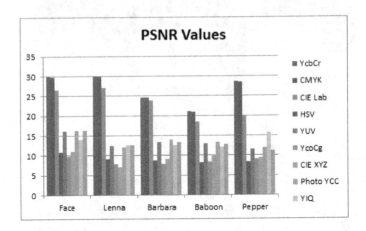

Fig. 7 Graph representation of PSNR values obtained for various color spaces on standard image database

Table 1 PSNR values obtained for various color space on standard image database

Color Spaces	Face	Lenna	Barbara	Baboon	Pepper
YCbCr	29.9639	30.0386	24.5285	21.0705	28.588
CMYK	29.8029	29.9144	24.6633	20.9491	28.4584
CIE Lab	26.6409	27.0881	23.9174	18.5227	20.0277
HSV	10.8418	9.1560	8.7289	8.1913	8.4664
YUV	16.1769	12.4860	13.2821	12.9519	11.5263
YCoCg	9.7821	7.9187	7.8571	8.4805	9.0324
CIE XYZ	11.0298	7.1170	9.0516	10.1207	9.4804
Photo YCC	16.3034	12.1421	14.0631	13.2981	11.8599
YIQ	13.9972	12.6303	12.6875	12.0666	15.756
YPbPr	16.2764	12.5932	13.3868	12.7577	11.232

$$PSNR = 10\log_{10}\left(\frac{255^2.N}{\sum_i (\hat{y}_i - y_i)^2}\right) \tag{18}$$

with $y, \hat{y} \in [0, 255]^N \subseteq R^N$, where N is the image size. Fig. 6 provides the visual comparison of the effect of YCbCr and CMYK color spaces on super-resolution. Graphical representation of PSNR values obtained for various color spaces on standard image database is shown in Fig. 7. From this representation, it is clear that YCbCr and CMYK color spaces gives better result compared to rest of the color spaces considered.

The PSNR values obtained by applying different color spaces prior to reconstruction in super-resolution are listed in Table 1. The data set used in this implementation is the same dataset used in single image scale-up implementation [25].

4 Conclusion

Super-resolution is one of the emerging techniques used in wide variety of applications. It has an inevitable role in many of the current image processing and machine vision applications. This work is done to determine the color space suitable for super-resolution applications. The effect of different color spaces like YCbCr, YCoCg, HSV, YUV, CIE XYZ, Photo YCC, CMYK, YIQ, CIE Lab and YPbPr on single image scale-up problem is analyzed by transforming the scale-down noisy image in to each of these color spaces prior to the reconstruction. The super resolution algorithm [25] used for this analysis already follows YCbCr color model, i.e. the input low-resolution RGB image in the reconstruction phase is transformed to YCbCr space prior to the reconstruction. From PSNR values, it is found that YCbCr and CMYK models give effective results. In addition to this, CIE Lab color space is also suitable for super-resolution application, to some extent.

References

1. Bruckstein, A.M., Donoho, D.L., Elad, M.: From sparse solutions of systems of equations to sparse modeling of signals and images. SIAM Review **51**(1), 34–81 (2009)
2. Candès, E.J., Fernandez-Granda, C.: Towards a mathematical theory of super-resolution. Communications on Pure and Applied Mathematics **67**(6), 906–956 (2014)
3. Cristani, M., Cheng, D.S., Murino, V., Pannullo, D.: Distilling information with super-resolution for video surveillance. In: Proceedings of the ACM 2nd International Workshop on Video Surveillance & Sensor Networks, pp. 2–11. ACM (2004)
4. Elad, M.: Sparse and redundant representations: from theory to applications in signal and image processing. Springer (2010)
5. Elad, M., Aharon, M.: Image denoising via learned dictionaries and sparse representation. In: 2006 IEEE Computer Society Conference on Computer Vision and Pattern Recognition, vol. 1, pp. 895–900. IEEE (2006)
6. Elad, M., Aharon, M.: Image denoising via sparse and redundant representations over learned dictionaries. IEEE Transactions on Image Processing **15**(12), 3736–3745 (2006)
7. Karimi, S.J.E., Kangarloo, K.: A survey on super-resolution methods for image reconstruction. International Journal of Computer Applications **19**(3), 0975–8887 (2014)
8. Farsiu, S., Robinson, D., Elad, M., Milanfar, P.: Advances and challenges in super-resolution. International Journal of Imaging Systems and Technology **14**(2), 47–57 (2004)
9. Farsiu, S., Robinson, M.D., Elad, M., Milanfar, P.: Fast and robust multiframe super resolution. IEEE Transactions on Image Processing **13**(10), 1327–1344 (2004)
10. Fernandez-Granda, C., Candes, E.J.: Super-resolution via transform-invariant group-sparse regularization. In: 2013 IEEE International Conference on Computer Vision (ICCV), pp. 3336–3343. IEEE (2013)
11. Ford, A., Roberts, A.: Colour space conversions, vol. 1–31. Westminster University, London (1998)

12. Hao, S., Lin, L., Weiping, Z., Limin, L.: Location and super-resolution enhancement of license plates based on video sequences. In: 2009 1st International Conference on Information Science and Engineering (ICISE), pp. 1319–1322. IEEE (2009)
13. Intel® integrated performance primitives for intel® architecture reference manual, vol. 2. Image and video processing
14. Keith, J.: Video Demystified. A Handbook for the Digital Engineer. Newnes (2004)
15. Kratochvíl, T., Melo, J.: Utilization of matlab for tv colorimetry and color spaces analysis
16. Kunter, M., Kim, J., Sikora, T.: Super-resolution mosaicing using embedded hybrid recursive folow-based segmentation. In: 2005 Fifth International Conference on Information, Communications and Signal Processing, pp. 1297–1301. IEEE (2005)
17. Li, F., Jia, X., Fraser, D.: Universal hmt based super resolution for remote sensing images. In: 15th IEEE International Conference on Image Processing, ICIP 2008, pp. 333–336. IEEE (2008)
18. Maintz, J.B., Viergever, M.A.: A Viergever. A survey of medical image registration. Medical Image Analysis 2(1), 1–36 (1998)
19. Maji, S.: Generative Models for Super-Resolution Single Molecule Microscopy Images of Biological Structures. PhD thesis, National Institutes of Health (2009)
20. Richardson, I.E.G.: Video Codec Design. Willey Interscience (2002)
21. Roohi, S., Zamani, J., Noorhosseini, M., Rahmati, M.: Super-resolution mri images using compressive sensing. In: 2012 20th Iranian Conference on Electrical Engineering (ICEE), pp. 1618–1622. IEEE (2012)
22. Schultz, R.R., Stevenson, R.L.: Extraction of high-resolution frames from video sequences. IEEE Transactions on Image Processing 5(6), 996–1011 (1996)
23. Marcollin, M.W., Sheppard, D.G., Hunt, D.R.: Iterative multiframe super-resolution algorithms for atmospheric turbulance- degraded imagery. In: IEEE International Conference on Acoustic Speech and Signal Processing, vol. 5, pp. 2857–2860, May 12–15, 1998
24. Yu, H., Xiang, M., Hua, H., Chun, Q.: Face image super-resolution through pocs and residue compensation. Visual Information Engineering (2008)
25. Zeyde, R., Elad, M., Protter, M.: On single image scale-up using sparse-representations. In: Boissonnat, J.-D., Chenin, P., Cohen, A., Gout, C., Lyche, T., Mazure, M.-L., Schumaker, L. (eds.) Curves and Surfaces 2011. LNCS, vol. 6920, pp. 711–730. Springer, Heidelberg (2012)
26. Zhang, Y., Mishra, R.K.: A review and comparison of commercially available pan-sharpening techniques for high resolution satellite image fusion. In: 2012 IEEE International Geoscience and Remote Sensing Symposium (IGARSS), pp. 182–185. IEEE (2012)

Encoded Scene Panorama Based Information Flow in a Multi Agent System

Sachin Meena and Jhilik Bhattacharya

Abstract The work presents a multi agent system using backchannels to interact and send environmental information in real time. The performance of the agent system in real time depends on its abilities to cooperate and communicate with each other. This has seen the evolvement of different architectures as well as Agent Communication Languages over time. The amount of data transfer and speed of processing are the two factors which control the real time response of Multi Agent Systems (MAS). Increase in the amount of information communicated between the agents will lead to improved functionality but at the cost of processing and communication. This paper utilizes an approach where an encoded panoramic mosaic is exploited to reduce visual data dimension communicated between agents in real time. Instead of transferring video data in bulk, the sender agent processes the frames to create a panorama. The sender updates the panorama only when there are changes in the scene. A single panorama hence encompasses the full 360 degree view at the t_0 instant as well as the dynamic information of the next t_1 to t_n instances. The value for n is set according to the amount of dynamic content of the scene. The panorama is refreshed when the Agent changes position. This reduces the data dimension transferred to other agents, which reconstructs the encoded panorama into individual image frames from t_1 to t_n. The amount of transfer is trimmed from $k \times n$ (k images per panorama and n panoramas) to a single panorama. Also if k images have a dimension $m \times n \times k$, a single panorama can be mapped as $\frac{m.k}{2} \times \frac{n.k}{2}$. The value of k depicts the number of pans used to acquire the 360 degree view and depends on the scene. The model is highly effective when there are minimal changes in the surrounding scene at fixed intervals.

S. Meena(✉) · J. Bhattacharya
Thapar University, Patiala, Punjab, India
e-mail: sachin.meena0@gmail.com, jhilik@thapar.edu

© Springer International Publishing Switzerland 2016
S. Berretti et al. (eds.), *Intelligent Systems Technologies and Applications*,
Advances in Intelligent Systems and Computing 384,
DOI: 10.1007/978-3-319-23036-8_47

541

1 Introduction

Multiagent Systems is a research area of distributed Artificial Intelligence where agents can be used for modelling any entity like web service, robot, or nodes of a grid for example. These artificial agents require to communicate and coordinate efficiently; retrieving, processing and exchanging expertise with other agents in-order to solve real time complex problems that cannot be solved individually [1] Research in this domain had a two-fold focus. One end concentrated on developing architectures of MAS while the other end worked with the communication language between the agents. This usually requires a common language, also known as an Agent Communication Language (ACL). Some widely used ACLs are KQML and FIPA-ACL [2] Although ACLs were popular in software agent systems, their use was limited in robot systems where low level data like video transfer was necessary. This continued until the introduction of backchannel for transferring the low level data [3].This work presents a MAS using the two tier architecture, i.e. ACL for high level data and backchannel for image data. The paper particularly deals with the information flow between the agents. The motivation behind this work was the requirement of high level i.e. semantic parsing of the environmental image data for analyzing the scene. One important factor observed while carrying out the work was the effect of data dimension on the system performance [4].Larger amount of environment data resulted in higher accuracy but at the cost of response time and storage space. Thus this research particularly focuses on setting an optimal trade off between the two. The use of data compression techniques like Run length encoding, Arithmetic encoding, MPEG, H.261, displaced frame difference, entropy encoding and many more, dates back to a long time. All these are used for removing the redundancy, common information thus reducing the effect on network bandwidth, storage capacity, speed and loss rate. The work introduces an approach where a panorama of the scene is used for encoding the moving parts in a scene. Individual images are reconstructed from the panorama using its encoded features at the receiver end. A single encoded panorama hence substitutes n panoramas at different time instances.This paper is organized in the following manner. Section 2 describes the system while section 3 gives its elaborate working. Experimental Results and Observations are presented in section 4. Section 5 briefs the conclusion and future work.

2 Brief System Description

A heterogeneous system with two agents is presented in the work. The system attempts to provide a high level environmental sensing. Generally Agents are equipped with sensors which acquire raw data from the environment. A low level parsing of this information is carried out for auto-control mechanism of the Agent. Agent 1 is a robot which is equipped with a pan-tilt camera, seven IR sensors and three ultrasonic sensors and a SBC. The sensors are connected to the SBC using a configured wireless network. The SBC gathers the sensor data and processes it before sending them to Agent 2. Agent 2 is a computer which communicates with Agent1 using ACL and backchannel. In the current work, Agent 1 perceives the environment and sends the gathered data to Agent 2 which performs a semantic parsing of the data. The semantic output is send back to Agent 1 on the basis of which it performs certain actions. This is further elaborated using figure 1and 2. This paper mainly focuses on camera data collection, processing for panorama creation,sending encoded panoramic mosaic and reconstruction at the receiver.

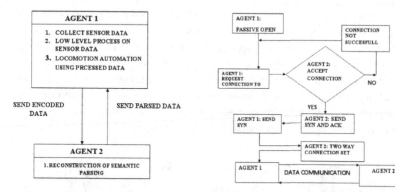

Fig. 1 Action performed by agent 1 and agent 2

Fig. 2 Flowchart of communication between agent1 and agent 2

3 Work Methodology

This section elaborates the procedure of data dimension reduction using panorama image. Agent 1 is equipped with a pan-tilt camera which is programmed to exhibit a pan-tilt motion continuously and grab picture. These images are to be send to Agent 2 continuously for semantic parsing. However, the current work creates a panorama of the scene at the t_0 instance. The dynamic areas of the scene are extracted using frame differencing. These are then encoded in the panorama using steganographic techniques before transferring it. At the receiver end the encoded dynamic areas are extracted and individual images are reconstructed. The approach is further described in the following subsections.

3.1 Panorama Creation Approach

For successfully constructing a panorama from two or more individual images, two adjustments are required. First is geometric adjustment and second is photometric adjustment [5].Geometric adjustment transforms the images into the same coordinate or computing surface [5],[6].Various methods mainly using SIFT or other interest point based feature matching and motion based registration techniques are reported in literature for panorama creation [7],[8]. Removing seam after stitching or photometric adjustment is another subtask of panorama creation. There are two kinds of image stitching methods: transition smoothing [9] and optimal seam finding [10].

This research uses SURF (Speeded Up Robust Feature [11]) for panorama creation which computes Hessian matrix as shown in equation 1.

$$H(p, \sigma) = \begin{bmatrix} L_{xx}(p, \sigma) & L_{(xy)}(p, \sigma) \\ L_{xy}(p, \sigma) & L_{(yy)}(p, \sigma) \end{bmatrix} \tag{1}$$

Where $L_{xx}(p, \sigma)$ is image convolution of second derivative $\frac{dx}{dx^2} g(\sigma)$. Once feature detection points and matching points of both the images are obtained, the two images are stitched and the stitching point position is stored for further processing. This approach is applied for all images and finally a panorama of k different images is obtained. This research uses a color balancing technique for photometric adjustment. The fundament frequency components of the different images are replaced with that of the reference image. Figure 3 shows the result of color balancing approach. Here, one image is taken as input image and other as reference image. According to the reference image, changes are applied to the input image.

(a) Input Image   (b) Reference image   (c) Input image after color balancing

Fig. 3 Color Balancing Approach

3.2 Moving Object Detection Approach

Let F_1^1, F_2^1, F_3^1,..,F_k^1 be the individual images with which panorama is created at t_1 using k different pans of the camera. At instance t_2 the set is represented by F_1^2, F_2^2, F_3^2,.., F_k^2. The two sets may differ due to the presence of moving regions/objects or lighting changes. The latter is already handled using color balancing. Dynamic content or moving object information is retrieved by computing the difference between F_i^j and $F_i^j - 1$. Position (in the image frame) and content (entire bounding box) of these dynamic regions are stored for future use.

3.3 Insert Collected Data into Panorama

Individual images in panorama are registered with reference to the first image. Hence for individual image reconstruction the transform value is required. Also the image boundaries in the panorama, the dynamic content and position are required. For reconstruction later at the receiver end, the dynamic content is inserted in the panorama using steganography techniques, whereas rest of the information is send in a structure. Suppose sizes of images are $R \times C$ pixels. Positions of images in panorama are calculated using equation 2 .

$$\begin{cases} P_i^s = JP_{i-1,i} - MP_i, \; P_i^e = JP_{i-1,i} + (C - MP_i) & i > 1 \\ P_i^s = 0, \; P_i^e = JP_{i,i+1} + (C - MP_i) & i = 1 \end{cases} \tag{2}$$

where P_i^s represents the starting position of image i and P_i^e represents last position of image i in the panorama. $JP(i-1, i)$ is the joining point between image $i-1$ and i in the panorama, MP_i gives the point in image i which corresponds to $JP(i-1, i)$ in the panorama. Different steganography techniques reported in literature [12],[13] vary according to the number of channels used for data hiding (one or all three), particular pixels selected (all pixels or edge pixels for example) and number of bits used in each pixel (one or two). In this research every channel is used to store the message. Message bits are distributed to every pixels last two bits in the corresponding three channels. Suppose first pixel values of bounding box information is $234, 243, 213$ then 4 pixels of the panorama will be required to hide the same. This is further ellaborated in the following example. Panorama pixel values $p1 = 194, 192, 123$, $p2 = 213, 123, 222$, $p3 = 111, 123, 213$, $p4 = 198, 231, 09$ transform into $p1 = 195, 195, 123$, $p2 = 214, 123, 221$, $p3 = 110, 120, 213$, $p4 = 214, 231, 09$ after data encoding. Example shows that maximum variation in pixel values of panorama is 3, which is not affected so much in original image.

3.4 Data Communication

Starting and ending points of each image in the panorama and their transformation matrices are send along with the encoded panorama to the receiver agent. The panorama is send using backchannel whereas other information is send in the

12-12 bit field for position of moving object			(576 bit field) 288-288 bit field for transform values			
MOVING OBJECT POSITION AT TIME t1	MOVING OBJECT POSITION AT TIME t2	MOVING OBJECT POSITION AT TIME t3	TRANSFORM VALUE TFROM1 OF IMAGES F_2^2 WITH RESPECT TO F_2^1		TRANSFORM VALUE TFROM1 OF IMAGES F_3^2 WITH RESPECT TO F_3^1	
STARTING POSITION OF IMAGE 1	ENDING POSITION OF IMAGE 1	STARTING POSITION OF IMAGE 2	ENDING POSITION OF IMAGE 2	STARTING POSITION OF IMAGE 3	ENDING POSITION OF IMAGE 3	

(72 bit field) 24-24 bit field for image position in panorama which is further divided into starting and ending position

Fig. 4 Structure send with panorama to agent2

structure as shown in the figure 4 . This is benificial when compared to Multi- agent system which uses KQML for communication.

Structure attached with panorama is 684 bit long. Panorama size is 1031x2751 pixels, maximum number of columns in panorama is 2751, and thus any column is represented by 12 bit. First 36 bit field is used for position of moving object which is further divided according to different images. Next 576 bit field is used for transform value of images with respect to image F_1^1. Transform values are in 3x3 matrixes each value is represented in 4bytes. So 288 bit field is used for transformation value. Last field is Image position in panorama, so again starting and ending position of image is not greater than 2751 and thus represented in 12 bit each.

3.5 Reconstruct Image from Panorama

Image Reconstruction includes decomposing the panorama into the individual images using starting and ending point and transformation values. The moving regions (decoded from the panorama) are then superimposed using alpha blending [16].

4 Experiments and Result

The experimental result shows panorama creation, moving object detection and reconstruction of images from the encoded panorama. 3 images are taken as input for creating a panorama.

Suppose at time t_2 and t_3 moving objects are on second image and third image respectively. Figure 5e shows the moving object at time t_1, t_2 and t_3 superimposed on the panorama created at t_1. Bounding box information and other related information

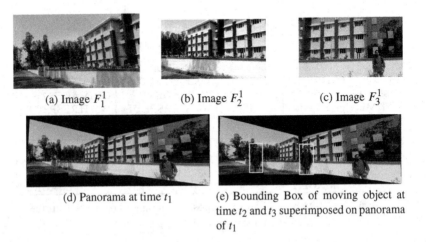

(a) Image F_1^1 (b) Image F_2^1 (c) Image F_3^1

(d) Panorama at time t_1 (e) Bounding Box of moving object at time t_2 and t_3 superimposed on panorama of t_1

Fig. 5 Three input images for panorama creation

(a) Extract image and bounding box position	(b) Original image transform and dynamic region super imposed

Fig. 6 Reconstructed Image from Panorama

with moving object are hidden into panorama created at time t_1. Size of panorama is 1031x2751x3=8508843 pixels, and size of moving object is 124x244x3=90768 pixels. Each pixel of moving object uses 4 pixel values of panorama for hiding the value,hence a maximum of 24 moving object can be encoded in the panorama of three images. Figure 6a shows the start and end position of F_2^2 image (rectangle in yellow color) as well as position of bounding box (circle in purple color). Figure 6b depicts the extracted image after applying inverse transform and moving object superimposition.

5   Conclusion and Future Work

The agent communication presented in this paper optimizes the visual data dimension by pre-analyzing the data and using steganographic techniques to encode it. It is observed that there is a reasonable improvement in communication speed as well as storage space using the approach (normally 3 image information requires 1920x1080x3 pixels but using these approach 1031x2751 pixels of panorama and 684 bit structure is used for same purpose). The quality of reconstructed data can be further improved by using other panorama creating techniques. The work currently uses a SURF feature matching for registration and image stitching. The color balancing technique, used to map the illumination differences between images is preferred over conventional seam removal techniques due to two reasons:- first being the speed efficiency of the former. Secondly, the panorama is further utilized for scene segmentation and semantic parsing which remains unhampered by presence of minor seams, if any. However work is in progress to compare the current application performances in terms of communication speed and reconstruction quality while using sparse models instead of SURF or other key point features. The experimental results deal with a single moving object, which will further be increased. Future work will also consider mapping the transformations between the same moving object from frame to frame. Thus instead of using the bounding box of the moving object in each frame, a single Gaussian mixture model (denoting the moving object) and a list of transformation mappings will be used.

References

1. Koes Berna, M., Nourbakhsh, I., Sycara, K.: Communication efficiency in multi-agent systems. In: 2004 IEEE International Conference on Proceedings of the Robotics and Automation, ICRA 2004, vol. 3. IEEE (2004)
2. Bordini, R.H., Dastani, M., Winikoff, M.: Current issues in multi-agent systems development. In: O'Hare, G.M.P., Ricci, A., O'Grady, M.J., Dikenelli, O. (eds.) ESAW 2006. LNCS (LNAI), vol. 4457, pp. 38–61. Springer, Heidelberg (2007)
3. Raphael, M.J., Deloach, S.A.: A knowledge base for knowledge-based multiagent system construction
4. Das, D., Gupta, S.: Communication, Cooperation, Coordination and Cognition of a Multi Agent System
5. Juan, L., Gwun, O.: Applied in panorama image stitching
6. Lowe, D.G.: Distinctive image features from scale-invariant key points
7. Leone, A., et al.: A fully automated approach for underwater mosaicking
8. Yu, Y., et al.: A novel algorithm for view and illumination invariant image matching
9. Pérez, P., Gangnet, M., Blake, A.: Poisson image editing
10. Agarwala, A., et al.: Interactive digital photomontage
11. Grimson, W.E.L.: Computational experiments with a feature based stereo algorithm
12. Nagpal, P., Baghla, S.: Video Compression by Memetic Algorithm. International Journal of Advanced Computer Science and Applications 2(6) (2011)
13. Laskar, S.A., Hemachandran, K.: Steganography Based on Random Pixel Selection for Efficient Data Hiding. International journal (2013)
14. Rossi, D.J., Willsky, A.S.: Reconstruction from projections based on detection and estimation of objects parts I and II: Performance analysis and robustness analysis. IEEE Trans. Acoust. Speech Signal Process ASSP–32, 886–906 (1984)
15. Gan, L., Qu, Z.: Research on 3-d reconstruction with a series of cross-sectional images. In: Innovative Computing, Information and Control, ICICIC 2006, vol. 1. IEEE (2006)
16. Uyttendaele, M., Eden, A., Szeliski, R.: Eliminating ghosting and exposure artifacts in image mosaics. In: Computer Vision and PatternRecognition, pp. 509–516. IEEE Computer Society (2001)

Steganalysis of LSB Using Energy Function

P.P. Amritha, M. Sreedivya Muraleedharan, K. Rajeev
and M. Sethumadhavan

Abstract This paper introduces an approach to estimate energy of pixel associated with its neighbors. We define an energy function of a pixel which replaces the pixel value by mean or median value of its neighborhood. The correlations inherent in a cover signal can be used for steganalysis, i.e, detection of presence of hidden data. Because of the interpixel dependencies exhibited by natural images this function was able to differentiate between cover and stego image. Energy function was modeled using Gibbs distribution even though pixels in an image have the property of Markov Random Field. Our method is trained to specific embedding techniques and has been tested on different textured images and is shown to provide satisfactory result in classifying cover and stego using energy distribution.

Keywords Markov random field · Steganography · Steganalysis · Gibbs distribution

1 Introduction

The natural images are those images which are the one to which the human vision is most sensitive. The statistical properties of the natural images are used for various researches and it has found a variety of applications in many areas like segmentation, feature extraction, edge detection etc. These statistical properties are used for probabilistic description and classification of different parts of an image and also for their quality estimation, to characterize the content of an image and its texture. Statistical methods can be further classified into first order, second order and higher order statistics. In this paper natural images are modeled using Gibbs distribution and this model is used as a prior to detect hidden messages in the images.

In any natural image there are inherent statistical properties that are identifiable and it can be made use for efficient representation of the image and also for recon-

P.P. Amritha(✉) · M.S. Muraleedharan · K. Rajeev · M. Sethumadhavan
TIFAC CORE in Cyber Security, Amrita Vishwa Vidyapeetham, Coimbatore, India
e-mail: {ammuviju,sreedivya.muraleedharan,rajeev.cys,m_sethumadhavan}@gmail.com

© Springer International Publishing Switzerland 2016   549
S. Berretti et al. (eds.), *Intelligent Systems Technologies and Applications*,
Advances in Intelligent Systems and Computing 384,
DOI: 10.1007/978-3-319-23036-8_48

struction. It is well known that the secret hiding technique of steganography is necessarily believed to be not causing any disturbance to the image visibly[5]. But it does affect some of the statistical properties of the image. The changes in any such properties can be visible by modeling the cover image and then comparing it with its stego model. The choice of model is also crucial to the success of the accurate detection. Targeted steganalytic methods are not capable of detecting general steganographic manipulations in images (universal steganalysis), since they are meant for specific stego algorithms[3]. Most universal steganalysis techniques use an image model to reconstruct an estimate of the original cover from the input[4]. Differences between reconstructed and input images are an indication of a steganographic manipulation.

Characterizing a prior distribution of natural images is a key problem of many low-level vision problems such as image denoising, super resolution etc. Markov Random Fields(MRFs) neighborhood have been extensively used for such problems and success have been shown in[10] and [14]. This ability of MRFs can also be used for detecting the hidden message in an image. However, a disadvantage of MRF models is that parameter estimation can become much more difficult due to the intractable nature of the partition function[1]. In this paper we are modeling pixels by Gibbs distribution using the concept of clique instead of MRF and then identify the changes due to embedding. The modeling of cover images are done first and then variations from this model for any stego image is identified. As an initial study the steganographic algorithm, LSB embedding is used.

The paper has been organized in a way to systematically cover all the areas of reference. Section 2 explains all the related works and existing systems that are using MRFs for various applications. These models are later incorporated for steganography detection. Section 3 explains the proposed system and its implementation details. Section 4 gives the experimental results with help of graphs for differentiating cover images and stego images with different payloads. Section 5 details out the conclusion and scope of the work.

2   Related Work

In image processing MRFs are now used extensively in several areas because of their ability to produce good, flexible, stochastic image models[8]. Image modeling is in itself done to represent the intensity of a given image. MRF image model has wide variety of applications in the areas of image and texture synthesis, image compression, image and texture segmentation, texture classification and surface reconstruction.

The information contained in the local, physical structure of images is sufficient to obtain a fair, global image representation. Most of the work on MRFs in image processing follows this remarkable guideline. Conditional probability distribution is used to capture the local structure and then to generate global image[2]. It follows by the important fact that the image intensity at any location of an image depends only on a neighborhood of pixels. Neighbors of any pixel $P(x,y)$ has two vertical and two horizontal neighbors, given by $(x + 1, y)$, $(x - 1, y)$, $(x, y + 1)$, $(x, y - 1)$, called the 4-neighbors of P denoted by $N_4(P)$ and four diagonal neighbors given by

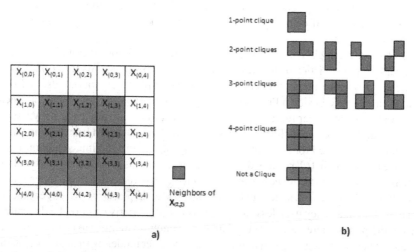

Fig. 1 An eight point a) neighborhood system, and b) its associated cliques[1].

$(x+1, y+1)$, $(x+1, y-1)$, $(x-1, y+1)$, $(x-1, y-1)$ denoted by $N_D(P)$. The points $N_D(P)$ and $N_4(P)$ are together known as 8-neighbors of the point P, denoted by $N_8(P)$. Fig. 1 shows an eight neighborhood systems with its associated cliques[1].

Let $X_s \epsilon \Omega$ be a random valued field defined on lattice S with neighborhood ∂s and with probability density function(pdf) $p(x)$. A random field X is a MRF if it has the property that for all $x \epsilon \Omega$

$$P(X_s = x_s | X_r, r \neq s) = P(X_s = x_s | X_{\partial r}) \qquad (1)$$

for all $x_s \epsilon \Omega$. Each pixel of the MRF is only dependent on its neighbors. But the definition of MRFs does not yield a natural method of writing down their distribution. This limitation was solved by using Gibbs distribution[1]. In order to define the Gibbs distribution, the concept of cliques must be discussed. Basically cliques are set of points which are all neighbors of one another. Examples of cliques for an eight point neighbornood system are illustrated in Fig.1 b).

Let the pdf of a discrete random field $X_s \epsilon \Omega$ be $p(x)$. $p(x)$ is a Gibbs Distribution if

$$p(x) = \frac{1}{z} exp\{-\sum_{c \epsilon C} \psi_c(x_c)\} \qquad (2)$$

where z is a normalization constant called partition function, C is the set of all cliques, x_c is the vector containing values of x on c and $\psi(x_c)$ is any functions of x_c. $\psi(x_c)$ is the potential function and the function

$$E(x) = \sum_{c \epsilon C} \psi_c(x_c) \qquad (3)$$

is the associated energy function. The Hammersely-Clifford theorem states that X is an MRF if and only if its pdf, $p(x)$ is a Gibbs distribution with the assumption that $p(x) > 0[1]$.

S. Roth et al., in [13] proposed an idea for a framework for learning image priors that capture the statistics of natural scenes. The approach extends traditional MRF models by learning potential functions over extended pixel neighborhoods. Field potentials are modeled using a Products-of-Experts framework that exploits non-linear functions of many linear filter response. In contrast to the already existing MRF approaches all parameters, including filters themselves were learned from learning data. This model were used for applications like image denoising and image inpainting.

U. Koster et al., in [6] explained the details of estimating the Markov random field potentials for a natural images. Here they have compared MRF with Independent Component Analysis and stated that MRF can be used in places where larger patches are to be handled. They used less samples in order to properly capture the dependencies of the pixels. Here they have used a Score Matching in order to estimate the potentials. They found that score matching produces better modeling than the ones using hand crafted Markov random field potentials.

Osindero et al., in [9] described an efficient learning procedure for multilayer generative models that combine the best aspects of Markov random fields and deep, directed belief nets[11]. The method described can learn models that have many hidden layers, each with its own MRF whose energy function is conditional on the values of the variables in the layer above. Detailed prior knowledge about the data to be modeled is not required. A similar method was used by Krizhevsky, A et al., in their model [7] to learn the filters in an unsupervised learning manner.

Marc'Aurelio Ranzato et al., in [12] proposed a model for natural images using gated MRF's. Markov Random Field was described for real-valued image modeling using two sets of latent variables. One set is used to gate the interactions between all pairs of pixels while the second set determines the mean intensities of each pixel. The conditional distribution over the input was found to be Gaussian with both mean and covariance determined by the configuration of latent variables, which is unlike previous models that were restricted to use Gaussian with either a fixed mean or a diagonal covariance matrix. This model has been used for various applications and one such is explained in where gated MRF is used to generate more realistic images.

3 Proposed System

In our system modeling of the images using Gibbs distribution is done in order to distinguish between cover and stego images. Our aim is to model a set of cover images that follows a similar distribution which is found to be altered when a message is embed into the cover. The existing systems have used many Markov models for applications like image denoising, segmentation, inpainting etc. Here we propose a method using energy function which follows Gibbs distribution. Gibbs distribution has already been used for above applications but not yet for steganalysis purpose.

The pixels x in the image that satisfies MRF property should follow Gibbs distribution. This means that the pdf, $p(x)$ for the image follows Gibbs distribution. The pdf has a partition function and an energy term depending on the vicinity of the random variable x. As we already stated, it is difficult to find $p(x)$. So the following procedure is followed. The input for the procedure is any natural image.

1. For a given image generate the energy function
2. In order to calculate the energy each and every pixel is changed iteratively. The energy is calculated for the changes in the pixel values. The change depends on the neighborhood of the pixel and this can be either four neighborhood or eight neighborhood. In our proposed system we use an eight neighborhood system and its associated cliques.
3. The change made has to be either accepted or rejected. This is done based on the energy value calculated. Acceptance of the change is done if the energy is small.

An uniform energy field is first considered and is later updated according to the energy calculation using equation(3). For each pixel value its energy is calculated based on its neighborhood. The new energy is updated only if it is smaller than the original energy. The Gibbs energy formula is used here. Equation (2) is used in order to define the probabilistic model and the energy is defined by the neighborhood of the pixel under consideration, where $E(x)$ is the energy defined over the neighborhood of the pixel using equation(3). In our proposed system we define function $\psi_c(x_c)$ using mean and median of the 8-neighborhood as follows.

$$\psi_C(x_c) = \sum_{x_i \in C} \frac{x_i}{\#C} \tag{4}$$

and

$$\psi_C(x_c) = median(C) \tag{5}$$

where C is the clique associated with x_c and $\#C$ is the number of the pixels in clique C.

The cover image was thus modeled using Gibbs energy. In order to identify the changes in an image due to embedding, these cover images where embedded with secret message of various payloads. A sequential LSB embedding algorithm was used in order to embed the message. Energy function was calculated for all the stego images created with various payload using above procedure done for cover images. The energy distribution for stego reveals the changes that occurred due to embedding. Even though exact payload cannot be recovered, the relative location of the embedding is easily visible from the energy plot.

4   Results and Discussions

To do the experiment a set of 100 natural images was selected with dimension 80×120.
All the images were in the jpg format. We have modeled the images using the energy
function in equation(3). The resultant model is described in the form of a graph, plotted the energy value of the pixel vs pixel position. We have taken the images with the
assumption that it is unaltered by any noise.

In Fig.2 original image and its different payload is shown. The figure explains the
changes brought in by performing LSB embedding on this image.The Fig.2 b) is the
result of 5 percentage embedding in the image. The Fig.2 c) is of 15 percentage embedding and Fig.2 c) for 30 percentage. It is clear from the image that there is no visible
changes one can identify after embedding. These are not good for identifying properly
whether the given image is stego or not.

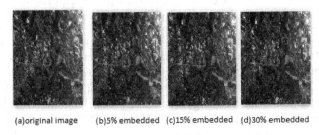

(a)original image   (b)5% embedded   (c)15% embedded   (d)30% embedded

Fig. 2 Image before and after embedding.

By using Gibbs distribution, the energy value is formulated and the slightest
changes in the pixel intensity values are identified. These changes thus help us to differentiate between cover and stego images. It also help us to locate the relative positions of embedding. The energy values corresponding to each pixel is plotted for
various payloads. From the graph it is clear that as the embedding capacity is increased
the energy value captured by the model varies.

We have defined the energy function using the mean and median of eight neighborhood of a pixel. From the graph it is clear that the changes have occurred at initial pixels denoting the sequential LSB embedding. The changes become more visible
as the embedding capacity increases. The energy plot with the median filter applied
has captured the change more accurately when compared to that of energy with mean.
Fig. 3,4,5 and 6 shows the energy value vs pixel positions plot for cover, 5 percentage
embedded image, 15 percentage embedded image and 30 percentage embedded image respectively. In all the Fig. 3,4,5 and 6(a) denote the plot with mean energy and (b)
median energy. When comparing Fig.3 (cover image) with Fig. 4,5 and 6 (stego images with different embedding range), between the intervals 0 and 30, we were able
to differentiate the changes in energy value.

Further observations can be made from plotting the count of the energy value of
the pixel. The plots for energy functions using mean and median reveal changes in
the count of energy values upon embedding. It can thus be seen that histogram attack

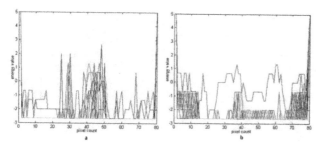

Fig. 3 Energy plot for cover image (a)with mean (b)with median

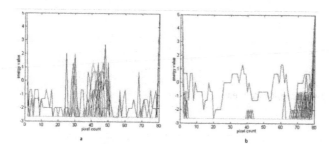

Fig. 4 Energy plot for 5 percentage embedded image (a)with mean (b)with median

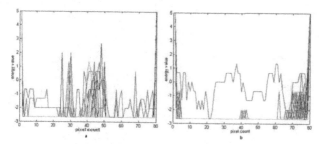

Fig. 5 Energy plot for 15 percentage embedded image (a)with mean (b)with median

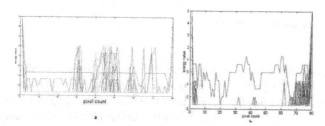

Fig. 6 Energy plot for 30 percentage embedded image (a)with mean (b)with median

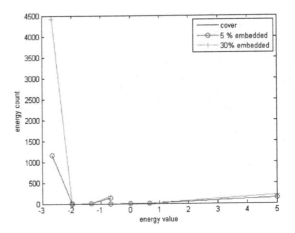

Fig. 7 Energy count plot for mean

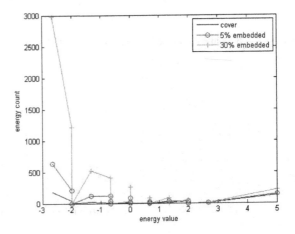

Fig. 8 Energy count plot for median

is possible not only on the pixel intensities but also on the energy values calculated. Fig.7 and 8 shows the plot of energy count for energy values with mean and median respectively. It is clear that from Fig.7 and 8 that when embedding increases the count of energy value varies. Using mean, energy function could not differentiate cover and stego image with 5 percentage embedding. But energy function with median was able to differentiate cover against stego with 5 percentage and 30 percentage embedding.

We have calculated energy values for a pixel using 2-point, 3-point and 4-point cliques. It is seen that energy values corresponding to these cliques is not good enough to find a deviation between cover and stego for different percentage of embedding. Clique concept was not showing satisfactory results because of first order statistics which we have used to model the energy function.

5 Conclusion

MRF modeling has been extensively used for a variety of image processing problems. The MRF modeling of the image is done in order to get a prior of the natural images. This prior will be used to identify whether a given image is stego or not. We have modeled the natural images using energy function which follows Gibbs distribution. Using this model we were able to identify the changes that happened to image on embedding a message with varying payloads. We formulated the energy function using mean and median of the eight neighborhood of a pixel and were able to identify the changes due to the embedding. The change were more clearly captured by energy function with median.

Although the proposed method is done only for LSB embedding, this method can be applied on other steganographic methods done in spatial domain also. This is because natural images follow a non-Gaussian distribution. The results can be further improved by modifying the energy function. Moreover we can use the concept of cliques to efficiently model cover by changing the energy function in terms of higher order statistics.

References

1. Bouman, C.: Model based image processing. Purdue University (2013)
2. Chellappa, R., Jain, A.: Markov random fields. theory and application, vol. 1 (1993)
3. Fridrich, J., Goljan, M.: Practical steganalysis of digital images-state of the art. In: Proceedings of SPIE, vol. 4675, pp. 1–13
4. Goljan, M., Fridrich, J., Cogranne, R.: Rich model for steganalysis of color images. In: IEEE Workshop on Information Forensic and Security, Atlanta, GA (2014)
5. Johnson, N.F., Jajodia, S.: Exploring steganography: seeing the unseen, vol. 31, pp. 26–34. IEEE (1998)
6. Köster, U., Lindgren, J.T., Hyvärinen, A.: Estimating markov random field potentials for natural images. In: Adali, T., Jutten, C., Romano, J.M.T., Barros, A.K. (eds.) ICA 2009. LNCS, vol. 5441, pp. 515–522. Springer, Heidelberg (2009)
7. Krizhevsky, A., Hinton, G.E., et al.: Factored 3-way restricted boltzmann machines for modeling natural images. In: International Conference on Artificial Intelligence and Statistics, pp. 621–628 (2010)
8. Li, S.Z.: Markov random field modeling in computer vision. Springer-Verlag New York, Inc. (1995)
9. Osindero, S., Hinton, G.E.: Modeling image patches with a directed hierarchy of markov random fields. In: Advances in Neural Information Processing Systems, pp. 1121–1128 (2008)
10. Rangarajan, A., Chellappa, R.: Markov random field models in image processing. Citeseer (1995)

11. Ranzato, M., Hinton, G.E.: Modeling pixel means and covariances using factorized third-order boltzmann machines. In: 2010 IEEE Conference on Computer Vision and Pattern Recognition (CVPR), pp. 2551–2558. IEEE (2010)
12. Ranzato, M., Mnih, V., Susskind, J.M., Hinton, G.E.: Modeling natural images using gated mrfs. IEEE Transactions on Pattern Analysis and Machine Intelligence 9, 2206–2222 (2013)
13. Roth, S., Black, M.J.: Fields of experts: a framework for learning image priors. In: IEEE Computer Society Conference on Computer Vision and Pattern Recognition, CVPR 2005, vol. 2, pp. 860–867. IEEE (2005)
14. Wang, C., Komodakis, N., Paragios, N.: Markov random field modeling, inference and learning in computer vision and image understanding: A survey. Computer Vision and Image Understanding 117(11), 1610–1627 (2013)

Real-Time Hand Gesture Detection and Recognition for Human Computer Interaction

Kapil Yadav and Jhilik Bhattacharya

Abstract This paper presents a gesture based system to interface Microsoft Word document. The system was developed using a two state discrete temporal model for gestures which works with distinct poses. The model fuses the state information along with individual pose recognition to activate the interfacing mechanism. It can be inferred from the experimental results that the model facilitates both accurate gesture recognition as well as promptness in response. The decision fusion of SURF and wavelet features prove to be robust for the current model. The selected features show a 96 percent accuracy when tested with gestures having varying background, scale, transformation and illumination conditions. The response time which varies between 3 to 3.5 second can be further improved by implementing the feature detection steps in VC++ environment instead of Matlab.

1 Introduction

Development of new technologies for Man Machine Interface hardware is in tandem with the corresponding advancement of software algorithms for data interpretation and processing. Technology enhancement thus saw a leap from Swept Frequency Capacitive Sensing techniques used in conventional touch screens, to interactive hardware displays like large displays, flexible displays and wearable displays for mixed reality. Current applications of MMI include smart homes, collaborative working environments, advanced information visualization and many more. The interfacing parameters range from multimodal interaction like touch, speech and gesture, to physiological factors such as ECG, EOG, Heart Rate, Eye blinking, facial expression for example. In particular, research focussed on intelligent human and machine interfaces to obtain a more intuitive and adaptive capability for advanced informa-

K. Yadav (✉) · J. Bhattacharya
Thapar University, Patiala, Punjab, India
e-mail: {kapilyadav1204,bjhilik}@gmail.com

© Springer International Publishing Switzerland 2016   559
S. Berretti et al. (eds.), *Intelligent Systems Technologies and Applications*,
Advances in Intelligent Systems and Computing 384,
DOI: 10.1007/978-3-319-23036-8_49

tion visualization that analyses and comprehends multi-dimensional information. To mention a few, considerable work has been done in voice based system control [1]. A vision-based system was presented in [2] for voluntary eye blink detection and pattern interpretation for HCI. Sign language development for deaf and dumb people has been a major drive for the advent of gesture based interfaces. Initially, work was done to automate communication between visually impaired and deaf people[3]. This later motivated exploration of gesture based interaction, whose primary application target was gaming and home entertainment. Work done by various researchers in this field also includes gesture based robot locomotion[4],[5] surgical systems. G-Speak spatial computing operating system, offers movement of data between different computing systems and displays through a gesture interface[6]. A Virtual keyboard was proposed [7] which provides a detectable surface on which user can move fingers that replicates the act of key pressing. The current work proposes a system which utilizes vision based hand gestures to interface Microsoft Word document in real time. The work uses gestures for opening, closing, changing font size and color, scrolling up and down in the word document. Gesture based research previously concentrated on offline gesture recognition mainly. Realtime gesture recognition requires accurate gesture modelling techniques for prompt response to the HCI. A probabilistic framework was presented by Ying Yin et. al. [8] for real-time gesture recognition. A taxonomy for Gesture modelling was presented in [9] where gesture was modelled as form, flow, temporal gestures. This work contributes a two state discrete temporal model for gestures which works with distinct poses. The model facilitates both accurate gesture recognition as well as promptness in response.

2 System Description

The system consists of a camera which will acquire images in real time. The software consists of three modules (as seen in figure 1). (i) A feature database which was

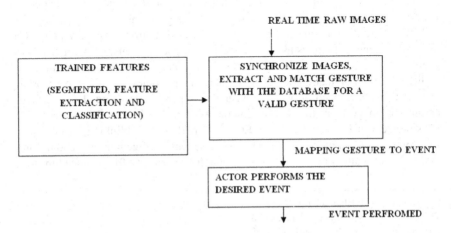

Fig. 1 Gesture based HCI system for Word Document Handling

trained offline (ii) A module called synchronize which detects gestures from captured images and matches it with the database for a valid gesture.(iii) The third module actor performs the desired event based on the matched gesture. The feature database generation and synchronization is done using Matlab. The actor needs to interface Microsoft OLE for the corresponding events to be handled in the word document. Document opening, closing, scroll up, scroll down, font color and size are the different events handled by the interface currently.

3 Methodology

The entire work can be divided in two phases training and testing. In the first phase, a set of gesture images are trained so that each gesture corresponds to a particular pattern. This includes hand region segmentation from an image, extracting features from the segmented region, using the feature vector for training. In the second phase a hand gesture has to be mapped to an event in realtime and the corresponding event needs to be performed.

3.1 Image Sequence by Camera and Acquizition

Images in realtime are acquired using the webcam provided in the laptop. The image acquisition program is looped to continuously capture snapshots every second. Any variations in gesture representation due to snapshot intervals are handled by the synchronization module.

3.2 Segmentation

In this phase, hand regions need to be extracted from the background for feature extraction and recognition purposes. The performance of the feature extractor largely varies on the segmentation algorithm selected. The task becomes challenging with cluttered background. Also, the algorithm should be robust against scene illumination and skin variations. Background modelling with K Gaussian distributions [10], connected-component labelling algorithm [11], Automatic Seeded Region Growing Algorithm (ASRGC) [13] with YCbCr model, meanshift filters are some of the mostly used segmentation algorithms. This paper uses k-means on RGB color image for skin color segmentation.

3.3 Features Extraction

Previous work on Gesture recognition has used shape based , keypoint based as well as region based algorithms. Shape based algorithms used different shape signatures (fourier descriptors for example) to detect hand shape while SIFT [12] is an example of keypoint based techniques. Region based algorithms like wavelet descriptors,

PCBR are also used by many researchers. The current method uses SURF and Wavelet descriptors as feature vectors.

3.3.1 SURF

SURF detector uses box filters to approximate the Gaussian and can be computed in constant time using the integral image [15]. At scale σ, Hessian matrix $H(p, \sigma)$ of a point p (x, y) in the image $f1$, is defined as shown in equation 1 .

$$H(p, \sigma) = \begin{bmatrix} L_{xx}(p, \sigma) \ L_{(xy)}(p, \sigma) \\ L_{xy}(p, \sigma) \ L_{(yy)}(p, \sigma) \end{bmatrix} \tag{1}$$

where $L_{xx}(p, \sigma)$ is image convolution of second derivative $\frac{dx}{dx^2} g(\sigma)$.

3.3.2 Wavelet Features

Input hand images are decomposed through the Wavelet Packet Decomposition using the Haar (Daubechies 1) basis function. The image is divided into four bands: LL(left-top), HL(right-top), LH(leftbottom) and HH(right-bottom). The HL band indicated the variation along the x-axis while the LH band shows the y-axis variation[12].

3.3.3 Canonical Correlation Feature

A feature fusion of of SURF and Wavelet Packets is obtained by computing the Canonical Correlation [14] of the two features. CC creates a new feature vector for each set of SURF and wavelet vectors such that the correlation between these variables is maximized and independent of affine transformation. Equation 2 gives the canonically correlated variable Z.

(a) K-Means Clustered image (hand color imposed on the segmented hand cluster)

(b) SURF features on hand image

(c) 2nd level wavelet decomposition

Fig. 2 SURF and Wavelet features extracted from K-means clustered image.

$$Z_i = \begin{bmatrix} A & 0 \\ 0 & B \end{bmatrix}^T \begin{bmatrix} X_i \\ Y_i \end{bmatrix}$$

$$C_{xy} = \frac{1}{L} \sum_{i=1}^{L} x_i y_i^t \tag{2}$$

Where x_i and y_i denote the SURF and wavelet feature vector of the ith image respectively. A and B are the eigenvectors of $C_{xx}^{-1} C_{xy} C_{yy}^{-1} C_{xy}^T$ and $C_{xx}^{-T} C_{xy} C_{yy}^{-1} C_{xy}$ respectively. C_{xy} gives the covariance matrix of x and y and L denotes the total number of training images.

3.4 Classification

Two types of classification are generally used by recognition applications. One category uses various distance metrics like Euclidean and Mahalanobis to compute the difference between the test vector with the different classes of vectors and assigns the test vector to the class having the least distance. Another category uses machine learning algorithms like SVM, NN, AdaBoost, HMM to classify the data. This work uses a combination of neural network and Euclidean distance classifiers. The feature vectors of SURF and Wavelet features and CC vector Z are tested with neural network (equation 3), whereas the decision fusion (equation4) is tested with Euclidean classifiers.

$$x = [-1, x_1, x_2, x_3, x_4, \ldots\ldots x_p]^T$$
$$\text{weights } w = [-1, w_1, w_2, w_3, w_4, \ldots\ldots w_p]^T$$
$$\text{output } odx = \sum_{i=0}^{P} x_i * W_i$$
$$\text{error } e_x = ox - odx \text{ odx is the actual output}$$
$$\text{updated weights } \Delta w_i = -\rho * \frac{\delta e}{\delta W_i} \tag{3}$$

$$d_i^{comb} = \frac{1}{d_i^y} + \frac{1}{d_i^x} \tag{4}$$

where d_i^x is computed for all image vectors x_i and test vector x^t. The i for which d^{comb} is the maximum is considered as the correct match.

3.5 Synchronization

A software-based system for the real-time synchronization of images captured by a lowcost camera framework is presented (as seen in figure 3). It is highly recommended for cases where special hardware cannot be used. Every gesture is identified in two steps. A start symbol denotes the start of the gesture, which is followed by appropriate document open, close or other gestures. As shown in the figure 4, interim invalid gestures may be captured in snapshots while the user is forming the particular gesture. All these gestures will not find any match in the database, however the recognition module will run unnecessarily wasting processor time. Hence a frame is passed for recognition only after it becomes static. Thus only when last three captured frames have no change, it is forwarded for segmentation, feature detection and gesture recognition. The start state is maintained as long as a valid gesture is not recognized. Once a gesture is recognized and the corresponding event is invoked, the cycle is complete. The next event will be marked by another start state.

3.6 Document Handling

The work uses existing MS Office Automation (OLE) modules and interfaces it with corresponding gesture recognition events. This is shown in figure 5. For example when a gesture denoting document open OLE is invoked, it is recognized as 001 which in turn calls the corresponding module to open the document, and the document is opened. Same is the case for the font color, size or scroll gestures.

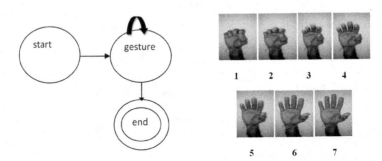

Fig. 3 A start gesture and another corresponding gesture is required for an event invocation.

Fig. 4 Transition between 2 gestures captured by the camera.

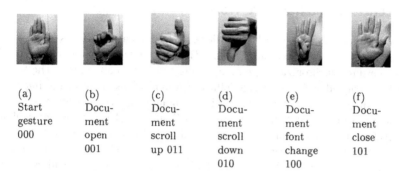

(a)	(b)	(c)	(d)	(e)	(f)
Start gesture 000	Docu-ment open 001	Docu-ment scroll up 011	Docu-ment scroll down 010	Docu-ment font change 100	Docu-ment close 101

Fig. 5 The different gestures used for particular events

4 Experimental Results

A dataset of 300 images are taken as test samples. Three types of test sample images are considered for experimental purpose for all the methods and procedures explained in this paper namely clear background (C1) , slightly cluttered background (C2), and slightly cluttered background with changing lighting conditions (C3). All the images are captured in home environment without any special lighting using a consumer quality web camera. The resolution of the images considered for processing after segmentation is 128 X 128. The accuracy and performance of the proposed system was further verified using realtime test cases. Hundred gestures , randomly selected and performed, by different people at different illumination and backgrounds were used for generating the test results shown in table 1 and 2. Table 1 gives a comparison of different feature vectors used. The technique which gives the best performance (decision fusion approach in the present case) is further used for word document interfacing. The time and performance accuracy for each event is shown in table 2.

It was seen that the decision fusion approach with wavelet and SURF features gives a satisfactory performance on the current dataset.

Table 1 Performance comparison of feature detection methods.

No	Method	Percentage
1	wavelet	93
2	SURF	90.5
3	feature fusion	87
4	decision fusion	96

Table 2 Performance of event invocation on gesture (in figure 5) recognition

Gesture	Percentage	Time(s)
5b	95	4.5
5c	92	4
5d	91	3.5
5e	93	3
5f	90	3.25
5a	98	2

5 Conclusion and Future Work

The work proposes a gesture based Microsoft Document handling system which operates using a two state gesture model. The performance of the system depends on the gesture synchronization and the gesture recognition algorithms. The former have been handled by a temporal modelling of gestures. This can further be improved by combining temporal model along with form and path. The gesture recognition shows a performance collation between SURF and wavelet transformation along with feature fusion and decision fusion approaches. The decision fusion of Wavelet and SURF features show a 96 percent accuracy on images with varying background, scale, transformation and illumination conditions. The number of features chosen for decision or feature fusion can be increased based on the available hardware for implementation. It was limited to two in the present work, implemented, on a laptop with Intel core i3 processor (CPU 2.27GHz) on a 64 bit windows platform. A number of extensions of the current work are under progress. A) The work is being implemented on other software. B) Two dimensional image information processed by the recognizer is being extended to include the depth dimension. This is further used by the algorithm to estimate user distance from the screen. Thus there will be an automatic zoom in or out depending on this distance. C) Dynamic gestures are being considered instead of static ones. As a result the same gesture will work for scroll up or down depending whether the hand is moving upwards or downwards.

References

1. Cui, B., Xue, T.: Design and realization of an intelligent access control system based on voice recognition. In: ISECS International Colloquium on Computing, Communication, Control, and Management. CCCM 2009, vol. 1. IEEE (2009)
2. Sumathi, S., Srivatsa, S.K., Uma Maheswari, M.: Vision based game development using human computer interaction (2010). arXiv preprint arXiv:1002.2191
3. Ghotkar, A.S., et al.: Hand gesture recognition for indian sign language. In: 2012 International Conference on Computer Communication and Informatics (ICCCI). IEEE (2012)
4. Lee, C., Xu, Y.: Online, interactive learning of gestures for human/robot interfaces. In: 1996 IEEE International Conference on Proceedings of the Robotics and Automation, vol. 4. IEEE (1996)
5. Nosowitz, D.: Video: MIT's Kinect Hack Tracks All Ten Fingers Simultaneously (2010)
6. Malik, S., Laszlo, J.: Visual touchpad: a two-handed gestural input device. In: Proceedings of the 6th international conference on Multimodal interfaces. ACM (2004)
7. Samanta, D., Sarcar, S., Ghosh, S.: An approach to design virtual keyboards for text composition in Indian languages. International Journal of Human-Computer Interaction 29(8), 516–540 (2013)
8. Yin, Y., Davis, R.: Real-time continuous gesture recognition for natural human-computer interaction. In: 2014 IEEE Symposium on Visual Languages and Human-Centric Computing (VL/HCC). IEEE (2014)
9. Wobbrock, J.O., Morris, M.R., Wilson, A.D.: User-defined gestures for surface computing. In: Proceedings of the SIGCHI Conference on Human Factors in Computing Systems. ACM (2009)

10. Lucchi, A., et al.: Supervoxel-based segmentation of mitochondria in em image stacks with learned shape features. IEEE Transactions on Medical Imaging **31**(2), 474–486 (2012)
11. Shapiro, Stockman, G.: Computer Vision. Prentice Hall (2002)
12. Kook-Yeol, Y.: Robust hand segmentation and tracking to illumination variation. In: 2014 IEEE International Conference on Consumer Electronics (ICCE), pp. 286–287, January 10–13, 2014
13. Yang, G., et al.: Research on a skin color detection algorithm based on self-adaptive skin color model. In: 2010 International Conference on Communications and Intelligence Information Security (ICCIIS). IEEE (2010)
14. Qi, F., Weihong, X., Qiang, L.: Research of Image Matching Based on Improved SURF Algorithm. TELKOMNIKA Indonesian Journal of Electrical Engineering **12**(2), 1395–1402 (2014)
15. Suaib, N.M., et al.: Performance evaluation of feature detection and feature matching for stereo visual odometry using SIFT and SURF. In: 2014 IEEE on Region 10 Symposium. IEEE (2014)

Effect of Distance Measures on the Performance of Face Recognition Using Principal Component Analysis

Sushma Niket Borade, Ratnadeep R. Deshmukh, and Pukhraj Shrishrimal

Abstract We examined effect of distance measures on the performance of face recognition using Principal Component Analysis. We tested commonly used 4 distance measures: City block, Euclidean, Cosine and Mahalanobis. The study was done on 400 images from ORL face database. We achieved the best recognition performance using Cosine and City block distance measures. It was observed that fewer images need to be extracted for achieving 100% cumulative recognition using Cosine metrics than using any other distance measure. The performance of Mahalanobis distance measure was poor compared to other distance measures. Our results show the importance of using appropriate distance measure for face recognition.

1   Introduction

Face recognition methods are categorized into two types: feature based and appearance based [1]. In the appearance-based approaches, whole face image is considered rather than just local features. Recognition is performed by comparing intensity values of whole face with all the other faces in the database. This approach is computationally very expensive as it performs classification in the high dimensional space. Principal Component Analysis (PCA) [2] is widely used for reducing dimensions. Sirovich and Kirby [3, 4] used PCA to efficiently represent face images. They showed that any face can be approximately reconstructed by using small number of eigenfaces and the corresponding weights. The face images are projected on the eigenfaces and these weights are obtained.

S.N. Borade(✉) · R.R. Deshmukh · P. Shrishrimal
Department of Computer Science and Information Technology,
Dr. Babasaheb Ambedkar Marathwada University, Aurangabad 431004, India
e-mail: {sushma.borade,pukhraj.shrishrimal}@gmail.com, rrdeshmukh.csit@bamu.ac.in

© Springer International Publishing Switzerland 2016   569
S. Berretti et al. (eds.), *Intelligent Systems Technologies and Applications*,
Advances in Intelligent Systems and Computing 384,
DOI: 10.1007/978-3-319-23036-8_50

Mathematically, the eigenfaces are nothing but the eigenvectors of the covariance matrix of all the face images. Based on the findings of Sirovich and Kirby, PCA was used by Turk and Pentland [5, 6] for detection and identification of face images.

This paper presents well known Principal Component Analysis technique for performing face recognition. We compare performance of face recognition using following four commonly used distance measures: City block, Euclidean, Cosine and Mahalanobis. We used following characteristics to measure the performance of the system: area above cumulative match characteristic curve (CMCA), recognition rate and the number of images (in percent) need to be extracted for achieving 100% cumulative recognition.

The organization of our paper is as follows: Section 2 discusses the PCA algorithm in detail. Section 3 describes the distance measures. Section 4 presents experiments and results of the research. Section 5 offers the conclusion.

2   Principal Component Analysis

In this paper, face recognition using PCA is implemented as proposed by Turk and Pentland. Consider a set of M face images as $\Gamma_1, \Gamma_2... \Gamma_M$, each of size $N \times N$. The average face of the set is defined as

$$\Psi = \frac{1}{M} \sum_{i=1}^{M} \Gamma_i \tag{1}$$

The difference between face image and average face, Ψ, is given by the vector $\phi_i = \Gamma_i - \Psi$. A covariance matrix C is found where

$$C = AA^T, \text{ where matrix } A = [\phi_1 \phi_2 \cdots \phi_M]. \tag{2}$$

As the size of matrix C is N^2 by N^2, its intractable task to find its N^2 eigenvectors. Therefore as proposed by Turk and Pentland, construct matrix $L = A^T A$ which is of size M by M, and find its M eigenvectors, v_i. These v_i give most significant M eigenvectors (i.e. eigenfaces) of C as:

$$u_l = \sum_{k=1}^{M} v_{lk} \phi_k, \quad l = 1,...,M \tag{3}$$

From these eigenfaces, M' ($< M$) eigenfaces corresponding to M' highest eigenvalues are selected. All face images in the training set are projected into the eigenspace by the operation:

$$w_k = u_k^T (\Gamma - \psi), \tag{4}$$

for k = 1, ..., M'.

A projection vector, $\Omega = [w_1, w_2, ..., w_M]$, formed using these weights, is stored in the database. The probe image to be identified is projected on the eigenspace. It is assigned the identity of the closest training image by comparing its projection vector with known projections of gallery images.

3   Distance Measures

We used City block, Euclidean, Cosine and Mahalanobis distance measures [7, 8, 9] for comparing performance results. Let x and y be n-dimensional feature vectors, and x_i and y_i be the ith components of the vectors. Distance measures between these two vectors $d(x, y)$ can be described mathematically as:

- City block distance:
  It is also called as Manhattan or L1 distance. It is the sum of the magnitudes of the differences along each dimension.

$$d(x, y) = \sum_{i=1}^{n} \left| x_i - y_i \right| \tag{5}$$

- Euclidean distance:
  This is also called as L2 or Pythagorean distance. Squared Euclidean distance between vectors x and y is the sum of squared differences of their i^{th} components.

$$d(x, y) = \sqrt{\sum_{i=1}^{n} \left(x_i - y_i \right)^2} \tag{6}$$

- Cosine:
  It measures the Cos of the angle between two vectors. For vectors x and y, the cosine similarity is represented using their dot product scaled by the product of their magnitudes as:

$$d(x, y) = -\frac{x \cdot y}{\|x\| \|y\|} = -\frac{\sum_{i=1}^{n} x_i y_i}{\sqrt{\sum_{i=1}^{n} x_i^2 \sum_{i=1}^{n} y_i^2}} \tag{7}$$

- Mahalanobis distance:

$$d(x, y) = -\sum_{i=1}^{n} z_i x_i y_i, \tag{8}$$

where $z_i = \sqrt{\dfrac{\lambda_i}{\lambda_i + \alpha^2}} \simeq \dfrac{1}{\sqrt{\lambda_i}}$ and $\alpha = 0.25$. Here λ_i is the ith eigenvalue corresponding to the ith eigenvector.

In this paper we are performing identification task. The cumulative match characteristic (CMC) curve is used to measure the system performance. CMC curve is a plot of rank values on the X-axis and the identification rate at or below that rank on the Y-axis [10].

4   Experiments and Results

To perform face recognition, the ORL face database [11] is used. It has face images taken between April 1992 and April 1994 at the AT and T Laboratories. The database has 400 grayscale face images with 10 images each from 40 persons. Size of each image is 112 x 92 pixels. In our experiments, each face image is resized to 50 x 42 pixels. System is trained using 200 images from 40 persons. For training, we used first 5 images per person and remaining five for testing. Thus there are 200 images in probe set. PCA based face recognition is implemented using MATLAB® R2013a.

The experimental results are summarized in Tables 1-5. It can be seen how different distance measures affect the performance of the system. We calculate the area above cumulative match characteristic curve (CMCA) to measure the overall performance of the distance measure. Smaller CMCA indicates better identification. We summarize number of images (in percent) extracted to achieve cumulative recognition rate in range of 80 to 100%. Smaller values indicate that fewer images need to be extracted for achieving the desired identification rate. We also present first one recognition rate. Here, for the given probe, the most similar image is extracted from the gallery set. Larger values indicate better recognition. Subscripts are used to mark the best results in Tables 1-5. CMC curve and the other characteristics used are shown in Fig. 1.

Table 1 Recognition using 10% of features (20)

Distance measure	Rank (%) of images					CMCA 0-10000	First1 rec
	80	85	90	95	100		
City block	2.38	3.63	4.88	12.5	42.5_3	419.38_3	161_2
Euclidean	2.81	3.85	4.90	11.25	37.5_2	390.63_2	157_3
Cosine	2.11	3.42	4.74	10.63	30_1	364.38_1	163_1
Mahalanobis	3.64	4.77	8.75	20	70	486.25	150

Table 2 Recognition using 20% of features (40)

Distance measure	Rank (%) of images					CMCA	First1
	80	85	90	95	100	0-10000	rec
City block	1.62	3.09	4.56	13.75	50_3	401.25_3	166_2
Euclidean	1.84	3.16	4.47	8.75	37.5_2	369.38_2	165_3
Cosine	0.94	2.5	4.06	8.75	30_1	336.25_1	170_1
Mahalanobis	3.90	5.83	13	30	60	583.13	151

Table 3 Recognition using 30% of features (60)

Distance measure	Rank (%) of images					CMCA	First1
	80	85	90	95	100	0-10000	rec
City block	0.19	2.12	4.04	11.25	40_3	360_2	172_1
Euclidean	1.53	2.92	4.31	8.75	37.5_2	363.13_3	167_2
Cosine	0.5	2.17	3.83	7.5	30_1	327.50_1	172_1
Mahalanobis	6.14	10	20.63	42.5	80	788.75	140_3

Table 4 Recognition using 60% of features (120)

Distance measure	Rank (%) of images					CMCA	First1
	80	85	90	95	100	0-10000	rec
City block	-0.21	1.88	3.96	9	35_2	344.38_2	173_1
Euclidean	1.32	2.80	4.26	8.75	35_2	358.75_3	168_3
Cosine	0.88	2.35	3.82	7.5	30_1	325.63_1	171_2
Mahalanobis	16.25	26.5	38.75	67.5	92.5_3	1215.6	127

Table 5 Recognition using 90% of features (180)

Distance measure	Rank (%) of images					CMCA	First1
	80	85	90	95	100	0-10000	rec
City block	-1.25	1.25	3.75	9	35_2	340.63_2	175_1
Euclidean	1.18	2.65	4.12	8.75	35_2	356.88_3	169_3
Cosine	0.88	2.35	3.82	6.25	30_1	323.13_1	171_2
Mahalanobis	9.46	12.5	25	34.17	77.5_3	838.13	97

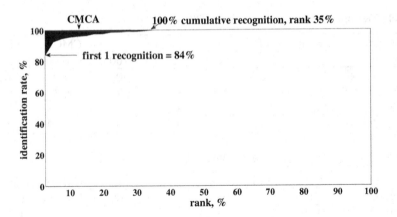

Fig. 1 Cumulative match characteristic

Distance measures can be sorted with respect to their performance using following characteristics: i) percent of images extracted to achieve the 100% cumulative identification (Cum 100), ii) first one recognition rate, iii) area above cumulative match characteristic curve (CMCA). The results are shown in Table 6.

Table 6 Sorted distance measures with respect to the system performance

Number of eigenvectors	Cum 100	First one recognition rate	CMCA
10%	Cosine	Cosine	Cosine
(20)	Euclidean	City block	Euclidean
20%	Cosine	Cosine	Cosine
(40)	Euclidean	City block	Euclidean
30%	Cosine	Cosine, City block	Cosine
(60)	Euclidean	Euclidean	City block
60%	Cosine	City block	Cosine
(120)	City block, Euclidean	Cosine	City block
90%	Cosine	City block	Cosine
(180)	City block, Euclidean	Cosine	City block

The graphical representation of relation between the eigenvectors used, and the system performance are presented in Fig. 2. Using Cosine distance measure, fewer images need to be extracted for achieving 100% cumulative recognition if we use 10 – 90% of eigenvectors. We achieved best results using Cosine distance with respect to CMCA (323.13 – 364.38). The largest first one recognition rates are achieved using Cosine distance measure (81.5 – 86%) if 10 – 30% of eigenvectors are used, and City block distance (86 – 87.5%), if 30 – 90% of eigenvectors are used.

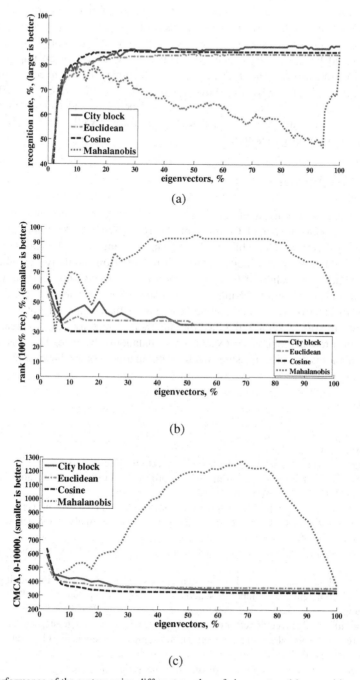

Fig. 2 Performance of the system using different number of eigenvectors (a) recognition rate, (b) cumulative 100% recognition, (c) CMCA

Now let us see comparison of our results with that of other researchers. Phillips et al. [12] showed that face recognition using PCA performs better with Cosine distance measure than using Euclidean, City block or Mahalanobis. According to Yambor and Draper [7], Mahalanobis distance showed significant improvements over City block, Euclidean or Cosine, if more than 60% of eigenvectors are used. Ahuja and Chhabra [13] also showed that better results were obtained using Mahalanobis distance than using City Block, Euclidean and Cosine on FERET dataset. Draper et al. [14] showed that PCA performed well with City block and Mahalanobis distance measure.

5   Conclusions

In this paper, we examined effect of distance measures on the performance of face recognition using Principal Component Analysis. We tested commonly used 4 distance measures: City block, Euclidean, Cosine and Mahalanobis. The experimental work was conducted on 400 images from publicly available ORL face database. We achieved best recognition performance using Cosine and City block distance measures. Using Cosine distance, we needed to extract fewer images to achieve 100% cumulative recognition than using any other distance measure. It was observed that as the number of eigenvectors increases, the recognition rate also increases except for Mahalanobis distance. The performance of Mahalanobis distance measure was poor compared to other distance measures.

References

1. Zhao, W., Chellappa, R., Phillips, P., Rosenfeld, A.: Face recognition: a literature survey. ACM Computing Surveys 35, 399–458 (2003)
2. Duda, R., Hart, P., Stork, D.: Pattern classification. Wiley (1973)
3. Sirovich, L., Kirby, M.: Low-dimensional procedure for the characterization of human faces. J. Opt. Soc. of America 4, 519–524 (1987)
4. Kirby, M., Sirovich, L.: Application of the karhunen-loève procedure for the characterization of human faces. IEEE Trans. Pattern Anal. and Mach. Intell. 12, 831–835 (1990)
5. Turk, M.A., Pentland, A.: Eigenfaces for recognition. J. Cogn. Neurosci. 3, 71–86 (1991)
6. Turk, M.A., Pentland, A.: Face recognition using eigenfaces. In: Proceedings of the IEEE Conference on Comput. Vis. and Pattern Recognit. pp. 586–591 (1991)
7. Yambor, W., Draper, B., Beveridge, J.: Analyzing PCA-based face recognition algorithms: eigenvector selection and distance measures. In: Christensen, H., Phillips, J. (eds.) Empirical evaluation methods in computer vision. World Scientific Press, Singapore (2002)
8. Miller, P., Lyle, J.: The effect of distance measures on the recognition rates of PCA and LDA based facial recognition. Technical rep., Clemson University (2008)

9. Sodhi, K.S., Lal, Madan: Comparative analysis of PCA-based face recognition system using different distance classifiers. Int. J. of Appl. or Innovation in Eng. & Manag. **2**, 341–348 (2013)
10. Biometrics Testing and Statistics. http://www.biometrics.gov/documnets/biotestingandstats.pdf
11. ORL Face Database. http://www.cl.cam.ac.uk/research/dtg/attarchive/facedatabase.html
12. Phillips, P.J., Moon, H., Rizvi, S., Rauss, P.: The FERET evaluation methodology for face recognition algorithms. T-PAMI **22**, 1090–1104 (2000)
13. Ahuja, M.S., Chhabra, S.: Effect of distance measures in PCA based face recognition. Int. J. of Enterp. Computing and Bus Systems **1**(2) (2011)
14. Draper, B.A., Baek, K., Bartlett, M.S., Beveridge, J.R.: Recognizing faces with PCA and ICA. Comput. Vis. and Image Underst. **91**, 115–137 (2003)

Author Index

Printed in the United States
By Bookmasters